HISTORY
OF PHYSICS

THE AMERICAN INSTITUTE OF PHYSICS is a not-for-profit membership corporation chartered in New York State in 1931 for the purpose of promoting the advancement and diffusion of the knowledge of physics and its application to human welfare. Leading societies in the field of physics and astronomy are its members. AIP's activities include providing services to its Member Societies in the publishing, fiscal, and educational areas, as well as other services which can best be performed by one operating agency rather than dispersed among the constituent societies.

Member Societies arrange for scientific meetings at which information on the latest advances in physics is exchanged. They also ensure that high standards are maintained in the publication of the results of scientific research. AIP has general responsibility for the publication and distribution of journals. The Institute is expected to stay in the forefront of publishing technology and to ensure that the services it performs for its Member Societies are efficient, reliable, and economical.

The Institute publishes its own scientific journals as well as those of its Member Societies; provides abstracting and indexing services; serves the public by making available to the press and other channels of public information reliable communications on physics and astronomy; carries on extensive manpower activities; encourages and assists in the documentation and study of the history and philosophy of physics; cooperates with local, national and international organizations devoted to physics and related sciences; and fosters the relations of physics to other sciences and to the arts and industry.

The scientists represented by the Institute through its Member Societies number more than 60,000. In addition, approximately 7000 students in over 500 colleges and universities are members of the Institute's Society of Physics Students, which includes the honor society Sigma Pi Sigma. Industry is represented through some 115 Corporate Associate members.

HISTORY OF PHYSICS

Edited by

Spencer R. Weart
Center for History of Physics
American Institute of Physics

and

Melba Phillips
Emeritus Professor of Physics
University of Chicago

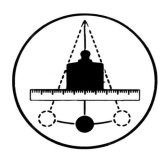

Readings from Physics Today

Number Two

American Institute of Physics
New York, New York
1985

Readings from Physics Today

PHYSICS TODAY, a publication of the American Institute of Physics, provides news coverage of national and international research activities in physics as well as government and institutional activities that affect physics. Both technical and nontechnical developments are covered by scientific articles, news, stories, book reviews, letters to the editor, calendars of meetings, and editorial opinion.

Articles in PHYSICS TODAY are intended to be of interest to—and understandable by—a broad audience of professionals from all subfields of physics as well as people with a general interest in physical science.

History of Physics is the second book in a series of volumes that contains reprinted articles and news material from PHYSICS TODAY in other areas and subfields of physics.

Cover and title design by Charles Grenner

Printed in the United States of America

Pub. No. R-315.1

Library of Congress Catalog Card No. 85-70236

ISBN 0-88318-468-0

Introduction

PHYSICS TODAY began publication in 1948, and for the first several years it contained no articles dealing with history of physics. There was nothing remarkable in this, for one would have had to look hard to find anywhere a journal article or a book dealing with history of physics. The chief exceptions to this rule were occasional scholarly studies of great early figures such as Galileo and Newton. The historical map of more recent times was mostly blank space, decorated here and there with prodigious figures (Maxwell, Kelvin, Planck,...) known less through direct investigation than through anecdotes that grew in the retelling, like the travellers' tales that populated early geographers' maps with tigers and sea serpents.

Historians did not notice that they lacked a history of modern science. Scholars who were not dedicated to the old narrow history of "kings and battles" were leaping to a history of people *en masse*: if presidents were not the key actors, then it must be labor unions or corporations. Yet evidence was accumulating that the center of modern history might lie somewhere between the leaders and the masses, and perhaps even in laboratories. Many would acknowledge that science had come to play a central role in the development of society, but few scholars investigated the question. To students of history, Einstein's and Schrödinger's equations seemed far more obscure than any medieval Latin parchment. As for the new generation of students of physics, they had little time to spare for literature, and were mostly satisfied with whatever colorful anecdotes they happened to hear about the past of their discipline.

Interest in history began among leading physicists

It was left to older physicists to notice that something important was being overlooked. Trained in the early decades of the century by professors who perhaps remained unconvinced of the value of relativity theory, these men and women had been young when quantum mechanics burst upon the scene. They had seen a mighty and complex intellectual process shake and transform physics in a way that no generation had known for centuries. Then, through the Great Depression, the Second World War, and the Cold War, they saw the physics community itself reshaped into a new form, while historical forces pulled physicists from their laboratories and placed them under the spotlights of the public stage—sometimes literally under spotlights. The older physicists wanted to understand what they had lived through. And they particularly wanted the next generation to understand this history, so that the physics community might brace itself to withstand and profit from the equally great transformations that might be expected at any time.

In 1952 PHYSICS TODAY carried its first two historical pieces, both reprinted in this book: an article by Karl T. Compton on the founding of the American Institute of Physics during the Depression, and an article by Edward U. Condon on the postwar relations between science and the federal government. Both men had been at the very center of the events they recounted, but they were not simply reminiscing about old times. Their articles carried lessons—which to this day are worth close attention—not only

about how the American physics community acquired its present shape, but also about how in their opinion the community ought to be shaped.

Meanwhile efforts were underway to bring an improved understanding of the past into physics education. A historical viewpoint is inevitably present in teaching, whether as explicit stories about scientists of the past, or implicitly. Many teachers had long recognized that teaching physics as an abstract and perfected intellectual structure, without a history, implicitly gives a distorted and perhaps even damaging picture of the nature of scientific research. And some teachers found history helpful in teaching physics concepts.[1] Such interest in the educational value of history became more widespread after the Second World War, as evidenced by symposia and, eventually, international conferences.[2] Physicists with a strong historical orientation, for example, Gerald Holton and Stephen Brush, worked with other historians of science not only to develop materials that could bring history into physics education, but also to make sure it was accurate history and not unverified folk tales. As more twentieth-century physics entered the curriculum, historical interest broadened still further. It was particularly during the 1960s that this movement took hold.

Professional history of physics rose in the 1960s

The feeling that the history of modern physics merited close attention inspired a few younger physicists to gamble their careers by turning from physics research to historical research. Senior physicists came to their support, and an institutional base for the research was gradually laid down. One important example of this work was the Sources for History of Quantum Physics project, launched in 1961 and led by Thomas S. Kuhn with the aid of several eminent physicists and a major grant from the National Science Foundation.[3] The original impulse had come with the realization that people like Einstein and Schrödinger, who would be honored so long as science was remembered, had died before anyone asked them for the full details of how they had made their epochal discoveries; but it was not too late to ask others and even to get their recollections down on tape. (As it happened, it was late indeed for Niels Bohr, who died after only two of several planned interview sessions.) In the course of the project the physicist–historians realized that tape-recorded recollections would not be enough to secure a complete and accurate history, and they set to microfilming files of correspondence as well. The resulting collection of interviews and microfilms has already served as "raw data" for a large number of scholars, and is used more frequently every year. People would have attempted to study the history of atomic and quantum physics even without this project, but the past two decades have seen publications on this subject (some of them included in this book) in a quantity, and on a level of accuracy, that would otherwise have been impossible.

At the same time as the Sources for History of Quantum Physics project was getting underway, other eminent physicists were working to create a more permanent institution. They were particularly concerned by the fact that physics, or at least physics after Newton, was simply not mentioned in history textbooks and other standard sources of culture. In the Smithsonian Museum, for example, physics was subsumed under "electrical engineering." In 1960 the concern over this situation helped establish a history program at the American Institute of Physics; in 1965 it became a permanent Center for History of Physics. The AIP Center has worked steadily to conduct oral history interviews and preserve documents, and has also worked to make the history known through projects such as preparation of exhibits and aid to scholars visiting its Niels Bohr Library. Many of the more recent articles in this book drew directly or indirectly upon the AIP Center's resources.

Such efforts within the borders of the physics community were reinforced during the 1960s by an outside movement. All around the United States and also to some extent overseas, historians of science and even entire history of science departments grew up in the universities. This was part of a general wave of university expansion and diversification, but it had a special relationship to physics. At many places, physicists played a role in getting the new departments established, and history of physics became by far the most popular specialty within the history of science (possibly excepting the history of medicine, which has its own unique traditions).

History of modern science largely the history of physics

Why was the history of modern science, so far as it has been written down, largely the history of physics? Perhaps it was for the same reasons that, ever since I. I. Rabi got together with Dwight Eisenhower, nearly all of the Presidents' science advisers have been physicists. One reason might be that physics is a master-key to all the twentieth-century sciences; another, that nuclear weaponry has made scientists and the public especially watchful of physics; yet another, that physicists have often had broader viewpoints than other technically oriented people, with an interest in everything from music to social relations. Whatever the causes, the result was a rising generation of professional historians of physics.

In the pages of PHYSICS TODAY, occasional articles began to appear recounting historical stories that had nothing to do with the personal reminiscences of the author. The articles by E. Mendoza and C. S. Smith, reprinted here, show that such writings could be not only interesting but also sophisticated history based on direct investigation of evidence, even though the authors were primarily scientists rather than professionally trained historians. In 1967 appeared the first articles by historians of science, Martin Klein and Lawrence Badash. Even these two, however, were originally trained as physicists and were employed in university physics departments. It was only in the late 1960s and especially in the 1970s that there came a significant number of people trained from the outset as deeply in history as in modern physics, and hired explicitly as historians of science; the first to appear in these pages is Charles Weiner, then Director of the AIP Center for History of Physics. Since the mid 1970s a number of articles by such people have appeared in PHYSICS TODAY. The author of the most recent article reprinted here, Robert Rosenberg, is trained as a mainline historian more than as a specialist in science.

A continuing dialogue between scientists and historians

The new type of article did not displace, but added onto, historical writings by full-time physicists. Leaders of the profession have continued to write articles based on their own experience—sometimes recounting their personal struggles and discoveries, sometimes telling of figures they knew during their careers, most often writing a combination of both. Some of these physi-cists have taken a leaf from historians, gathering reminiscences from colleagues and searching for documentary evidence, aspiring to a high standard of scholarly accuracy. This points to a remarkable characteristic of the history of physics, not only in the pages of PHYSICS TODAY but more generally: any scholar or general reader who would pursue the subject will end up reading a mixture of first-person reminiscences and retrospective historical accounts, mixed together with no clear boundary dividing them.

This close relationship between the people who are studying history of physics, and the people who are the subjects of that study, has helped to mobilize continuing support for the work. In the 1980s this support is stronger than ever. The number of full-time professional historians of physics continues to increase, while physicists themselves are more than ever writing historical articles and books, cooperating in oral history interviews, and aiding in the permanent preservation of correspondence and other valuable unpublished papers in archival repositories. Many physicists give personal, cash support directly to the Friends of the AIP Center for History of Physics. Through The American Physical Society they support a Division of History of Physics, created in 1980 and already gathering together more members than some of the Society's older divisions; the Division is very active in arranging sessions of historical papers at meetings and in a number of other areas. Through government agencies such as the National Science Foundation and the Department of Energy, and through grants and donations by industrial corporations and private foundations such as Bell Laboratories, IBM, and the Sloan Foundation, physicists and their friends are supporting a number of important projects. Examples are a project to publish all of Einstein's papers and correspondence; an American Institute of Physics study of preservation of historical documentation at government-contract laboratories; an International Project in the History of Solid State Physics; and a Laser History Project.

A sturdy institutional base guarantees the continuation of such activities. To be sure, the years since 1970 have seen a severe weakening in universities of many academic fields and particularly the humanities, and this has affected all fields of history. Important history of science departments and groups have been weakened or even disbanded. But outside or alongside the universities, American history of science as a whole has been strengthened in the past half-dozen years by the creation of new institutions such as the Charles Babbage Institute for the History of Information Processing, the IEEE Center for History of Electrical Engineering, and the Center for History of Chemistry. Modelled initially on the AIP Center for History of Physics but supported by their own respective disciplines, these centers not only complement but reinforce work in the history of physics itself. Meanwhile, at such places as the University of California, Berkeley, Office for History of Science and Technology, and the Smithsonian Institution, groups interested in the history of modern physics have grown vigorously.

There remain some intellectual weaknesses that have been present from the outset. Study of the history of modern physics has concentrated overwhelmingly on the theories of relativity and quantum mechanics, perhaps because of these theories' philosophical interest, and on nuclear physics, perhaps because of its social implications. Other fields such as solid state physics, which may be even more important in the long run of history, are only recently beginning to attract intensive study. Another weakness is that most historians, coming from a literary and theoretical tradition, have written far more about the history of theory than of experiment. Yet another problem is that nothing has been written about the history of physics in industry, except by the very few historians who have themselves worked as physicists in industry.

These deficiencies are rarely made good in first-person accounts by physicists, most of which also tend to center more on theory in the universities than on experiments or industrial research.

Meanwhile, the history of physics that gets written is still read mainly by people trained in physics. A few pioneering books, museum exhibits, and public television programs have reached a wider audience, so that general historians and the public at large are beginning to appreciate some features of the rise of modern physics—but only beginning. Much more research and writing must be done before nonphysicists can get a good feeling for the history of physics, both in its own right and as an integral part of modern history as a whole.

PHYSICS TODAY articles give an overview

We reprint here a selection of articles from the American Institute of Physics magazine PHYSICS TODAY. The magazine, like the AIP itself, was founded partly in hopes of providing common services that would help keep the physics community, with its highly diverse interests, from suffering fragmentation into subdisciplines. History articles, which are perhaps the most generally popular of all the types published in the magazine, have served especially well in binding the community together, if only by giving what one physics student described as a feeling for the shared "lore and traditions" of the discipline.

No sampler of writings could give a comprehensive picture of the history of modern physics; such a comprehensive picture has indeed never been attempted by any author. The articles reprinted here are more like pieces of a mosaic, with much blank space in between. Yet by looking over the scattered pieces the reader can get an idea of the mosaic as a whole, that is, of what has happened in physics over the past two or three generations. These pieces by their very heterogeneity may give a truer impression than could be found in a single synthetic work.

We have had space for less than half of the history articles that were available, and anyone looking over back issues of PHYSICS TODAY will find other articles of a quality as high as those included here. Reasons of balance, no doubt somewhat arbitrary, have dictated hard choices. Also not included here, but very useful for historical purposes, are the PHYSICS TODAY obituaries. Through these writings the physics community maintains a tradition of respect for its past members, a tradition once shared by all scientific disciplines but which most other fields have allowed to lapse. Finally, the magazine's staff-written news columns and particularly its "Search and Discovery" section have always contained much of interest. In articles such as Gloria Lubkin's annual pieces on Nobel Prize winners, these columns contain as much historical investigation as current journalism.

Different ways to read this book

We advise readers that a history collection like this one should not be approached as you would approach a physics textbook—not with that grim determination to read through from the start until you reach the end of the book, or your patience, or the semester. Read this book more as you would read a physics journal: skim the titles to find one that sounds interesting, dip into the article to see if it is appropriate to your interests (there are pieces here which are suitable for high school students, and others that assume knowledge around the graduate student level), and then read all or perhaps only parts of the article. The pieces can be read in any order, although we have put them in a rough sequence by way of offering suggestions.

Bear in mind that this book mixes together two types of historical writing which should be read in different ways. The difference is a traditional one, noted, for example, in 1891 by the great historian Frederick Jackson Turner. "The antiquarian," he wrote, "strives to bring back the past for the sake of the past; the historian strives to show the present to itself by revealing its origin from the past." Turner was more concerned with lessons for the present than with what he called the "dead past."[4] He and many later historians have striven to find general rules that might guide us—if only the famous rule that the one thing we learn from history is not to be surprised by anything that happens. Many of the articles here do aim "to show the present to itself," using historical evidence to uncover the patterns of human action that shaped the physics community and that continue to shape it. The form of a bird's wing can be understood only if you know the evolutionary history of birds.

Much of the writing in these pages, however, has been done "for the sake of the past." One thing we have learned from historical and sociological studies of physicists is that most people in the discipline work less for material rewards such as wealth or leisure, which few scientists can expect, than for the privilege of putting their life's effort into an imperishable structure. Whether as discoverer or teacher, the goal is to leave a part of oneself within the ever-growing and immortal entity that is physics. Physicists are therefore specially concerned that their discoveries be justly remembered, and that their colleagues and predecessors likewise be remembered for what they did. Only through such a tradition of memory can they feel themselves firmly placed in time, whether past, present, or future. One purpose of reading and writing history is to confirm this sense of identity within the community.

History has important lessons for today

Yet even reminiscences designed as a simple memorial to past events are at the same time lessons in the traditions of the community. These lessons are aimed at the present: what the great figures of older days did, the author may imply, we in our own lives should emulate (or if the result was bad, avoid). The wise reader will therefore inspect every writing, however much it seems to stay in the past, for the advice it may imply; historians and sociologists will even use such writings as evidence for standards set up for scientific behavior. Of course, the wise reader will also notice that articles which seem to analyze the past only in order to reveal patterns of present concern, are reciprocally invaded by an interest in the past for its own sake. Nobody can be a good historian, or for that matter a good physicist, who does not respect, as individuals in their own right and in their own times, the people who laid the foundations for our present world.

1. See Florian Cajori, "The Pedagogic Value of the History of Physics," School Rev., 278–285 (May 1899); Lloyd W. Taylor, Physics: The Pioneer Science (Houghton Mifflin, Boston, 1941).

2. Symposia: "Use of Historical Material in Elementary and Advanced Instruction," Am. J. Phys. 18, 332 (1950); Proceedings of the International Working Seminar on the Role of History of Physics in Physics Education (University Press of New England, Hanover, NH, 1972).

3. Thomas S. Kuhn, John L. Heilbron, Paul Forman, and Lini Allen, Sources for History of Quantum Physics. An Inventory and Report (American Philosophical Society, Philadelphia, 1967).

4. F. J. Turner, "The Significance of History," The Varieties of History, edited by Fritz Stern (Vintage, New York, 1972), p. 201.

HISTORY OF PHYSICS

Edited by

Spencer R. Weart
Center for History of Physics
American Institute of Physics

Melba Phillips
Emeritus Professor of Physics
University of Chicago

Table of Contents

Author Affiliations

Luis W. Alvarez, holder of the Nobel Prize in Physics, is Emeritus Professor of Physics at the University of California at Berkeley. (p. 198)

Philip W. Anderson, holder of the Nobel Prize in Physics, has been a staff member of the AT&T Bell Laboratories, and is professor of physics at Princeton University. (p. 194)

Lawrence Badash is professor in the History Department of the University of California at Santa Barbara. (p. 103)

Felix Bloch (1905–1983), holder of the Nobel Prize in Physics, taught in Zürich and Leipzig and was professor of physics at Stanford University. (p. 319)

Ferdinand G. Brickwedde is Evan Pugh Research Professor of Physics Emeritus in the Department of Physics at Penn State University, University Park. (p. 208)

Laurie M. Brown is professor in the Department of Physics and Astronomy at Northwestern University. (pp. 340, 346)

Stephen G. Brush is professor in the Department of History and the Institute for Physical Science and Technology at the University of Maryland at College Park. (p. 42)

Karl T. Compton (1887–1954) was a professor of physics at Princeton University and then President of the Massachusetts Institute of Technology; he served on many important boards and committees. (p. 74)

Edward U. Condon (1902–1974) taught physics at Princeton and Washington University, St. Louis; he was Associate Director of the Westinghouse research laboratories, Director of the National Bureau of Standards, and Director of Research and Development for the Corning Glass Works. (pp. 130, 310)

David H. DeVorkin is Chairman of the Department of Space Science and Exploration in the National Air and Space Museum of the Smithsonian Institution. (p. 50)

Albert Einstein (1879–1955), holder of the Nobel Prize in Physics, was professor of physics at the University of Berlin and member of the Institute for Advanced Study, Princeton. (p. 243)

Anne Eisenberg teaches science writing at the Polytechnic Institute of New York. (p. 234)

Paul P. Ewald, now in retirement in Ithaca, New York, was professor of physics at Stuttgart Polytechnic University, the Queen's University in Belfast, and the Polytechnic Institute of Brooklyn. (p. 68)

Enrico Fermi (1901–1954), holder of the Nobel Prize in Physics, was professor of physics at the University of Rome, at Columbia University, and at the University of Chicago. (p. 282)

Anthony P. French is professor in the Department of Physics at the Massachusetts Institute of Technology. (p. 138)

Otto R. Frisch (1904–1979) worked in Germany, the Niels Bohr Institute in Copenhagen, and in England, where he was professor at the Cavendish Laboratory of Cambridge University. (p. 272)

Richard K. Gehrenbeck is associate professor in the Department of Physics and Astronomy of the University of Rhode Island. (p. 324)

Samuel A. Goudsmit (1902–1978) studied at Leiden University; he was professor of physics at the University of Michigan, senior scientist at Brookhaven National Laboratory, visiting professor at the University of Nevada, and editor of the *Physical Review*. (p. 246)

John L. Heilbron is professor in the Department of History and Director of the Office for History of Science and Technology of the University of California at Berkeley. (pp. 14, 303)

Lillian Hartmann Hoddeson is a member of the Physics Department of the University of Illinois at Urbana–Champaign and the historian of physics at Fermilab. (pp. 61, 346)

Charles H. Holbrow is professor of physics and Chairman of the Department of Physics and Astronomy at Colgate University. (p. 86)

Vera Kistiakowsky is professor of physics at the Massachusetts Institute of Technology. (p. 149)

Martin J. Klein is Eugene Higgins Professor of the History of Physics at Yale University. (p. 294)

M. Stanley Livingston, now in retirement in Santa Fe, was professor of physics at the Massachusetts Institute of Technology, then Director of the Cambridge Electron Accelerator, and subsequently Associate Director of the laboratory now called Fermilab. (p. 255)

Edwin M. McMillan, holder of the Nobel Prize in Physics, is former Director of the Lawrence Berkeley Laboratory and Emeritus Professor of Physics at the University of California at Berkeley. (p. 261)

E. Mendoza has taught physics in the Physical Laboratories of Manchester University, England. (pp. 20, 25)

John D. Miller is professor of education at the University of California in Berkeley. (p. 29)

Mark L. Oliphant, now in retirement in Canberra, worked at the Cavendish Laboratory and directed physics laboratories at the University of Birmingham, England, and subsequently at the Australian National University. (p. 173)

Melba Phillips (*editor*), now in retirement in New York City, is Emeritus Professor of Physics at the University of Chicago. (p. 78)

Robert Rosenberg is a research associate in the Edison papers project at Rutgers University. (p. 108)

Robert G. Sachs is professor in the Physics Department of the University of Chicago and Director of the Enrico Fermi Institute there. (p. 228)

Roland Schmitt is the General Electric Company's Senior Vice President for Corporate Research and Development, directing the GE Research and Development Center in Schenectady, New York. (p. 354)

Robert S. Shankland (1908–1982) was Ambrose Swasey Professor of Physics at Case Western Reserve University. (p. 36)

Alice Kimball Smith is Dean Emeritus of the Bunting Institute at Radcliffe College. (p. 221)

Cyril Stanley Smith is Institute Professor Emeritus at the Massachusetts Institute of Technology. (p. 2)

Grace Marmor Spruch is professor of physics at Rutgers University. (p. 214)

George P. Thomson (1892–1975), son of J. J. Thomson and holder of the Nobel Prize in Physics, was professor at the University of Aberdeen and the Imperial College of Science, and Master of Corpus Christi College, Cambridge. (p. 289)

George E. Uhlenbeck studied at Leiden University; he was professor of physics at the University of Michigan and is Professor Emeritus of Physics at the Rockefeller University, New York. (p. 246)

Charles Weiner, former Director of the Center for History of Physics at the American Institute of Physics, is Professor of History of Science and Technology in the Program in Science, Technology, and Society at the Massachusetts Institute of Technology. (pp. 115, 221, 332)

Spencer R. Weart (*editor*) is Manager of the Center for History of Physics at the American Institute of Physics. (pp. 123, 159)

Victor F. Weisskopf, a former director of CERN, is Institute Emeritus Professor of Physics and Senior Lecturer at the Massachusetts Institute of Technology. (p. 358)

John A. Wheeler is Ashbel Smith Professor and Blumbert Professor of Physics and Director of the Center for Theoretical Physics at the University of Texas at Austin. (p. 272)

—Chapter 1—
Before Our Times

The main subject of PHYSICS TODAY is the subject declared in the magazine's name, but we all recognize there is much to learn from the past as well as from the immediate present. For many people the past is simply what they remember themselves, perhaps supplemented by what acquaintances remember of their own lives. But since time changes neither physical law nor human nature, there can be an equal fascination in stories of events long vanished from living memory. This section gives some of those stories, arranged in roughly chronological order.

In various writings Cyril Stanley Smith has shown how, long before science began, people were working to appreciate the order of nature with both aesthetic sensitivity and ingenious logic—a type of work that has only become more important over the centuries. It was not until the time of Galileo, however, that a few people began to organize observations by means of laws whose validity all serious thinkers could acknowledge. Great figures like Galileo and Newton are the subject of numberless scholarly articles and books, but our PHYSICS TODAY authors have preferred to write about matters less familiar to the average physicist. Some of the articles in this section,

especially the pair by E. Mendoza, summarize a broad area with particular attention to correcting historical myths that are still all too prevalent. The articles on Franklin and Rowland go further, showing how mid eighteenth and late nineteenth century "natural philosophers," or at least these particular two individuals, approached physics as a whole—an intellectual enterprise with aims somewhat different from what most physicists claim today.

The articles on Michelson, Poincaré, and the Hertzsprung–Russell Diagram take a still more focussed approach. Each shows a particular scientific subject as it developed over a few years or decades. The difficulties that researchers encountered in each case are good examples of the sort of problems all physicists and astronomers must face, and it is worth noting how the problems were (or were not) surmounted. It is also worth noting that in none of these cases did science advance by the fully modern mode with its extremes of hasty competition and teamwork.

Incidentally, these three articles are the only ones in this book that deal with astronomy; articles on the history of modern astronomy are included in the first volume of this reprint series, *Astrophysics Today.*

Contents

The prehistory of
SOLID-STATE PHYSICS

PHYSICS TODAY / DECEMBER 1965

By Cyril Stanley Smith

Introduction

Prehistory implies the selection of a date when history begins. In solid-state physics this is very recent, dating, perhaps, from Debye's specific-heat theory of 1913, but most of all from the famous diffraction experiment of Friedrich, Knipping, and Von Laue in March 1912. It was this tool of perfection which laid the ground for imperfection to become of interest to physicists. The growth of solid-state physics marks, I think, a basic change in the attitude of physicists toward matter. Virtually all the development of mechanics, marvellous though it was, was based on a treatment of matter that was essentially structureless and whose measured elastic constants and densities gave the constants to put into equations that became ever more elaborate. When physicists at last paid attention to the structure of real crystals, they soon became aware of imperfections, both theoretically and experimentally, and the great flourishing of solid-state physics in the last three decades has been mostly based on the elucidation of the role of mechanical, ionic, and electrical imperfections in a crystal, accompanied, of course, by a continued development of understanding of bonding and dynamics of the ideal lattice.

There would be no physics at all if it were not possible to find models ideal enough to compute and sufficiently close to reality to be meaningful: this has meant selecting areas of study one after another in which this approach would be most fruitful at a given time and ignoring others. It is nevertheless interesting to read nineteenth-century treatises on physics, whether research papers or

Cyril Stanley Smith is Institute Professor at the Massachusetts Institute of Technology. His article is based on a lecture at the meeting of the American Physical Society in New York on January 29, 1965, which began by the author's remarking: "Those who know me will suspect that the title is a disguise for a talk on the history of metallurgy. They will be partly right, though a subtitle might be *The interplay of mathematics and aesthetic empiricism in science.* If here I overemphasize empiricism, it is because I am talking to physicists—a talk to practical metallurgists would, conversely, overemphasize the value of mathematical theory."

textbooks, and to note the avoidance of the real structure of matter. Despite the development of good crystallography early in the nineteenth century and despite the development of an essentially valid ball-stacking model of ionic crystals as early as 1812, virtually all nineteenth-century physics, when it dealt with any structural concepts at all, was based on the molecule. This is not, perhaps, surprising, since the molecule had such a magnificent quantitative success in the kinetic theory of gases and in explaining the composition of chemical compounds. (It is notable, however, that chemists studied only those compounds that fitted the theory, and Bertholet and others who insisted that analyses frequently did not agree with the law of simple multiple proportions were ignored.) Then Cauchy's model of crystal elasticity based on a simple lattice failed to agree with measurement, and all crystalline properties were referred to the anisotropy of the molecule as a unit, not to the arrangement of the units. Von Laue remarks in his *History of Physics* that "no physical phenomenon [of the nineteenth century] required the acceptance of the space lattice hypothesis." I think he should rather have said that physicists refused to accept the concept, for the phenomena themselves certainly depended on lattices, while physicists overexploited the adjustable flexibility of the molecule to explain all anisotropic behavior, whether optical, thermal, elastic, or electrical. Perhaps the most revealing index of this blindness is that the great Von Laue himself, a month before he had the epoch-making idea of the diffraction of x-rays from the three-dimensional crystal grating, had to be told by a graduate student that some people supposed that atoms might be arranged in a regular array in a crystal. It was a measure of his greatness how quickly he saw the significance of the relationship to his theory of crossed optical gratings; and it is a measure of greatness again, and of the times, that the graduate student, Paul Ewald, went on to write the first text intended for physicists in which the properties of matter are realistically discussed on the basis of their real structural and mechanical behavior. This was his section in the eleventh edition of Müller and Pouillet's *Lehrbuch der Physik,* written in 1927-28. Ewald drew heavily upon the experimental work of Mark, Polanyi, and Schmid, on the metallurgist's study of grain growth and the properties of single crystals. It was symptomatic that this was an edited book with chapters by different specialists.

There is something about the very nature of physics itself that has produced this late development: one cannot simultaneously have two views of the world, a broad and a narrow one. Perhaps, indeed, physics could turn to real solids only after some centuries of concern with simple mechanics, and perhaps solid-state physics could only result from a fusion of two streams of knowledge which had to have time for development in isolation before they impinged on each other with exciting results. In the seventeenth century, when qualitative speculation was still permitted, a natural philosopher could enjoy the diversity of properties of solids, which were explained in terms of the interaction of imaginary corpuscles or parts; rigorous physics following Newton quite rightly discouraged such speculation, but unfortunately the discouragement served also to exclude any interest in the phenomena.*

However, concern with the real behavior of matter, if not a physicist's characteristic, is certainly a human one. The evolutionary advantage that accompanied the ability to exploit the cracking of stone gave rise to man himself. Studies, or perhaps I should say enjoyment, of the plasticity, crystallization, and vitrification of silicates and the selective absorption of certain wavelengths of light by metallic ions in an appropriate environment gave rise to the magnificent art of ceramics. The making of jewelry, tools, and weapons involved knowledge, if not atomistic understanding, of virtually every property now being studied by physicists except electrical conductivity and the effects of irradiation. There is something about man's relationship to matter through his senses

that inspired him to experiment empirically with the effect of heat on natural substances, singly and in mixture, at the same time that he was experimenting with social organization and long before he began to develop the more intellectual mechanical arts. Virtually not until the twentieth century did the engineer outstrip the materials that had been discovered 4000 years earlier; and progress in metallurgy had been mostly that of making more of the old metals and alloys more cheaply.[1] Thanks largely to recent discoveries of physics—at first electricity and lately nuclear fission—the metallurgist is now forced to be more qualitatively creative than he has been for many centuries. I use the word "quality" intentionally, for I believe that quality (in both of its meanings) has inspired human advance far more than has numerical quantity.

Philosophy—Aristotelian and corpuscular

Greek philosophy was much concerned with qualities, culminating in Aristotle's theories of matter, in which the four elements carried the elemental qualities—hot, cold, dry, and moist—in various combinations in a body to give rise to all of the properties that were perceptible to the senses. These ideas dominated most thinking until the seventeenth century, and most explanations of the nature of bodies lay in purely ad hoc suggestions as to the relative amounts of the qualities, with an ingeniousness but disregard for verifiability that we find shocking today. Nevertheless, it should be noticed that it is precisely the qualities that concern the solid-state physicists that were then regarded as central to understanding of matter—conductivity, plasticity, fusibility, color, texture, and hardness. The seventeeth century saw the end of this. Physics—mathematical physics in the pattern that was nucleated in the Middle Ages, began to crystallize around Galileo, and reached marvellous maturity with Newton— changed all this, for qualities could not be calculated, and even when it became possible to measure "properties" something had to be left out, everything dependent on the interaction of many parts. Mechanics and optics alone proved amenable to mathematical treatment.

Virtually every advance since the seventeenth century has stemmed from the unwillingness of the physicist to talk vaguely about things that cannot be reduced to computable models whose inaccuracies can be exposed and removed by continual interaction with experiment. Science is in very essence both mathematical and experimental, but at times one or the other viewpoint has grown

---

*I don't wish to accuse physicists of being particularly perverse in refusing to look at crystals. The most recently published history of the constitution of matter bears the promising title, *The Architecture of Matter,* but it is concerned almost entirely with atomic and subatomic concepts. Even historians seem to be unable to see beyond atomic or molecular bricks to the magnificently diverse structures that are composed of them, unless they go the whole hog and study cosmology at the other end of the scale, equally intangible and so equally capable of being oversimplified for the purpose of thought.

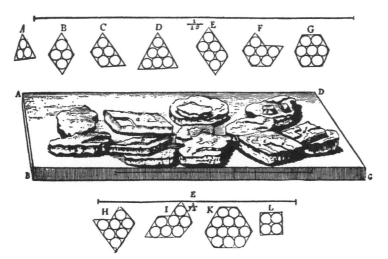

Drawing by Robert Hooke showing packing of spheres to match polyhedral shapes in alum and salt crystals. (*Micrographia*, London 1665). This drawing and illustrations on pages 21, 26, and 29 were taken from C. S. Smith, *A History of Metallography*, Chicago 1965.

beyond balance. That most marvellous of physicists, Robert Hooke, wrote in 1665: ". . . and here the difficulty is . . . least by seeking to inlarge our Knowledge, we should render it weak and uncertain; and least by being too scrupulous and exact about every Circumstance of it, we should confine and streighten it too much."

The idea that many properties were somehow related to the interaction of smaller units of structure was developed by Democritus and other early Greek thinkers and might have reached fruition by interaction with the Pythagorean emphasis upon form had they not been rejected by the most authoritative Greek philosopher, Aristotle. How different the history of science might have been had he been an atomist, or had his work called forth constructive criticism instead of adulation! Really creative thinking occurred again only in late medieval times after the revival of the forgotten atomism. Marshall Clagett at the recent Montreal meeting of the History of Science Society discussed Nicholas Oresme's remarkable fourteenth-century ideas in which he makes the qualities themselves depend on form. He says: "The ratio of intensities is not so properly or so easily attainable by the senses as is the ratio of extensions," and then describes how to plot the intensity of a quality normal to the extension of the substance, and discusses a kind of resonance between adjacent bodies depending upon the conformity and difformity of the arrangements of their representations in quality space. Remarking that experience and philosophy alike show that all natural bodies determine their shapes in themselves, he says they also determine in themselves the qualities that are natural to them, and that, "In addition to the shape that

these qualities possess in their subject, it is necessary that they be figured with a figuration that they possess from their intensity," and, "It is necessary that qualities of this sort have diverse powers and action depending on the difference in figurations previously described." He does not quite go on to describe a Brillouin-zone polyhedron, but his remarks on the mutually conformable configuration of qualities in seeds would not startle a modern biologist.

Oresme's ideas were based on an intuitive feeling for form. His realization that the intensity of a quality could be plotted so as to make it appreciable to the senses was a great inspiration, but it led to no immediate development. Everyone knows of the great developments of astronomy that occurred in the sixteenth and seventeenth centuries; few people have studied the equally interesting but less fruitful studies on the properties of matter that occurred at the same time, for the practical consolidation of knowledge in this area was not accompanied by a theory of the kind that could become part of the mathematical mainstream of science.

In the seventeenth century, natural philosophy reached its prescientific height and this was the last time for three centuries that respectable thinkers concerned themselves with the properties of real solids. Atomism, or at least corpuscular philosophy, was invoked to explain everything; but the shape of the parts, like the proportions of the preceding Aristotelian qualities, were adjustable ad lib, and could not be expressed in the soon-to-be-mandatory mathematical form or related to experiment. Nevertheless, there are some seventeenth-century writings that are entrancing for a twentieth-century solid-state physicist to read. In a purely qualitative way, physicists and philosophers deduced models of behavior based upon shape, size, and interaction of parts which (if we properly select for each occasion the appropriate unit as an atom, molecule, subgrain, microcrystal, or crystal) are qualitatively as we would have them today. Molecules are formed by parts of different shapes sticking together, and metals are plastic because the parts can slide over each other and change neighbors without losing coherence. Descartes, who had watched wrought iron coming to nature in the molten bath of a fiery hearth, saw that there was something about particles on one scale which enabled them to be joined into grains within which cohesion was greater than with other grains, though oddly he failed to see that the grains were crystalline. The most popular Cartesian physicist, Rohault, in his *Traité de Physique* (1671), supposes that plastic materials are made of parts with complicated textures intermixed with each other, hooked together like the rings of a chain or entwined like the threads of a cord, while brittle bodies are of simple texture with particles touching one another at only a few places. He talks about the preferred orientation of particles after hammering or drawing, and the preferential clumping of particles into grosser particles under the influence of heat, structures which in steel can be preserved by quenching and are responsible for its hardness. Somewhat later (1722), these ideas in the mind of the great Réaumur led to the inversion of the ancient belief that steel was a purified iron (logical enough, since steel resulted from prolonged treatment in fire, which does usually purify) and he suggested that it arose from the addition of some particulate matter ("sulfurs and salts") which could be distributed or segregated by heat treatment within a hierarchy of structures of iron particles with accompanying hardening or softening.

Another Cartesian physicist, Hartsoeker (1696), let his imagination run wild. He cooked up all kinds of amazing contraptions to explain the properties of matter. Corrosive sublimate becomes a ball of mercury with, stuck all over it, particles of salt and vitriol shaped like needles and cutting blades; air is a hollow ball built of wirelike rings to give it the necessary elasticity. He conjectured that the particles of a substance like iron, which is hard when cold but malleable when hot, must have teeth which slide over each other when the particles of heat have sufficiently separated them; the parts of mercury, being spherical, can slide

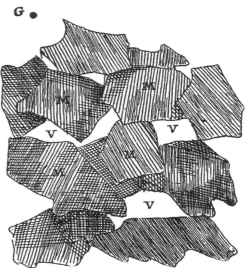

Drawing by R. A. F. de Réaumur (1772) showing "A grain of steel as it would look if it were vastly enlarged. Its natural size is shown in G. MMM are the molecules of which the grain is composed; VV the voids left between them". Réaumur explained the conversion of iron into steel by diffusion into the iron of particles of reducing and saline matter from the cementing compound. He explained the hardening of steel by the redistribution of this matter between the grains and intergranular spaces. He had no concept that there was crystalline order within the grains.

easily between polyhedral particles of gold (is not this indeed the basis of liquid-metal embrittlement?) and so on. After numerous specific examples he ends, "But I do not wish to deprive the reader of the pleasure of himself making the search following the principles that have been established above." It is precisely this element of uncontrolled imagination in the speculation that made respectable physicists turn their back on this kind of thinking. Yet the particle, of course, usually without

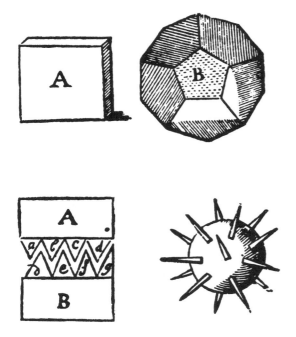

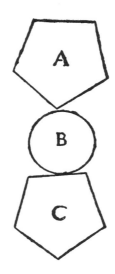

Conjectural shapes of the particles of matter according to the corpuscular physicist Nicholas Hartsoeker (1696). The spherical ball with attached spikes represents mercuric chloride; the toothed pieces are iron, which is hard when cold because the particles interlock, but is easily forged when heat particles distend the parts so that they can slide over each other.

any such specific remarks as to its shape and packing, was accepted by virtually everyone after the
middle of the seventeenth century.[2] As in so many
things, Newton provided (in the notes to the second edition of his *Opticks* [London: 1718]) a summary of a viewpoint beyond which it was unwise
to go:

> There are therefore Agents in Nature able to make
> the Particles of Bodies stick together by very strong
> Attractions. And it is the Business of experimental
> Philosophy to find them out.
>
> Now the smallest Particles of Matter may cohere by
> the strongest Attractions, and compose bigger Par-
> ticles of weaker Virtue and many of these may co-
> here and compose bigger Particles whose Virtue is
> still weaker, and so on for divers Successions, until
> the Progression end in the biggest Particles on
> which the Operations in Chymistry, and the Col-
> ours of natural Bodies depend, which by co-
> hering compose Bodies of a sensible Magnitude. If
> the Body is compact, and bends or yields inward to
> Pression without any sliding of its Parts, it is hard
> and elastick, returning to its Figure with a Force
> arising from the mutual Attraction of its Parts. If
> the Parts slide upon one another, the Body is mal-
> leable or soft. If they slip easily, and are of a fit
> Size to be agitated by Heat, and the Heat is big
> enough to keep them in Agitation, the Body is
> fluid. . . .

This is not the Newton of the *Principia* speaking,
but it was the *Principia* that set the tone for physics. Virtually all speculation on the nature of solids
disappears thereafter from the writings of good
physicists for two centuries. The physics of solids
was limited almost exclusively to idealized elasticity, a favorite subject with mathematicians as well
as physicists, but one which, except for the mathematical atomism of Boscovich, was divorced from
any concepts as to ultimate structure. This is not
to say that there were not speculations on the nature of crystals and even some marvellous mathematics of crystallography to which I will return
later, but both of these were outside the mainstream of physics. But I wish to return to the
theme of qualities and take up another thread.

The alchemists

The nuclear physicist can laugh at the alchemist's
misguided attempt at transmutation, but the solid-
state physicist shouldn't. Transmutation has not always had today's connotation of a change in the
nucleus of an atom. Looked at qualitatively, the
change from a mixture of sand and ashes into
glass, from a mixture of malachite, calamine, and
charcoal into gleaming brass, or from a white
fabric into an Emperor's purple robe *is* a most
spectacular and fundamental change. To Aristoteli-
ans, the whole difference between substances lies in

their particular combination of qualities, and since
it is clearly possible to produce some of these at
will, why not others? The modern materials engineer is producing new qualities all the time, but
he does not call it transmutation.

As has been argued especially well by Hopkins
in his *Alchemy, Child of Greek Philosophy* (1934),
alchemy began reasonably enough on the basis of
the well-known changes in color and nature which
had long been exploited by artisans for decorative
purposes in goldwork, in enamels and in dyeing.
It was supported by the belief that somehow behind these changes there lay a key to the relationships and transformations in the larger world (a
view that anyone with a spark of the artist in him
must admire) but it failed eventually simply because the adepts came to have too great a belief
in the premature theory, and they became too preoccupied in the observable qualities rather than
their compositional causes, and so were unable to
benefit from the innumerable experiments that
were done. If the yellow matter that came from
heating copper with certain substances was regarded as only an inferior gold, the experiment
was a failure: it could have been regarded as a
more castable, harder, and resplendent form of
useful copper. Yet what wonderful physical changes
the alchemists produced, and how fervently and
how rightly they believed in the significance of
the difference between the qualities of a shiny
ductile metal; a black, brittle sulfide; a crumbly
crystalline salt; gleaming, hard diamond; infusible
earths; and the vapors, phlegms, and tars that
came from distilling animal and vegetable matter.
These properties are the subject of solid-state
physics, but there were no solid-state physicists in
those days.

The beauty of alchemical mysticism attracted ad-
herents long after it was obvious that it was not a
fruitful guide (obvious in retrospect, that is). It
was slowly replaced by the belief that eventually
became the mainstream of chemistry that the qualities were dependent upon composition and that
they were not dependent only on the units but also
sometimes on the manner of combination. At first,
however, the qualities needed an embodiment, and
perhaps largely under the influence of miners, mercury and sulfur (the philosophical kinds, not the
ordinary materials) were thought to account for most
substances by their varied combination. Sulfur represented the inflammable principle, the soul, the
fire of Aristotle, while mercury was the materializa-
tion of the fluidity principle. Paracelsus early in
the sixteenth century methodized this viewpoint and
added a third principle, salt; he also directed chem-

An eighteenth-century metallurgical laboratory with apparatus for determining the physical and chemical properties of metals. (William Lewis *Commercium philosophico-technicum*, **London 1763**)

istry toward a useful practical purpose, medicine, and away from its domination by mystic philosophy. Salt, sulfur, and mercury—excellent examples of ionic, Van der Waals, and metallic bonding; had diamond with its covalent bonding been added, all types of today's quantum theory of solids would have been represented. The problem was in the realm of solid-state physics, but there were no solid-state physicists in those days.

No physicist arose to meet the challenge, but chemists had to do something and so did practical smelters and assayers of ore. The inflammable principle, the reducing principle, the sulfur of Paracelsus, was supposed to be transferred from charcoal to a metal ore when the latter was converted to metal. It became the terra pinguis, the unctuous earth of J. J. Becher in 1667, and was elevated to that important chemical principle, phlogiston, by J. H. Stahl, a metallurgical chemist, in 1703. Very much of eighteenth-century chemistry revolved around the phlogiston theory and the degree to which this evanescent material was transferred from one substance to another in reaction. But the study of reactions was now being done systematically, and tables of affinity appeared—the first in 1718—putting substances in order of their affinity for each other, each being able to displace those above it from compounds. Though phlogiston

had some of the chameleon-like variability of the alchemist's elusive elixir, it was responsible for metallicity and its loss left a calx (an oxide in today's terminology). The presence of an excess of it changed iron into steel, and still more into cast iron. Parallel with the phlogiston studies went an intensive study of the composition of matter, sparked to some extent by the desire to duplicate Chinese porcelain. The definition of element became something that could not be chemically broken down and it appeared that there were many elements, though not an infinite number. Analytical chemistry evolved from the assayer's ancient technique of extracting the noble metals in weighable metallic form, usually on the basis of ingenious pyrochemical reactions, and became broadly applicable when it was found in the eighteenth century that compounds of definite composition could be precipitated reproducibly by reaction in aqueous solution and weighed. This was accompanied by a growing interest in the role of gases and the rather sudden appreciation of the chemical role of atmospheric oxygen, which quickly demolished the phlogiston theory. The new chemical nomenclature of Lavoisier and his associates tied together all of the analytical data into a clear listing of the elements and their relationships in numerous natural and artificial compounds, and there-

15. Depositing Arrangement No. 6.—*Deposition by Magnet and Coil* (Fig. 14).— We may produce deposition in the separate liquid by connecting the two pieces of immersed metal with any other source of depositing power—for instance, if a long copper wire A, covered with silk or cotton, is coiled upon a large bar of pure soft iron B, and its ends C and D are immersed in a solution of sulphate of copper E, and the poles of a powerful horse-shoe magnet F are brought in contact very many times with the end of the bar, and every time before removing the magnet from the bar one of the ends

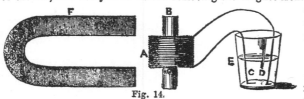

Fig. 14.

of the wire is taken out of the liquid, and re-placed before returning the magnet, one end of the copper will slightly dissolve, and the other receive a thin copper de-posit; but if each of the ends is allowed to remain constantly in the liquid, no such effects will occur.

An early phlogiston pump (not so named!). From G. Gore, *Theory and practice of Electrodeposition*, **London 1856**

after composition alone became the chemist's explanation for all of his phenomena. The chemist, the mineralogist, and the metallurgist were still almost the only people seriously interested in the nature of solids.

Looked at from today's viewpoint, it is obvious that the phlogistonists were right. The difference in properties between black brittle cuprite and shiny malleable copper *is* due to phlogiston: phlogiston is simply the valence electron in the conduction band of today's quantum theory. The phlogistonists did overlook the oxygen atom which trapped the electron, and this is a pretty large thing to overlook, but they were right physically if not chemically. They had to use other atoms (composition) to manipulate the phlogiston; today we simply pump phlogiston through an electrolytic cell, add it to ions, and get metal. A ton of aluminum, it turns out, needs just about two ounces of phlogiston for its preparation!

After the development of analytical chemistry in the 1780's, very many of the age-old properties of metals and other materials were found to be associated with specific compositions, and even very minor amounts of impurities such as phosphorous or sulfur in iron were found sometimes to be associated with great physical changes. One of the first triumphs—again under inspiration from the Orient, in this case in the form of the Damascus sword—was the discovery that it was minute but varying amounts of carbon, a real material substance now classed as an element, that was responsible for the striking differences between wrought iron, steel, and cast iron.

After this, composition per se was, for a time, regarded as a sufficient explanation of the wondrous diversity of properties of substances. Analysis provided the basis for the classification of substances. After the atomic theory of Dalton (which was no more of an atomic theory than had existed for centuries but was a really fine quantitative theory of simple molecules) chemists' eyes were for a

long time closed to compounds that were not simple. The reactions of metallurgy, which largely involve solid solutions, lost interest to the chemist, who now worked mostly with ionic compounds or aqueous solutions of them (or with organic molecules) and interpreted the simple ratios of atoms found by analysis as representing molecules. Superb quantitative proof of the existence of molecules was provided by the combining volumes of gases and by the kinetic theory of their PVT relations, but most of the chemist's precipitated compounds were actually in simple ratios only because of the geometric requirements of the crystal lattice. Physicists were of no help. If nineteenth-century physicists were interested in solids at all, they too talked about the relations of the molecules, though molecules were often supposed to be spatially oriented (not on lattice points but sometimes within unit cells) to account for the anisotropic properties of crystals.

Crystallography

The introduction of the crystal makes me take another leap back in time. Crystals initially were simply bodies with a certain geometric external shape, and quartz was the archetype. They were brittle, commonly transparent.

There are few subjects better adapted to elegant treatment by the mathematical physicist than is crystallography, yet, although physical properties of crystals were often measured, crystallography did not really become part of physics until after x-ray diffraction. Nineteenth-century physicists showed an almost incredible restraint in speculating on the details of the atomic, or as they would call it, molecular, arrangements responsible for the symmetrical anisotropy of the shapes and properties of crystalline matter. The mineralogists, however, fairly early realized the value of crystal measurement in the identification of minerals. Though much had been done before, it was Linnaeus' desire for classification in the realm of natural history that gave the real impetus to the collection of data on crystal faces and their angles, and the seeking of a satisfactory model that would explain them in their diversity. The great Haüy who was the first to develop the mathematics of the angular relationships did this on the basis of an earlier supposition that crystals were composed of aggregates of tiny polyhedra (called integrant molecules), with all faces that did not correspond to the plane faces of the unit arising from the removal of polyhedra in a simply stepped array of building blocks. Incidentally, he remarks that the similarity between different individual crystals of the same species is less evident than the similarity between different individuals of a biological species—a view that we find astonishing today with our mind on the perfect regularity of the space lattice as the main characteristic.

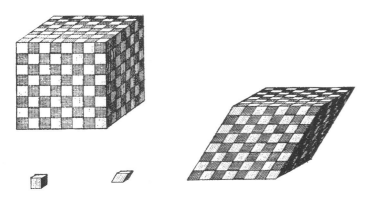

The assembly of polyhedral parts to give cubic and rhombohedral crystals (Grignon, *Essai de Physique sur le fer*, **Paris 1775). A model of this kind was used by Haüy as the basis of the first calculations of the angles between crystal faces in 1784.**

Models of crystal structures made by W. H. Wollaston in 1812. (Proceedings of the Royal Society, 1813)

Haüy explicitly disclaims the possibility of knowing the ultimate structure of matter, though he considered the structural units definitely to be polyhedra within which the molecular interactions were different from those outside; it was a kind of geometric package, and had been arrived at by Haüy as by others before him simply on the basis of observations on the disparity between the cleavage and the growth faces of crystals. To our minds, the stacking of balls seems to provide a more physically meaningful model of simple crystals, though mathematically, of course, there need be no difference between the two. It is therefore particularly interesting to see that the first thoughts about the nature of crystals involved exactly this model. It was suggested by Thomas Harriot about 1599 though first published by Kepler in 1611 and developed particularly by Hooke and Huygens later in the seventeenth century. Hooke, for instance, showed that all of the surfaces of alum crystals could be matched by stacks of globular bullets arranged in close packing, and he suggested that sea salt is built of globules placed in a cubic arrangement. He saw the relation of stacking to the sixfold dendrite formation in snow, though in characteristic Hookian fashion he merely outlined a program of study and did not follow it through. Huygens used a similar model with spheroids to explain the cleavage and optical properties of calcite, but after him the ball model disappears, to be replaced with stacks of polyhedra. Even more astonishing is the fact that when the stacking-of-spheres model is resurrected by the great Wollaston in 1812 and used to explain the nature of several bodies, including the alternate regular stacking of large and small spheres to account for rock salt, again it is rejected in favor of Haüy's approach. The polyhedra somehow seemed to lend themselves more readily to mathematics, and they were mathematically, though not physically, replaced somewhat later in the century by the more ideal point-group model of the mathematical crystallographers. Stacks of ball-shaped atoms came back again in a paper by Barlow in the 1880's—notice that it is the more pragmatic approach of the English, not the elegant mathematics of Continental physicists, that produced it. It was being well developed by Barlow and Pope, Sohncke and others, all of a chemical turn of mind, when x-ray diffraction suddenly provided the experimental handle to enable both the symmetry and the chemistry to be combined in a properly scientific scheme. The first surprise, indeed it was a shock, was the realization that no molecules existed in simple ionic crystals. The relationship between the ball-like atoms and the stacking polyhedra of the unit cell was thereafter clear to every freshman. It is ironic that just as the ball model was vindicated, the atom itself lost all reality and we have now turned to the neat polyhedra of the Brillouin zone as the most reasonable model of the unit of the crystal. For the first time, the model is one which is neither a determining unit nor a dominating array, but results from the two-way interaction between unit and arrangement.

This brings me to the subject I am mainly interested in, metallurgy, for nineteenth-century metallurgy is virtually a qualitative preparation for twentieth-century solid-state physics. Here, for the first time, the earlier qualitative speculation on the relation between structure and properties begins to take definite useful form. First, it was realized that the essence of crystallinity lies in internal order not in external form, and more important, that most solid inorganic bodies are composed of hosts of microcrystals. The knowledge that metals had a granular texture, of course, goes back to the earliest broken piece of metal, and the fracture test was the principal basis of selection and quality control for millennia. In the eighteenth century, Réaumur used experiments on fracture combined with Cartesian corpuscular philosophy to give the first good theory of steel, but it was not until the middle of the nineteenth century that the granular structure was experimentally shown to be microcrystalline. The nucleating observations occurred appropriately enough in the steelmaking town of Sheffield in England, almost exactly a hundred years ago, when Henry Clifton Sorby, for the first time in history, prepared the surface of a sample of steel carefully enough so that the structure could be seen under a microscope without the distortions that had rendered the structure invisible to early microscopists. The background of Sorby's use of acid to develop the structure is itself an interesting bit of history, for it has roots not only in the artist's etched prints and decoration of armor but also in the oriental "Damascus" sword, the etching of which led to the etching of meteorites. Sorby saw that metals did not crystallize under vibration—a long-lived myth—but were always finely polycrystalline. He saw that they could be distorted while maintaining crystallinity and would recrystallize either as a resut of an allotropic transformation as in steel, or simply on heating after straining by cold work. He also identified most of the phases now known in steel, but he did not continue in the field very long, and it was left to other workers who took up the subject

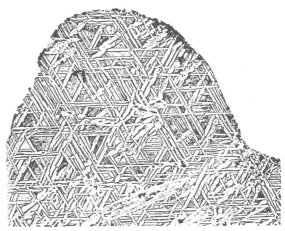

Print made directly from the etched and inked surface of the Elbogen iron meteorite by Schriebers and von Widmanstätten in 1813. Slightly enlarged

The earliest photomicrograph of a piece of wrought iron. Made by Henry Clifton Sorby in Sheffield in August 1864. Sorby's work showed conclusively that deformation did not destroy crystallinity.

after 1880 to reveal the richness of structure in metals and alloys, and to associate the changes of structure with the properties that had been empirically discovered and long used. Slip bands were seen in 1896, and in 1900 their nature and significance were appreciated by Ewing and Rosenhain. Slip interference soon became the metallurgist's theory of hardening.

Long before physicists began to get interested in problems of deformation and the nature of grain boundaries, metallurgists knew the phenomena intimately though empirically, and had developed their own naive little models to account for the behavior. Though through most of history the metallurgist's closest association has been with chemists, by the second decade of the twentieth century they were thinking in physical terms if not as physicists. Chemists had grown and studied metal crystals as curiosities for over a century, but it was the metallurgist, H. C. H. Carpenter, who first did significant mechanical tests on single crystals of metal, and it was the report of his work which triggered off G. I. Taylor's renewed interest in deformation that culminated in the invention of the dislocation. Almost a century earlier, collaboration with a practical cutler in work on the alloys of steel (partly aimed at duplicating oriental Damascus steel) had helped to give Michael Faraday the sense of structure which so dominated his thinking.[3]

I don't mean to say that metallurgists in the nineteenth century did not benefit from physics; indeed, their whole approach was always based upon a knowledge of college physics, the tamped-down general level of science which Derek Price properly regards as being the route through which science mainly influences technology. But it must be admitted that physicists were usually unable to work up interest in the complicated problems that concerned metallurgists. The metallurgist tends quite literally to enjoy the wide range of the behavior of metals, while the physicist will look only at those aspects that are ripe for understanding.

Faraday soon lost interest in metals, and it was a very good thing for science that he did. A few other physicists tried to look realistically at solids, but they had little following. There is the French physicist, Louis Savart, who in 1829, to explain the details of Chladni figures on vibrating plates, made some very acute comments on the structure of metals. He realized that normally there were assemblages of a vast number of little crystals packed together at random, but that preferred orientation would develop under special conditions of casting, working, and annealing. He observed the difference between the static and dynamic modulus of elasticity, and attributed changes of internal friction to structural relaxation. The elastic aftereffect attracted experimentalists, while Boltzman and Maxwell provided characteristic theories, the former purely mathematical, the latter based on changing molecular aggregations. Others who were concerned with complicated structures in relation to physical properties were M. L. Frankenheim and particularly O. Lehmann, whose *Molekular Physik* (1889) is a fine museum of phenomena that depend upon crystalline perfection and imperfection, and he had a sense of form that was more that of a biologist than a physicist. This viewpoint, however, did not find its way into textbooks, not even the advanced ones which decided what things the discipline of physics should be concerned with.

Finally, no one who, like myself, has experienced the wonderful stimulation that came to metallurgy from the impingement of physics in the 1920's, and especially right after World War II, can be blind to results of the joining of two streams of development that had been to some extent separate. One cannot deplore the earlier separation, for neither field was ripe for profitable interaction. Recently, however solid-state physics has advanced to the point of becoming a separate profession, and physical metallurgy has become metal physics. Though both fields have gained competence and immense utility, they have perhaps become less exciting, for the diversity of material behavior has been reduced to unitary phenomena that are well understood, at least "in principle." The framework for studying complexity is still lacking, and, deplorably, the study of it is not encouraged in most universities.

Metallurgists trained in the 1920's, as I was, saw in the richness of visible microstructure a key to the understanding of most of the phenomena that their predecessors had discovered and used. Most developments since then have been on an atomic scale, especially flowing from the application of that marvelous tool, x-ray diffraction. As a microscopist, however, I have been delighted to see the recent return to direct observation of the structures of irregular aggregates of imperfections with the electron microscope. Great things are certainly stirring, but I have a little feeling that with metallurgy and physics now so close together, the new viewpoint that will trigger off the next wave of excitement and advance will have to come from outside. Somehow, I think it must be a concern with far more complex things than have been allowed in the domain of respectable physics in the past. I wouldn't be entirely surprised if it comes from biology when the high fashion of biology returns from the molecule to the organism. It will certainly have some of the old natural historian's view in it, and it may even have a big dose of something as unscientific as art, for of all people the artist seems to be best able to make significant, if not always precise, statements about very complex interrelationships.

References

1. C. S. Smith, Materials and the Development of Civilization and Science, Science, **148**, 908 (1965).

2. For an excellent history of corpuscular philosophy, see Marie Boas, The Establishment of the Mechanical Philosophy, Osiris, **10**, 412 (1952). The metallurgical aspects are mentioned in C. S. Smith, *A History of Metallography*, chapter 8 (University of Chicago Press, Chicago, 1960 and 1965).

3. L. P. Williams, Faraday and the Alloys of Steel, *The Sorby Centennial Symposium on the History of Metallurgy*, C. S. Smith, ed., pp. 145-162 (Gordon and Breach, New York, 1965).

Franklin's physics

"Poor Richard's" ability to extract the heart from the matter
and express it plainly, evident in his work with electricity, led to the international
scientific reputation that preceded his political missions.

John L. Heilbron

PHYSICS TODAY / JULY 1976

Benjamin Franklin usually receives good marks for his physics from those who have taken the trouble to study it. To contemporaries he was the "Kepler of Electricity" (Volta being the Newton), the "Modern Prometheus," the "Father of Electricity." Among moderns, Robert Millikan credits him with the discovery of the electron and brackets him with Laplace as the two greatest scientists of the 18th century. Millikan, whose promotion of Franklin was perhaps intended to facilitate a reappraisal of the relative contributions of himself and J. J. Thomson to the investigation of electrons, went too far. But one does not have to consider Franklin a Kepler, Newton, Prometheus or Millikan to perceive that he was one of the most important natural philosophers of the Age of Reason.

Franklin's international reputation derived from his work on electricity, done primarily in the late 1740's and made public in the early 1750's. The reputation preceded and assisted his political missions to England and France. (The portrait shown in figure 1 was engraved for sale in Paris.) The relation between his electricity and his embassy may be taken as a symbol of the coherence of his life's work. The same cast of mind and habits of thought appear in his science, in his social and political writings, and indeed in the conduct of his printing business.

Plus and minus electricity

Franklin took up electricity in the winter of 1745–6, in his fortieth year, when his business no longer needed his full attention and yielded an income that could support learned leisure. Printing was by no means an inappropriate prep-

aration for an Enlightenment experimentalist; it taught some of the requisite qualities, the coordination of head and hand, familiarity with wood and metal, exactness, neatness, dispatch. The productive English electricians contemporary to Franklin also came from the higher trades: William Watson, an apothecary; John Ellicott, a clockmaker, and Benjamin Wilson, a painter. And printing, as practiced by Franklin, brought not only manual skills but also practice in straight and accurate thinking. In editing or composing, the successful printer had to be clear, economical and pertinent; everything was set by hand, and paper cost as much as labor. These experiences helped to frame Franklin's style. The same power of extracting the heart from the matter and expressing it plainly that delights us in the sayings of Poor Richard was ready to serve Franklin when he began to unscramble the phenomena of electricity. He also drew upon his experience of men and institutions. His success in building up his business and shaping his community, in mastering people and machines, no doubt supported his characteristic optimism, the expectation that he could control or cajole his environment.

The standard electrical demonstrations of the early 1740's employed the odd apparatus shown in figure 2. It had been introduced a decade earlier by Stephen Gray, formerly a dyer but then a resident of the London Charterhouse, where charity boys were always available for use as capacitors. One caught an urchin, hung him up with insulating cords, electrified him by contact with rubbed glass, and drew sparks from his nose. Franklin witnessed such sport in 1744, when a travelling lecturer in natural philosophy, one Dr Spencer of Edinburgh, visited the middle colonies. It was not Spencer's

operations, however, that made Franklin an active electrician, but the gift of a glass tube to the Library Company of Philadelphia (of which Franklin was a founding member) and the simultaneous appearance in the *Gentleman's Magazine* of an article describing the latest amusements procurable by electricity.

The *Gentleman's* was a lively monthly of political and intellectual news, published in London. The Library Company subscribed to it, and Franklin probably read it regularly. He was often the first to see it, in his capacity as postmaster of Philadelphia, and he had tried to introduce a colonial version, the *General Magazine for all the British Plantations in America*. This publication had run for six months in 1741, filled out, as was the *Gentleman's*, with bits and pieces taken from books and other magazines. The article on electricity that caught Franklin's fancy in 1745 was just such a filler, a translation of a piece published anonymously in a literary review called *Bibliothèque raisonnée*. Although written in French, the review was conducted by Dutch professors and published in Amsterdam (figure 3). The anonymous contributor of electrical news was Albrecht von Haller, the celebrated Swiss biologist, litterateur and all-round polymath, then a professor at the University of Göttingen. Franklin's first steps in electricity were guided not, as has been thought, by the works of Watson and Wilson or by his untutored imagination, but by a popular report in a Dutch journal of the latest findings of German electricians.

Haller's account includes a thought-provoking experiment in which the usual boy now stands upon insulating supports of pitch. He grasps or is tied to a chain electrified by the tube or by a globe spun by a machine like a cutler's wheel (figure

The author is professor of history and director of the Office for History of Science and Technology, University of California, Berkeley.

4); should anyone approach the boy, a spark will jump between them, "accompanied with a crackling noise, and a sudden pain of which both parties are but too sensible."

Franklin seized on this experiment, extended and simplified it, and made it the basis of a new system of electricity. Let two persons stand upon wax. Let one, A, rub the tube, while the second, B, "draws the electrical fire" by extending his finger towards it. Both will appear electrified to C standing on the floor; that is, C will perceive a spark on approaching either of them with his knuckle. If A and B touch during the rubbing, neither will appear electrified; if they first touch afterwards, they will experience a spark stronger than that exchanged by either with C, and in the process lose all their electricity. Gentleman A, says Franklin in explanation, the one who collects the fire from himself into the tube, suffers a deficit in his usual stock of fire, or electrifies minus; B, who draws the fire from the tube, receives a superabundance, and electrifies plus; while C, who stands on the ground, retains his just and proper share. Any two, brought into contact, will experience a shock in proportion to their disparity of fire, that democratic element forever striving to attach itself to each equally.

The form of this analysis appears to have been habitual with Franklin. His first published work, *A Dissertation on Liberty and Necessity, Pleasure and Pain*, which he printed up himself in 1725, considers the problem of freedom of the will in much the same terms as he later used to classify electrical sparks. Since God is omniscient, omnipotent and all good, our world, His creation, must be arranged for the very best: there is no room for liberty of action. The only cause for motion in the universe is pain, or

Scientist, philosopher, diplomat. This portrait of Franklin by F. N. Martinet was offered for sale in Paris with an inscription reading in part ". . .America has placed him at the head of scholars; Greece would have numbered him among the Gods." Note the lightning conductor visible through the window and the electrostatic apparatus behind the chair.   Figure 1

rather its avoidance; fortunately we do not lack sources of uneasiness, and keep busy seeking surcease. "The fulfilling or satisfaction of this desire, produces the sensation of pleasure, great or small in exact proportion to the desire." Franklin makes much of the exact proportion, or rather equality, between the stimulating pain and the relieving pleasure. The one supposes the other; should the pain last until the end, death will bring proportionate relief. Consider A, an animate creature, and B, a rock. Let A have ten degrees of pain. Ten degrees of pleasure must therefore be credited to his account; "pleasure and pain are in their nature inseparable." Let him then have his pleasure; he thereby returns to the neutral state, which B has enjoyed throughout. One cannot miss the analogy between the animating pain, the inseparable, equal compensating pleasure, and the inert rock, on the one hand, and negative electricity, positive electricity, and the neutral state, on the other.

The chief result of this analysis, as it pertained to electricity, was the discovery of contrary electrical states. The originality of the discovery is perhaps best gauged by the extreme reluctance of Europeans to accept it. Eventually they did so, largely on the strength of Franklin's analysis of the Leyden jar, which other contemporary theories could not explain.

The Leyden jar

The Leyden jar charges by accumulating electrical fire on its internal coating. The accumulation is made possible by grounding the external surface; for as positive electricity develops inside, the answering negative must be able to establish itself outside. Franklin believed that the charging continued until the outer surface of the bottle was exhausted: "no more can be thrown into the upper part when no more can be driven out of the lower." He demonstrated the equa-

lity by arranging a cork to play between wires attached to the coatings, as in figure 5; the cork swings to and fro, carrying fire from the top to the bottom, until the original state has been restored.

How does the accumulation produce a deficit, the plus yield a minus? Franklin supposes that the bottle's glass is absolutely impermeable to the electrical matter; that the particles of electric matter repel one another; that the repulsion operates over distances at least as great as the thickness of the jar, and that this macroscopic force, arising from the accumulation within the bottle, drives out the electrical matter naturally resident in the exterior surface of the jar. Most of these suppositions were peculiar to Franklin, particularly the odd notion that the bottle contained no more electricity when charged than when normal (its pleasure and pain separate but equal) and the revolutionary concept of the impenetrability of glass. Earlier electricians, arguing from the exercise of electrical attraction across glass screens, had concluded that the (material) agent of electricity could penetrate glass. Since, as they also knew, glass could insulate charged bodies (that is, prevent the flow of electrical matter to ground) they understood that glass could not transmit electrical matter very far. But everyone believed that transmission could occur over distances of the order of bottle thicknesses. Electricians were accordingly perplexed to discover, in the case of the condenser, that a very thin glass, grounded on one side, could preserve a very large charge. Several of them escaped from their dilemma by the sort of argument made familiar by the quantum physicist: Glass is either transparent or opaque to the electrical matter according to the experiment tried.

Characteristically Franklin cut through the paradox by firmly choosing one alternative and ignoring or downgrading the

phenomena that supported the other. Together with the impenetrability of glass he perforce admitted action over macroscopic distances, a proposition that, despite the success of the gravitational theory, was still a bugaboo even among Newtonian physicists. But—and this is also characteristic—Franklin made no effort to relate what he took to be the macroscopic results of the charging to the primitive repulsive forces that he understood to generate them.

For example, his proposition that the positive charge on the inner coating equals the negative on the outer conflicts with his charging mechanism. The condition for the cessation of charging must be the vanishing of the force driving electrical matter into the grounding wire, and—if the primitive force decreases with distance, as Franklin supposed it to do—the macroscopic force can only be annulled when the farther accumulation exceeds the nearer deficit. The discovery by later Franklinists of the inequality of the charges on the two surfaces of the condenser was to mark a substantial advance over the theories of the founder. Of course the advance had been set up by the acceptance of Franklin's approach. He always applauded such improvements in his theory, it being more important, he said, that science advance than that he be considered a great philosopher. He generously left the second, and sometimes also the first, approximation to others.

In disregarding the hypothetical dynamics of the electrical matter Franklin distinguished himself from the leading contemporary European electricians, from J. A. Nollet in France, Watson and Wilson in England, and the Germans mentioned by Haller. Franklin did not know his colleagues' habits when he began; he had not read their papers, and his cicerone Haller had omitted their intricate theories as imperfect and premature. He went his way until he met with the electrical mechanics of Watson and Wilson. He then devised an alternative scheme, a specimen of which is illustrated in figure 6.

In this scheme the positively electrified spike holds its redundant electrical matter as a conformal atmosphere. Note that the portions HABI and KLCB are held by the large areas AB and BC, respectively, while HAF, IKB, and LCM rest on much smaller surfaces. The spike retains its atmosphere by an attractive force between electrical and common matter; hence, Franklin says, it is easy to draw off electricity from a corner or a point, where there is little attracting surface. This scheme, which is not natural to Franklin, did not earn him high marks for physics. He unwittingly introduced two inconsistent sets of forces, one to establish the conformal atmosphere, the other to preserve it; for if the forces that maintained it determined its shape, it would be shallow opposite points and deep opposite

Stephen Gray's charity boy. Suspended by insulating cords and charged with a rubbed glass rod, the boy could provide diversion for onlookers by having sparks drawn from his nose or (as in the illustration above) by attracting bits of leaf brass. (From J. C. Doppelmayr, *Neu-entdeckte Phaenomena*, Nuremburg, 1744.)
Figure 2

Bibliothèque raisonée. The title page of the first volume (1728) shows one of its enlightened reviewers at work. Franklin's attention was drawn to electrical experiments by an article, originating in this review, translated for *Gentleman's Magazine.*

Figure 3

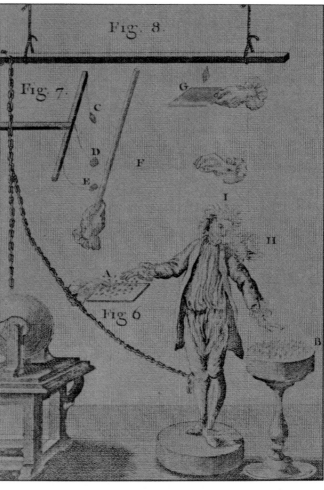

Another human capacitator, this time erect; he again attracts leaf brass, as at A and B. His flying hair and the fluff on his shoulder demonstrate the repulsive force of electricity. (From J. A. Nollet, *Essai sur l'électricité des corps,* Paris 1746).

Figure 4

plains. Again, the primitive forces proposed—attraction between the elements of common and those of electrical matter, repulsion between particles of the electrical—conflict with the fact that neutral bodies do not interact electrically. Let the quantities of electrical and common matters in the first body be E and M, those in the second e and m. Then E will be attracted by m and repelled by e, but M will be drawn by e without compensating repulsion. Similarly there is an unbalanced attraction on m. Franklin's electrical mechanics, the result of an attempt to copy continental physicists, require unelectrified bodies to run together.

Lightning

The same indifference to the exigencies of the forces he introduced appears in Franklin's theory of the lightning rod. So does the same bold process of simplification that discovered the contrary electricities and insisted upon the electrical opacity of glass. That lightning and electricity agreed in many properties was a commonplace when Franklin took up the subject; Haller, for example, emphasized the parallel between the transmis-

sion of electricity along an insulated string and the direct path of lightning along the "wire of a steeple clock from top to bottom" (so the *Gentleman's* mistranslates *"un fil d'archal qui servoit à faire sonner une clochette sur le haut d'une tour"*). In 1748 the Bordeaux Academy of Sciences offered a prize for an essay on the relation between lightning and electricity. It was won by a physician who took as his byword an old conceit of Nollet's: *"l'électricité est entre nos mains ce que le tonnerre est entre les mains de la nature."*

One of Franklin's collaborators discovered that a grounded metallic point could quietly discharge an insulated iron shot at some distance, while a blunt object could extract the electricity only when very near, and then suddenly, noisily, and with a show of sparks. Whence the difference? According to Franklin's homemade physics, the point acts only upon the small surface of the shot directly facing it; it therefore can pull away the redundant electrical matter a little at a time, as the excess redistributes itself to compensate for the loss of the portion just removed. "And, as in plucking the hairs

from a horse's tail, a degree of strength not sufficient to pull away a handful at once, could yet easily strip it hair by hair; so a blunt body presented can not draw off a number of particles [*of the electrical matter*] at once, but a pointed one, with no greater force, takes them away easily, particle by particle."

Franklin's bold and optimistic imagination immediately assimilated the shot to a thunder cloud and the pointed punch or bodkin to an instrument capable of robbing the heavens of their menacing electricity. To illustrate and confirm his thought, he invented a straightforward but misleading laboratory demonstration, easily reproduced. Take a pair of brass scales hanging by silk threads from a two-foot beam; suspend the whole from the ceiling by a twisted cord attached to the center of the beam, maintaining the pans about a foot above the floor; set a small blunt instrument like a leather punch upright on the ground. Now electrify one pan and let the cord unwind; the charged pan inclines slightly each time it passes over the punch until the relaxation of the cord brings it close enough for a spark to jump between them.

If, however, you mount a pin, point up-permost, atop the punch, the pan silently loses fire to the point at each pass and, no matter how close it approaches, never throws a spark into the punch.

It was no doubt his faith in the power of points that allowed Franklin to go beyond the European electricians who had con-tented themselves with suggesting an-alogies between lightning and electricity. Why not catch a little lightning and make it do the usual electrical parlor tricks? The probe would have to be insulated, but Franklin apprehended no danger, prob-ably because he believed that a sharply pointed rod would bring down lightning slowly enough to allow the observer to prevent dangerous accumulations. He accordingly proposed, in 1750, that a sentry box containing an insulating stand A (figure 7) be mounted on a tower or steeple; an iron rod, projecting 20 or 30 feet above the box, would fetch the light-ning, which the sentry would draw off in sparks. The sight, sound, smell and touch of the sparks would confirm the identity of lightning and laboratory elec-tricity.

Franklin did not try this dangerous experiment himself; much of Poor Rich-ard's caution was native to his inventor. The drama was first staged in France, in 1752, by a clique headed by G. L. Leclerc, later comte de Buffon, who had found the electrical theories of the unknown printer from Philadelphia useful ammunition in his bitter feud with R. A. F. de Réaumur, who supported the traditional approach to electricity of his protégé Nollet. Buf-fon's agents, sharing Franklin's caution, did not expose themselves to thunderbolts either. They engaged a retired dragoon to draw the sparks; fortunately for the old soldier the rod picked up only minor electrical disturbances in the lower at-mosphere. The first to perform Frank-lin's experiment as initially conceived was a member of the Petersburg Academy of Sciences, G. W. Richmann. The thun-derbolt that he enticed into his home killed him instantly.

Richmann had known that he ran a risk. "In these times [*he said*] even the physicist has an opportunity to display his fortitude." There was good evidence that Franklin's sentry might be in peril. As Haller had observed, lightning liked to run down wires and ropes attached to church steeples. Often enough the other ends of these ropes were held by men en-gaged in the standard early-modern de-fense against lightning—ringing bells. Now the destruction of bellringers by lightning had been remarked, and Franklin's sentry stood in the same rela-tion to his box and rod as the bellringer did to his church and steeple. In fact Franklin tacitly admitted the danger of his sentry in his consequential proposal to replace bellringers by *grounded* rods as protection against lightning. (The practice of breaking up thunder clouds by

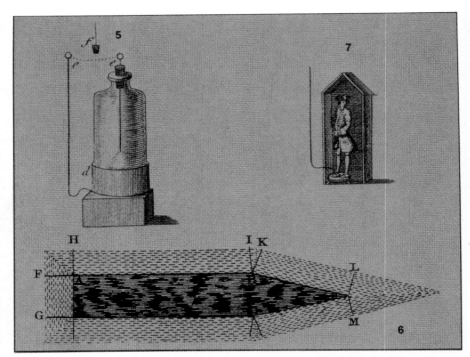

Experiment and observation. Franklin's most persuasive demonstration **(figure 5)** of the qualitative difference between the electrifications of the two coatings of a Leyden jar: the cork *f* oscillates between the points *e*, fetching the surplus electricity from the interior of the bottle to make up the deficit of the outside coating *d*. His analysis **(figure 6)** of the binding of electrical atmospheres postulates an attractive force between electrical and common matter, and the sentry box **(figure 7)** illustrates his design for bringing down lightning from the clouds. (From Franklin, *Experiments and Observations on Electricity,* London, 1751–54.)   Figures 5,6,7

sounding bells was outlawed in several places in the 1770's and 1780's; not so much on physical or humanitarian prin-ciples, it must be confessed, but for noise abatement.)

That Franklin did not conclude, from the protective role of the grounded rods, that insulated ones might be very dan-gerous agreed with his usual sanguinity. His confidence rested, as already sug-gested, on his belief in the analogy of na-ture, on his extrapolation from the silent discharges effected by points in the lab-oratory to the operations of iron rods on thunder clouds. The same sort of ex-trapolation may perhaps be seen in his optimistic political and social philosophy; he appears to have considered organiza-tion at the federal level to be analogous to local combinations, without regard to scale.

The Franklinist faith in the power of points may be illustrated by the mock-heroic battle of the knobs and spikes, which broke out in the late 1770's when a British power magazine, defended by sharp grounded rods as directed by Franklin, suffered minor damage from lightning. Wilson immediately located the trouble in the points. In elaborate experiments conducted in a London dance hall grandly named "The Pan-theon" (figure 8) he showed, what no one doubted, that. pointed conductors dis-charged electrified bodies at greater dis-tances than blunt ones. Since, he said, Franklin's points evidently do not draw

down lightning silently, but are struck just like blunt rods, it is only prudent to ter-minate lightning conductors obtusely; for loaded clouds that would strike to pointed rods might, if high enough, pass harm-lessly over blunted ones.

Wilson's large-scale experiments had been made possible by George III, to whom he had access through aristocrats whose portraits he had painted. Franklin had represented the disobedient colonies which, at the time of the Pantheon dem-onstrations, were in full revolution. The shape of lightning conductors became a matter of politics. The King (according to a fine story perhaps invented by the French) instructed the President of the Royal Society, Sir John Pringle, that lightning rods would henceforth end in knobs. Pringle, a great friend of Frank-lin's, replied that the "prerogatives of the president of the Royal Society do not ex-tend to altering the laws of nature," and, according to the story, forthwith resigned.

Fortunately neither the cause of the Revolution nor the efficacy of lightning rods rested upon the supposititious ad-vantage of points over knobs. The anal-ogy that Franklin trusted does not hold: on Nature's scale, on the scale of thun-derclouds, points and knobs appear about the same; "obtuse Wilson" (as the Fran-klinists called him) was quite right in in-sisting that pointed rods cannot quietly despoil clouds of lightning. And yet, even though the analogy does not hold, the optimism expressed by it, the expectation that experience with puny effects of our own creation can guide us to the control of the great powers of Nature, was not misplaced. Lightning rods work. The aristocratic hanger-on, Wilson, warned that we must not expect "anything like absolute security" in such matters. The optimistic republican, Franklin, trusted that Nature could be mastered.

Utility

Several passages in Franklin's writings suggest that he cultivated science chiefly with an eye to its utility. In a report of electrical experiments dated 1748 he declared himself "chagrined a little" that his work on electricity had not yet produced anything "of use to mankind." The best he could offer were imaginary improvements on electrical games described by Haller—a picnic on an electrocuted turkey, roasted on an electrical jack before a fire ignited by an electric spark; a toast to the electricians of Europe, drunk from electrified bumpers (small, thin, nearly-full wine glasses charged as a Leyden jar) "under the discharge of guns from the electrical battery." This playfulness disappeared from Franklin's account of experiments undertaken to show that light-colored cloths "imbibe" the heat of the Sun less readily than dark ones. "What signifies philosophy [*he then said*] that does not apply to some use?" He goes on to recommend white clothes for the tropics. In another place he writes as if the practical implications of natural laws are for him the main objective. "It is of real use to know that china left in the air unsupported will fall and break; but *how* it comes to fall, and *why* it breaks, are matters of speculation. It is indeed a pleasure to know them, but we can preserve our china without it." To this evidence may be added the testimony of his inventions, the Pennsylvania fireplace, bifocals, the glass harmonica and, above all, the lightning rod.

Yet, for all his emphasis on utility, Franklin cultivated science primarily for intellectual pleasure. The message about china plates follows immediately upon the highly conjectural analogy between the power of points and the stripping of a horse's tail. Franklin constantly built up, and as often discarded such "pretty systems"; the principal use of which, he said, was the discarding, for that might "help to make a vain man humble." He spoke of his work in electricity not as a hopeful inventor, but as an eager savant. "I never before was engaged in any study that so totally engrossed my attention and my time as this has lately done," he wrote in 1747. "What with making experiments when I can be alone, and repeating them to my friends and acquaintances, who, from the novelty of the thing, come continually in crowds to see them, I have, during some months past, had little leisure for anything else." Franklin may have hoped that something practical

The Pantheon experiment. A model of the powder magazine (shown at the right), armed with grounded points or knobs, was drawn on rails under the huge cylinders electrified by the machine in the center background. The cylinders represented clouds, the motion of the model their drift over the magazine. (From Benjamin Wilson's paper in *Philosophical Transactions* of the Royal Society of London, **68:1**, 239–313, 1778.) Figure 8

would emerge from his studies, but he did not study primarily for utility. In the case of electricity he set the principles of the subject and developed them in analyzing the condenser before he sought practical applications.

Perhaps Franklin's most frivolous study was magic squares. He indulged his taste for these useless toys for several years, until he acquired such a "knack . . . that I could fill the cells of any magic square, of reasonable size, with a series of numbers as fast as I could write them, disposed in such a manner as that the sums of every row, horizontal, perpendicular, or diagonal, should be equal." But this by no means satisfied him; he invented supererogatory tasks, the creation of grids with bizarre additional symmetries, such as the great 16 × 16 table published in the *Gentleman's Magazine* in 1768. It was, as Franklin allowed in terms far from utilitarian, "the most magically magical of any magic square ever made by any magician."

Poor Richard nonetheless felt obliged to apologize for time wasted on number magic. He took up the squares, he said, to pass the time ("which I still think I might have employed more usefully") when, as clerk of the Pennsylvania Assembly, he was obliged to sit through much tedious government business. Filling in squares therefore did have some utility: it kept Franklin awake, and made him appear alert, at meetings he would have preferred to miss. For our era, the Age of the Committee, Franklin's application of the ancient magic square as an antidote to boredom could be a most useful invention. ☐

For further reading . . .

Most of the quotations from Franklin come from his *Experiments and Observations on Electricity*: the data about the history of electricity are taken from John Heilbron's *Electricity in the 17th and 18th Centuries: A study of early modern physics*, Berkeley (1979). Additional pertinent information may be found in I. B. Cohen, *Franklin and Newton*, Philadelphia (1956) and in Carl van Doren, *Benjamin Franklin*, New York (1938).

A Sketch for a History of EARLY THERMODYNAMICS

By E. Mendoza

PHYSICS TODAY / FEBRUARY 1961

ACCOUNTS of the origins of the first and second laws of thermodynamics follow a fairly standard pattern. The caloric theory of heat, we are told, assumed that heat was a fluid endowed with a number of properties, among them indestructibility. The cannon-boring experiments of Rumford (1798) and the ice-rubbing experiment of Davy (1799) destroyed the basis of the caloric theory because they showed that heat could be created by the expenditure of work. A full half-century elapsed, however, before Joule repeated and extended Rumford's experiments and measured the conversion factor J accurately with his paddle wheels. In the meantime (in 1824) Carnot formulated the second law of thermodynamics and drew many valid conclusions about the efficiency of heat engines though his ideas were based on the caloric theory. Kelvin came across Carnot's work, as rewritten by Clapeyron; he became convinced of its truth and because it was based on the caloric theory he found it difficult to accept Joule's results. However, by 1850 both Kelvin and Clausius had formulated the first and second laws as we know them now. In retrospect, the caloric theory of heat seemed to have been slightly ridiculous.

It seems to me that the pattern just sketched out is incorrect in many ways. It is particularly unfortunate that it should be so, for the discovery of the first law is an episode in the history of physics which can be studied by students as an example of the way that the great ideas of science have evolved.

The facts seem to be that the caloric theory did not reach its highest state of development till *after* the work of Rumford and Davy had (in our modern view) destroyed its very basis—indeed these same experiments were regarded by the physicists of the time as enriching the caloric theory, as filling in some of the missing details. Further, at its highest point, the caloric theory was sophisticatedly mathematical; the properties of the caloric fluid—the model behind the abstract mathematics—were rarely stressed and were indeed usually regarded as irrelevant. The mathematics predicted most of the correct results, and where the equations differed in essential ways from our own correct ones, there were reputable experimental results to support them. Finally, when the modern two laws of thermodynamics were formulated, the whole of the mathematical apparatus of the caloric theory was taken over. The attitudes of modern thermodynamics, with its jargon of perfect differentials and of partial differential coefficients, were inherited from the previous epoch. Perhaps this account implies that science does not progress tidily, but I think it is worthwhile giving.

The Two Theories of Heat

THE two hypotheses—that heat was a mode of motion of the particles of bodies, and that heat was a substance—had their origins in two quite different sets of observations. The obvious production of heat by friction gave rise to the one; indeed the mechanical theory of heat is by far the more ancient of the two. On the other hand, the idea of the conservation of heat in calorimetric experiments was only conceived in the eighteenth century. Joseph Black had defined several interlocking quantities—temperature, specific heat, latent heat, and quantity of heat—and had at the same time postulated the conservation in a thermal mixing process. Then with the rise of the atomic theory and the discovery of oxygen, many quantitative things could be explained by the idea that heat was a gas of indestructible atoms. The conservation of heat was assured on this model; further, the atoms of caloric could enter into chemical combination with the atoms of a sub-

E. Mendoza is senior lecturer in physics at the Physical Laboratories of Manchester University in England.

stance (when the heat was latent) or be free (when the heat could affect a thermometer). In Lavoisier's view, the caloric atoms were an essential constituent of oxygen and their release gave rise to the heat of combustion. Thus, in contrast to the old-fashioned dynamical theory, the caloric theory of heat used a few basic ideas of the up-to-date atomic theory and could explain beautifully the facts of combustion and calorimetry.

Yet the French physicists and chemists always kept it firmly in mind that there were two hypotheses which at the time were equally valid. Every statement of the theory of heat invariably placed the two theories side by side, usually with a statement that the two, though seemingly quite different, must be only varied aspects of the same underlying cause. There was no obvious contradiction between the two hypotheses. One of the earliest statements of this kind comes from the *Memoir on Heat* written by Laplace and Lavoisier in 1786. They state:

> We will not decide at all between the two foregoing hypotheses. Several phenomena seem favourable to the one, such as the heat produced by the friction of two solid bodies, for example; but there are others which are explained more simply by the other—perhaps they both hold at the same time. . . . In general, one can change the first hypothesis into the second by changing the words "free heat, combined heat and heat released" into "*vis viva*, loss of *vis viva* and increase of *vis viva*".

Here we may note that the words "heat" and "caloric" were always regarded as interchangeable and that the *vis viva*—the living force—of a system of particles was twice the kinetic energy. The identity of the two theories is therefore explicitly stated. This statement, though an early one, is typical of all those written by French scientists for the next sixty years.

This means that the French scientists did *not* consider that the issue was straightforward—that either the caloric theory was true or the dynamic theory; on the contrary, they held that both were true. Thus it was that Rumford's work had very little impact on them. For example, one of his papers described how he measured the density of caloric by weighing some ice and then reweighing it after it had melted, concluding that the density of caloric, if it existed at all, was negligible. Subsequent accounts of the caloric theory therefore incorporated the additional statement that the mass of the caloric atoms was very small—like electricity. Further, in his other experiments, Rumford showed that the supply of heat produced by friction was apparently inexhaustible. Subsequent statements of the caloric theory therefore included the additional statement that the number of caloric atoms which could be rubbed off by friction was negligible compared with the number actually inside a body—like frictional electricity.

It is usually said that the first symptom of the inadequacy of any theory is observed when each new experiment demands that a new hypothesis be added. From our modern viewpoint these additions to the caloric theory were of just this kind. But from the con-

Pierre Simon Laplace, who dominated the French Academy of Sciences in his later years.
(*Culver Pictures, Inc.*)

temporary point of view they were extremely reasonable statements. Far from killing the caloric theory, Rumford's experiments added to the understanding of it.

The British scientists, in contrast to the French, were mostly interested in chemistry and atomic theory and therefore adopted the caloric view uncritically. Rarely were the two theories placed side by side for fair comparison in their writings. Even Davy used caloric concepts when he found them convenient. But it was in France that the most significant developments were made, in the decade from 1810 to 1820.

Perfect Differentials

THE mathematical version of the caloric theory gradually evolved in a series of papers by Laplace and Poisson. By 1818, the theory of heat was usually cast in the following form—the quotation is from a brief introductory paragraph in a paper by Poisson:

> Let ρ be the density of a gas, θ its centigrade temperature, p the pressure which it exerts on unit area, the measure of its elasticity: then one has
>
> $$p = a\rho \, (1 + \alpha\theta)$$
>
> where a and α are two coefficients. . . . The total quantity of heat contained in a given weight of this gas, in a gram for example, cannot be calculated: but one can consider the excess of this quantity over that

contained in a gram of gas at an arbitrarily chosen pressure and temperature. Designating this excess by q, it will be a function of p, ρ and θ, or simply of p and ρ since these three variables are connected by the preceding equation; thus we have

$$q = f(p,\rho)$$

where f indicates a function whose form must be found.

By defining q as the excess quantity of heat over an arbitrary zero, Poisson avoided the difficulty that the absolute quantity of heat was much greater than what could be rubbed off by friction. By stating that q was a unique function of the thermodynamic coordinates—for this is the significance of the second equation—he summarized tersely many experimental facts, for example the equality of the latent heats of boiling and condensation, or what we should now call the uniqueness of the enthalpy as a function of pressure and temperature.

We may put this analytical statement into perspective by stating for comparison the starting points of elementary modern thermodynamics. In such treatments, we first restrict ourselves to systems which have single-valued equations of state, and then we postulate that there are two independent heat-like quantities which are single-valued functions of the thermodynamic coordinates—we usually choose the internal energy U and the entropy S, which can be expressed as $U(p,V)$ and $S(p,V)$. In short, the caloric theory differed from our own approach in that it recognized only one law of thermodynamics—one heat function $q(p,V)$—where we have two.

Laplace and Poisson then used this analytical law to calculate the temperature rise of a gas when it was compressed adiabatically, to explain the experimental results of Clément and Désormes. Since q was a unique function of p and V, dq could be expressed (in modern notation) as

$$
\begin{aligned}
dq &= (\partial q/\partial p)_v dp + (\partial q/\partial v)_p dv \\
&= (\partial q/\partial T)_v(\partial T/\partial p)_v dp + (\partial q/\partial T)_p(\partial T/\partial v)_p dv \\
&= C_v \cdot V \cdot dp/R + C_p \cdot p \cdot dV/R \qquad (1)
\end{aligned}
$$

putting the specific heats as dq/dT with suitable subscripts, and substituting $pV = RT$. Assuming that the specific heats were constant with temperature the equation was then integrated to give

$$q = f(pV^\gamma). \qquad (2)$$

In an abiadatic change the total quantity of heat did not alter; hence such a change was governed by the law

$$pV^\gamma = \text{constant}.$$

It is well known that Laplace corrected Newton's expression for the velocity of sound, assuming that the wave motion was adiabatic instead of isothermal; this was his method of calculation. Thus the assumption that the quantity of caloric was a unique function of the pressure and volume of a gas allowed the velocity of sound to be correlated with direct measurements of the ratio γ. It was something of a triumph and was ob-

Delaroche & Bérard's apparatus. Gas contained in B and B' was driven through apparatus by heads of water in vessels A and A' in the room above. Normally it exchanged heat in the little spiral in the other half of diagram; the apparatus is, however, shown arranged for finding the heat capacity of the spiral by forcing hot water through it.

viously proof of the correctness of that basic assumption.

It took later scientists many years to realize that this same result, that the ratio of adiabatic and isothermal elasticities of a gas is equal to the ratio of two suitably defined specific heats, follows straight from the definitions, and results from any physical model of heat whatever.

Pistons and Cycles

THE mathematical approach to thermodynamics is essentially the same as that which we use today. The other approach, using cycles of operations with frictionless pistons, was evolved by Sadi Carnot. He was capable of an extraordinary precision of thought and was no mean mathematician. But his single published work, *Reflections on the Motive Power of Fire* (1824), was conceived as a popular book for engineers, to stimulate them into designing better heat engines. Thus all his proofs and theorems are based on the actions of engines, however idealized. His concept of the cycle of operations was consciously based on the assumption of the uniqueness of the quantity of heat as a function of coordinates; he had probably been taught that theorem at his Army Engineering School, the Ecole Polytechnique, where Laplace and Poisson were instructors.

In perspective, we can see that this pictorial approach had a comparatively short life. After Clausius used it in 1865 to derive the concept of entropy and thereby show that the two laws of thermodynamics could be expressed in the same way as the old caloric theory, the more mathematical approach became dominant once more; pistons and cycles were relegated to teaching textbooks.

Experimental Proofs

THE rise of temperature of a gas when it was compressed suddenly would be easily explained on the model that caloric itself was atomic—the heat atoms were squeezed out from the gas atoms "like water from a sponge". This qualitative idea was however given quantitative expression; it followed from equation (2) above.

Laplace made the *assumption* that the function f was the simplest possible—that it was linear. Thus the heat content of a gas could be written

$$q = A + B \cdot T \cdot p^{(1-\gamma)/\gamma}$$

where A and B were constants, p and T being chosen

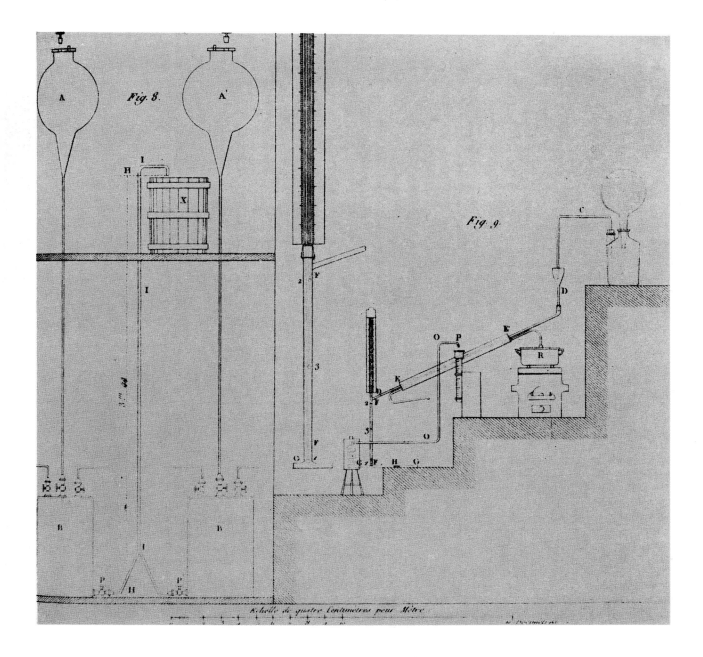

here as the appropriate variables. The specific heat C_p followed by differentiating with respect to T, showing at once that it was proportional to the pressure raised to the power $(1 - \gamma)/\gamma$. Putting $\gamma = 1.4$, the specific heat of air should decrease approximately as the cube root of the pressure.

Carnot on the other hand deduced a number of theorems leading to a slightly different result—his method gave the form of the function explicitly and showed that the heat content and the specific heat decreased with the logarithm of the pressure. But both Carnot's and Laplace's expressions, though different in detail, predicted decreases of specific heat with pressure, showing that a rise of pressure should release heat and so cause a rise of temperature. They were the quantitative expressions of the "squeezing out" process.

The experimental measurements of Delaroche and Bérard of the specific heats at atmospheric pressure of a large number of gases were performed in 1812 and deservedly won a prize award by the Institut de France. Their apparatus was beautifully designed, their techniques were highly developed, and most of their results were accurate. Unfortunately they also performed two measurements of the specific heat of air at one value of the pressure slightly above atmospheric—to be precise, at 1006 mm pressure. They found that for this 30% increase the specific heat of unit mass of air was reduced by about 10%, which agreed almost exactly with Laplace's prediction. This observation remained for years one of the cornerstones of the whole caloric theory.

Carnot later compared the same observations with

his own expression and concluded that the coefficient of the log p term was small. In 1837, von Suerman in Germany performed measurements on air at *reduced* pressures, finding that Carnot's formula (or more precisely Clapeyron's version of the same expression) fitted better and that Laplace's assumption was not correct. But everyone was agreed that there *was* a variation of specific heat, in conformity with the predictions of the caloric theory.

Thus by the late 1830's a considerable body of experimental results had been accumulated and an advanced mathematical technique had been evolved in support of the caloric theory. At the same time, these decades were alive with speculation about the dynamical theory of heat. Claims have been advanced on behalf of several people as the real originators of the First Law—but few of these ever wrote down an equation or quoted numbers other than isolated estimates of J which proved nothing. Even Mayer's brilliant intuitions were largely concerned with qualitative speculations about the conservation of energy in different forms; there was little that was quantitative and even that could be explained on existing theories. In Poisson's phrase, the undulatory theory of heat was sterile.

Carnot and the First Law

THE dynamical theory implied that the heat content q was not a unique function of pressure and temperature and that the single law of thermodynamics was wrong. But this essential point was still not recognized by all physicists. Perhaps they took refuge in the

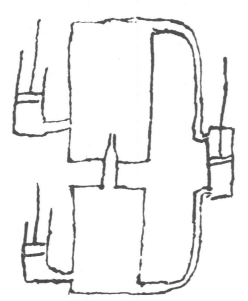

A sketch taken from Carnot's private manuscript notes (the original is one inch high), showing a proposed experiment on free expansion of gases. It was not until 25 years had elapsed that Joule and Thomson proposed and performed this experiment.

postulate that the quantities of heat so evolved were small compared with the total so that the error of the assumption was small; perhaps they did not believe that the supply of heat produced by friction was really inexhaustible. At any rate, it is astonishing to find a person as critical as Clapeyron writing (in 1834) only two or three pages before explicitly stating the uniqueness of the heat function:

> It follows that a quantity of mechanical action and a quantity of heat which can pass from a hot body to a cold body are quantities of the same nature, and that it is possible to replace the one by the other; in the same manner as in mechanics a body which is able to fall from a certain height and a mass moving with a certain velocity are quantities of the same order, which can be transformed one into the other by physical means.

Clapeyron was discussing the functioning of heat engines, not the nature of heat, when he wrote this paragraph, but the implication was nevertheless quite clear. The opinion of Laplace and Lavoisier, that there was no conflict between the two theories of heat, was still held.

Possibly the only person who grasped the essential conflict was Carnot himself. In fact he occupies a special position in any history of the subject because, though he only published the one short book on heat engines, some notebooks of his have been preserved in which he mused about the shortcomings and improbabilities of the caloric model, and gradually groped toward the equivalence of heat and work. These notes constitute a revealing record of the objections which could at that time be raised against the dynamical theory. Mostly they stem from the fact that there was no clear picture of the structure of atoms or of solids, so that the nature of the thermal agitation of atoms in solids could not be imagined. For example, Carnot states that if heat is what we now call energy then the fact that the whole universe cannot be imagined to run down must imply (on the dynamical theory) that atoms cannot touch one another; for if they did touch there would be friction and the heat vibrations would die down. In that case he was unable to visualize what forces could hold the atoms in position in a solid if they were not touching. Any forces would have to act through an ether; since an ether had to be a fluid, it too had to be atomic in structure, so the difficulty could not be solved. Finally, however, he explicitly stated the equivalence of heat and work, leaving the question of the microscopic picture unsolved. He estimated J quite accurately.

A careful examination of these notebooks together with the *manuscript* of Carnot's book on heat engines and the published version of it shows that he had started on this train of speculation about the First Law at the same time as he was writing about the Second. Certainly by the time he came to correct the proofs of his book he had realized that the very basis of all his theorems and demonstrations was wrong. For example,

had begun to doubt his own methods. We can only be thankful that this is what he did.

(It is unfortunate that something of a "mystique" has grown up around Carnot's writings. From his use of the word "caloric" it has been deduced that he had a prevision of the concept of entropy. However, the words he used were merely interpretations of the equations he wrote down, and it is clear that together with those written down by all other contemporary physicists, these equations were only true by coincidence.)

Joule's Experiments

JOULE'S first research (started when he was aged 19) was on the design of electric motors. Though these early machines were spidery little affairs hardly recognizable as the forerunners of those familiar to us, Joule envisaged them as the prime movers of the future. At first he thought of them as possible perpetual motion machines, but the i^2R formula for the heating effect of a current was an early result of the investigations. He also found that the attractive force of an electromagnet was proportional to i^2, and the similarity of the formulas led him to think of a connection between mechanical and heating effects. Eventually he was led to do a remarkable experiment with a simple dynamo whose armature was immersed in a rotating vessel full of water. With the armature stationary and connected to a battery he measured the heating; by rotating the armature he superimposed a second current and found that he could create or *destroy* heat according to the sense of rotation. The change of heating was proportional to the work done in rotating the armature. This experiment, in Joule's view, showed conclusively that the accepted theory of heat was wrong and he started at once on a series of experiments of great variety to prove his point of view.

The electrical experiments had given $J = 4.60$ joules/calorie in our modern units. The heating of water forced through narrow holes in a piston gave 4.25 units; heating by the friction of two solid surfaces rubbing beneath water or mercury gave the same value. He pumped air into a cylinder to 22 atmospheres and measured the heat produced in the cylinder; comparing this with the pV term, J emerged as 4.60 units. Then he allowed the gas to escape slowly—the cylinder cooled and J was found to be 4.38 units. But when the gas escaped slowly from the high-pressure cylinder into another, without performing external work, the cooling of one cylinder was equal to the heating of the other so that there was no net production of heat. These experiments took him five years to do—from 1843 to 1848.

After these experiments were finished Joule allowed himself to speculate on the philosophical and other aspects of the theory. It was, however, the quantitative aspect of his work which eventually carried conviction. The conversion factor was the same within 15% however the work was performed: electromagnetically, by solid or liquid friction, or by the changes of volume of a gas. This could not be plausibly explained on any

James Prescott Joule as he appeared at the time of his classic experiments.

concerning his theorem that the motive power of heat is independent of the working substance, he originally wrote:

> The fundamental law which we proposed to confirm seems to us to have been placed beyond doubt. . . . We will now apply the theoretical ideas expressed above to the examination of the different methods proposed up to now for the realisation of the motive power of heat.

But in the printed version he altered this to:

> The fundamental law which we proposed to confirm seems to us however to require new verifications in order to be placed beyond doubt. It is based on the theory of heat as it is understood today, and it should be said that this foundation does not appear to be of unquestionable solidity. New experiments alone can decide the question. Meanwhile we can apply the theoretical ideas expressed above, regarding them as exact, to the examination of the different methods proposed up to now for the realisation of the motive power of heat.

He had realized that the Law $q = f(p,V)$ was no longer true and this destroyed the idea of the cycle of operations. He had discovered the First Law to the exclusion of the Second. The essential step of postulating that there were two independent laws was too difficult to take.

The point of this episode is that we know that Sadi Carnot was a reserved and taciturn man, something of a perfectionist. It is therefore extraordinary that he allowed the publication of his book to proceed after he

caloric model. Two years after this series of experiments he measured J accurately by stirring water with paddle wheels, but these experiments were relatively unimportant.

Joule wrote a number of papers about his work but till almost the end of this important epoch he was intellectually quite isolated. The commonest objection to his theory was that it all depended on temperature rises of a few hundredths of a degree, which could hardly be significant enough. But two papers, Grove's "On the Correlation of Physical Forces" and Helmholtz' "On the Conservation of Force" helped to prepare the intellectual climate for the acceptance of Joule's theory.

Clapeyron's paper on the motive power of heat had been published in England in 1837, in Taylor's *Scientific Memoirs,* a journal which specialized in translations of foreign papers; Joule was familiar with it. By 1844 he was already confident enough to reject Clapeyron's description of the cycle of operations in the steam engine. He flatly contradicted the view that the passage of heat from boiler to condenser was sufficient to produce work. For the first time, the issue appeared to be clear—either Carnot or Joule was right.

Synthesis

WILLIAM THOMSON (Lord Kelvin) seems to have been the key figure in the synthesis of the two theories. He worked in Paris as a sort of research assistant in Regnault's laboratory in 1845 and there learned of Clapeyron's paper. He proposed the work scale of temperature wholly in caloric terms. Though he became a close friend of Joule, had a deep respect for his experiments, and always quoted his opinion, he could not accept the newer theory. His principal objection was that there were no examples of the reverse conversion of heat into work. Joule wrote to him that the Peltier effect could provide one such process, but it took Thomson four years to understand this remark. In 1849, Thomson published an account of Carnot's theory. There were many references to Joule's work but the "ordinarily received and almost universally acknowledged" principle that heat was conserved in a cycle of operations was still the accepted basis. Later in the year William's brother James published theoretical predictions based on Clapeyron's equation for the lowering of the freezing point of water by pressure; experiments confirmed the predictions—and hardened Thomson's conviction that Carnot's methods and the theory it was founded on were true.

The change of viewpoint happened quite suddenly. Probably Clausius was the first to see that there were two *independent* principles. In 1850 he wrote:

> It is not at all necessary to discard Carnot's theory entirely, a step which we certainly would find it hard to take since it has to some extent been conspicuously verified by experiment. A careful examination shows that the new method does not contradict the essential principle of Carnot, but only the subsidiary statement that *no heat is lost,* since in the production of work it

may well be that at the same time a certain quantity of heat is consumed and another quantity transferred from a hotter to a colder body, and both quantities of heat stand in a definite relation to the work that is done.

At about the same time, Thomson saw the light. Some theoretical work by Rankine on the adiabatic expansion of steam, together with the observation that high-pressure steam escaping from a safety valve does not scald because it comes out dry, abruptly convinced Thomson that steam could be heated by friction. It is difficult to see why this should suddenly have appeared so conclusive to him when Joule had been using the same concepts for seven years. However that may be, Thomson soon embarked on a long paper, stating the two laws explicitly and independently, one ascribed to Joule and the other to Carnot and Clausius. The introductory historical account was of course quite biased and incomplete; it was the forerunner of those which are usually written today. This paper, with the appendixes which were added at various times, included the thermoelectric relations and a discussion of elasticity.

In 1850 Clausius wrote that the "internal work U"

> has the properties which are commonly assigned to the total heat, of being a function of V and T and of being therefore fully determined by the initial and final conditions of the gas.

He treated U with the same mathematical techniques as Laplace and Poisson and Clapeyron had applied to q. The quantity $\Sigma dQ/T$ began to appear quite early in papers by Thomson and Clausius, but it was not till 1865 that Clausius deemed it worthy of special definition. He wrote:

> We can say of it that it is the *transformation content* of the body, in the same way as we say of the quantity U that it is the *heat and work content* of the body

and coined the name entropy for it. The mathematical methods of the caloric theory were finally recovered; thermodynamics today still bears the impress of Laplace and Poisson, just as surely as electrostatics.

Conclusion

THE conventional description of the caloric theory, as a qualitative model of heat processes which had to be abandoned as soon as Rumford did his cannonboring experiments, is obviously untrue. The difficulty encountered by the proponents of the dynamic theory of heat was that they had first to break the stranglehold of a glib mathematical formulation, a method which could make a sufficient number of correct predictions to give the illusion of being the whole truth. But probably this was a necessary stage in the development of the subject, since it did after all allow the formulations to be worked out. After that, there just remained the enormous intellectual difficulty of proposing two laws where instinct said that only one existed; when that was done the theory of heat was virtually complete.

A Sketch for a History of

THE KINETIC THEORY OF GASES

*By **E. Mendoza***

PHYSICS TODAY / MARCH 1961

THE ideas that solids are composed of compact arrays of atoms, while gases are composed of atoms or molecules in very rapid translational motion, are so obvious that we accept them nowadays without question; in teaching textbooks they are stated as if they were axioms. In its most elementary form, without any sophisticated calculations about the distribution of velocities, with only the one assumption that the impacts of the molecules on the walls of the containing vessel produce the pressure, a very simple calculation gives the equation

$$pV = \tfrac{1}{3} mNc^2 \tag{1}$$

where m and N are the mass of a molecule and the number per unit volume, and c is a velocity; p is the pressure and V the volume of the gas.

This formula poses something of a historical puzzle. For comparing it with $pV = RT$, there is the strong implication that the temperature—and therefore the heat content of a gas whose specific heat is constant with temperature—is proportional to the kinetic energy of translation of the molecules, and hence that heat is a form of motion. It is stated in every textbook that this kinetic theory originated with Daniel Bernoulli in the middle of the eighteenth century. But it is equally well known that the dynamical theory of heat was not accepted till a whole century later. On the face of it, therefore, scientists seem to have been singularly obtuse not to have recognized the straightforward implications of Eq. (1) for so long.

But in reality it seems that the kinetic theory of gases is quite a modern development. It was not at all obviously correct; it was not accepted into physics until it had overcome some formidable opposition. The outline of the story will be given here.

The Static Theory of Gases

IT is quite true that Bernoulli did give an excellent account of the kinetic theory in his book on hydrodynamics published in 1738. But I have never been able to trace a single reference to this theory in any paper or book published in France or England during

E. Mendoza is senior lecturer in physics at Manchester University, England. His "Sketch for a History of Early Thermodynamics" appeared in the February 1961 issue of *Physics Today*, p. 32.

the first half of the nineteenth century; it was piously disinterred in 1859. The influence that Bernoulli's kinetic theory had on other physicists during the critical period was nil; it might just as well never have been written.

Most scientists in France and Britain adopted instead the *static* theory of gases. According to this, the forces which held atoms together in a solid were attractive forces which gave the solid its cohesion, but in a gas these changed into repulsions. The atoms tried to get as far from one another as they could, and this purely static effect produced the pressure. A gas was therefore merely a highly expanded solid; except for accidental effects like convection, the atoms in a gas were quite stationary.

This theory originated with Newton but Laplace refined it in several authoritative papers published around 1824. The origin of the repulsive forces was taken to be the short-range repulsions of the caloric atoms inside the gas molecules. Lengthy calculations showed that whatever the law of force

$$\text{Pressure} = (\text{constant}) \, \rho^2 \, q^2$$

where ρ was the gas density, q the charge of caloric in each molecule. Considering the dynamic equilibrium of emission and absorption of the caloric and taking the temperature to be proportional to the density of caloric atoms in transit, he found

$$\text{Temperature} = (\text{constant}) \, \rho \, q^2$$

and hence the gas laws followed. These papers are deeply impressive, but they leave the nasty impression that the abstractness of the mathematics was a sign of decadence. The fact that the quantity of heat appears squared in both these formulae seems to conceal some basic confusion, hidden somewhere under the mathematics.

In England Newton's theory was widely taught, but not everyone was in agreement. When Davy wrote his *Elements of Chemical Philosophy* in 1812, he inclined quite strongly towards the dynamical theory of heat and proposed that in solids the motion was a vibration or undulation of the atoms, but that in gases the atoms also *rotated* about their axes. He seems to have had a glimmering of the idea of the partition of energy between the rotational and vibrational modes, and to have tried to explain the latent heat of boiling in this way. The idea that gas atoms revolved on their axes made a great impression on Davy's contemporaries. For combined with the orthodox static theory of gases it allowed a precise model to be made of the origin of the repulsive force between gas molecules—namely the centrifugal force of the revolving atomic atmospheres. This idea was later taken up by Joule and Rankine.

In this discussion it is important to realize that there were several possible concepts of atoms. They could be point centers of force—the forces could have a finite range or could extend an infinite distance—or they could be particles with definite shapes. There were difficulties in imagining the collisions between atomic *particles* to be perfectly elastic, however; for a body could deform elastically only if its parts moved relatively to one another, whereas an atom was usually held to be an indivisible elementary particle and therefore without substructure. For the same reason, Davy's idea of atoms with revolving atmospheres offended some purists. But I get the impression that different scientists had quite private views on such questions which they rarely bothered to state explicitly.

Herapath's Hypothesis

JOHN Herapath, a self-taught schoolmaster from Bristol, originated the kinetic theory of gases as we know it. He had a genius for distorting irrelevant facts to fit incorrect theories; but if we take the cruel but realistic criterion that the most important scientists are those whose ideas have influenced others, who have proposed theories which are in the main stream of scientific thought, then Herapath is among the most

important; the kinetic theory of gases is firmly founded on "Herapath's hypothesis".

He began by noting a small discrepancy in some observations on the motion of the moon, and proposed that Newton's constant of gravitation was not in fact a constant but varied with the temperature of the planet concerned. Thus he was led to a study of Newton's *model* of the gravific ether, the gas whose pressure produced the gravitational force—though Newton himself had of course stressed that the model was not very important. Hence Herapath was led to study the properties of gases in general. He tried to deduce them from the caloric theory and made no progress; then he accepted Newton's theory of static atoms with mutual repulsions but could not see

. . . how any intestine motion could augment or diminish this repulsive power. But it struck me that if gases were made up of particles or atoms mutually impinging on one another and on the sides of the vessel containing them . . .

the theory would be more simple, consistent, and easy. After very many pages of quite impenetrable verbal arguments, exhibiting an astonishing confusion about the meaning of the law of conservation of momentum, he eventually reached a set of propositions which are roughly equivalent to Eq. (1) above.

He then gave experimental proofs that his hypothesis was correct. In a thermal mixing experiment in a calorimeter, he said, quantity of motion was conserved; and quantity of motion, as everyone knew, was momentum. Since momentum depended on the first power of the velocity whereas Eq. (1) implied that the absolute temperature varied as the square, he predicted that when equal masses of the same substance at absolute temperatures T_1 and T_2 were mixed, they would reach equilibrium at T_3 where

$$\sqrt{T_1} + \sqrt{T_2} = 2\sqrt{T_3}.$$

Water at 0°C and 100°C should reach equilibrium not at 50°C but at 48°C. He could not perform the experi-

The first practical steam carriage to ply along English roads, between Bath and London, in 1829. John Herapath is up front. (Information kindly supplied by Spencer D. Herapath of London.)

Waterston as he appeared at the age of 46. (Courtesy Oliver & Boyd Ltd. from *The Collected Scientific Papers of J. J. Waterston*, edited by J. B. S. Haldane.)

ment himself for lack of good thermometers, so he searched the literature. Crawford, he found, had determined the equilibrium temperature to be 50.0°C; but this result, said Herapath, was the *expected* one, it was therefore suspect and should be rejected. (Actually it is within 0.05° of the correct value.) De Luc, on the other hand, had found 48.3°C. This confirmed Herapath's theory. But there was another proof, equally convincing, that his theory of gravity was correct. For it was well known that the acceleration of gravity at the earth's surface varied from equator to pole in a way which did not conform to the known ellipticity of the earth. But Herapath could now explain this in terms of the influence of the temperature at the two latitudes on Newton's "constant" of gravitation. Again his theory was in agreement with observation.

These papers were published in 1821, to be followed by long drawn-out disputes, attacks, refutations, and denials. But after they died down, there were *three* rival theories of gases—Newton's static theory, Davy's rotational model, and Herapath's hypothesis.

Joule and Others

IF Herapath remains a comic figure in spite of his real achievement, John James Waterston was a man whose genius was dogged by tragic ill luck. In 1843, while a schoolteacher for the East India Company in Bombay, he had a book published in Edinburgh—anonymously—entitled *Thoughts on the Mental Functions,* an attempt to explain human behavior in mathematical and physical terms. In a note at the end, he gave a full and accurate account of the kinetic theory

of gases. But nobody read the book. Two years later he sent a paper to the Royal Society on the physics of media composed of free and perfectly elastic molecules in a state of motion; he wrote that he hoped that "although the fundamental hypothesis [of perfect elasticity] is likely to be repulsive to mathematicians, they will not reject it without a fair trial". But the referee reported that it was "nothing but nonsense" and only a short abstract was published, in another journal. Waterston not only developed the basic ideas precisely and was the first to see the relevance of Graham's recently published law of the effusion of gases through small holes, but he also stated the principle of equipartition of energy, introduced the concept of the mean free path (the "impinging distance") and proposed modifications of the model to represent imperfect gases. But his work was passed over and his influence on the main stream of science was negligible compared with that of the gregarious Mr. Herapath.

Joule favored Davy's rotational hypothesis at first. In a paper on electrolysis (1844) he spoke of revolving atmospheres of electricity and later he used the same idea to explain radiation. In his paper on the rarefaction of air he said that the centrifugal force of the revolving atmospheres was the sole cause of the expansion of a gas when the pressure was removed. But his main interest was to calculate the specific heats of gases. In one of his notebooks is to be found the rough draft of a lecture in which he drew a block of a substance

> . . . containing a number of atoms each of which revolves rapidly on its axis in the direction of the hands of a watch. Suppose now a number of fine cords to be rolled round each of these atoms and to pass over a wheel. It is evident that the force of the atoms will be diminished in winding up the weight W. This diminution of the velocity of the atoms is what we generally call a diminution of temperature. . . .

But shortly afterwards he realized that both rotational and translational motion could give the result that the vis viva of the atoms was proportional to the heat content. In 1848 he wrote that since Herapath's hypothesis was simpler, he would use it in preference to Davy's. He calculated the molecular velocities in several gases and also some specific heats. These were the first definite numbers ever to emerge from the kinetic theory of gases (except for those in Waterston's papers). Thus the kinetic theory of gases and the dynamical theory of heat were developed at the same time and largely by the same people.

Rankine developed the rotational (or vortex) theory to its highest refinement shortly *after* this. The essence of his method was to divide up each revolving atmosphere into concentric shells, typically of area $4\pi r^2$

and acted upon by a centrifugal force of the type mc^2/r; the pressure p was therefore of the type $mc^2/4\pi r^3$. The total volume V of N such atoms was $\frac{4}{3}\pi r^3$ N. Substituting, one arrives again at Eq. (1). Rankine extended this to arbitrarily shaped vortices and again reached the same result—as must always be for any form of motion because of the implicit assumption of the equipartition of energy. This was an interesting situation, for no experiment could ever decide which was the correct model. But within a few years Herapath's hypothesis gained almost universal acceptance. Even Waterston managed to get a paper published on it in 1851. The German scientists Krönig and Clausius evolved just the same ideas independently in 1856 and 1857 (though Clausius certainly knew of Joule's results on molecular velocities). Two years later, a German translation of Bernoulli's old paper was published. Within a short time even British scientists were writing of "Bernoulli's theory lately revived by Mr. Herapath" and it was not long before Herapath's name was almost forgotten.

The vortex model persisted for a long time, however, in various guises. Maxwell dismissed it for an ordinary gas, in preference to Herapath's hypothesis, because he thought the rigidity would be too high. But he used it, as is well known, as the basis of his model of the electromagnetic ether. Later still, Kelvin made smoke-ring models of atoms to explain spectra. His calcula-tions of the modes of oscillation of such systems have a very modern sound.

Conclusion

THE outstanding feature of this story is that—like the dynamical theory of heat—the kinetic theory of gases had first to break the grip of an abstruse and authoritative mathematical theory before the simple basic physical ideas could be accepted. These difficulties should, perhaps, be presented in their proper perspec-tives in our teaching textbooks.

Above all, these episodes accentuate the problem of communication in science. The records of the Royal Society and the French Academy of Sciences are blotted again and again by the rejection of outstanding dis-coveries. On the other hand, a hypothesis like Hera-path's, published in a journal with less stringent refereeing, was embedded amid so much nonsense-writing that it took the instinct and genius of a man like Joule to uncover the one idea worth preserving. When the amateur historian descends into the stacks of a library he can contemplate the yards and yards of dusty volumes, records of decades of busy scientific activity. On the average, perhaps one short paragraph out of the huge output of any one year was *really* worth writing. Scientific researches are like fishes' eggs—only one in thousands ever reaches maturity. It is a chastening thought.

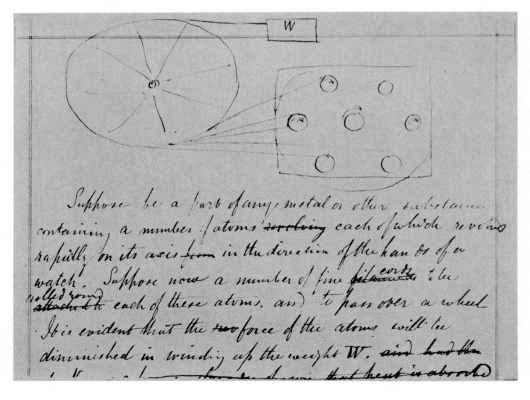

A sketch (drawn sideways to save space) from one of Joule's private notebooks. It illustrates his idea of the rotating atom theory.

Rowland's physics

Of the three most eminent US physicists of the late nineteenth century, Gibbs, Michelson and Rowland, it was the "doughty knight of Baltimore" who had the broadest impact, setting the pace for the golden age of US physics.

John D. Miller

PHYSICS TODAY / JULY 1976

"Those were the days," reminisced Daniel Gilman, the President of The Johns Hopkins University, "when scientific lecture-rooms in America gloried in demonstrations of 'wonders' of Nature— 'the bright light, the loud noise, and the bad smell.' Rowland would have none of this." The Johns Hopkins physicist thus characterized was Henry Rowland, whose contributions—particularly those in spectroscopy and electromagnetism— secured him a high place in the ranks of nineteenth-century physicists.[1]

Neither experimental nor theoretical physics was widely practiced in the United States during the middle of the nineteenth century. The popular study was natural history, a subject for which the bountiful Nature of a young and largely unexplored country offered a collage of unknown plants, animals and geological patterns to investigate. The description of such novelties required neither mathematical nor other formal training, and although professionals appeared, the field was particularly attractive to dilettantes and amateurs.[2] But the logic and mathematics of exacting physical investigations lay beyond the reach of amateurs. It was not surprising, therefore, that papers on natural history filled the *American Journal of Science,* the leading science publication in the United States and one read extensively abroad—or that the mathematics so seldom encountered in the *Journal's* pages appeared puerile in comparison to European standards.

However, in the latter half of the century at least three physicists in America

John D. Miller is an associate professor at, and associate director of, the Lawrence Hall of Science of the University of California, Berkeley.

practiced their mathematics as well as their science at a level equal to the best of their European colleagues: J. Willard Gibbs, Albert Michelson and Rowland. All three men were anomalies in nineteenth-century America.

Of these three it was Rowland who, in terms of laboratory investigations, set the most influential standards for physics research, particularly through his precise instrumentation. Although Michelson became well known, particularly when his ether experiments were found to corroborate the Lorentz–Fitzgerald contraction hypothesis, this did not come until much later. Furthermore, recent findings have refuted the idea of a generic relationship between Michelson's work and Albert Einstein's postulates of 1905.[3]

Rowland, in contrast, had completed major experiments in magnetism and electricity, and his diffraction gratings and Sun-spectrum photographs were, in the 1880's, distributed and acclaimed throughout the world. Later, in 1894, Michelson himself even turned to Rowland for advice concerning the suitable outfitting of a new physical laboratory at the University of Chicago. As for Gibbs, although his work was of the highest quality, it was all theoretical—it is doubtful that he had ever set foot inside a laboratory. In the end it was Rowland who set the pace in US laboratory physics in the last quarter of the century.[4]

The education of a civil engineer

When Rowland was born in 1848 he became the only son in a family of five children. When he was eleven, his position of responsibility was underscored by the death of his father, a Protestant theologian. This occupation had been followed by three generations of Rowland

males, all of whom are reported to have possessed exceptional intellects and dominant personalities. In fact the earliest, David Rowland (1719–94) was from his Providence pulpit such a zealous defender of his country against foreign oppression that during the Revolutionary War he was forced to flee the city, escaping with his family up the Connecticut River in darkness through the midst of a surrounding British fleet.

Rowland's mother Harriet expected him to follow family tradition and enrolled him, at thirteen, in classical studies at New Jersey's Newark Academy. But Henry was more interested in mechanics and electricity, as a small pocket notebook that he began in 1862 records. Found here are accounts of the things he made— electromagnets, induction coils, galvanometers and electric motors. Scientific studies had not yet won prominence in American education, however, and according to Samuel Farrand, the Academy headmaster, a scientific education to Harriet Rowland "seemed like throwing away her boy."

Rowland suffered through three years of Latin and Greek, finally writing in July 1865 that classical studies were "horrible," declaring " *'Non feram, non patiar, non sinam'* [I will not bear, I will not suffer, I will not tolerate] is a sentence which just expresses my condition." He petitioned his mother to study science. Eventually she relented, and when Rowland was 17 he enrolled in the Rensselaer Technological Institute in Troy, New York.

The "scientific course" was largely oriented toward practical applications and led to a civil-engineering degree. The school also trained mechanical and hydraulic engineers as well as architects and superintendents of gas and iron works.

ROWLAND

At Rensselaer Rowland studied mathematics through the calculus of variations, but spent most of his time with apparatus that he constructed and operated in his boarding-house room. His letters home were filled with descriptions of galvanometers, electrometers and a Ruhmkorff coil that would, as he wrote his sister Jennie, "charge and discharge a Leyden jar twenty times a second or more." He was very active in the school's scientific club, reading papers in 1868 on spectroscopes, the mechanical equivalent of heat and tests of his induction coils.

In the fall of 1876, his third year, Rowland arranged his classes so as to spend all his mornings uninterrupted on his experiments. He also began to keep records of his work in bound notebooks. Entries included sketches of gold-leaf electroscopes, the origin of electricity produced by the contact of water and heated metal, the upward force of wind necessary to support a man in flight with "wings" twenty feet in length, and the change in the position of the apparent poles of a horseshoe magnet when an armature was placed across its legs.

The next year brought records of much more serious studies—particularly in numerous references to one special source of ideas: The *Experimental Researches in Electricity* of Michael Faraday is the first entry in a list of scientific books contained in one notebook. Another two-volume notebook of 1868 contains more than a dozen references such as "Notes from Faraday's Experimental Researches in Electricity" and "Thoughts

suggested by the reading . . . of Faraday."

Faraday's ideas, as recorded in these volumes, eventually provided the basis for two of Rowland's major experimental investigations: the magnetic analogy to Ohm's Law and the magnetic effect of a moving electrical charge. His interests were thus not limited to civil engineering; he wrote to his mother in May 1868:

"You know that from a child I have been extremely fond of experiment; this liking, instead of decreasing, has gradually grown upon me until it has become a part of my nature and it would be folly for me to attempt to give it up . . . I intend to devote myself hereafter entirely to science—if she gives me wealth I will receive it as coming from a friend but if not I will not murmur."

The magnetic analogy to Ohm's law

After his graduation from Rensselaer in June 1870 with the degree of Civil Engineer, Rowland could not find a position in experimental science. He spent his time on a series of magnetic researches conducted at his mother's Newark home.

These experiments were originally intended to determine the distribution of magnetism in several iron and steel bars. Rowland soon found it difficult, however, to interpret his measurements. At this time little was accurately known about the effects of various media and geometric configurations on the transmission of magnetic forces. Experimenters such as Sir William Harris, William Sturgeon, Sir James Joule, Heinrich Lenz and Joseph

Henry had studied isolated factors, but their experiments had produced no single model of magnetic action that simultaneously took into account the shape of the core, its material composition and the arrangement of the energizing coil.

By using detection coils that could be quickly reversed by physical means, wired to a galvanometer of his own design, Rowland began a very accurate mapping of the magnetic fields produced by direct electric currents, checking alignments with his rifle sights. The method depended upon obtaining a reversal time that is a small fraction of the natural oscillation period of the galvanometer.

As he traced Faraday's lines of force in various materials he experienced difficulty in interpreting his measurements, for the configuration of the lines appeared to shift as the current in the electromagnets was varied. The shift was only a few per cent but it was enough to upset Rowland's venerated sense of precision. His interest was aroused to seek a theoretical, mathematical model of the phenomena.

For a physical model to analyze, Rowland turned to the analogy between electricity and magnetism that Faraday had postulated in his diary. The idea involved *Gymnotus electricus,* an eel having electric organs concentrated along its body and tail. Using his bare hands to estimate intensities, Faraday in 1838 had studied the distribution of electricity surrounding the eel during its moment of discharge, sketching the action as shown in figure 1. Faraday pictured the physical lines of electric force as continuous through the cells of the eel and its sur-

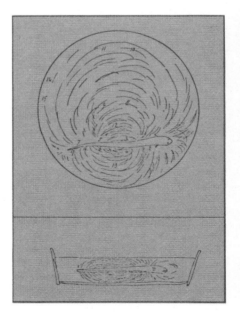

The electric action of an eel, as Michael Faraday sketched it in his notebook in 1838. Estimating intensities with his bare hands, Faraday found at the moment of discharge that "every part of the water" is filled with a current from the front to the rear of the eel. From this Rowland drew an analogy to the lines of force surrounding a bar magnet.   Figure 1

rounding medium, "for they form continuous curves like I have imagined within and without the magnet." The three-dimensional aspect of the analogy represented characteristic Faraday brilliance.[5]

In a notebook of 1872 Rowland cited Faraday's *Gymnotus* work, but evolved the idea further, using the well known nineteenth-century telegraph circuit. The imperfect insulators of a telegraph line correspond to lines of magnetic force in a medium surrounding a magnet; the galvanic cell represents the source of the lines of force, and the analog of the horizontal wire is the conduction of the lines within the magnetic material itself.

Having pictured the circuit for one electric cell of the eel, Rowland imagined a distributed set of such cells, as shown in figure 2a, corresponding to a series of short bar magnets laid end to end as shown in figure 2b. Rowland in 1873 added the effects of the driving force of each cell to obtain the following equations, which describe the distribution of Faraday's lines of force in and near a long magnetic bar:

$$Q_\epsilon = \frac{M}{2\sqrt{RR'}} \frac{1-A}{(A\epsilon^{rb}-1)} (\epsilon^{rx} - \epsilon^{r(b-x)})$$

$$Q' = \frac{\epsilon^{rb}-1}{(A\epsilon^{rb}-1)(\sqrt{RR'}-s')} \frac{M}{r}$$
$$- \frac{M}{2R} \frac{1-A}{A\epsilon^{rb}-1} (\epsilon^{rb}+1-\epsilon^{rx}-\epsilon^{r(b-x)})$$

In these equations $r = (R/R')^{1/2}$,

$$A = \frac{\sqrt{RR'}+s'}{\sqrt{RR'}-s'}$$

and

R = resistance to lines of force of 1 m of length of bar
R' = resistance of medium along 1 m of length of bar
Q' = lines of force in the bar at any point
Q_ϵ = lines of force passing from the bar along a small distance l
ϵ = base of Napierian logarithms
x = distance from one end of the helix
b = length of helix
s' = resistance, at the end of the helix, of the rest of the bar and the medium
M = magnetizing force of the helix.

These complex exponential forms were obviously far from Ohm's simple proportionality law. However, Rowland went on to study the predictions of his equations for the center of a long thin magnet, for which, from his observations of the lines in the media surrounding the magnet and from symmetry considerations, he expected the lines to assume homogeneous paths and his equations to reflect this simplification. Rowland did not publish the details of the algebraic transformation of these equations under these limiting conditions. At the center, $x = b/2$, of any magnet, the number of lines, Q_ϵ, passing along the bar at some small distance clearly vanishes, since the factor on the right of the first expression becomes zero

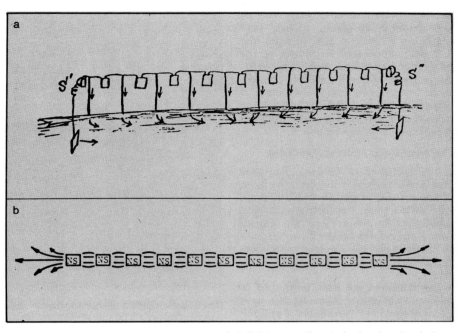

A telegraph circuit provided Rowland with a heuristic link between Faraday's electric eel and a bar magnet. Sketch **a**, taken from one of Rowland's notebooks, shows a nineteenth-century telegraph circuit with distributed galvanic cells added. The series of short bar magnets shown in **b** completes the analogy, which led to a magnetic equivalent of Ohm's law. Figure 2

there. On the other hand, the number of lines in the bar does not vanish at this point but equals

$$Q' = \frac{1-\epsilon^{-rb}}{(A-\epsilon^{-rb})(\sqrt{RR'}-s')} \frac{M}{r}$$
$$- \frac{M}{2R} \frac{1-A}{A-\epsilon^{-rb}} (\epsilon^{-rb}+1-2\epsilon^{-rb/2})$$

For the case of the infinitely long bar, $b \to \infty$, Rowland found that this equation reduces to

$$Q' = \frac{M}{R}$$

Thus in the center of the magnet the number of lines passing through the magnetic medium was proportional to M, the magnetizing force in the magnet, and inversely proportional to R, the resistance to these lines of force. He had therefore found a magnetic analogy to Ohm's law for electric circuits.[6]

To use this simplified formula for measuring the magnetic properties of different metals Rowland constructed toroidal magnets. In this geometry the lines of force closely approximate those observed at the center of a long bar. An early Rowland ring is shown in figure 3.

The above equations were the product of several experiments begun three years earlier. Rowland had repeatedly received rejection notices from the editors of the *American Journal of Science,* who finally admitted that they simply did not understand his mathematics. But before receiving this explanation Rowland sent his 1873 paper directly to James Clerk Maxwell, whose treatise of that year had contained a general theory of magnetism from which Rowland's equations could be derived. This congruency could have been expected: Both men had started with Faraday's ideas. Maxwell was much taken by Rowland's work and arranged for immediate publication in the *Philosophical Magazine* in England.

In 1875 Rowland found an appointment at Rensselaer, but was given no de-

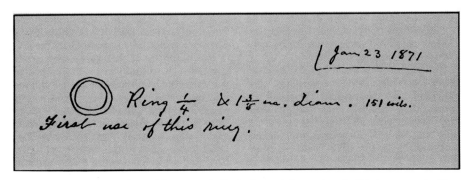

In a toroidal magnet the lines of force are similar to those at the center of a thin bar magnet. The first Rowland ring is shown in a dated entry from Rowland's notebook. Figure 3

cent laboratory in which to carry out exact experimental research. That year through a relative he met Gilman, who had been appointed President of the newly founded Johns Hopkins University. Gilman saw Maxwell's letters, thought them "worth more than a whole stack of recommendations" and hired the young Rensselaer engineer to organize a physics department at Johns Hopkins.

The charge-convection experiment

That summer of 1875, Gilman took Rowland to Europe to inspect institutions of science and to visit instrument shops with a view toward outfitting a physical laboratory for Johns Hopkins. On his own, Rowland was Maxwell's house guest in Scotland and then crossed to the Continent, arriving in Berlin in late October.

Rowland had not been impressed by much of what he saw, reporting that many shops seemed like "museums of antiquity" and the laboratories looked as though the "architect had got the best of the physicist." But in Germany he wrote Gilman,

"You were right when you said I would find no lack of scientific spirit here and the apparatus shows it. In America we have apparatus for illustration, in England and France they have apparatus for illustration and experiment, but in Germany, they have only apparatus for experimental investigation. Our country is hardly ripe for the latter course though I should like to see it pursued to the best of our ability."

Caught up in this spirit, Rowland applied for a general course of study in the university laboratory at Berlin under Hermann Helmholtz. The distinguished German physicist's reply was prompt and negative, citing crowded conditions.

Not giving up, Rowland wrote to Helmholtz again, this time proposing as specific experiments either an extension of his magnetic researches or a plan that Rowland had recorded in his Rensselaer notebook of 1868:[7]

"The question I first wish to take up is that of whether it is the mere motion of something through space which produces the magnetic effect of an electric current, or whether those effects are due to some change in the conducting body which, by affecting some medium around the body, produces the magnetic effects."

Rowland and Maxwell had discussed these ideas the previous summer, and Rowland told Helmholtz, "Maxwell assumes that the last case will produce magnetic effects although he has since told me he had no reason for the assumption." In the 1873 *Treatise,* Maxwell did in fact indicate his "supposition that a moving electrified body is equivalent to an electric current" but did not give his reasons.

Rowland–Hutchinson charge-convection apparatus seen from above, in an 1889 photo. The question Rowland asked was whether the mere motion of a charge could generate magnetic effects similar to those of a current in a wire. His answer, yes, proved hard to confirm.   Figure 4

This time Helmholtz was interested and had a storage room cleaned out in the basement for Rowland. Months earlier Helmholtz had been investigating the possibility of convection currents related to his and Franz Neumann's potential theory of magnetic actions, which he described as "open" circuits. Did the discharge of arc observed between disconnected wires, Helmholtz wondered, actually complete the circuit, accompanied by magnetic effects?

Rowland set up a single gilded ebonite (vulcanite) disk, 21 centimeters in diameter, to revolve about a vertical axis 60 times a second. He reversed the polarity of the electrification while observing the reflection of a beam of light reflected from a mirror attached to a delicate magnetic astatic-needle system. The mirror was placed on a thread between two compass needles aligned with poles opposed to cancel the effect of the Earth's magnetism. The disk revolved on a plane between the needles. Figure 4 shows an 1889 version of the apparatus. After several weeks of trials he reported a distinct deflection of the beam by several millimeters, noting that this "qualitative effect . . . once being obtained, never failed." He was reporting a magnetic force only about 1/50 000 of that of the Earth's horizontal component in Berlin.

Not stopping at this qualitative measurement, Rowland went on to compute the expected magnetic force to compare it to measured values. To do this he had to assume some value for Maxwell's v, the ratio of electromagnetic to electrostatic units—the constant Maxwell had postulated to be equal to the velocity of light. Rowland assumed a value of 288 million m/sec, as measured by Maxwell, to produce the Table on page 43 from 62 readings of individual deflections.

The difference between expected and measured values of force, Rowland noted, was 3, 10, and 4 per cent respectively with Maxwell's value for v. He observed, however, that the "value v = 300 000 000 meters per second, satisfies the first and last series of the experiments best."

This mechanically brillant experiment cost only about fifty dollars. Many were to attempt to repeat it in a variety of forms—and be frustrated—in the next 25 years. It was Maxwell himself who bestowed the laurels, writing in serio-comic verse,

The mounted disk of ebonite
   Has whirled before, nor whirled in vain;
Rowland of Troy, that doughty knight,
   Convection currents did obtain
In such a disk, of power to wheedle,
From its loved North the subtle needle.

'Twas when Sir Rowland, as a stage
   From Troy to Baltimore, took rest
In Berlin, there old Archimage,
   Armed him to follow up this quest;
Right glad to find himself possessor
Of the irrepressible Professor.

But wouldst thou twirl that disk once more,
   Then follow in Childe Rowland's train,
To where in busy Baltimore
   He brews the bantlings of his brain . . .

Back to Baltimore

When Rowland returned to Baltimore from Europe in the spring of 1876 he told Gilman, "Give me time and apparatus and if our University is not known, it will

Rowland in a self-portrait, about 1882. Centrally placed in his bachelor apartment was the bronze horse Rowland bought with the prize money that was awarded to him for his precise measurement, in 1880, of the mechanical equivalent of heat. Figure 5

not be my fault." Rowland wanted research apparatus "*not* for the illustration of lectures." He argued that, although it was *sometimes* possible to produce good work with poor apparatus—just as it is possible to cut down a tree with a pen-knife—there is work that can not possibly be done without calling to our aid all the resources of mechanics. To this class, he asserted, belong "many of the higher questions in mathematical physics."

Gilman showed him two former boarding houses in downtown Baltimore that were to serve as temporary laboratories. Rowland said that all he needed in one of the buildings was the back kitchen and a solid pier built up from the ground "to sustain such instruments as require steadiness."

His intentions were to carry out a series of measurements of basic physical constants. The philosopher, logician and meteorologist Charles Peirce visited Baltimore in 1878 and was critical of Rowland's plans. But Rowland went ahead anyway, beginning a new determination of the mechanical equivalent of heat. This massive project led in 1880 to a 125-page report, including subsidiary

investigations in thermometry and calorimetry. The research, a paradigm of precise measurement, won Rowland the Venetian Prize in 1881 and, recommended by Peirce, an honorary doctoral degree. In the background of Rowland's self-portrait, figure 5, the bronze horse that he purchased with the prize money can be seen.

Rowland also carried out precise measurements of Maxwell's ratio of units at this time, again as a test of the electromagnetic theory of light. He used a spherical condenser that had been machined with great precision to provide a known capacitance. Its stored charge in turn was passed through a calibrated galvanometer.

The first of these measurements looked promising, Rowland wrote to Maxwell in April 1879: "I believe the experiment is a link in the proof of your theory seeing that the result is, by the first rough calculation, 299 000 000 meters per second, though the corrections may amount of ½ per ct. or so." But subsequent measurements produced a v of 297 900 000 m/sec, a value that decreased with the number of discharges employed; this discouraged

Rowland initially from publishing his results.

A third constant related to v was the value of the standard resistance, the ohm. In the electromagnetic system of units evolved in the latter half of the century, length/time were interestingly also the dimensions of resistance. Many measurements of v ultimately depended on knowledge of the ohm. Rowland was critical of measurements made by the British Association and a German group headed by Friedrich Kohlrausch, discovering arithmetical errors and inconsistencies through dimensional analysis. His criticism proved valid, and he served on several international committees through the 1880's, presiding over the International Electrical Congress in Chicago in 1893.

It is little known but Rowland had assembled the most elaborate and extensive set of equipment to be found anywhere in the world in the late 1870's and early 1880's. Physicists at Harvard had inventoried US collections,[8] and Rowland's student Edwin Hall told Gilman that Johns Hopkins "would be the loser" if it exchanged its apparatus with that at Cambridge University's Cavendish Laboratory, for example, particularly if "what belongs to Prof. Rowland personally was included."

In 1877, by using superb galvanometers and an experimental configuration devised by Rowland, Hall measured an electric potential acting perpendicular to a line of current flow and to a magnetic field. A fluid model of electricity was used throughout Hall's work, perhaps the last productive use of this model leading to fundamental electrical laws. In 1894 Rowland disclosed the extent to which he had been involved in Hall's work, telling George Fitzgerald that the convention experiment, ". . . together with that of Mr. Hall [Hall effect] which was really my experiment also, were made to find the nature of electric conduction. Indeed I had already obtained the Hall effect on a small scale before I made Mr. Hall try it with a gold leaf which gave a larger effect. My plate was copper or brass and I only obtained 1 mm deflection. Mr. Hall simply repeated my experiment, according to my direction, with gold leaf." Rowland's colleague Joseph Ames wrote in an account of the period, "There have been several striking cases where it might have seemed to an impartial observer that Rowland's name should have appeared on the title page."

"Magnifique" gratings

Another major line of Rowland's study was spectroscopy. Before 1881, the problem of ruling an optical grating of high resolution, yet free from large periodic errors in spacing, had been solved only partially. With his sense of precision mechanics, Rowland became interested in the critical worm screw that advances

Magnetic force due to a rotating charge.

	Series I	Series II	Series III
Measured magnetic force (horizontal component)	0.000 003 27	0.000 003 17	0.000 003 39
Computer magnetic force (horizontal component compound using Maxwell's value of v)	0.000 003 37	0.000 003 49	0.000 003 55

the metallized glass plates under an oscillating diamond scriber. Periodic errors in the screw resulted automatically in grating errors. Rowland invented a method of grinding screws, submerged in water, over a three-week period.[9] A ruling engine employing the new screw design was completed in 1882. The figure on page 26 of this issue of PHYSICS TODAY shows Rowland with one of his ruling engines. At about this time Rowland also invented the concave grating, which eliminated the need for auxiliary telescopes or other optical accessories for observing the spectrum under study.

Rowland, accompanied by his colleague John Trowbridge, took sample gratings to the Paris electrical conference of 1882. Trowbridge reported on the reactions of French physicist E. E. Mascart, Sir William Thomson and Kohlrausch:

"It is needless to say that they were astonished. Mascart kept muttering 'superb,' 'magnifique.' The Germans spread their palms, looked as if they wished they had ventral fins and tails to express their sentiments ... We left [Paris] with the feeling that there was little to be learned there in the way of physical science, and having sent for the above scientists as heralds to proclaim the preeminence of American diffraction gratings ..."

In England Rowland told an equally enthusiastic audience, "I have ruled 43 000 lines to the inch and I can rule one million to the inch, but what would be the use, no one would ever know that I had really done it." Trowbridge wrote that there was much laughter at this: "This young American was like the Yosemite, Niagara, [the] Pullman parlor car; far ahead of anything in England ..."

There was great demand for Rowland gratings, which Johns Hopkins distributed at cost throughout the world. One of special note went to Pieter Zeeman, who used it in 1897 to observe the magnetic widening of the two D lines of the sodium spectrum. Figure 6 shows a spectrometer in use in Rowland's own laboratory.

Distractions from pure science

In 1890 Rowland, then 42, was married and discovered through a life-insurance examination that he had diabetes, which was incurable at that time. He was given ten years to live.

Until his marriage Rowland had rarely worked on any commercially related scientific work. (Once, in 1879, he had tested the efficiency of Edison's newly invented electric light.) Nor had he filed patents on any of his laboratory-apparatus inventions. But when Rowland's children were born in the early 1890's all this changed.

By 1896 he had filed or received confirmation of at least nineteen patent claims dealing with commercial electrical equipment. He also spent an immense

Using one of Rowland's gratings in his own laboratory at Johns Hopkins, about 1885. This photograph, printed from a glass negative, was taken by gaslight.   Figure 6

amount of time on a complicated multiplex telegraph system. However, the telegraph with its delicate synchronization system never proved commercially practical and the company went bankrupt shortly after Rowland's death.

A commercial consulting project also absorbed much of Rowland's time during 1892–93. He was retained as chief design consultant for the Cataract Construction Company, which was involved with the design of a power-generating plant at Niagara Falls. The generation and transmission of electric power on such a

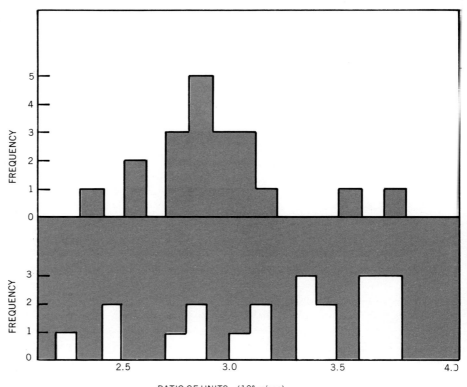

Histograms of historical data. The upper diagram shows the frequency of charge-convection measurements for values of v, the ratio of electromagnetic to electrostatic units, from Rowland's 1876 Berlin experiment. The lower histogram, of data taken by Rowland and Cary Hutchinson in Baltimore in 1889, is much less satisfactory, due to electrical noise.   Figure 7

scale had never been attempted, and Rowland spent most of his time during that period on the project. When his fee of $10 000 was rejected by Cataract he brought suit. The jury awarded in his favor, but it had been an inopportune time for the physicist to be occupied in court.

This was the year that Philipp Lenard's paper on cathode rays in the free atmosphere appeared; the following year brought Wilhelm Röntgen's announcement of a hitherto unknown and mysterious form of radiation. To Rowland it was a disappointing period, in which he published only two minor electrical papers.

It was not until 1899 that he again directed basic researches into the nature of electricity and magnetism on any large scale. Since his Berlin convection experiments of 1875 there had been numerous attempts to repeat the investigation, with mixed results. In fact, in 1889 Rowland and one of his students attempted a repetition and obtained much less satisfactory results than those from Berlin in 1876. In 1970 I summarized the two sets of data in figure 7 by reconstructing the ratio of units v from the raw data of each set of experiments. The spurious effect of trolley lines and other technologies of an electrically noisier age are apparent in the 1889 data.

The most disturbing of these attempts was a series of researches conducted late in the century by Victor Crémieu at the University of Paris, who could not find any magnetic effect.

But, for Rowland, the effect discovered by Zeeman of the sodium D line splitting could be explained by the convection equipment. Vibrating, electrified "matter" within a molecule gripped the ether. This might produce a magnetic effect, which interacted with Zeeman's externally applied magnetism. Perhaps

the rotating matter of the Earth likewise retained "a feeble hold on the ether sufficient to produce the Earth's magnetism . . ." Late in the decade he therefore decided to undertake new experiments in an attempt to measure directly an interaction with the ether. At the same time he directed a new series of charge-convection experiments.

By Christmas 1900, results from a series of ether experiments in which a cylinder wound with 80 meters of wire and revolved with great velocity in air appeared promising and Rowland wrote to reserve space in the *American Journal of Science*. But when the commutator leads were reversed, the galvanometer failed to reverse; he never again attained a steady deflection. Yet positive results were obtained from a new series of convection experiments and were reported to Rowland shortly before his death on 16 April 1901.

The decades of the 1870's and 1880's had been the most productive for Rowland, but the picture is also clear in the 1890's of a dying physicist torn between commitments to science and to family. Only once did he appear to refer publicly to his diabetic condition, but then it was with considerable bitterness and frustration:[10]

"What blasphemy to attribute to God [death] which is due to our own and our ancestors' selfishness is not founding institutions for medical research in sufficient number and with sufficient means to discover the truth. Such deaths are murder. Thus the present generation suffers for the sins of the past and we die because our ancestors dissipated their wealth in armies and navies, in the foolish pomp and circumstance of society and neglected to provide us with a knowledge of natural laws."

It was not until 1921 that Frederick Banting and John Macleod discovered

insulin, sharing the Nobel Prize in medicine in 1923.

My historical research on Rowland began in 1967 through a grant from the Smithsonian Institution. Archivist Frieda C. Thies (retired) assisted me in organizing Rowland's scientific notebooks, which I recovered from uncatalogued storage at the Johns Hopkins University in 1968. The interested reader can find additional technical references in the Isis *articles in reference 1. For a copy of Rowland's letter to Helmholtz, reference 7, I am obliged to Christa Kirsten, Archiv Direktor, Deutsche Akademie der Wissenschaften.*

References

(Unless otherwise noted quotations and notebook citations refer to materials contained in the Rowland and Gilman manuscript collections at Johns Hopkins University.)

1. J. D. Miller, Isis **63**, 5 (1972); **66**, 230 (1975).
2. *Science in Nineteenth Century America,* (N. Reingold, ed.), Hill and Wang, New York (1964).
3. G. Holton, Isis **60**, 2 (1969).
4. *Selected Papers of Great American Physicists,* (S. R. Weart, ed.), American Institute of Physics, New York (1976).
5. *Faraday's Diary,* 1820–62 (T. Martin, ed.), G. Bell and Sons, London (1933), volume III, page 354.
6. H. Rowland, Phil. Mag. **46**, 140 (1873).
7. Rowland to Helmholtz, 13 Nov. 1875, in the Archives of the Deutsche Akademie der Wissenschaften, (East) Berlin.
8. J. W. Gibbs, E. R. Wolcott, E. C. Pickering, and J. Trowbridge, list of [scientific] apparatus, Harvard College Library Bulletin, volume 11, pages 302, 350 (1879).
9. H. Rowland, "Screw," reference 4, page 85.
10. H. Rowland, Presidential address to The American Physical Society, 28 Oct. 1899, reference 4, page 91.   □

Michelson and his interferometer

Pioneering applications in such diverse fields as astronomy, atomic spectra and mensuration followed the initial disappointment over the failure to detect a luminiferous ether.

Robert S. Shankland

PHYSICS TODAY / APRIL 1974

Albert Abraham Michelson was the first American scientist to win the Nobel Prize, and his career is one of the most fascinating in the entire history of physics. His earliest work was firmly based on the classical physics of geometrical optics—in a precise determination of the velocity of light by an improved Foucault method. But then he mastered wave optics and invented his interferometer, and from that point on he proceeded to dazzle the scientific world with a display of the applications he found for his invention during a career that exhibited throughout a unique pattern of originality and dedication to physics.

The interferometer came into being for the specific purpose of measuring the Earth's motion through the luminiferous ether, a project familiar to generations of physics students as the "Michelson–Morley experiment." Although this single undertaking has proved important enough to guarantee Michelson's place in history, the unexpected negative result caused response at the time to be lukewarm, and this is not the work for which the Nobel Prize was awarded in 1907. He was honored instead for the other applications of his invention—particularly for his work on the determination of the length of the International Standard Meter in terms of the wavelength of light, but also for such diverse and pioneering achievements as the discovery of fine and hyperfine structure in atomic spectra and the first application of interference measurements in astronomy.

The birth of a concept

Michelson's invention of this remarkable instrument, the interferometer—which to the present day plays important roles in Fourier spectroscopy, laser-beam interferometers and the ring-laser gyro—came suddenly with but little relationship to his earlier researches on the speed of light.

He had been born in 1852 at Strzelno in the Prussian province of Posen and travelled with his parents to frontier towns of California and Nevada. Then he made his way with the greatest determination to the Naval Academy at Annapolis, where he excelled in science and made his first precise measurement of the speed of light. One will search in vain in his Annapolis textbook[1] and in his papers and correspondence for clues as to what inspired his great invention. At Annapolis, and later, when Simon Newcomb invited him to collaborate with him at the Naval Observatory in Washington, Michelson's velocity-of-light determinations employed exclusively the methods of ray or geometrical optics, with heliostats, mirrors and lenses to produce intense beams of light; there is no indication in this period of his concern or interest in the wave properties of light or in optical interference.

But in a few weeks in 1880 between his last velocity of light determinations with Simon Newcomb in Washington and his first work in Helmholtz's laboratory at Berlin (where he had gone on leave from the Navy for special study and research), he clearly had mastered the basic principles of the wave nature of light and then invented his interferometer, which is one of the most powerful and elegant applications of the characteristic interaction between light waves.

However, two events had occurred in Washington that bear closely on the invention of his interferometer. The first was a letter, dated 19 March 1879, which James Clerk Maxwell[2] had written to David Peck Todd at the Nautical Almanac Office inquiring about astronomical observations on Jupiter's satellites suitable for a determination of the speed of light but which more importantly, might reveal the Earth's motion through the ether of space. In this letter, which was also studied by

Newcomb and Michelson, Maxwell had asserted that no terrestrial method was capable of measuring the speed of light to the one part in a hundred million that would be necessary in any laboratory experiment to detect the Earth's motion through the ether. Maxwell's statement appears clearly to have been the challenge that the young Michelson accepted for developing his interferometer specifically to carry out a laboratory ether-drift experiment, which he first conducted in Germany and later in its final form with Edward W. Morley at Cleveland.

A second clue showing Michelson's shift in interest from ray optics to wave optics after his study of Maxwell's letter is suggested by a short paper he presented to the Philosophical Society of Washington on 24 April 1880. It is entitled "The Modifications Suffered by Light in Passing Through a Very Narrow Slit."[3] This report gives a brief but accurate account of his observations on the already well known diffraction phenomena produced by a narrow slit. However, the subject seems to have been new to Michelson, and he reported his keen observations on the color and polarization of the light as he narrowed the slit width while using sunlight for the source. This early paper is certainly not one of his major contributions, but it does reveal his remarkable observational ability as he describes precisely the colors, polarization, and diffraction patterns produced. This paper strongly suggests he had already appreciated that the key to meeting Maxwell's challenge for precision optics was essentially to find a method of measurement that would directly employ the extremely short wavelengths of light and not depend on the macroscopic length and time measurements of ray optics that he had employed exclusively in his earlier work.

When Michelson arrived at Helm-

Albert A. Michelson in 1927 at his desk in the Ryerson Physical Laboratory, University of Chicago. This is one of two photographs, taken by H. P. Burch, that Michelson often said he liked better than any others. (Courtesy of the Michelson Museum). Figure 1

Robert S. Shankland is Ambrose Swasey Professor of Physics at Case Western Reserve University, Cleveland, Ohio.

The Michelson–Morley experiment as used in Cleveland in 1887, with its optical parts mounted on a five-foot-square sandstone slab. This photograph was found in 1968 by D. T. McAllister in a Michelson notebook at Mount Wilson Observatory. (Courtesy of the Michelson Museum and the Hale Observatories.)   Figure 2

Holder for optical-flat "beam-splitter" of the Michelson–Morley interferometer used in 1886–87 at what is now Case Western Reserve University.   Figure 3

holtz's laboratory in Berlin in the fall of 1880, he experienced for the first time the thrill of a well equipped and active research center, for at that time this was probably the outstanding laboratory in Europe for physics research. There also he was suddenly brought in touch with the best apparatus available for experiments in optics, for Helmholtz himself was already world famous for his researches in physiological optics. The questions that had been raised in Michelson's mind by the phenomena of his narrow-slit experiment in Washington had "sensitized" him to react strongly and appreciate fully the many new stimuli of Helmholtz's laboratory. In any event, soon after his arrival in Berlin his pondering and search for an optical method that would meet the severe requirements posed by an ether-drift experiment aroused his natural creative instincts and he invented the Michelson interferometer. (But it is possible that he had already conceived the essential elements of the instrument while still in Washington, where Newcomb had introduced him to Alexander Graham Bell who later, on Newcomb's recommendation, supplied the necessary funds to have the first interferometer built by Schmidt and Haensch in Berlin.)

In later years he always stated that the interferometer was devised specifically for the ether-drift experiment. It is, of course, impossible to trace precise paths in the creative thinking of a scientist and conclusively demonstrate how he finally arrived at his goal, and there are discontinuities in the process that even the man, himself, cannot explain. But it seems clear that Maxwell's letter and the narrow-slit experiment in Washington were essential spurs to Michelson's genius for his invention of the interferometer.

This instrument is a classic example of symmetry, and apparent simplicity. He dispensed with the narrow slits that physicists had employed since the days of Thomas Young to produce interference between coherent light beams, and instead used a large glass optical flat silvered just enough on one face to half reflect and half transmit the entire wavefront of the light impinging on it, thus giving much greater intensity and permitting a wide range of experiments that had been impossible with all earlier optical apparatus. Once the two coherent light beams were produced at the optical "beam splitter," they could then each be directed by mirrors and lenses in a variety of ways (through moving water for example) and then be reunited to add and subtract their vibrations to produce the beautiful patterns of bright and dark interference fringes that Michelson studied in one experiment after another for the rest of his life. He spent the last forty years at the University of Chicago (figure 1 is a photograph dating from this period).

Ether drift

We will note here only a few of the great experiments he carried out with his interferometer. As already stated, it was specifically devised to measure the motion of the Earth through the ether, a medium that in those days was universally believed to be essential for the propagation of light. In this experiment, first tried unsuccessfully at Potsdam in 1881, and then after Michelson became the first professor of physics at Case School of Applied Science, it was conducted in its definitive form (see figures 2 and 3) by Michelson and Morley at Cleveland in 1887.

One of the two coherent light beams produced in the interferometer was caused to traverse a to-and-fro path along the direction of the Earth's motion, while the other light beam travelled along a path of exactly equal length in a perpendicular direction. On their return the two light beams were recombined to produce white-light interference fringes, so that the central white fringe could serve as a reference. Michelson had confidently expected from calculations that, when the apparatus was rotated so as to interchange the positions of the two light beams, the pattern of interference fringes would shift and thus reveal the Earth's motion through the ether. This procedure, in effect, compares with great precision the speed of light in the two arms of the interferometer. The ether theory predicted that this speed should be altered unequally by the Earth's motion, to a degree proportional to the square of the ratio of the Earth's speed to that of light. The apparatus was sensitive enough to have shown this extremely small effect discussed by Maxwell, but no significant shift of the interference fringes was observed. The scientific world generally, and Michelson in particular, were greatly disappointed by this result, which was in direct conflict with accepted theory at that time. It was many years before the work of George Fitzgerald, H. Antoon Lorentz, Joseph Larmor, Henri Poincaré and, finally, Albert Einstein carried theoretical physics to the point where Michelson and Morley's result could not only be explained, but served as an essential basis for our modern concepts of space and time.

It is a curious fact that for many years Michelson seldom mentioned this result. It did not appear in his Vice-Presidential Address to the American Association for the Advancement of Science, delivered at Cleveland in 1888; his students at Case School of Applied Science never heard of it in his physics classes there, and it is absent from his Nobel Prize lecture in 1907. After many years Michelson did discuss it in his optics courses at the University of Chicago, but only after the relativity theory was fully established; even then it was described primarily in its relation to the ether theory of Augustin Fresnel and Lorentz, rather than for its importance to relativity.[4] But, in Einstein's words, Michelson had "led the physicists into new paths, and through his marvelous experimental work paved the way for the development of the theory of relativity. He uncovered an insidious defect in the aether theory of light as it then existed, and stimulated the ideas of H. A. Lorentz and Fitzgerald out of which the special theory of relativity developed. This in turn pointed the way to the general theory of relativity, and to the theory of gravitation."[5] As Robert A. Millikan emphasized in 1948 at the dedication of the Michelson Laboratory in California, the ether-drift trial has long been regarded as one of the two greatest physics experiments performed in the nineteenth century (the other being the Faraday-Henry discovery of electromagnetic induction).

Measuring the meter

But strangely enough this was not the work for which Michelson was awarded a Nobel Prize, the first such award to an American. Rather, the Prize was given primarily in recognition with Morley in Cleveland, and in 1887 they abruptly abandoned the search for the ether to prove the feasibility of their optical method for standardization of the meter.[6] An early form of interferometer built for this purpose is now at Clark University and is shown in figure 4. Michelson alone com-Paris which was jealously guarded against damage or loss. Clearly a reproducible standard of length was highly desirable—one that could be duplicated at any major laboratory in the world.

The solution of the problem was first undertaken by Michelson in collaboration with Morley in Cleveland, and in 1887 they abruptly abandoned the search for the ether to prove the feasibility of their optical method for standardization of the meter.[6] An early form of interferometer built for this purpose is now at Clark University and is shown in figure 4. Michelson alone com-

Early interferometer of the type developed by Michelson and Edward W. Morley and used by Michelson in Paris for measuring the standard meter in wavelengths of cadmium light, 1892–93. (Courtesy of the Michelson Museum and Clark University.) Figure 4

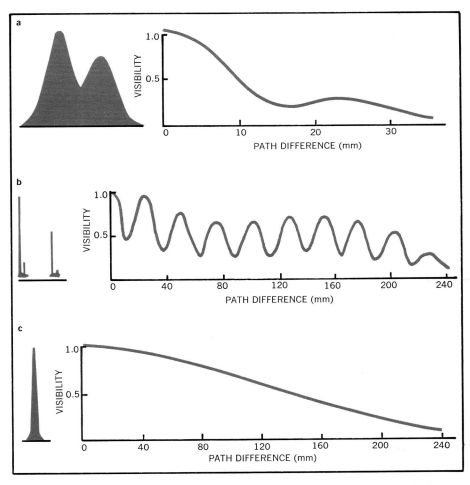

"Visibility curves" of interference fringes as a function of light-path differences in the two interferometer arms (solid color curves on right), with the analyzed structure of the spectrum lines (colored peaks on left). Part a: Fine-structure doublet of H-alpha line of hydrogen. Part b: Hyperfine structure in a line of thallium. Part c: The narrow red line of cadmium used to standardize the meter. (From A. A. Michelson, "Light Waves and Their Uses," University of Chicago Press, 1903.) Figure 5

pleted the determination in Paris, and since that time it has been a matter of little concern whether or not the standard meter bar continues to exist, for thanks to Michelson and the later development of the orange–red line of krypton[7] as a new primary standard, the length of a light wave is now the official standard of length.

Two major discoveries were made by Michelson and Morley in the course of their standard meter work in Cleveland.[8] To measure the meter in terms of light waves it was essential that the interference fringes in their special interferometer should be produced by light of an extremely narrow spectrum line, so that interference between light beams with a large difference in path was possible. In the course of their search for such a light source they analyzed many spectrum lines with the interferometer by observing the changes in the "visibility" of the fringes as the path difference was increased. Today this process is the basis of the large activity in Fourier spectroscopy.[9] They were surprised to find that nearly all spectrum lines are complex and thus discovered what is now known as "fine-structure" in the spectrum of hydrogen, and "hyperfine structure" in the spectra of mercury and thallium (see figure 5). It was many years before the full significance of these findings for atomic and nuclear physics was understood, and it is interesting to note that the detailed explanation of fine structure requires the relativity theory that owed so much to Michelson's other experiments. The discovery of fine structure and hyperfine structure will always ensure their work an important place among those experiments that were basic for the development of quantum mechanics and nuclear physics. They also were the stimuli that led Michelson to his invention of the echelon spectroscope, his harmonic analyzer for more accurate Fourier spectroscopy, and his long program at the University of Chicago in the ruling of diffraction gratings.

The Earth's rotation

Michelson's experiments on the ether continued from 1881 until 1929, and "to the end he hoped to empirically prove there was this medium known as *the ether.*"[10] One of the most interesting applications of his interferometer for this search was in the Michelson-Gale-Pearson experiment conducted in 1924–25 on the Illinois Prairie at what is now the Clearing industrial area west of Chicago. As early as 1904 Michelson had proposed an interferometer experiment to reveal the Earth's rotation through the ether. During 1921–23 he had made preliminary trials at the Mount Wilson Observatory, for after Eddington's successful

Part of the system of evacuated pipes used in the Michelson–Gale–Pearson experiment at Clearing, Illinois, 1924–25. This photograph shows, left to right; Charles Stein, Thomas J. O'Donnell, Fred Pearson, Henry G. Gale, J. H. Purdy and an unidentified worker. (Courtesy of the Michelson Museum and J. H. Purdy.)　　　Figure 6

solar-eclipse expeditions in 1919 had found the deflection of starlight by the sun, as predicted by general relativity, there had been a great revival of interest in all related experiments.

Michelson was in ill-health at the time, but with the active collaboration of Henry G. Gale, Fred Pearson and Tom O'Donnell a large system of 12-inch-diameter pipes for the light beams was set up on a rectangle (300 meters by 600 meters) on level ground. Figure 6 shows part of the large rectangle of pipes employed. The two light beams from a Michelson interferometer were reflected, (one in each direction) around the circuit of evacuated pipes. The Earth's rotation affected the times of travel of the two beams unequally and thus would be revealed by the interference fringes when the two beams were re-united. A second system of fringes from light-beams travelling in a smaller rectangle of pipes established the fiduciary point of the interference pattern. Michelson's poor health and the excessive newspaper publicity that attended this experiment cooled his enthusiasm for the work, but it was carried through successfully. This experiment is the optical analogue of the Foucault pendulum, and as such "only shows that the Earth rotates on its axis," as Michelson caustically remarked. The results were definite, giving a shift of 0.25 fringe in the larger optical circuit.[11] However, since this result was in agreement not only with both the special and the general theories of relativi-

ty but also with Fresnel's old fixed-ether theory, it did not give the decisive test that had been hoped for.

The techniques of this experiment were of great interest to Einstein, and the following letter from him accurately describes its relation to relativity:

September 17, 1953

Dear Dr. Shankland:

The Michelson–Gale experiment does, of course, concern the relativity question but, as you mentioned yourself, not insofar as relativity theory differs from Lorentz' theory based on an ether at rest. My admiration for Michelson's experiment is for the ingenious method to compare the location of the interference pattern with the location of the image of the light source. In this way he overcomes the difficulty that we are not able to change the direction of the earth's rotation.

Sincerely yours,
Albert Einstein (signed)

However, modern applications of this method in the ring-laser gyro have proved to be of great value for measuring and guiding rotations in the navigation of satellites, missiles, and aircraft.

The diameter of a star

One final application of the interferometer should be emphasized—in this case to astronomy. As shown in figure 7, Michelson adapted his instrument for use with large telescopes to mea-

Michelson's twenty-foot stellar interferometer mounted on top of the 100-inch Hooker telescope at Mount Wilson Observatory in 1920, as used to measure the angular diameter of Betelgeuse. The outer two (movable) mirrors collect the starlight and the two inner ones direct it to the eyepiece. (Courtesy of the Hale Observatories.) Figure 7

sure the diameters of heavenly bodies. First, at the Lick Observatory in 1891, he determined the sizes of Jupiter's satellites;[12] later, in 1920 at the Mount Wilson Observatory, Michelson and Pease measured for the first time in history the angular diameter (0.047 seconds of arc) of a star (Betelgeuse).[13] This latter feat was one of the greatest triumphs of his life-long devotion to precision measurements with light waves, and extensions of his method are now an essential element for much work in long baseline radioastronomy. After explaining the technical details of the stellar measurement to a joint meeting of the American Physical Society and the AAAS, he then urged his children "to always remember the wonder of it."

In closing this account we should also realize that Michelson's other continuing scientific interest in addition to his interferometer was the measurement of the speed of light. He pursued this for over half a century from his first determination along the old sea-wall at Annapolis (1877–79), then across the Potomac in Washington, then along the railroad tracks in Cleveland (1882–84) and, finally, at Mount Wilson and at Santa Ana in California

until the end of his days (1931).[14] The accuracy of his results improved steadily over the years and the continuing importance of this fundamental constant for science has fully justified the care that he lavished on its determination.

* * *

This article has been adapted from an address given 21 October 1973 at the New York University Hall of Fame Meeting at Town Hall, New York City.

References

1. A. Ganot, *Treatise on Physics*, (Atkinson's translation), New York (1873).
2. J. C. Maxwell, reprinted in Nature **21**, 314 (1880).
3. A. A. Michelson, Smithsonian Misc. Collections **20**, 119 and 148 (1881).
4. V. O. Knudsen, Notes of Michelson's University of Chicago Lectures made in 1917. Also correspondence with Michelson's students, Harvey Fletcher, Ralph D. Bennett and Richard L. Doan.
5. A. Einstein, Science **73**, 379 (1931).
6. A. A. Michelson, E. W. Morley, Amer. J. Sci. **34**, 427 (1887).
7. Natl. Bur. of Std. Publ. 232, April 1961.
8. A. A. Michelson, E. W. Morley, Amer. J. Sci. **38**, 181 (1889).
9. J. N. Howard, G. A. Vanasse, A. T. Stair, D. J. Baker, Aspen Conference on Fourier Spectroscopy, 1970.
10. Letter of T. J. O'Donnell (Michelson's instrument maker) to R. S. Shankland, 12 July 1973.
11. A. A. Michelson, H. G. Gale, F. Pearson, Astrophys. J. **61**, 137 (1925).
12. A. A. Michelson, Publ. Astron. Soc. Pacific **3**, 274 (1891).
13. A. A. Michelson, F. G. Pease, Astrophys. J. **53**, 249 (1921).
14. Dorothy Michelson Livingston, *The Master of Light*, Scribners, New York (1973). □

Poincaré and cosmic evolution

Among his other, better known, studies this nineteenth-century "mathematical naturalist" enquired into the origin and stability of the solar system, the fate of the universe and the shapes of rotating fluid masses.

Stephen G. Brush

PHYSICS TODAY / MARCH 1980

Henri Poincaré is well known today for his contributions to many areas of mathematics and his popular writings on science. His attempts to apply physical theories to the evolution of the solar system and the rest of the universe are largely forgotten, except by a few specialists. Yet the crisp lucid prose of this brilliant thinker[1] can still help the modern reader to appreciate the worldview of nineteenth-century science, and provides a useful introduction to a fascinating historical phenomenon that I will call "the mathematician as naturalist" (see the box on page 44).

Many of Poincaré's colleagues, I suppose, silently waved the flag of caution when he published his ideas on cosmic evolution. Speculations about the remote past and the distant future of the world should be avoided by a sensible mathematician, especially at a time when scientists are no longer confident that their fundamental theories are valid even

for phenomena that can be studied in the laboratory. It is obviously dangerous to extrapolate those theories to the indefinitely large domains of space and time variables needed to explain such hypothetical events as the origin of the solar system, the birth of the Moon, the long-term periodicity of planetary orbits, the attrition of the Earth's rotation, and the ultimate fate of the entire universe. In France, where the positivist influence was still strong at the end of the century, scientists were discouraged from theorizing about the nature of the world beyond their immediate observations.

Yet the temptation to study cosmic evolution, already irresistible for anyone with a modicum of curiosity about the world, is strengthened for a mathematician by the knowledge that refined reasoning and careful calculation have in the past produced some remarkable advances in physical astronomy. Beyond the obvious example of Isaac Newton, one re-

calls several successful applications of mathematics; three of the most spectacular happen to have been made by Frenchmen. In 1758 Alexis Clairaut predicted the return of Halley's comet, expected the following year, within 30 days, by taking account of the effect of the major planets on its orbit. In 1784 Pierre Laplace showed that the "long inequality" of Jupiter and Saturn was cyclic, not secular, thus eliminating one of the major reasons for doubting the stability of the solar system. In 1846 Urbain LeVerrier pinpointed the position of a previously unknown planet by analyzing anomalies in the orbit of Uranus, and the resulting discovery of Neptune demonstrated again the amazing power of Newtonian celestial mechanics.

As its hegemony crumbled in theoretical physics and other scientific disciplines during the 19th century, France retained its leadership (though certainly not a monopoly) in mathematical astronomy.

Poincaré, heir to this glorious tradition, could hardly ignore the classic problems that had elicited brilliant contributions but not definitive solutions from his predecessors; surely he could peer a little further by standing on the shoulders of those giants.

The problems, in order of Poincaré's most intense concern with them, were:
▶ equilibrium figures of rotating fluid masses (1885)
▶ stability of the solar system and ultimate fate of the universe (1889)
▶ origin of the solar system (1911).
These problems were of course closely interrelated; in particular, we will have to discuss aspects of Laplace's nebular hypothesis in connection with each of them.

Rotating fluids

The first problem goes back to Newton and Christiaan Huygens, who concluded that the centrifugal force associated with the Earth's rotation would cause it to bulge at the equator and flatten at the poles, taking the form of an oblate spheroid. Because two leading astronomers in Paris—Jean-Dominique Cassini and his son Jacques—and some theorists following the ideas of René Descartes came to the opposite conclusion, this problem was seen as a crucial test of the competing Newtonian and Cartesian systems of the world. The results of French expeditions, in the 1730's, to measure the length of the degree of latitude at high and low latitudes, confirmed the predicted flattening at the poles and thus helped to ensure the victory of Newton over Descartes.[2]

To calculate the precise geometrical form of the rotating fluid, including the quantitative relation between speed of rotation and deviation from sphericity, proved to be a much more difficult task, but the work was motivated by its geophysical significance. It was generally believed that the Earth had been formed as a hot fluid, which cooled, solidified (at least on the outside), and contracted. By conservation of angular momentum, contraction would increase the speed of rotation and hence the amount of equatorial bulge. Is the present oblateness just what one would expect for a fluid mass spinning at the present rotation speed of the Earth? If not, does it indicate that the Earth solidified while spinning faster or more slowly than it does now? A secular decrease in rotation rate might be attributable to dissipation by tidal friction, and would presumably be associated with transfer of angular momentum to the Moon.

Stephen G. Brush is a professor in the Department of History and the Institute for Physical Science and Technology, and a member of the Committee on History and Philosophy of Science, at the University of Maryland, College Park.

A possible objection to the hypothesis of a frozen-in equilibrium shape would be that a solid sphere the size of the Earth, composed of known materials such as rocks and iron, would lack the mechanical strength to maintain a non-equilibrium shape against distorting forces, and hence must have very nearly the same shape as a fluid mass rotating at the present rate.

The early history of this problem has been discussed in excruciating detail by Isaac Todhunter.[2] I will mention only the bare minimum needed to put Poincaré's work in context.

Ellipsoids

In 1742 Colin Maclaurin showed that a series of ellipsoids of revolution would be equilibrium figures for low rates of rotation. There was considerable further work by Laplace and others on the details of the solution. It was usually assumed, in mathematical treatments, that the fluid is ideal (no viscosity), homogeneous, and incompressible, conditions that are obviously not satisfied inside the Earth; yet it was apparently thought that if the problem could be solved for a more realistic model, only minor qualitative corrections would be obtained.

In 1834, C. G. J. Jacobi opened up a new aspect of the subject by showing that an ellipsoid with three unequal axes can be a figure of equilibrium. The possibility that a rotating fluid does not have complete rotational symmetry was apparently a shock to the intuition of some 19th-century scientists. The Jacobi ellipsoid was proposed as a model for variable stars, on the assumption that the asymmetrical bulges would emit more light than the flattened sides and thus rotation would produce an apparent change in brightness.

Poincaré's interest in the problem was stimulated by a section in the *Treatise on Natural Philosophy* by William Thomson and P. G. Tait; moreover, he had been teaching fluid mechanics at the Sorbonne starting in 1881, and was dissatisfied with the standard textbook treatments of rotating fluids.[3] His first papers do no more than supply explicit proofs of statements made by Thomson and Tait concerning the stability of annular surfaces of revolution. That problem is closely related to the question of the physical state of Saturn's rings, and this connection may explain how he came across Sonya Kovalevsky's memoir on Saturn's rings; he subsequently credited her with introducing the appropriate methods for such problems.

In his long memoir of 1885, Poincaré discussed a new series of equilibrium figures that branch off from the Jacobi ellipsoids with increasing angular momentum, just as the Jacobi ellipsoids branch off from the Maclaurin ellipsoids. The new figures were later called *piriform* (pearshaped); they can be described qualitatively by imagining an ellipsoid cut

in half, then letting one half flatten and approach a hemisphere while the other becomes more and more elongated. (See figures on pages 46 and 47.) A furrow develops around the elongated part, giving the impression that it is being "strangled" or "wants to separate" into a small and large part. Poincaré apologizes for using such non-mathematical language and cautiously points out that it is difficult to say whether this separation will indeed take place. Nevertheless he thinks it is possible that the next stage of evolution of the system will be a stable equilibrium state of a large and a small body revolving around each other, comparable to a planet and a satellite. He notes (as he did on several other occasions) that this process is not necessarily the one envisaged in Laplace's theory of the origin of the solar system, since according to that theory the primeval nebula is very strongly concentrated at the center, whereas the fluid masses considered by Maclaurin, Jacobi and Poincaré have uniform density.

Poincaré's 1885 paper "came as a revelation" to George Howard Darwin (son of Charles Robert Darwin). Darwin recalled, in awarding the Gold Medal of the Royal Astronomical Society to Poincaré in 1900,[4]

I had attempted to attack the question from the other end, and to trace the coalescence of two detached bodies into a single one—but alas! I have to admit that my work contained no far-reaching general principles—no light on the stability of the systems I tried to draw—nothing of all that which renders Poincaré's memoir one that will always mark an important epoch in the history not only of evolutionary astronomy, but of the wider fields of general dynamics.

In 1878 Darwin had traced the history of the Earth–Moon system, as influenced by tidal forces, back to a time, 54 million years ago, when the Moon was only 6000 miles from the surface of the Earth, and its time of revolution around the Earth was the same as the Earth's rotation period at that epoch, 5 hours 36 minutes.

These results point strongly to the conclusion that, if the moon and earth were ever molten viscous masses, then they once formed parts of a common mass.

We are thus led at once to the inquiry as to how and why the planet broke up. The conditions of stability of rotating masses of fluid are unfortunately unknown, and it is therefore impossible to do more than speculate on the subject.[5]

Since Poincaré, apparently unconcerned about the passage of time on a small scale, did not date his letters to G. H. Darwin, I cannot say just when their collaboration began, but it was in full swing by 1901. In that year each presented a long memoir to the Royal Society of London on pear-shaped figures. Poincaré thought these figures are probably stable, but this could be proved only by very complicated calculations, which he hoped to facilitate by reducing the stability condition to a convenient analytical form. Darwin performed the actual calculations for Poincaré's theory, and concluded that the pear-shaped figures are indeed stable. This would imply that as the fluid planet cools and contracts, a part of it gradually separates but remains in an orbit close to the primary body. The other alternative, if the pear-shaped figure is never stable, is that the body suddenly undergoes an enormous deformation and a series of oscillations, followed by catastrophic disintegration.

Darwin's conclusion was contradicted in 1905 by the Russian mathematician Aleksandr Mikhailovich Lyapunov, who determined by a different method that the pear-shaped figure is initially unstable. The disagreement was still unresolved at the time of Poincaré's death; he was more inclined to believe Lyapunov, having earlier been impressed by his incisive work on similar problems.

In 1915 James Jeans tackled the problem by another method, which enabled him to discover an error in Darwin's calculations; he then confirmed Lyapunov's conclusion that the pear-shaped figures are always unstable. An even stronger result in the same direction was obtained by Elie Cartan in 1924. "And at this point," the astrophysicist S. Chandrasekhar noted, "the subject quietly went into a coma." [6]

Chandrasekhar speaks rather harshly of Poincaré's influence on this field of research. He says Poincaré's "spectacular discovery" of the pear-shaped figures "channeled all subsequent investigations along directions which appeared rich with possibilities; but the long quest it entailed turned out, in the end, to be after a chimera. . . . The grand mental panorama that was thus created was so intoxicating that those who followed Poincaré were not to recover from its pursuit." [7]

Poincaré's pear-shaped figures are no longer believed to play any role in cosmic evolution. But the hypothesis that fission following rotational instability of a fluid mass could lead to the formation of double-star systems is still being investigated by at least a few astrophysicists; so Poincaré's general approach may come back into favor.[8]

Stability of the solar system

Poincaré's concern with "stability" showed up in another mathematical problem with greater relevance to astronomy than the evolution of homogeneous fluid masses: the effect of gravitational perturbations on the orbits of planets in the solar system. In its simplest form this is the famous "three-body problem." Whereas a single planet could

continue to move forever in a Kepler orbit around the Sun in the absence of frictional resistance and other forces, the presence of a second planet must disturb its motion and, over a sufficiently long period of time, might cause it either to spiral into the Sun or wander off to infinity.

According to the "clockwork universe" concept, or the "Newtonian world machine" as it is sometimes called by historians of ideas, the effect of perturbations is cyclic rather than secular: each planet remains in an orbit whose dimensions change back and forth between fixed limits. According to Newton himself, if nothing more than gravitational forces were involved, the perturbations would have a secular effect and occasional (divine?) intervention is needed to restore the system to its proper state. That statement was the occasion for the Leibniz–Clarke debate of 1715–16, in which Gottfried Wilhelm Leibniz accused Newton of disrespect for God through the implication that He was not competent enough to construct a clockwork universe that could run forever by itself, but rather has to wind it up from time to time. Newton stood his ground, arguing that to relegate God's actions to the indefinite past was the first step toward eliminating Him entirely from our conception of the world.

During the 18th century, research in celestial mechanics focused on three "inequalities," meaning, in this context, deviations from cyclic motion in Kepler orbits. Each inequality appeared to be a secular effect of the kind mentioned by Newton, and thus to endanger the long-term stability of the solar system:

▶ The secular acceleration of the Moon, noticed by Edmund Halley in 1693 and apparently confirmed by the detailed calculations of Tobias Mayer, implied that the Earth–Moon distance was decreasing; if it continued, the Moon would eventually crash into the Earth.

▶ The long inequality of Jupiter and Saturn, also first noted by Halley (1695), was a gradual acceleration of Jupiter and a retardation of Saturn. The ultimate result would be the loss of Saturn—one of the most interesting heavenly bodies, because of its rings—from the solar system, and gradual destruction of the inner planets as Jupiter fell toward the Sun.

▶ The decrease in the obliquity of the ecliptic, from about 23°51′ in the 3rd century BC to about 23°28′ in the 18th century AD, threatened to abolish seasonal variations of climate on Earth, if it led to a final state in which the Earth's axis of rotation is always parallel to the axis of its orbit around the Sun.

Following heroic but inconclusive work on these problems by Leonhard Euler and Joseph Lagrange, Laplace finally explained all three phenomena as cyclic rather than secular effects, and moreover proved some general theorems suggesting that the parameters of planetary orbits oscillate around fixed values. Thus by the standards of 18th-century celestial mechanics Laplace proved the stability of the solar system, and justified the clockwork universe philosophy. This is why he might have replied, when Napoleon protested that his book on the universe failed to mention its creator, "Sir, I have no need of that hypothesis." (So Newton was right when he warned that the clockwork view would lead scientists to atheism; yet his own theory of gravity started them on that path!)

This digression on 18th-century celestial mechanics may suggest why Laplace had such a high reputation in the 19th century—a reputation that 20th-century scientists may find hard to appreciate because the three inequalities mentioned above, and the other problems he solved such as the speed of sound, are familiar only to a few specialists. Moreover, we now know that his analysis of the first inequality was defective—the Moon is slowing down, not speeding up, so that it was much closer to the Earth in the past; hence the possibility of explaining its origin in the way G. H. Darwin suggested—and that his arguments for the stability of the solar system are not conclusive. But in the 19th century Laplace's authority in astronomy was so great that his nebular hypothesis for the origin of the solar system was widely accepted in spite of many serious defects and Laplace's own diffidence in presenting it.

Evolutionary philosophies

In a way it was Laplace who furnished the physical basis for the evolutionary philosophy that dominated science in the late 19th century. This came about by two rather different routes. First, the supposed proof of the stability of the solar system implied that the Earth had remained at more or less the same average distance from the Sun for an indefinitely long time in the past; hence the temperature at the surface of the Earth had been roughly the same for countless millions of years. Therefore geologists (Hutton, Playfair, Lyell) could assume that the same physical causes that we now see in action had been operating with the same intensity in the past, favoring a "uniformitarian" as opposed to a "catastrophist" approach in geological explanation. Charles Darwin was then able to invoke a geological time-scale on the order of hundreds of millions of years to permit a slow process of biological evolution by natural selection.

Second, the nebular hypothesis provided an example of evolution in the universe, which could be taken (as it was, for example, by Robert Chambers and Herbert Spencer) as the first stage of a comprehensive scheme of cosmic evolution leading to the emergence of plants, animals and humans. In fact, the nebular hypothesis was attacked on theological grounds (Laplace's reputation as an atheist may have played some part in this) just as biological evolutionary theories were castigated. The success of Laplace's supporters in overcoming this criticism, and the fact that intellectuals thereby became accustomed to talking about evolution in the cosmos, may have assisted the favorable reception of Charles Darwin's theory.[9]

Nevertheless there was a conflict between these two kinds of evolution, for the nebular hypothesis implied that the Earth had originally been a hot fluid mass that subsequently solidified and cooled. The time needed to cool to the present state could be estimated from Fourier's heat-conduction theory (assuming no present sources of heat inside the Earth) and the resulting "age" of the Earth was much less than the time geologists needed for their uniformitarian explanations. It was also much less than the time Darwin had suggested was available for biological evolution; hence arose the famous controversy on the age of the Earth. It was settled by the discovery of radium. Though Thomson's estimate of the age of the Earth turned out to be much too low, this problem did play an important role in the origin of his 1852 principle of the dissipation of energy.[10]

Thermodynamics and stability

When Poincaré became interested in the stability of the solar system in the 1880's, the problem had acquired physical as well as mathematical aspects. According to Thomson's dissipation principle, or the generalized second law of thermodynamics, irreversible processes in the solar system should push it toward a final equilibrium state, in which planetary and satellite orbits would not necessarily be the same as they are now. On the other hand the work of Rudolf Clausius, James Clerk Maxwell and Ludwig Boltzmann suggested that irreversible processes themselves might be explained in terms of the motions and collisions of molecules obeying Newton's laws. Irreversibility might be no more than a statistical effect resulting from our inability to keep track of the paths of the immense number of molecules in a macroscopic sample, as the counterexample of Maxwell's Demon suggested.

Poincaré's memoir on the stability of the solar system, or rather on the three-body problem and the equations of dynamics, was submitted in 1889 for a prize offered by King Oscar II of Sweden. His motivation was at first primarily mathematical—the problem involved the properties of solutions of differential equations near singularities—but by the time he had won the prize and published the memoir (1889) he had acquainted himself with the physical as well as the astronomical aspects of the problem. As in the case of rotating fluids, he acknowledged his debt to the results of "la savante mathématicienne" Sonya Kovalevski.

Stability, Poincaré points out, may have two different meanings in celestial mechanics. It may entail that the point P representing the position of the system in space (in general, in the n-dimensional phase space of positions and momenta), never goes beyond a fixed distance from its starting point. Alternatively, one may define "stability in the sense of Poisson" as the condition that P returns after a sufficiently long time as close as one likes to its original position. Poincaré's recurrence theorem states that almost all solutions of the equations of mechanics possess stability in the sense of Poisson, provided that P never leaves a fixed volume V. In general the system will return not once but infinitely many times to a configuration very close to its initial one.[11]

The claim that "almost all" solutions are stable is expressed by Poincaré as follows: there are an infinite number of *unstable* solutions, as well as an infinite number of *stable* solutions, but the former are the exception and the latter the rule—in the same sense that the rational numbers are exceptions while the irrationals are the rule. The probability that the initial conditions in any real problem correspond to an unstable solution is zero.

The historian of mathematics would presumably want to investigate the extent to which Poincaré was familiar with the work of Georg Cantor, Emile Borel, and Henri Lebesgue, and how much that work may have influenced his conception of the recurrence theorem in 1889 and later. I cannot go into this point here except to note that there is a brief remark in the 1889 paper to the effect that Cantor has shown that a set can be "perfect" but not "continuous"; hence one cannot, strictly speaking, draw conclusions about the infinity of trajectories near periodic solutions. It was my feeling, when I first wrote on Poincaré's recurrence theorem several years ago, that his proof would not be considered rigorous by a 20th-century mathematician because it lacked the notion of a set of measure zero, and indeed that was the view of Constantin Carathéodory who provided a measure-theoretical proof in 1919. However, Clifford Truesdell tells me that there is nothing really wrong with Poincaré's proof; the measure-theoretical reformulation by Carathéodory is merely cosmetic.

From my viewpoint one of the most interesting of all of Poincaré's writings is a short paper, "Le mécanisme et l'expérience," which he published in *Revue de Metaphysique et de Morale* in 1893. Here he alludes to the contemporary crisis of the atomistic philosophy in physical science, and discusses the cosmological implications of his recurrence theorem. First he describes the "reversibility paradox" that arises when one tries to reconcile the Second Law of Thermodynamics with any theory based on Newtonian mechanics. The fundamental equation of Newtonian mechanics, $F = ma$, is time reversible, and therefore any motion in one direction in time can be replaced by a motion in the opposite direction without violating Newton's laws—the entire system can run either "backwards" or "forwards." Yet all experience teaches that natural phenomena are irreversible. The "English kinetic theorists" (a curious omission of Boltzmann!) have made a valiant attempt to overcome this difficulty, through a statistical explanation—"the apparent irreversibility of natural phenomena is . . . due to the fact that the molecules are too small and too numerous for our gross senses to deal with them." [12] Maxwell's example of the fictional demon shows how the Second Law could be violated if this were not the case.

Poincaré then introduces his own contribution to the debate:[13]

A theorem, easy to prove, tells us that a bounded world, governed only by the laws of mechanics, will always pass through a state very close to its initial state. On the other hand, according to accepted experimental laws (if one attributes absolute validity to them, and if one is willing to press their consequences to the extreme), the universe tends toward a certain final state, from which it will never depart. In this final state, which will be a kind of death, all bodies will be at rest at the same temperature.

I do not know if it has been remarked that the English kinetic theories can extricate themselves from this contradiction. The world, according to them, tends at first toward a state where it remains for a long time without apparent change; and this is consistent with experience; but it does not remain that way forever, if the theorem cited above is not violated; it merely stays there for an enormously long time, a time which is longer the more numerous are the molecules. This state will not be the final death of the universe, but a sort of slumber, from which it will awake after millions of millions of centuries.

According to this theory, to see heat pass from a cold body to a warm one, it will not be necessary to have the acute vision, the intelligence, and the dexterity of Maxwell's demon; it will suffice to have a little patience.

In 1896 the mathematician Ernst Zermelo used Poincaré's recurrence theorem to attack the atomistic theory of heat and the mechanical worldview in general. He had not seen Poincaré's 1893 paper quoted above—he says that Poincaré had not noticed the applicability of his own theorem to the mechanical theory of heat—and that paper seems to have been missed by all the other participants in the debate about the "recurrence paradox." Zermelo's position is that the Second Law does have absolute validity, and therefore entropy can never decrease. According to the mechanical theory of heat a physical system is represented by a collection of atoms obeying Newtonian mechanics; it must obey the recurrence theorem and therefore its entropy will eventually decrease in order to return to its initial value. The only acceptable way to avoid the contradiction is to abandon the mechanical theory.

Boltzmann, the major defender of the mechanical theory at this time, conceded that Poincaré's recurrence theorem is valid; moreover, he claimed that it is completely in harmony with his own sta-

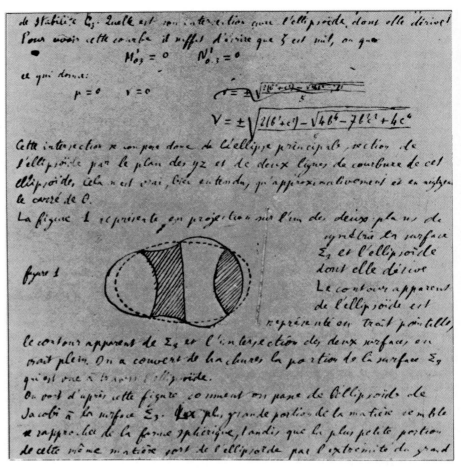

Poincaré's pear-shaped figure of equilibrium is postulated in an 1885 memoir, illustrated in this handwritten version in the *Ouvres* (ref. 1, vol. XI, page 283). The shaded areas in the sketch indicate those parts of the surface that lie within the ellipsoid Poincaré has drawn with a dashed line.

tistical viewpoint, which leads one to expect that there is a small but finite probability that the system will be in any possible state, including the initial one—hence it will eventually reach that state if you wait long enough. But he rejected Zermelo's assertion that there is a contradiction between recurrence and the Second Law; the time for the predicted return to the initial state is many orders of magnitude greater than the times for which the Second Law has been verified.[14]

Poincaré did not comment on Boltzmann's statistical interpretation of the Second Law until a few years later, but in a popular article on the stability of the solar system (1898) he stated that entropy always increases. When it has once changed from its original value, "it can never return again . . . The world consequently could never return to its original state, or to a slightly different state, so soon as its entropy has changed. It is the contrary of stability." [15]

Poincaré clearly regards the periodic solutions demanded by his recurrence theorem as artefacts of the mathematical idealization employed in treating the solar system as a collection of mass points moving in a vacuum, interacting only with

an inverse-square attractive force. But the problem of the stability of the solar system is different from that of solving the set of equations usually considered by mathematicians:

Real bodies are not material points, and they are subject to other forces than the Newtonian attraction. These complementary forces ought to have the effect of gradually modifying the orbits, even when the fictitious bodies, considered by the mathematician, possess absolute stability.

What we must ask ourselves then is, whether this stability will be more easily destroyed by the simple action of Newtonian attraction or by these complementary forces.

When the approximation shall be pushed so far that we are certain that the very slow variations, which the Newtonian attraction imposes on the orbits of the fictitious bodies, can only be very small during the time that suffices for the complementary forces to destroy the system; when, I say, the approximation shall be pushed as far as that, it will be useless to go further, at least from the point of view of application, and we must consider ourselves satisfied.

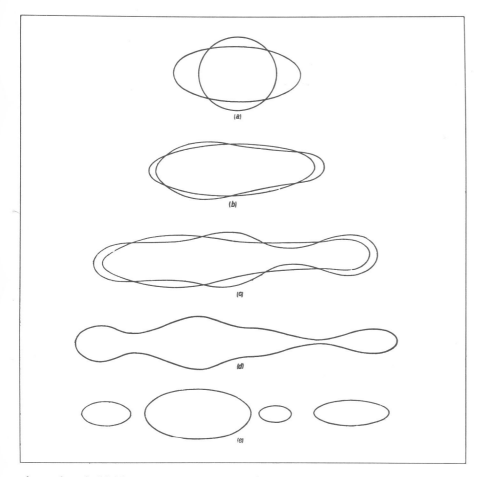

James Jeans's tidal theory to explain the formation of planets begins with a distortion of the Sun by close encounter with another star. Successive diagrams in the sequence above include increasingly "pear-shaped" figures, (b and c), and separation (e). (From Jeans, ref. 19, page 127).

But it seems that this point is attained; without quoting figures, I think that the effects of these complementary forces are much greater than those of the terms neglected by the analysts in the most recent demonstrations on stability.

Let us see which are the most important of these complementary forces. The first idea which comes to mind is that Newton's law is, doubtless, not absolutely correct; that the attraction is not rigorously proportional to the inverse square of the distances, but to some other function of them. In this way Prof. Newcomb has recently tried to explain the movement of the perihelion of Mercury. But it is soon seen that this would not influence the stability . . .

The Second Law of Thermodynamics does, however, destroy the stability of any real physical system, through the action of irreversible processes. In particular:

▶ The existence of a resisting medium in interplanetary space seems to be indicated by anomalies in the motion of Encke's comet. This would eventually cause the planets to fall into the Sun; but the estimated effect is very small.

▶ Tidal forces acting on deformable bodies (liquid or solid) dissipate energy at a significant rate. As Charles-Eugéne Delaunay and G. H. Darwin have shown, the effect of these forces has been to slow the rotation of the Earth, and to force the Moon to keep the same face toward the Earth. In the future, the rotation of the Earth will also become synchronous with the motion of the Moon around it, and the orbit of the Moon will become precisely circular. Both the month and the day will become equal to about 65 of our present days.

Such would be the final state if there were no resisting medium, and if the earth and the moon existed alone.

But the sun also produces tides, the attraction of the planets likewise produces them on the sun. The solar system therefore would tend to a condition in which the sun, all the planets and their satellites, would move with the same velocity round the same axis, as if they were parts of one solid invariable body. The final angular velocity would, on the other hand, differ little from the velocity of revolution of Jupiter. This would be the final state of the solar system if there were not a resisting medium; but the action of this medium, if it exists, would not

allow such a condition to be assumed, and would end by precipitating all the planets into the sun . . .

This is not all: the earth is magnetic, and very probably the other planets and the sun are the same. The following well-known experiment is one which we owe to Foucault: a copper disc rotating in the presence of an electromagnet suffers a great resistance, and becomes heated when the electromagnet is brought into action. A moving conductor in a magnetic field is traversed by induction currents which heat it; the produced heat can only be derived from the *vis viva* of the conductor. We can therefore foresee that the electrodynamic actions of the electromagnet on the currents of induction must oppose the movement of the conductor. In this way Foucault's experiment is explained. The celestial bodies must undergo an analogous resistance because they are magnetic and conductors.

The same phenomenon, though much weakened by the distance, will therefore be produced; but the effects, being produced always in the same direction, will end by accumulating; they add themselves, besides, to those of the tides, and tend to bring the system to the same final state.

Thus the celestial bodies do not escape Carnot's law, according to which the world tends to a state of final repose. They would not escape it, even if they were separated by an absolute vacuum. Their energy is dissipated; and although this dissipation only takes place extremely slowly, it is sufficiently rapid that one need not consider terms neglected in the actual demonstrations of the stability of the solar system.

Poincaré's confidence in the absolute validity of the Second Law survived the efforts of Maxwell, Boltzmann, and J. W. Gibbs to establish a statistical interpretation, but finally succumbed to a much less sophisticated argument. In 1904 he announced to the Congress of Arts and Sciences at St. Louis that the phenomenon of Brownian movement is a visible violation of the Second Law, as Leon Gouy had suggested fifteen years earlier:[16]

. . . we see under our eyes now motion transformed into heat by friction, now heat changed inversely into motion, and that without loss since the movement lasts forever. This is the contrary of the principle of Carnot.

If this be so, to see the world return backward, we no longer have need of the infinitely subtle eye of Maxwell's demon; our microscope suffices us.

Gouy's suggestion had been generally ignored by physicists, and it was not until a year after Poincaré's St. Louis address, when Albert Einstein published a quantitative theory of Brownian motion, that the integrity of the Second Law was seriously compromised. After the experimental confirmation of Einstein's theory by Jean Perrin, scientists could no longer doubt the essential truth of the Maxwell–Boltzmann statistical theory—or for that matter the existence of atoms.[17]

But can the Second Law really be reversed on the astronomical scale? Can the heat death of the universe be avoided? Poincaré's last pronouncement on this subject was elicited by a new cosmological hypothesis, developed beginning in 1903 by the Swedish physical chemist Svante Arrhenius. According to Arrhenius, the universe is like a giant heat engine, operating by heat flow between high-temperature stars and low-temperature nebulae. The latter behave like automatic Maxwell demons, because molecules that escape from nebulae eventually are captured by stars and contribute energy to them as they fall in, thus helping to maintain the high temperatures of the stars. To explain why stars like the Sun do not seem to be gaining mass by this process, Arrhenius postulated in addition that they eject molecules by radiation pressure.

In his paper on the "Arrhenius demon," Poincaré showed that a more careful analysis of the physical effects involved in Arrhenius's scheme leads to the opposite conclusion: the Second Law cannot be violated in this way. But he left open the possibility that some other mechanism could be found to accomplish the same purpose.

Origin of the solar system

By this time (1911), Poincaré had turned his attention from endings to beginnings. His book on cosmogonical hypotheses is generally regarded as a classic by workers in this field; it contains the first serious attempt (aside from that of Edouard Roche) to give a comprehensive analysis of the properties of models based on Laplace's nebular hypothesis. As a review of theories of the origin of the solar system it is well worth reading even today, but somewhat unsatisfactory because it ignores some of the most important theories proposed at the beginning of the 20th century, in particular the tidal-planetesimal hypothesis of the American cosmogonists T. C. Chamberlin and F. R. Moulton, published in 1905. Poincaré does not take seriously the major objections to the nebular hypothesis, and pays little attention to the binary-collision or tidal-disruption theories that were popular at the time he wrote. (Lest the reader think this is a Whiggish criticism of Poincaré, I should remark that the binary theory was rejected around 1935, and recent theories again postulate a primeval nebula.)

If a homogeneous nebula contracts and spins off rings that condense into planets, the result should be a central body that rotates much more rapidly than does our Sun at present. In other words, if angular momentum is conserved in the process, it is difficult to understand how the major planets rather than the Sun came to have most of the angular momentum of the system. That difficulty had been mentioned by Jacques Babinet in 1861 and by Maurice Fouché in 1884, but they considered it an argument for assuming that the nebula was initially highly condensed toward the center rather than as a decisive objection to the nebular hypothesis itself. Poincaré seems to have adopted their conclusions without realizing that the condensed-nebula model was vulnerable to other serious objections. He does not even mention the 1900 papers of Chamberlin and Moulton, which persuaded most astronomers to abandon the nebular hypothesis, or their alternative theory, proposed in 1905, which was favorably received by many American and British astronomers before 1911. (The Chamberlin–Moulton theory postulated a close encounter of the Sun with another star, drawing material out of the Sun by tidal forces; the material first solidified to small particles, which then formed planets by accretion.[18])

Poincaré suggested that a homogeneous nebula, rather than forming a planetary system by Laplace's process, would evolve through the pear-shaped figures he had investigated in 1885, and then split into a double-star system as proposed by T. J. J. See and G. H. Darwin.[19] A similar process might also be responsible for the birth of the Moon from the Earth. While he was aware of Lyapunov's proof of the instability of the pear-shaped figures, he did not realize that this proof made them irrelevant to astronomical evolution.

Ironically one of Poincaré's results on the stability of a fluid ring, which he interpreted as proof that such a ring could be formed in the fashion suggested by Laplace, was used to reach exactly the opposite conclusion by James Jeans.[19] (Jeans favored a tidal theory similar to that of Chamberlin and Moulton.) I

Henri Poincaré

Jules Henri Poincaré (1854–1912) came from a prosperous middle-class family in Nancy. His cousin Raymond Poincaré was several times prime minister of France and its President during World War I. Though sometimes regarded as the world's greatest mathematician during his own lifetime, Henri was not a child prodigy and always had difficulty with arithmetic. He took a degree in mining engineering but soon established himself as a mathematics professor in Paris, where he remained from 1881 until his death. His eyesight was bad and his handwriting terrible; he didn't bother to revise or polish his hastily written lecture notes before publishing them; yet Poincaré was one of the most successful scientists of all time in communicating his ideas to the public. (Four paperbacks of his essays were available until recently from Dover.)

Poincaré's pathbreaking researches in complex-variable theory, differential equations and combinatorial topology earned him an undisputed and continuing high reputation in pure mathematics. The value of his contributions to modern physics is less certain; recognizing the crisis that threatened to undermine nearly every previously-accepted law of nature at the turn of the century, Poincaré was reluctant to propose radical solutions, and preferred to modify the existing theories, which he sometimes regarded as no more than conventions. When he reported Wilhelm Röntgen's work on x rays to the Paris Academy of Science, he offered speculations that now seem pedestrian but apparently inspired Henri Becquerel to begin the research that led him to discover radioactivity. When he undertook an elaborate development of H. A. Lorentz's theory of electrons, he derived much of the mathematical structure of relativity theory but retained the ether hypothesis, relinquishing to Albert Einstein the glory of discovering the physical significance of the relativity principle. Poincaré's masterpiece on celestial mechanics was not translated into English until 1967, when the needs of the space program made it a valuable reference work for NASA.

For students of the psychology of science, Poincaré's most memorable publication is the chapter on "Mathematical Discovery" in *Science and Method*. Recalling his own research on what he called "Fuchsian functions" in honor of the German mathematician Lazarus Fuchs, he described three episodes of intensive effort leading to an impasse, followed by a period when his conscious mind was occupied by non-mathematical thoughts. In each case an important new idea suddenly came to him with great clarity and certainty, obviously the result of an unconscious process in which many possible combinations have been tried, and a single fruitful one had emerged as a candidate for detailed calculation and verification. The process of unconscious manipulation and selection, he argued, could not be purely mechanical but must depend on a "special aesthetic sensibility" that recognizes the most beautiful or harmonious mathematical entity from among billions of possible alternatives.

In contrast to his cousin who pursued a vindictive policy against Germany after World War I, Henri Poincaré demonstrated a special appreciation for the works of German mathematicians. When Felix Klein pointed out that he had considered Poincaré's "Fuchsian functions" though Fuchs himself had not, Poincaré graciously gave the name "Kleinian" to the next class of functions which he discovered. Thus Fuchsian functions are those not studied by Fuchs while Kleinian functions are those not studied by Klein; the properties of both were in fact determined primarily by Poincaré.

suppose the moral of this example, and of Poincaré's excursions into cosmic evolution in general, is that sound mathematical work can indeed have an impact on science, but not necessarily in the way anticipated by the mathematician himself.

In summary, I think Poincaré's view of cosmic evolution was characteristic of the late 19th century: processes in the physical world are gradual and irreversible; discontinuous changes obviously occur, but only when really necessary and then not in a catastrophic manner. The results of mathematical calculation could be interpreted to support such a view, but could not provide a proof strong enough to withstand the onslaught of the revolutionary events and theories of the 20th century. As I have suggested elsewhere,[20] the year 1905 was the turning point in several areas of science, heralding radical changes. To a lesser extent one may claim that Poincaré's concept of cosmic evolution was undermined by developments in that year: Lyapunov's proof of the instability of pear-shaped figures, Einstein's theory of Brownian movement, and the Chamberlin–Moulton theory of the origin of the solar system. What these three have in common is the idea of catastrophe or random collision, administering a shock treatment to the 19th-century idea of a stable, slowly evolving universe. The mathematician who wants to be a naturalist must now assimilate a new set of physical concepts; the need for mathematical expertise is greater now than ever before.

This paper is based on research supported by the History and Philosophy of Science Program of the National Science Foundation. I am indebted to Arthur I. Miller for suggestions on an earlier draft, to John Blackmore for sending me copies of the Poincaré–Darwin correspondence, and to William K. Rose for information on current research in astrophysics.

References

1. The basic source for Poincaré's technical articles is *Oeuvres de Henri Poincaré*, Gauthier-Villars, Paris (1951–1956); see also *Figures d'Equilibre d'une Masse Fluide*, Naud, Paris (1902), and *Leçons sur les Hypotheses Cosmogoniques,* second edition, Hermann, Paris (1913).

2. T. B. Jones, *The Figure of the Earth*, Coronado Press, Lawrence, Kansas (1967); H. Brown, *Science and the Human Comedy*, University of Toronto Press, Toronto (1976), chapter 8; I. Todhunter, *A History of the Mathematical Theories of Attraction and the Figure of the Earth, from the Time of Newton to that of Laplace*, reprint of the 1873 edition, Dover Publications, New York (1962).

3. W. Thomson, P. G. Tait, *Treatise on Natural Philosophy*, Clarendon Press, Oxford (1867); second edition, Cambridge University Press, Cambridge (1879–1883). J. Levy, "Poincaré et le Mécanique Celeste," lecture at The Hague, 1954, published in *Oeuvres de Henri Poincaré*, Vol. 11, pages 225–232.

4. G. H. Darwin, Mon. Not. Roy. Astr. Soc. **60**, 406 (1900), page 411.

5. G. H. Darwin, Phil. Trans. Roy. Soc. London **170**, 447 (1879), pages 535–536.

6. S. Chandrasekhar, *Ellipsoidal Figures of Equilibrium*, Yale University Press, New Haven (1969), page 12; see also R. A. Lyttleton, *The Stability of Rotating Liquid Masses*, Cambridge University Press (1953).

7. Chandrasekhar, ref. 6, page 11.

8. J. P. Ostriker, in *Stellar Rotation*, A. Slettebak, ed., Gordon & Breach, New York (1970), page 147, and in *Theoretical Principles in Astrophysics and Relativity*, N. R. Lebovitz et al., eds, University of Chicago Press, Chicago (1978), page 59; N. R. Lebovitz, Astrophys. J. **175**, 171 (1972); R. C. Fleck, Jr, Astrophys. J. **225**, 198 (1978).

9. See R. Numbers, *Creation by Natural Law: Laplace's Nebular Hypothesis in American Thought*, University of Washington Press, Seattle (1977).

10. J. D. Burchfield, *Lord Kelvin and the Age of the Earth*, Science History Pubs., New York (1975). S. G. Brush, *The Temperature of History*, Franklin, New York (1978), Chapter III; *The Kind of Motion We Call Heat*, North-Holland Pub. Co., Amsterdam (1976), Chapter 14. L. Badash, Proc. Amer. Philos. Soc. **112**, 157 (1968).

11. For the relation of the recurrence theorem to Nietzsche's "eternal return" and other aspects of 19th-century culture, see Brush, *The Temperature of History*, Chapter V.

12. H. Poincaré, Rev. Metaphys. Mor. **1**, 534 (1893); quotation from the translation in S. G. Brush, *Kinetic Theory*, Vol. 2, Pergamon Press, New York (1966), page 205.

13. Brush, ref. 12, page 206.

14. For translations of the Zermelo and Boltzmann papers see Brush, *Kinetic Theory*, Volume 2.

15. *Oeuvres de Henri Poincaré*, Vol. 8, page 538; translation in Nature, **58**, 183 (1898). [The printed text says, twice, entropy always *decreases*]. The long quotations in the text are from Nature, **58**, 184–185 (1898).

16. H. Poincaré, *Congress of Arts and Science, Universal Exposition, St. Louis*, Vol. I, Houghton, Mifflin & Co., Boston (1905), pages 604–622, quotation from page 610. Reprinted in The Monist, **15**, 1 (1905).

17. See Brush, *The Kind of Motion We Call Heat*, pages 669–700.

18. For further details on this theory and its history see S. G. Brush, J. Hist. Astron. **9**, 1, 77 (1978).

19. H. Poincaré, *Leçons sur les Hypotheses Cosmogoniques*, pages 22–23; J. H. Jeans, *Problems of Cosmogony and Stellar Dynamics*, Cambridge University Press, London (1919), pages 147–153; see also G. P. Kuiper, J. Roy. Astron. Soc. Canada **50**, 105 (1956).

20. S. G. Brush, in *Rutherford and Physics at the Turn of the Century* (M. Bunge, W. R. Shea, eds.), Science History Pubs., New York (1979), page 140. □

See also S. G. Brush, "From Bump to Clump: Theories of the Origin of the Solar System 1900–1960," in P. A. Hanle and V. D. Chamberlain, eds., Space Science Comes of Age (Washington, 1981) pp. 78–100; Brush, "Nickel for Your Thoughts: Urey and the Origin of the Moon, Science 217 (1982), pp. 891–898; Brush, Statistical Physics and the Atomic Theory of Matter from Boyle and Newton to Landau and Onsager (Princeton, 1983), Ch. II.

Steps toward the Hertzsprung–Russell Diagram

In the late nineteenth century, astronomers seeking to classify stars by their spectra using then-current concepts of stellar evolution found a temperature–luminosity plot that revolutionized the subject.

David H. DeVorkin

PHYSICS TODAY / MARCH 1978

Every student of stellar astronomy encounters the fundamental relationship expressed by the Hertzsprung–Russell Diagram. One cannot effectively discuss stars—how they are born, live and die, how they are distributed in space and how our Sun fits amongst them—without using this relationship as a fundamental tool of communication.

The Diagram, now almost seventy years old, is today seen in many forms. Basically it is a plot of stellar energy output against stellar surface temperature (see figure 1). The majority of stars plotted occupy a well-defined diagonal band, with a secondary grouping along the top. The observation, first made unambiguously by Ejnar Hertzsprung in 1905 and then by Henry Norris Russell in 1910, was that fainter stars are, on the average, redder than bright ones—except for those prominent stars grouped at the top of the diagram. Astronomers were on the verge of discovering this relationship for quite some time, effectively from the early 1890's. What kept this discovery from being realized and exploited earlier? We will see that the observations necessary to identify stars of similar spectral type, but of vastly differing luminosities—today identified as "giants" and "dwarfs"—were not available until after the turn of the century. As I shall show, 19th-century astronomers were unable to detect the existence of giants and dwarfs among stars of the same spectral type, which led them seriously astray.

But to say that astronomers needed only to produce adequate data before the diagram was possible is an oversimplification. In fact neither Hertzsprung nor

David DeVorkin is presently on leave from Central Connecticut State College as consultant to the Center for History of Physics, American Institute of Physics.

Russell looked directly for the relationship. Each came to it from independent directions, and with different interests. But both required very much the same data base—the brightnesses and spectra of stars—and so both had to turn to a single critically important source: Harvard College Observatory and E.C. Pickering.[1]

The meaning of stellar spectra

The origins of the Hertzsprung–Russell Diagram have one common theme: the understanding of the meaning of the different spectra seen amongst the stars. Since the 1860's and the time of Gustav Kirchhoff, astronomers engaged in spectral classification, including Angelo Secchi, Hermann Carl Vogel, J. Norman Lockyer and William Huggins, all held to the same basic observation that of all the stars examined (which by the 1880's had amounted to several thousands) only a few basic types were to be found, though variants existed. Astronomers then as now were fascinated by the variants— stars that had variable spectra or stars with bright-line spectra. But on the whole, the meaning of the variation of spectra among the few normal groups was the primary question. Throughout the late 19th century the possibility that composition differences were the cause was a persistent theme but the pervading uniformitarian philosophy of Nature, and the fact that the stars did arrange themselves into so few fundamentally different groups, were strong arguments for some other explanation. Secchi in the 1860's and 70's, and Lockyer after him, worked hard to establish temperature as the primary variable causing changes in spectral type. To most, however, the simple correlation of spectrum with stellar color was somehow at the base of the differences seen in spectra.

But there was a problem with this apparently simple picture. The trouble was that at the time, in what was a highly empirical subject, this problem was itself far from being empirical. Very few astronomers of the late 19th century could approach the question of the meaning of stellar spectra without being influenced by the idea that stars were mechanisms that radiated energy from a finite store and hence experienced a continual process of aging. This process became known as the "evolution" of a star, terminology inspired by the Darwinian revolution but in its usage somewhat misleading. And since astronomers had concluded that all sources of energy— chemical, electrical or meteoritic—were inadequate or impossible, only the process of the cooling of an incandescent sphere undergoing continual gravitational contraction, thereby converting mechanical energy into heat, was thought possible.

The cooling process was thought to be directly visible through the spectral differences seen among stars. Thus when astronomers set about examining stars for their spectra, and began looking for an appropriate system for their classification, just about all the systems devised began with blue stars. Blue stars were apparently the hottest, and had the simplest spectra. These stars were also most closely associated with gaseous nebulae in space, and had spectra quite similar to nebulae (exhibiting dark-line spectra with the same groupings and sequences found in parts of the bright-line nebular spectrum. The process of contraction of blue stars from nebulae was supposed to continue to the yellow stars (solar type), and finally, in the general cooling process, to the red stars and then to extinction. This order from blue to red was followed by all the major and popular classifications, with but a few exceptions, which we shall

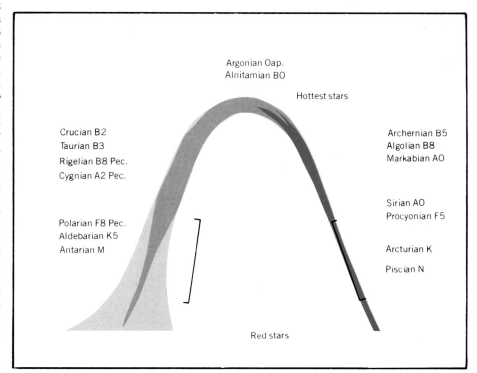

The Hertzsprung–Russell Diagram with the observed stellar spectral classes on the Harvard System plotted against stellar luminosity, or total energy output (Sun = 1). There is a well-defined relationship between a star's surface temperature and its energy output. But, as can be seen from the diagram, each of the redder stars may have one of at least two possible luminosities for its spectral class. The failure to recognize this feature led 19th-century astronomers astray in their attempts to search out empirical relationships among the various parameters that describe the physical characteristics of stars. Figure 1

identify later in the course of this article.

As spectra became better identified, and especially when larger telescopic apertures allowed for an increase in the dispersive powers of the attached spectroscopes, many peculiarities in individual spectra became apparent. One of the most important of these was the early recognition by Secchi of two distinct types of spectra among red stars. In these two red classes, banded structure that appeared in the same positions exhibited different structure. Secchi thought that the differences were enough to warrant a separate class, which he tacked on to his system as a fourth class. Thus his first class were the blue stars, his second were the solar type yellows, and his third and fourth were the reds. Vogel, however, believed that this separation was too great. In his system of classification, which he developed from 1874 through 1895, he retained only the three major classes, very much as defined by Secchi, but used subdivisions for what he considered to be secondary spectral distinctions. Vogel based his classification system directly on stellar evolution, and felt that the two classes of red stars were explained by minor variations in composition. Lockyer, however, advocated Secchi's original separation. He believed that Secchi's first red class, class III, exhibited bright lines and hence made them closer in evolutionary stage to nebulae than were the blue stars. Secchi's class IV stars, in Lockyer's minority view, were furthest removed from nebulae, and hence occupied the classic evolutionary place of red stars.

Lockyer favored Secchi's system because he was one of the very few who did not follow the popular concept of evolution. In Lockyer's view stars passed through the temperature sequence twice, ascending in temperature from a cold

nebular state and then, after attaining a maximum temperature as a blue star, cooling to extinction through the normal color and spectral progression. Even though there was much in Lockyer's scheme that can be seen at subtler levels in the evolutionary schemes of the majority of astronomers of the time—those represented best by Vogel and Huggins—Lockyer's zeal in connecting his temperature arch (see figure 2) with

his belief that nebulae were swarms of meteors in collision, and that all stars on his ascending branch were condensing meteoric swarms, kept his views quite unpopular throughout the latter half of the 19th century.[2]

From the theoretical side, Lockyer's model for stellar evolution was anticipated by the studies of J. Homer Lane of Washington, D.C. and of August Ritter of Potsdam who, between 1870 and 1883,

Norman Lockyer's "Temperature Arch," which first appeared in the late 1880's. Lockyer's elaborate classification system is represented by generic archetypes. Within the arch we have bracketed the region where a significant number of stars examined for parallax (and hence for luminosity) by Russell and Hinks were also included on Lockyer's later lists. Clearly these stars, while exhibiting similar temperatures, must differ greatly in other characteristics. Figure 2

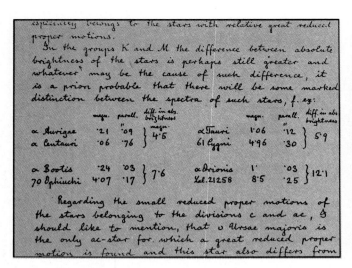

Part of a letter from Hertzsprung to Pickering dated 15 March 1906. Here Hertzsprung shows that in the redder classes, the two sequences of stars (Antonia Maury's spectral classifications "c" and "non-c," proposed in 1897) differed by greater amounts in absolute magnitude.

The listing reads left to right in columns. Reproduction from the E. C. Pickering Collection, Harvard University Archives. The photograph of Hertzsprung on the right is reproduced by courtesy of Dorrit Hoffleit, Yale University Observatory.   Figure 3

discussed the behavior of contracting gas spheres in convective equilibrium. Their independent findings showed that such bodies, beginning their lives as perfect gases, first heated upon contraction and began a cooling process only when densities within their interiors reached levels that caused them to deviate from the perfect-gas laws.[3] These works began to be noticed generally by astronomers in the 1890's and caused considerable consternation in those, especially Huggins, who wished to reconcile them with the observed sequence of spectra.

Aside from difficulties reconciling theoretical arguments with observation, the observations themselves left much to be desired. Lockyer's belief in the presence of bright lines in some red stars was symptomatic of the great difficulty of interpreting visual stellar spectra. In the 1880's photography began to rectify the situation, but even then, astronomers found that the highly limited sensitivity and poor reproducibility of photographic emulsions kept the new technique from causing an overnight sensation. It was also painfully evident that most of the prevalent classification schemes could well be fortuitous, and fraught with selection effects. The persistence of these crucial limitations in technique and completeness was in keeping with the status of the rest of the astronomical data base. Systems for determining the brightness of stars were far from standardized, and very few stars had reliable trigonometric parallaxes to determine their distances.

Such was the situation in stellar astronomy when Pickering, then a young physicist at MIT, accepted a post as the new director of Harvard College Observatory in 1877. In the 1880's Pickering began to organize two large projects with generous support from the family of Henry Draper and others, to improve the

situation. He established an objective-prism survey of the spectra of stars visible from Cambridge and Arequipa, Peru, and developed an accurate and consistent scheme for the determination of the apparent brightnesses of stars.

The use of objective prisms, thin prisms placed in front of the objective lenses of telescopes, was not new to Pickering. Both Joseph Fraunhofer and Secchi had used this efficient means of securing spectra. But Pickering attached these prisms to wide-field photographic astrographs, and thereby was able to secure spectra of hundreds of stars in one exposure. These exposed plates yielded the spectral classes of thousands of stars through direct eye examination in the rooms of the Harvard College Observatory, a comparatively mild environment compared with the cramped and often frigid confines of the telescope dome.

Pickering's projects were made possible by the enthusiastic and untiring assistance of a corps of women, headed by Wilhaminia P. Fleming. By 1890, he and Mrs Fleming brought out the first *Henry Draper Catalogue of Stellar Spectra* containing some 10 000 stars.

Pickering devised a simple alphabetic scheme of classification based upon the visibility of the hydrogen lines. Class A showed hydrogen lines strongest, and the series ranged on to O, P and Q. Refinements followed with more and better spectra. By 1898 he and a new assistant, Annie J. Cannon, who worked with Fleming, decided that the order must be reversed to O,B,A,F,G,K,M, primarily because O and B stars had similar helium spectra and both were closely associated in space with nebulae. This was a most important reversal in the classification system (which appeared in 1901) for it demonstrates the influence of evolution upon the Harvard classifiers. The fact that the Harvard classification has since

turned out to be a highly accurate temperature classification appears therefore to be fortuitous. It has, however, remained standard to this day.

But Pickering had long realized that the average quality of each of the tens of thousands of stars his team had been classifying was at best rough. Stellar spectra were far more complicated than the single objective prism could reveal, and therefore warranted closer attention along the lines advocated by the pioneers Huggins and Vogel. Pickering therefore designed, as a corollary project, an examination of a few bright stars under higher dispersion. Antonia Maury, Henry Draper's niece and one of the few women at that time actually trained in astronomy and physics, was delegated the task, and through the 1890's from a small but high quality sample of stellar spectra she devised an extremely sophisticated system of classification.

In 1897 Maury proposed her new system of classification. It had 22 numerical groups and identified differences within many of these groups in terms of relative line strengths and line widths. In brief, she detected two primary subdivisions: stars with normal spectra (hydrogen lines broad) designated a and b; and stars with especially sharp hydrogen lines and with metallic lines somewhat enhanced, designated c and ac. She later identified these two subdivisions as "c" and "non-c" in character. Maury noted that the existence of the subdivisions within several of her numerical groups suggested the existence of "parallel courses of development." Her provocative words were not heeded at the time. Indeed, her work remained unnoticed until Hertzsprung decided to find out if the two subdivisions she had detected represented anything uniquely interesting in the physical properties of the stars themselves.

But Hertzsprung's work can only be

appreciated within its context, the statistical examination of the spatial distribution of spectra, which was a growing interest in astronomy since 1890.

Spatial distribution of spectra

Secchi and a few others had long realized that stars of his first class (blue stars) tended to be concentrated more towards the Milky Way plane than were stars of other classes. By 1890, this concentration had been noticed also for bright-line stars, but significant statistical studies began to appear only with the availability of the *Henry Draper Catalogue*. These studies had two major themes: The analysis of the structure of the sidereal system, and the nature of the stars themselves. Which were the most luminous stars, and which were the least? Which were the largest in radius and which were the smallest? Clearly, an analysis of the mean distances of the different spectral classes, when compared to their mean relative apparent brightnesses, would yield statistical information about their relative actual brightnesses and sizes (neglecting by necessity any differences due to relative emittance as a function of color or spectral class). It must be remembered that relative size implied relative age, because the only conceivable direction of evolution, on whatever scheme, was contraction. To astronomers of the turn of the century therefore, such studies could yield information about the evolutionary status of the different spectral classes.

The first distribution studies were also the simplest, correlating spectral type with position on the celestial sphere. But later studies correlated apparent motions too, and when examined statistically, these yielded mean distances.

W.H.S. Monck, an Irishman about whom little is known, was one of the first, along with the legendary J.C. Kapteyn, to examine stars in this manner, correlating proper motions with spectra to determine mean distances and hence mean intrinsic brightnesses. When Monck sat down to the task of comparing the data at hand he found, in 1892, that proper motions increased with advancing (blue to red) spectral type, except that the reddest stars did not, as a group, have the largest motions; the yellow stars did. Monck concluded that the yellow stars must, as a class, be the closest to the Sun. He went so far to suggest that the Sun might be in a small cloud of solar-type yellow stars. But the fact remained that when he compared their distances and mean apparent brightnesses to the corresponding quantities for the other spectral classes, the solar-type stars came out the least luminous intrinsically. This meant that there were red stars brighter, and possibly larger, than the Sun. By 1893, not only Monck but also Kapteyn had come to this conclusion. Monck thus altered the "normal" course of evolution by placing

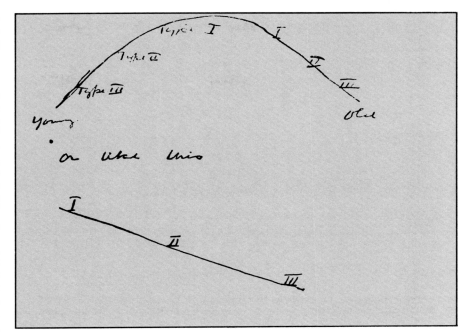

In Russell's lecture notes dated 14 March 1907 the upper curve represents Lockyer's view, where Type I are blue stars, Type II, yellow, and Type III, the red stars. Russell's second curve represents the classical cooling line, and is strikingly similar to what one would expect from a main sequence. Russell commented on these two curves noting "... we cannot be sure at present though some things look as if the first hypothesis is correct..." Reproduction from the Henry Norris Russell Papers, Princeton University Library.

Figure 4

the red stars of classes K and M before the solar-type G stars. In this he clearly was following the dictum of contraction, in spite of the fact that it played havoc with the accepted temperature history of stars.

In the next year, J.E. Gore, a friend of Monck's, showed in a popular text titled *The Worlds of Space* that there were red stars of great dimension. Using proper motions and brightness and neglecting colors, Gore found that these "giant stars" as he called them (borrowing terminology initiated by R.A. Proctor) were, as in the case of Arcturus, some 80 times the Sun's diameter or about the size of Venus's orbit. Thus, if Gore's calculations were anywhere near correct, how could solar stars cool and contract into red stars when there were red stars far larger than the Sun? These bright red stars at least could not have succeeded the yellow or blue stars, and must necessarily be quite young in their life histories. Gore did not mention the theories of Lockyer or Lane and Ritter, which would have supported his findings, but Monck did consider briefly the possibility of giant stars in his later work. He could not press the question, for he considered his data base too weak. Though spectra had become plentiful enough by that time, consistent measures of distance (and hence luminosity) from proper motions were still lacking, and far too few direct trigonometric parallaxes were available for any proper statistical analysis.

Monck therefore was not able to take the step taken by Hertzsprung just a few

years later, when for the first time it was found that the reason red stars had greater mean brightnesses was the inclusion among them of giant stars. It is quite clear today that Monck's and Kapteyn's samples were affected by the great visibility of red giant stars. Though they are quite rare in space, they were the only red stars bright enough to be easily photographed for spectra.

In 1900 the problems just discussed in statistical astronomy were very much open. The proliferation of spectral classification schemes, and their interpretation, frustrated many. Further confusion came from a general lack of consensus over the physical meaning of spectra. It had long been believed that the normal spectral sequence revealed the temperature history of stars, with slight variations due to composition. But by the late 1890's, other physical variables such as density and atmospheric pressure demanded serious consideration as the primary causes. There were those, including the illustrious William Huggins, who felt that extensive masking by the stellar atmosphere, dependent upon atmospheric density, was the primary factor. In 1900 he suggested that selective absorption by an extensive stellar atmosphere might be great enough to cause a blue star to appear red. He saw the red stars as the most advanced in life, and therefore the densest. By Lane's law, they should also be the hottest, thought Huggins, who for some reason seemed to think that the stars remained perfectly gaseous throughout their lives. Thus, he intro-

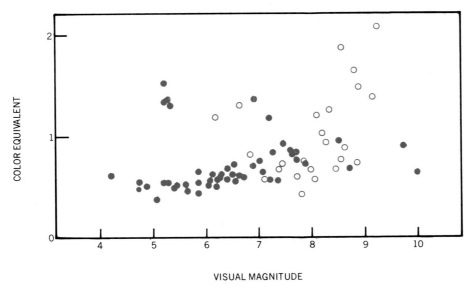

One of the first diagrams published by Hertzsprung for the Hyades star cluster, adapted from Potsdam Publications **22** (1911), page 29. The horizontal coordinate represented apparent magnitude, the vertical one Hertzsprung's "color-equivalent," a measure of stellar color.   Figure 5

duced masking to argue that the red stars were, in fact, the hottest, but appeared the coolest due to their selectively absorbing atmospheres. Huggins's ideas helped to confuse the interpretation of spectra, and hence it remained quite difficult to apply the radiation laws of Josef Stefan, Wilhelm Wien and Max Planck to the stars.

Ejnar Hertzsprung

Happily this situation did not stop the young Danish photochemist Ejnar Hertzsprung. As one of his earliest interests in astronomy, Hertzsprung applied the laws of radiation to find, in 1906, that Arcturus (the same star singled out by Gore) was the size of Mars's orbit. At the same time he also revived the statistical studies of Monck and Kapteyn, and entered directly upon the work that led him to construct the first "Hertzsprung–Russell" Diagram.

But what was it that allowed Hertzsprung to rediscover what Monck and Kapteyn had found but could not exploit? Hertzsprung stated at the outset of his first statistical study in 1905 that it was Maury's classification system that stimulated his interest in searching for what determines the differences in spectra among the stars. In particular he wanted to know why there were subdivisions among her spectra (a,b; c and ac). Hertzsprung began, as had Monck and Kapteyn, using proper motions statistically to derive relative distances and relative brightnesses for the different spectral classes. But, unlike the others, he had Maury's classification as a guide.

In his first analysis he found, as did the others, that the major spectral classes exhibiting greatest proper motion were the solar classes and not the red classes. But after a detailed analysis of the various groups defined on Maury's system he found that for all stars brighter than magnitude +5 the red ones had vanishingly small proper motions, and only a very few had parallaxes. But among stars of large proper motion or parallax he found mostly faint red and yellow stars. Hertzsprung was particularly intrigued that the former group contained c stars, and the latter, non-c stars. What made Hertzsprung's analysis extremely difficult was that, for the redder classes, the subdivisions were not distinct at all; so he had to construct an elaborate indirect process of analysis that allowed him to come to this conclusion.

Nevertheless, triggered by Maury's classification, Hertzsprung had found a filter by which he could distinguish intrinsically bright and faint red stars, depending upon which proper motion and brightness group they fell into. After he established the technique, he concluded that the total sample of yellow stars appeared fainter because there was a greater proportion of dwarf yellows relative to giant yellows in it. This was due to the fact that the dwarf yellows were just a bit brighter than the dwarf reds, and therefore appeared more frequently in general surveys.

After the publication of his first paper on the subject, in an obscure German photographic journal, Hertzsprung wrote to Pickering in March 1906 discussing his work and the resulting significance of the Maury system of subdivisions, which could now be used to detect luminosity differences wherever the subdivisions were distinct. Within this letter was a descriptive table outlining how he felt the c and non-c spectra should be examined so as to illustrate the great luminosity differences between them (see figure 3).

During 1906 Hertzsprung continued his work, and in 1907 he published a second paper with a slightly different selection of stars. Here he was concerned with refining the magnitude differences between the c and non-c stars by incorporating reliable parallax data, where available. He also discussed the space densities of stars in each class, finding correctly that giants of all classes were rare. With this second paper, Hertzsprung's local reputation grew. He had become a close friend of Karl Schwarzschild[4] and as a result followed Schwarzschild from Göttingen to Potsdam as a staff astronomer when the latter became the director there. But Hertzsprung's international reputation had not yet been made, even though his papers and letters were in Pickering's hands.

In 1908 when Hertzsprung received a copy of the latest *Harvard Annals* he realized with some surprise that Pickering had not taken his 1906 letter and 1905 paper seriously, for Maury's spectroscopic notation and subdivisions had not been reinstated in the publication (they were dropped in the 1901 *Harvard Annals* in Cannon's extension of the original alphabetic system). Hertzsprung wrote[5] to Pickering in July 1908 to voice his concern over the apparent neglect of so important a discovery:

It is hardly exaggerated to say that the spectral classification now adopted is of similar value as a botany, which divide the flowers according to their size and color. To neglect the c-properties in classifying stellar spectra I think, is nearly the same thing as if the zoologist, who has detected the deciding differences between a whale and a fish, would continue in classifying them together.

Hertzsprung wished that Maury's classification system would be reinstated so that stars of great luminosity could be identified. In early August Pickering responded cordially but skeptically, noting that he did not have enough faith in his own spectra to believe in Maury's subdivisions. He felt that the objective prism spectra she had used did not have the resolution or standardization one would need to be able to determine real differences in line structure, since slight instrumental changes could easily change the appearance of the lines. Pickering believed that her line differences could only be confirmed by the use of high-quality spectra taken with slit spectrographs—a conclusion he had first voiced in print in 1901.

It is understandable that Pickering would be very cautious about using Maury's subdivisions. At the time, his main concern was putting the general Harvard Spectral Classification system on a sure footing in the astronomical community. At the time no system was generally preferred, and many of Pickering's colleagues, for example George Ellery Hale, continued to use the earlier systems of Secchi and Vogel. Pickering was all too aware of the inconsistencies found in many classification systems that had tried to say too much in the past, and strove to keep his own as simple and unambiguous as possible. Still, after 17 years and tens of thousands of stars classified, there was no generally accepted standard.

But this did not comfort Hertzsprung. He had also noticed that in addition to line-width variations, line ratios in the two subdivisions were different. Further, as he wrote back to Pickering arguing his point:[6]

> The fact that none of the stars called c by Antonia Maury has any certain trace of proper motion is, I think, sufficient to show that these stars are physically very different from those of divisions a and b.

By October 1908 Hertzsprung sent a new manuscript to Pickering before he submitted it to the *Astronomische Nachrichten*. In private to Schwarzschild he had expressed his bitter disappointment over Pickering's attitude[7] but to Pickering he maintained a diplomatically firm air, noting that his paper was intended for publication with or without Pickering's approval. If, however, Pickering wished to provide commentary, Hertzsprung would find it most welcome. The *Astronomische Nachrichten* paper did appear in 1909, and was a partial restatement and expansion of Hertzsprung's earlier work. These three papers contained tabulated data sufficient for a Hertzsprung–Russell Diagram, but no diagrams appeared. These only came in 1910 and 1911.

To place Pickering's skepticism into proper context, we have to provide a fuller picture of his involvement in the development of the diagram, which centered upon support for the work of Henry Norris Russell.

Henry Norris Russell

It is not rare in the history of science to find the most pivotal and crucial discoveries and studies made independently by different people at about the same time. In many cases the time was right, the need was apparent, and the discovery was "in the air." Though there is good evidence that this is true here, the universal nature of the diagram allowed for its discovery by workers interested initially in different goals.

The influences upon Russell causing him to come to the diagram, or to the relationship behind it, are quite different from those upon Hertzsprung. While Hertzsprung was intrigued by Maury's classifications, and attempted to unravel their meaning hoping for a better understanding of the apparently anomolous statistical behavior of the red stars, Russell came to the problem primarily from an interest in evolution stimulated by Lockyer's writings.

After a brilliant student career at

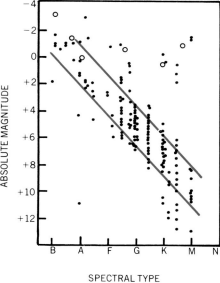

Russell's 1914 diagram. The vertical coordinate is absolute magnitude derived from his parallax work. The horizontal coordinate is spectral class on the Harvard System. The large open circles along the upper part of the diagram represent mean absolute brightnesses for bright stars whose parallaxes were on the order of their probable errors. All these stars had very small proper motions, indicating a statistically distant sample. Adapted from H. N. Russell, "Relations Between the Spectra and Other Characteristics of the Stars," in *Popular Astronomy* **22** (1914) page 285, figure 1. Figure 6

Princeton, Russell spent several postgraduate years (1902–05) studying at Cambridge University and developing, with A.R. Hinks, Chief Assistant at the Cambridge University Observatory, one of the first photographic parallax programs ever attempted. Though this was clearly a pilot program, the 55 stars selected for study included 21 common to a recently published (1902) list of stellar spectra by Lockyer. Lockyer had selected only the brightest stars for his listing, while Hinks and Russell said their criteria for choosing parallax stars included brightness only as a minor consideration. Understandably, they preferred the more fruitful criteria of large proper motion and previous parallax measurement in choices of parallax candidates. Thus it is surprising that half their stars were on Lockyer's list. Further, when one examines the distribution of stellar types they chose, it is obvious that the Lockyer stars chosen were just those that could best test his double-branched temperature arch (see figure 2).

Russell's interest in Lockyer's hypothesis can be seen in lecture notes he prepared for a course in 1907 at Princeton. For a lecture in March 1907 on stellar evolution he first reviewed spectral classification, then the two possible courses for evolution, clearly preferring Lockyer's (see figure 4). Of great interest, though,

is how he chose to represent the classical theory due primarily to Vogel. He showed it as a descending line quite like what one would expect from a rudimentary representation of the main sequence in a Hertzsprung–Russell Diagram. Unfortunately, since Russell did not label his axes, we cannot say that he knew in 1907 that for main sequence stars, brightness diminished with increasing redness. At best, this sketch represents Russell's keen intuitive powers.

To exploit his parallax work fully, Russell needed to reduce his parallaxes to account for the probable parallactic motions of the reference stars. For whereas parallaxes based upon visual meridian circle measures yielded fundamental positions and motions relative to the terrestrial observer, photographic parallaxes revealed only motion relative to the selected background reference stars. These background stars could also have their own parallactic motions, which would have to be taken into account before the actual parallactic motion of the program star could be determined. To do this Russell resorted to Kapteyn's established technique of statistically derived proper motions based upon brightness and spectral class. Thus Russell needed spectra and magnitudes, best available from Harvard and Pickering.

It was actually Pickering who approached Russell, having heard of his needs.[8] This is of interest because, by the time Russell and Pickering were in contact, Pickering had already received Hertzsprung's early paper and arguments for why the K and M red stars did not, as a group, have the largest proper motions. Yet Pickering suggested to Russell in late April 1908 that Harvard would produce the spectra of the parallax and reference stars, and added[9]: "The material would perhaps be sufficient to determine which were the most distant, stars of Class A or Class K."

After Russell sent Pickering identifications for the stars in need of spectra and magnitude, a long gestation period set in. By September 1909 Russell had received most of the data from Pickering and found at the outset that:

> ...the fainter stars average redder than the brighter ones. I do not know of any previous evidence on this question . . . I would not now risk reversing the proposition and saying that the red stars average intrinsically fainter—some of them certainly do; but Antares and α Orionis are of enormous brightness, and the average may be pretty high.

These conclusions[10] are strikingly close to Hertzsprung's and so should have prompted Pickering to reply with mention of Hertzsprung's work, if only to state that Russell had come to the same conclusions but from a much more direct and reliable data base. But Pickering remained silent, quite possibly so skeptical

	O Oa-Oe	B B0-B5	A B8-A5	F A8-F2	G F5-G0	K G5-K2	M K5-Ma	N Na-Nb	R R0-R5
-6.0		2	1						
-4.5	1	5	2	2	1	1	1		
-3.0		5	5	1	1	5	3		
-1.5	3	2q	29	4	7	17	7	1	
0.0	5	74	83	20	18	65	24		
+1.5	2	122	201	31	40	167	49	7	
+3.0	2	140	522	89	72	418	120	11	2
+4.5	5	58	809	271	199	817	189	10	4
+6.0	3	50	855	469	448	1123	196	11	3
+7.5	1	25	465	396	806	844	92	7	
+9.0		7	210	140	726	566	26	1	1
+10.5			68	16	277	331	9		
+12.0			7	6	47	116	12		
+13.5					5	16	7		
+15.0						1	4	2	
+16.5						1	2		
+18.0			1						

Unpublished Russell diagram clipped to a note from Lockyer to Russell dated June 1913, while Russell was in London. Russell evidently had this diagram with him when he visited Lockyer. Russell attempted here to indicate the number of stars found in each magnitude and spectral class range. The photograph on the left shows Henry Norris Russell as he appeared prior to World War I. The diagram is reproduced from the Russell Papers, Princeton University Library; the photograph is in the American Institute of Physics Margaret Russell Edmondson Collection.   Figure 7

of Hertzsprung's use of Maury's data that he had decided to keep the matter to himself for fear of misleading Russell.

The earliest diagrams

It should now be clear that the fundamental empirical relationship between the spectra or colors of stars and their intrinsic brightnesses was established independently by Hertzsprung and by Russell well before it was ever put into the form of a diagram. Russell had the model as early as 1907—if we are allowed to read between the lines—and could have produced a diagram easily by 1909. A.V. Nielsen has shown that Hertzsprung, as early as 1908, had created a diagram of an open cluster of stars, but kept it from publication because of instrumental errors.[11]

The first diagram to see print was for the Pleiades cluster, in a paper written in June 1910 by H. Rosenberg, Hertzsprung's assistant at Potsdam. Hertzsprung's own diagrams of the Pleiades and Hyades clusters came soon after (see figure 5).

Russell first heard of Hertzsprung's work from Schwarzschild during a meeting of astronomers at Harvard in August 1910, and in 1911 he wrote to Pickering suggesting that they might follow up Hertzsprung's cluster diagrams with spectra of the stars he included, instead of the color-equivalents Hertzsprung was

using. The primary reason for the lapse of time between 1910 and late 1913—when Russell became capable of producing a diagram and when he actually did so—was his own concern for the meaning of the great luminosity difference found between "giants" and "dwarfs" (terminology he had created while attempting to describe his findings in correspondence with Pickering). The differences could be due to mass or to volume. Russell's chief activity in this interval was to establish that it was a volume difference, from studies of binary stars he had initiated and that were carried out by his graduate student Harlow Shapley. Russell had earlier developed a method for determining the densities of eclipsing binaries, had maintained considerable interest in stellar densities and binary reductions all the while, and by 1910 had strong evidence that there were stars of extremely low density and hence enormous volume—giant in size but not in mass. While Shapley continued his own binary star orbit calculations through 1912, Russell began to realize that the mass range among all stars was quite small compared to variations in other physical characteristics. Shapley's examination of about 90 binary systems helped confirm Russell's first results, which then only awaited the proper opportunity for presentation. This arose in June 1913, while Russell and a small band of American astronomers

stopped briefly in London en route to the summer meetings of the International Solar Union held in Bonn.

While in London, the Americans were invited to present results of recent research to the Royal Astronomical Society. Russell presented his discussion "Relations Between the Spectra and Other Characteristics of the Stars"—a title he had kept prepared for several years. His paper was brief, due to the usual time limitations, and did not appear in print for a few months. Its first appearance was without the diagrams, though his text referred to them.

Reactions to Russell's work

Initial reactions to Russell's work were positive. Arthur Stanley Eddington worried a bit at first about Russell's thoughts on evolution, which went against the established grain, but in correspondence he admitted a deep interest and fascination. Whatever Eddington felt about Russell's evolution, which was literally a revival of Lockyer's old ideas, he was sure of the great value of the diagram (see figure 6) and wanted to publish one in a book about to see print. While Russell was in London, he met and discussed his ideas with Lockyer, who for obvious reasons was delighted with the turn of events this American had brought. A note from Lockyer to Russell (found in Russell's papers at Princeton) discussing this meeting in 1913 was clipped to three rudimentary Diagrams, and a histogram picturing the mean apparent brightnesses of the various spectral classes. I include one here, for it may be Russell's earliest attempt to represent his findings graphically (see figure 7).

In the years following Hertzsprung's and Russell's presentations the Diagram became better refined. Its primary function to picture the vast differences between giants and dwarfs was strongly supported by the invention and application of the technique of spectroscopic parallaxes, which allowed absolute luminosities to be determined by a means independent of trigonometric parallaxes. In 1920 the angular diameter of a giant star was measured by A.A. Michelson and Francis Pease at Mount Wilson and was found to be very close to predicted values.

Thus, while the diagram itself remains as an empirical fact, its interpretation has changed radically in past years.[12] Russell saw the giant branch and main sequence as a continuous series of homologously contracting gas spheres. While they were giants they behaved as perfect gases and thus heated upon contraction. But they turned into relatively incompressible fluids once on the main sequence—causing further contraction to result from cooling only.

The main sequence persisted as an evolutionary track until the mid-1920's, and the position of the giants in evolution remained unsolved until the early 1950's. Many aspects of theoretical astrophysics had to develop and mature before our present interpretation of the Hertzsprung–Russell Diagram became possible. In Russell's time, stars were purely convective, fully mixed and capable of contraction only. The many advances needed to change these 19th-century views represent the mainstream of progress in stellar astronomy over the past sixty-five years. It is a tribute to Russell's memory that he had something to do with almost all of them.

I would like to thank the archivists at Princeton and Harvard Universities, and at the Lick Observatory Archives, for aiding me in my research. Material made available by the AIP Center for History of Physics has been central to this work. I would particularly like to thank A.J. Meadows of the University of Leicester for his interest and support for my studies of the history of the Hertzsprung–Russell Diagram.

References

1. See: D.H. DeVorkin, "The Origins of the Hertzsprung–Russell Diagram" in *In Memory of Henry Norris Russell,* A.G. Davis-Philip, D.H. DeVorkin, eds. (Dudley Observatory Report No. 13, Proceedings of IAU Symposium **80,** 1977). This book includes recollections of Russell's scientific life by his students, colleagues and historians. For general background information on the topics discussed in this paper see: B.Z. Jones, L.G. Boyd, *The Harvard College Observatory,* Harvard (1971); A.V. Nielsen, "The History of the HR Diagram" *Centaurus* **9** (1963), page 219; D. Hermann, "Ejnar Hertzsprung—'Zur Strahlung der Sterne'" *Ostwalds Klassiker* no. 255, Leipzig (1976); O. Struve, V. Zebergs, *Astronomy of the 20th Century,* Macmillan (1962).

2. See: A.J. Meadows, *Science and Controversy—A Biography of Sir Norman Lockyer,* MIT (1972).

3. See: S. Chandrasekhar, *Stellar Structure,* Dover (1957), pages 176–179.

4. See: Nielsen, ref. 1.

5. Letter, Hertzsprung to Pickering (22 July 1908) Harvard Archives, E.C. Pickering Collection.

6. Letter, Hertzsprung to Pickering (17 August 1908) Harvard.

7. Letter, E. Hertzsprung to K. Schwarzschild (26 August 1908) Schwarzschild Papers Microfilm, American Institute of Physics Niels Bohr Library.

8. See: Jones and Boyd, ref. 1.

9. Letter, E.C. Pickering to H.N. Russell (22 April 1908) Princeton University Library, Henry Norris Russell Papers.

10. Letter, H.N. Russell to E.C. Pickering (24 September 1909) Harvard.

11. Nielsen, ref. 1, page 241.

12. For an excellent review of the history of the Hertzsprung–Russell Diagram since its discovery see: B.W. Sitterly, "Changing Interpretations of the Hertzsprung–Russell Diagram, 1910–1940: A Historical Note," in *Vistas in Astronomy* **12,** Pergamon (1970), page 357. ◻

── Chapter 2 ──────────────────────
Institutions of Physics

Most scientists begin their careers fascinated with pure scientific knowledge and only gradually come to understand that discoveries are made by real people. The history of science likewise had to mature before it could fully recognize the importance of institutions in the growth of knowledge. The most important of these institutions, of course, are the great universities and laboratories, and they are so important that we can take it for granted that physicists have at least a rough understanding of how they came into being and how they function. For other institutions, understanding is much less widespread.

The organization of the discipline itself often seems like dull stuff to the average physicist, and usually attracts the attention of only a few leaders of the field. It is precisely in attending to such things that they *are* leaders. The scientific enterprise would instantly collapse without its own self-created institutions, which for centuries have been astonishingly democratic, durable, and unobtrusive. The articles in this section describing, for example, the founding of the American Institute of Physics, the American Association of Physics Teachers, and institutions in the field of crystallography, should be read with the thought in mind that without these institutions, such vital everyday matters as journal publication and conferences would look quite different and might not even be possible.

The French entrepreneur of science Jean Perrin remarked that science may be done with brains, but "brains, annoyingly enough, are attached to stomachs." Feeding those stomachs takes money. The articles in this section on the Kellogg Radiation Laboratory, on Bell Labs (a piece which is one of the few detailed case studies ever written on industrial physics), and on the Office of Naval Research, show how physics has raised funds for research by proving its usefulness in medicine, business, and war. In each case, however, a simple utilitarian appeal was not the whole story; everyone seems to have recognized that physics has an appeal and an importance that goes beyond anything it can immediately deliver.

Contents

The roots of solid-state research at Bell Labs

The impact of science on industry—and of industry on science—
is nowhere better illustrated than by the origins of the solid-state group
at Bell Laboratories, which gave the world the transistor.

Lillian Hartmann Hoddeson

PHYSICS TODAY / MARCH 1977

Solid-state physics has experienced a dramatic growth in the last four decades; whereas in the 1920's the term "solid-state physics" was not yet in use, this is now the single most populated sub-field of physics. Much of this growth has taken place in industry, so that today a small number of industrial laboratories are producing a substantial fraction of contributions in the field.

In this article I explore the roots and beginnings of basic solid-state research in one industrial setting, Bell Laboratories, where crucial advances were made, such as those leading to modern semiconductor electronics. By focussing on these developments we may hope to gain insight into the mechanisms of the contemporary impact of basic physics on industry, as well as into the complementary role that industrial policies have in turn played in shaping specific areas of modern research.

The roots of Bell's solid-state program developed gradually, in a series of stages generated by internal technological needs of the expanding telephone industry. The stages show a striking reciprocal interplay between science and technology in the context of corporate expansion. Let us examine four stages:

1875–1906 A newly invented device establishes an industry.
1907–24 Technological needs called for by the growth of the industry lead to in-house research.
1925–35 Interactions with scientific research outside the Laboratories help focus some of its basic technical studies on even more fundamental scientific issues.
1936–45 The intensified focus on scientific underpinnings of technological

problems leads to proliferating scientific and technological developments, among them the formation of the famous solid-state group that in 1947 would demonstrate the first transistor.

Let us recapitulate these stages in more detail, starting at the time of the invention of the telephone.

Establishing a telephone industry

In 1876 Alexander Graham Bell received a patent for his method of transmitting sounds by electrical undulation, and in 1877 he patented his "magneto-telephone," a device that could actually transmit speech. The telephone industry began several months later when the first telephones were leased to subscribers.

The manufacture, installation and maintenance of telephones in the growing business raised new technological problems.[1] However, the earliest of these did not require scientific training or fundamental research and, understandably enough, the infant company supported neither scientific education nor research. Not even Bell's "mechanical assistant," Thomas Watson, had any formal scientific training. When the technical staff expanded during the next few years, Watson was joined by inventors, not scientists.

To be sure, many telephone problems of the 1880's and 1890's—attenuation and distortion of telephone signals, crosstalk, switching, interference from other electrical devices such as street lighting or electric railways—were caused by electromagnetic phenomena that were just receiving scientific explanation. James Clerk Maxwell's *Treatise on Electricity and Magnetism* had only recently been published (in 1873), and it had limited experimental support. (Heinrich Hertz's experimental confirmation of electromagnetic waves came in 1888.)

The first decisive step towards in-house research occurred in 1885 when Hammond V. Hayes, the first PhD in the Bell System (and holder of the second physics doctorate awarded by Harvard) became chief of the technical staff. But the engineers on Hayes's small staff in the 1880's were not trained in mathematics and could not readily apply electromagnetic theory to the engineering problems they encountered. Approaching immediate practical problems by the cut-and-try approach seemed more promising than taking staff time out to comprehend and develop the scientific underpinnings.

Yet even before the turn of the century, Hayes had hired a handful of university-trained scientists to work on technical problems. In 1890 he recruited John Stone Stone, trained at Johns Hopkins University in advanced mathematical theory, to work on sound transmission; in 1897, George Campbell, an MIT-trained physicist with five years of postdoctoral study, to clarify the role of the inductive impedance in telephone communications, and in 1899, Edwin Colpitts from Harvard, to study alternating electrical currents and inductive interference due to electric trolley cars and power-transmission systems.

Frank Baldwin Jewett (later to become the first president of Bell Labs) was hired in 1904 to work under Campbell as a transmission engineer. Jewett, who was the first member of the technical staff to have some close experience with the atomic physics then being developed, was teaching physics and electrical engineering at MIT at the time he was hired. While in the doctoral program at the University of Chicago, Jewett had been a research assistant to A. A. Michelson and a close friend of Robert Millikan. The latter, then a young physics instructor, exposed Jewett to the new discoveries

Lillian Hartmann Hoddeson is an assistant professor of physics at Rutgers University, New Brunswick, New Jersey.

being made in electron physics. Jewett's association with Millikan would soon contribute crucially to the beginnings of basic research within the Bell System.

Thus by 1907 several trained scientists were working in the company, but primarily as engineers, not as part of an organized basic-research program.

Spanning the continent

In 1907 Theodore Vail, who had left the company in a dispute twenty years earlier, was rehired as President. Two decisions Vail made then had major impact on the movement towards establishment of basic scientific research.

First, he brought together all technical workers into a single department. The new engineering department—which ultimately evolved into the Bell Telephone Laboratories—was established at 463 West Street in New York City as a division of Bell's manufacturing arm, Western Electric. Hayes retired and Vail appointed John J. Carty to head the new department. Carty, who at first sight seems a throwback to an earlier era—he had joined the company in 1879 as a boy operator, and had no formal scientific training—actually proved to be closer to the new style. He was a research enthusiast who had by then made an impressive series of technical contributions to the art of telephony, including application of the two-wire metallic circuit, the first multiple switchboard, the bridging bell and the repeating-coil phantom circuit.

Vail's second decision was to build a transcontinental telephone line from New York to San Francisco, in time for the 1914 Panama–Pacific Exposition. It was soon recognized, however, that no such line could be achieved unless a "repeater"—a device that could amplify telephone signals attenuated by dis-

HAMMOND V. HAYES, 1907

tances—could be developed. But to design a usable amplifier for coast-to-coast service would require a detailed understanding of the new electron physics, a subject beyond the working knowledge of anyone then in the company.

Attenuation had become a progressively more obtrusive problem as the company's lines lengthened—from approximately two miles between Boston and Cambridge in 1876 to 900 miles between New York and Chicago in 1892, and then to 2100 miles between New York and Denver in 1911. As early as 1899, Campbell had developed a "loading coil," which cut energy losses dramatically by increasing the inductive impedance of the lines; the New York to Denver line could not have been built without it. (Michael Pupin, at Columbia University, also invented the loading coil at this time and won the patent fight against Campbell.

The company, however, then bought Pupin's patent, and Campbell went on further to develop the loading coil for telephone application.) But to go farther than Denver it would be necessary to add an amplifier to the system.

A mechanical amplifier designed by Herbert Shreeve, based on a vibrating diaphragm, had been tested as early as 1904. The amplified signal was similar to the original one but it was typically quite badly distorted; Shreeve's repeater tended to favor some frequencies and discriminate against others. When used on lines with loading coils, the signal was all but destroyed.

Something less sluggish than a vibrating diaphragm was needed, such as electrified gas particles, or free electrons. The development of this idea required knowledge of the most recent electron physics. Therefore in 1910 Jewett discussed the problem with his graduate-school friend Millikan, who later recalled that Jewett asked him to recommend "one or two, or even three, of the best young men who are taking their doctorates with you and are intimately familiar with your field. Let us take them into our laboratory in New York and assign to them the sole task of developing the telephone repeater." Millikan recommended his best graduate student, Harold Arnold, who in January 1911 joined Western Electric's engineering department.

The first research branch

Three months later the Bell System established its first research branch as a division of this department. Headed by Colpitts, the new group had as its specific directive to produce "the highest grade research laboratory work." Jewett was given responsibility for directing research on the most immediate problem, the repeater.

A pattern was developing that would deepen throughout the following five decades: The Bell System would support increasing programs of basic research within an expanding engineering effort. In-house research directly pertinent to communications needs would circumvent the necessity of buying patents from other institutions or individuals or, as in the case of the radio research, would protect existing Bell patents.

The trend towards more fundamental studies was reinforced by what was apparently a new, if unwritten, policy: The directors of research were chosen from among the scientists who were trained in the Bell System's own laboratories. Such men understood that creative scientists need freedom to speculate and explore intellectually and to communicate with researchers working on similar problems—even if these were employed outside the company. In short, the scientists required latitude comparable to that available in academic laboratories. Ac-

THEODORE N. VAIL, 1915

GEN. JOHN J. CARTY, WORLD WAR I

tive competition in the larger scientific community would also be recognized in time as the most effective means for Bell to achieve the awareness of scientific frontiers it deemed necessary to maintain its market advantage.

The solution to the amplifier problem began with the triode offered in 1912 to American Telephone and Telegraph by Lee De Forest. The triode could amplify weak signals; however, due to the relatively large telephone currents required, the gas inside the tube would ionize. As John Mills recalled, "[the tube] would fill with blue haze, seem to choke, and then transmit no further speech until the incoming current had been greatly reduced."[3] That problem was eventually solved by Arnold's development of a high-vacuum version of De Forest's triode.

The first transcontinental line opened in time for the Exposition; in January 1915 Alexander Graham Bell in New York reissued his famous command to his former assistant in San Francisco: "Mr Watson—come here—I want you." To this Watson replied, "It would take a week to get there."

Basic research takes root

In the third stage (1925–35) basic research took firm root in company policy. In 1925 a new corporation headed by Jewett—the Bell Telephone Laboratories—took over Western Electric's engineering department. But the organizational changes of 1925 did not alter the new research policy.

Basic research continued to expand and diversify in the Bell System. The trend may be illustrated by the work of the vacuum-tube department, an organiza-

tion that originally had evolved out of the earlier research on repeaters. By 1930 this department was staffed by almost 200 scientists and co-workers organized in subgroups focussing on specialized aspects of vacuum-tube phenomena. These included thermionic emission and the interaction of electrons with solids. Examples of fundamental research that grew out of such investigations during the later 1920's are the well known studies on thermionic noise by J. B. Johnson and Harry Nyquist, Harold Black's important study of negative feedback, and the famous experiments by Clinton Davisson and Lester Germer that provided experimental verification of the wave behavior

of electrons. (It is of interest that Davisson and Germer were not initially aware of the relation their experiments had to quantum mechanics. Instead these experiments were in part an outgrowth of Arnold's desire to understand fully the issues raised in his patent fight with Irving Langmuir over the development of the high-vacuum tube.[4,5]) Among other examples of fundamental research in this period was that carried out by Richard Bozorth on magnetic materials.

In the following decade interactions increased between researchers at the Laboratories and those in universities both here and abroad. The new quantum physics entered Bell Laboratories re-

JOSEPH A. BECKER AND C. J. CALBICK, 1927

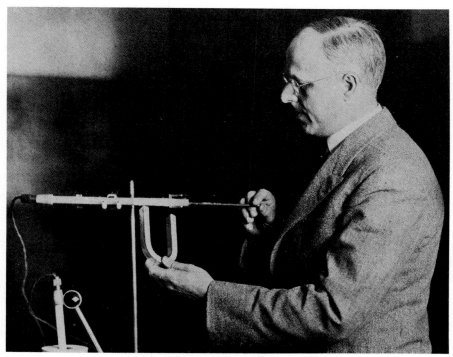

HAROLD D. ARNOLD, 1931

search and contributed towards still more intensive focus on fundamental questions. The quantum theory of solids, developed between 1926 and 1933 by Wolfgang Pauli, Werner Heisenberg, Arnold Sommerfeld, Felix Bloch and others would create a context for Bell's innovations of the subsequent decades in solid-state physics.

The quantum theory of solids was soon recognized as relevant to technical studies at Bell such as thermionic emission, photoelectricity and conduction. Walter Brattain and Joseph Becker, for example, drew upon the classic work of Arnold Sommerfeld and Lothar Nordheim in 1928 on the electron theory of metals to compute thermionic emission formulas.[6]

New ideas

The quantum theory entered through a number of avenues, some of them uncommon for an industrial laboratory of that period. One of these was Bell's lively colloquium series, organized in 1919 "to review scientific progress by means of contributed papers and general discussions of current scientific literature." In the early years, most of the talks were given by Bell Labs scientists; during the 1920's, however, researchers from all over the world spoke there on recent advances in physics and chemistry. Prominent European visitors during the period included Sommerfeld, from Munich, who spoke in 1923 on "Atomic Structure" and in 1929 on "The Photoelectric Effect in a Single Atom and in a Metal;" Ernest Rutherford, from the Cavendish Laboratory, who in 1924 spoke on "Recent Researches Concerning Atomic Nuclei;" Erwin Schrödinger, from Zurich and

Berlin, who in 1927 spoke on "The Undulatory Theory of the Electron;" Eugene Wigner, of Berlin and Princeton, who in 1932 discussed "Applications of Quantum Mechanics to Chemistry," and Paul Ewald, from Stuttgart, who in 1936 spoke on "Crystal Growth and Crystal Perfection."

Distinguished American scientists from other institutions who delivered colloquia at Bell in the same period included Robert Millikan (Cal Tech) in 1925, Robert Mulliken (New York University) in 1927, Edward Condon (Princeton) in 1928, Harold Urey (Columbia University) in 1932, I. I. Rabi (Columbia) in 1933 and John Van Vleck (Harvard) in 1936. In December 1933, there was a symposium on the recently discovered "Positive Electron." Speakers included Bell's Karl K. Darrow, who gave an historical review, and Gregory Breit, then at NYU, who presented P. A. M. Dirac's theory of holes; Rabi led the discussion.

Much of the impetus behind Bell's colloquium came from Darrow, who had been on Bell's staff since 1917. Particularly during the summer months, Darrow would visit major European and American research centers and attend physical-society meetings. Scientists often would accept Darrow's invitation to visit Bell Labs and give colloquia there. During the period, Darrow also helped transmit new ideas in physics by writing a semipopular series, "Some Advances in Contemporary Physics," for the *Bell System Technical Journal.* The topics included "Waves and Quanta" (1925); "The Atom-Model" (1925); "Statistical Theories of Matter, Radiation and Electricity" (1929), and "The Nucleus" (1933). The series was widely read and often

evoked strong response; Brattain, for example, claims his awareness of Bell Labs was stimulated by Darrow's articles written, as Brattain put it, "in his gorgeous language." [7]

Individual study and self-education provided another path of entry for new ideas at Bell, aided ironically by the Depression, which caused a reduction in 1932 of the work week for Bell's staff from $5\frac{1}{2}$ to 4 days. (In 1934 the staff went back to a $4\frac{1}{2}$-day-week and in 1936 to five days.) In a number of cases the extra time was devoted to individual study of quantum physics or to course work at Columbia and elsewhere. Some study efforts were disseminated more widely in the Laboratories; Brattain, on his return to Bell after attending Sommerfeld's lectures on the electron theory of metals at the 1931 Michigan Summer Symposium, gave a series of informal lectures on that theory.

By this time other industrial firms were also making strides in the application of research at their institutions. The extent to which leaders of research saw such activity as a common enterprise is illustrated by the joint monthly luncheon meetings of some twenty industrial laboratory leaders including Charles Kettering of General Motors, Kenneth Mees of Eastman Kodak, Willis Whitney of General Electric and Jewett of Bell. They discussed shared problems and issues, such as organization, personnel, patents and the relation of industrial research to economic conditions. Sometimes the group of "directors of industrial research," as they called themselves, would visit each others' laboratories. A tradition of individual visits to other research laboratories also evolved in this period.

Industrial researchers were frequently included in programs of the academic community. For example, during the late 1920's and early 1930's, MIT ran a colloquium series within their electrical-engineering department in which members of various manufacturing, operating and engineering companies, including Bell Laboratories, were invited to lecture on how fundamental science could be applied to engineering problems. In 1928 Bell's Mervin Kelly spoke in this series on "Thermionic Filaments of Vacuum Tubes used in Wire Telephony;" in 1936, Bozorth reported on "Recent Research in Magnetic Alloys." By the mid-1930's the problems, approaches and atmospheres of fundamental research at Bell Labs were remarkably similar to those in university laboratories.

Establishing solid-state research

The fourth stage, the establishment of basic research in solid-state physics culminating in the development of the transistor, began in 1936, when Kelly was appointed director of research. Kelly, like Arnold and Jewett before him, had taken his doctorate in physics at Chicago (where he had worked with Millikan on the oil-drop experiment), and had for a period (1928–34) led the vacuum-tube department. Kelly had become very much aware of the potential value of an amplifier without vacuum tubes—which were large, expensive, fragile, slow, relatively noisy, and often unreliable and short-lived. He is said to have manifested an interest in the early 1930's in developing an amplifier based on the properties of solid materials.

Some researchers on Bell's staff were already exploring the amplification properties of semiconductors. For example, Becker and Brattain were studying the properties of copper oxide but they did not fully understand the physical basis for their observations. Raymond Sears, who worked closely with Becker during the 1930's, recalls:[8]

"Becker all along felt that there was something in a copper-oxide rectifier that ought to have an analogy to the vacuum tube. There was a nonlinearity of the conduction in the forward and in the reverse direction. And so Joe himself would try to imbed a wire mesh in the oxide layer of copper oxide, in order to almost try to make a grid, like in a vacuum tube. I do well remember that. And Brattain and I would tell him, 'Look, that's not the way to go about it. You've got to understand how things work.' "

Brattain describes[9] his original motivation for attending the Michigan Summer Symposium in 1931 as his desire to obtain "a thorough knowledge" of the work function in thermionic emission and the photoelectric effect.

Kelly became convinced that the route to a solid-state amplifier was a deeper understanding of the basic physics of solids. By the mid-1930's he began to indicate a desire to create a new kind of research team to be composed of chemists, physicists and metallurgists who would focus on basic solid-state physics. This interest, which according to Bozorth was expressed even earlier by Oliver Buckley, director of research at Bell from 1933 to 1936, probably motivated Kelly's hiring of theoretical physicist William Shockley in 1936. (In 1952, the year after Buckley retired as president of the Laboratories, Bell Labs and The American Physical Society established the Oliver E. Buckley Prize in solid-state physics, thus commemorating Buckley's long-standing interest in fundamental solid-state physics research.)

Shockley's thesis adviser at MIT, John Slater, was head of one of the two major US training centers of that period for young solid-state physicists. The other was Princeton; several of Eugene Wigner's graduate students, e.g., John Bardeen, Conyers Herring and Frederick Seitz, became part of the first generation of physicists to refer to themselves as "solid-state" physicists. Connections

CLINTON S. DAVISSON AND MERVIN J. KELLY, 1951

were close during the early 1930's between the physics departments at MIT and Princeton; Bell's three leading solid-state theorists during the mid 1940's—Shockley, Bardeen and Herring—had known one another during their graduate-school days.

At Bell, Shockley first worked on vacuum-tube phenomena but soon joined a new research group under the direction of Harvey Fletcher, the well known acoustics researcher who was director of physical research at that time. This group was recently described[9] by Joseph Burton, who later became one of its members, as ". . . a group of fairly new people. Wooldridge, Townes, Shockley and Nix. All had been brought up to some degree in modern solid-state physics." In this group, Foster Nix engaged in a series of studies of phenomena in metals and alloys, and interested Shockley in the order–disorder phenomenon in alloys. Shockley thus moved closer to basic solid-state physics, in which he had been trained at MIT. Dean Wooldridge, who had joined the Labs in 1936, was at this time working on the theory of secondary emission, magnetic sound recording and television; Charles Townes, who joined in 1939, soon became involved in radar bombsight research.

Nix recently described[10] his impressions of that group: "When Kelly created this little group of independent people—there were Shockley and I and Wooldridge—under Fletcher, we were told, 'You do whatever you please; anything you want to do is all right with me' . . ."

Shockley and Nix were central to the organization in 1936 of an informal study group approved by Kelly that came to function as another important avenue for

entry of the quantum theory of solids. The group—including Shockley, Townes, Nix and Wooldridge, as well as Brattain (chiefly working on copper-oxide rectifiers), Alan Holden (whose speciality was crystals), Addison White (working on dielectrics), Bozorth (researching magnetic materials) and Howell Williams (in magnetics under Bozorth)—met weekly for more than four years to discuss the then recent works on quantum solid-state physics, including the books by Nevill Mott and H. Jones, Mott and Ronald Gurney, Richard Tolman and Linus Pauling. James Fisk (initially studying nuclear fission with Shockley) and Burton (working on photoelectron emission) joined the study group in 1939.

The transistor

Meanwhile, an advance took place at Bell's radio lab in Holmdel, which would contribute fundamentally to the invention of the transistor. Several researchers noticed that some samples of the semiconductor silicon were effective detectors of high-frequency microwaves. One of the researchers, Russell Ohl, became interested in obtaining pure silicon samples and involved several of Bell's metallurgists in the problem. During the cooling of hot silicon ingots, Jack Scaff and Henry Theurer produced the first silicon p–n junction; a substantial photovoltaic effect was produced when the silicon was illuminated (this was in 1940).

When Kelly learned of this he recognized that here might be the key to the solid-state amplifier. Brattain recalls,[6]

"Becker and I were invited to a conference in Kelly's office to discuss the meaning of this phenomenon. We were presumably the physicists who

OLIVER E. BUCKLEY, 1952

were supposed to know something about semiconductors . . . [and] we were completely flabbergasted at Ohl's demonstration. The effect was apparently at least two orders of magnitude greater in room light than anything we'd ever seen . . . I even thought my leg maybe was being pulled."

The beginnings of solid-state physics at Bell Labs and the first steps towards the transistor were therefore definitely under way at the advent of World War II. War research at Bell and elsewhere led to new advances, such as resonance techniques, thermal-neutron scattering and improved computing methods, which would in the postwar period contribute in fundamental ways to solid-state physics. And perhaps most important to the advancement of solid-state electronics, a large effort was invested in development on an expanded scale of materials with very small quantities of impurities. Silicon and germanium became prototypes for the study of solid-state physics after the war, in part because the technology for producing them had become well developed.

The war also led to nationwide recognition of both industrial and academic research as a national resource, contributing to Bell's growing support of in-house basic research. President Buckley expressed his attitude in a letter to *The New York Times* of 25 August 1949:

"One sure way to defeat the scientific spirit is to attempt to direct inquiry from above. All successful industrial research directors know this, and have learned from experience that one thing a 'director of research' must never do is to direct research, nor can he permit direction of research by any supervisory board. Successful research goes in the direction in which some inquiring mind finds itself impelled. True, goals are set, goals of understanding in the case of fundamental research . . . . The director of research does his part by building teams and seeing that they are supplied with facilities and given freedom to pursue their inquiry. He also insures for them contacts essential to their work, but at the same time protects them from interference or diversion arising from demands of immediate operating needs . . ."

As to solid-state physics proper, the long discussions between Buckley and Kelly on Bell's basic research during the late 1930's and throughout the war years resulted in formal authorization in January 1945 of the mixed group of researchers that Kelly had envisioned for so long—the group of physicists, chemists, physical chemists and metallurgists, jointly directed to pursue basic research in solid-state physics. The solid-state research group was co-headed by the chemist Stanley Morgan, who had been at Bell since the mid-1920's, and the physicist Shockley. The authorization reflects the vision Kelly had during the 1930's, of a unified approach to all solid-state problems.

Two other basic research groups were also established at the same time; one, headed by James Fisk, to pursue fundamental studies in electron dynamics, and another, headed by Wooldridge, devoted to basic research in physical electronics. Fisk suggested to Kelly that he invite Bardeen, by this time recognized as one of the outstanding solid-state theorists in the country, to join the new solid-state group. Bardeen joined in 1945, and in the following year Herring was hired into the new physical-electronics group. With the addition of Bardeen and Herring, Bell Labs became more than able to hold its own as a leading research institution, in theoretical as well as experimental solid-state physics.

A subgroup of the new solid-state division under Shockley's direction—Bardeen, Brattain, experimental physicist Gerald Pearson, physical chemist Robert Gibney and circuit expert Hilbert Moore—began to focus on the semiconductors silicon and germanium. In December 1947, Bardeen and Brattain demonstrated the first point-contact transistor and in the following year, Shockley developed the first junction transistor. In 1956, Bardeen, Brattain and Shockley received a Nobel Prize for invention of the transistor.

First by necessity, then by design

The cycle was now complete: Bell's program of basic research, which had evolved out of technical concerns of an industry initially generated by one device, had given birth to another device. And soon the cycle would expand dramatically, for the transistor would increase the financial base—and the size—of solid-state physics and begin the age of solid-state electronics.

A highly successful union had been achieved in the Bell System of two traditionally distinct—now proven complementary—approaches to the physical world, the more particular approach of the technical worker and the more abstract approach of the research scientist. It was a union initiated by necessity and only later welded by design.

This article summarizes research subsequently published as three articles: "The Emergence of Basic Research in the Bell Telephone System 1875–1915," Technology and Culture 22, 512 (1981); "The Entry of the Quantum Theory of Solids into the Bell Telephone Laboratories, 1925–40: A Case-Study of the Industrial Application of Fundamental Science," Minerva 18, 423 (1980); and "The Discovery of the Point-Contact Transistor," Historical Studies in the Physical Sciences 12, 41 (1981). The research, based on documents (including notebooks, letters and technical memoranda) and tape-recorded interviews with Bell Laboratories scientists, was made possible by the cooperation and support of several institutions, the most prominent among which are Bell Laboratories and the Center for History of Physics of the American Institute of Physics.

References

1. See, for example, *A History of Engineering and Science in the Bell System, The Early Years (1875–1925)* (M. Fagan, ed.), Bell Telephone Laboratories, Murray Hill, N.J. (1975).

2. *Autobiography of Robert A. Millikan*, Prentice-Hall, New York (1950), page 117.

3. J. Mills, Bell Tel. Quart. **19**, 5 (1940).

4. L. Germer, "The discovery of electron diffraction," (unpublished memorandum), reel 66, Archives for the History of Quantum Physics (available at AIP, New York; Amer. Philos. Soc., Philadelphia; Niels Bohr Inst., Copenhagen; University of California, Berkeley).

5. R. Gehrenbeck, *Clinton Davisson, Lester Germer and the Discovery of Electron Diffraction*, doctoral thesis, University of Minnesota (1973).

6. W. Brattain, J. Becker, Phys. Rev. **45,** 696 (1934).

7. Interview with W. Brattain by A. Holden and W. J. King, January 1964, Oral History Collection, AIP Niels Bohr Library, New York.

8. Interview with R. Sears by L. Hoddeson, 14 July 1975, Oral History Collection, AIP Niels Bohr Library, New York.

9. Interview with W. Brattain by C. Weiner, 28 May 1974, Oral History Collection, AIP Niels Bohr Library, New York.

10. Interview with F. Nix by L. Hoddeson, 27 June 1975, Oral History Collection, AIP Niels Bohr Library, New York.

11. Interview with J. Burton by L. Hoddeson, 22 July 1974, Oral History Collection, AIP Niels Bohr Library, New York. □

Paul P. Ewald heads the department of physics at the Polytechnic Institute of Brooklyn and is editor of *Acta Crystallographica*, the international journal of crystallography. He came to the United States four years ago from Ireland, where he had served as professor of mathematical physics at the Queen's University in Belfast.

Some personal experiences in
of CRYSTAL

By **P. P. Ewald**

PHYSICS TODAY / DECEMBER 1953

An article based on Professor Ewald's address as Retiring President of the American Crystallographic Association at its meeting in Ann Arbor, Michigan, June 24, 1953.

X-RAY CRYSTALLOGRAPHY, like any good crystallization, grew from a few distinct nuclei. The first nucleus was the Laue-Friedrich-Knipping experiment in Munich. Hardly had the news of this new effect been given at the spring 1912 meeting of the Bavarian Academy of Science and found its way into the papers, before a second nucleation was induced in England. While Laue had explained the effect as one of diffraction of very short light waves by the regular lattice arrangement of scattering atoms, W. L. Bragg concluded from the shape of the Laue spots that they should be explained as an effect of reflection of waves on the internal atomic planes, an idea that led him at once to what is now known as Bragg's Law. Thus it was the focusing property which gave the first clue to the Bragg version of the phenomenon, as published in the *Proceedings of the Cambridge Philosophical Society* in November 1912.

Soon after this W. H. Bragg applied this principle in the construction of the x-ray spectrometer, an instrument which led to the fundamental discovery of the K and L series of characteristic line spectra as distinct from the continuous "white" spectrum of the general Bremsstrahlung. With this discovery the wide field of x-ray spectroscopy was opened up precisely in time to give convincing support to the Bohr theory of the atom in its first infancy and subsequently throughout the stages of increasing refinement and complexity. The second wide field opened up by W. H. Bragg's discovery was x-ray crystal analysis, for which the characteristic wavelengths provided the yardstick for measuring the distances between atoms or atomic net planes in a crystal. The use of this yardstick was, however, only obtainable by first determining a crystal structure without its application. This was achieved by W. L. Bragg by comparing the Laue pictures of NaCl, KCl, KBr, and KI; the changes produced in replacing a lighter atom by a heavier one of greater reflecting power led to the confirmation of the "spatial chessboard" structure which had been postulated for these salts by Barlow and Pope. Once the relative arrangement of the atoms was known, the absolute scale of their distances followed from the density of the crystal.

What exciting years were these last pre-war years 1912 and 1913. They belong to those periods of eruptive development that occur when an entirely new impact hits and unites fields of science which for many years had not yielded to the most strenuous external pressures. This had been the case with x-rays prior to 1912, with optical spectroscopy and with the interpretation of the first quantum phenomena in the theories of radiation and of the photo-electric effect. In these same years a revival of interest in the theory of the solid state took place; in Born-Kármán's paper on specific heat (1912) the first application of quantization to the lattice model of solids was made, and shortly after that, in 1915, appeared Born's *Dynamik der Kristallgitter* which marks the nucleation of the modern theory of solids. Immediate as the impact of the new discoveries was on physics, it was a delayed one for chemistry. The fact that in simple inorganic salts the concept of a molecule should no longer hold did not please the chemists. Ephraim's book on inorganic chemistry was, as far as I am aware, the first textbook fully to accept this fact, but it did not appear until 1921. Progress in x-ray diffraction came from many European countries in those early years. Maurice de Broglie in Paris was quick in developing his own spectroscopic methods and in training co-workers like Trillat and Thibaut. Some members of this audience may remember the unique setup of his laboratory in his private *hôtel* in the rue Lord Byron where cables for the current came in by holes cut in the Gobelins adorning the walls. In Holland Lorentz developed the Lorentz-factor in his lectures and Debye,

the international coordination

DIFFRACTOMETRY

at that time in Utrecht, ventured to tackle the theory of diffraction by a lattice in thermal vibration—a problem which appeared superhuman to anyone but a Debye. In England Moseley made the first systematic survey of the K- and L-series throughout the periodic system and Darwin discussed the absolute intensity of x-ray reflection by setting up the first dynamical theory far ahead of all others; in order to account for the difference between the theoretically expected, and the observed intensities, he developed the idea of the mosaic crystal which proved indispensable for all later work. The crystallographer G. Wulff in Russia showed the advantages of crystallographic projection techniques; Nishikawa obtained the first fibre diagrams and Terada, also in Tokyo, was the first to observe the sudden appearance and disappearance of the diffracted spectra on the fluorescent screen. Remember that all this happened at a time when the identity of the Bragg reflection and the Laue diffraction theories was not yet generally understood.

It is hard nowadays, especially for the younger among you who have been taught x-ray diffraction in a well organized university course, to imagine how crystal analysis then appeared to those engaged in it. It may be illustrated by a post card I received from W. L. Bragg on which he wrote that he had measured the spectra of pyrites and had been trying to obtain the structure. *"But it is terribly complicated"*, he wrote. It was the first example of a cubic crystal in which the trigonal axes do not intersect.

THE WAR of 1914–18 brought not only the interruption of international relations, it even brought the actual x-ray diffraction studies very nearly to a standstill. The application of these studies to chemical and technical problems had not yet been discovered. Only one advance of great importance was made in 1917, and that independently in Göttingen and in Schenectady by Debye and Scherer and by A. W. Hull, respectively. While all previous measurements required fairly large well-formed single crystals, which were not always easy to obtain, the powder method was applicable to practically all solid substances. I first heard of this method from the crystallographer A. Johnsen, then professor in Kiel, and keeper of a fine collection of minerals. His words, which I remember well, are significant for the enthusiasm with which the powder method was acclaimed: Who would still want to take single crystal pictures, painstakingly adjusted and hard to index? We just powder our crystals in a mortar and get the powder lines to fit into a quadratic form and that gives us all the information.

In the period after 1918 the retarded development flared up afresh. The two Bragg schools, at the Royal Institution and in Manchester, were the leading centers for structure analysis and for the training of the next generation of physicists in this art. They were also an international meeting ground of crystallographers. The spectrometer remained for a long time the main instrument. Apart from giving a direct indication of the strength of reflection, it offered the great convenience of showing exactly from what plane the reflection came, an advantage that was lost in the rotation diagrams of Polanyi (1921) and only regained in the Weissenberg x-ray goniometer method (1924). The deciphering of the experimental data was achieved by frontal assault in each case. A "normal" decline of intensity with increasing order of reflection was established by W. H. Bragg and the deviations from the normal sequence were attributed to the halvings or similar subdivisions of the sets of reflecting planes and later to the structure factor. The usefulness of space group theory in providing a framework for the atomic positions was stressed by Niggli in his book (1919), but given the practical test by Wyckoff, together with his numerous co-workers, in determining a great number of structures with the help of his *Analytical Expression of the Results of the Theory of Space Groups* which appeared in 1922. The first English adaptation of the theory of space groups followed in 1924 (Astbury and Yardley).

About the same time the first books on the new subject appeared. The book by the Braggs, *X-Rays and Crystal Structure*, had already been published in 1915; it gave mainly a coordinated account of their own investigations and is still a fundamental book, unique for the simplicity of its reasoning and the beauty of its style. It gave the direction to the English school of x-ray workers, but it was never meant, at the early date of its publication, to present more than the line of thought that had achieved the great first results. It was often reprinted but never expanded or revised.

My own book, *Kristalle und Röntgenstrahlen* (1923), represented the continental point of view and aimed at integration of the advances in the methods of x-ray diffraction and at discussing the implications of the results. It was sold out in two years and I never prepared a second edition, partly because the subject was growing so rapidly, but mainly because the two editions of the chapter I wrote for the *Handbuch der Physik* (1926 and 1933) gave a more succinct and modern presentation of the same matter.

Similar monographs on the subject were written in France by Ch. Mauguin (1924) and in the U.S.A. by Wyckoff (*The Structure of Crystals*, 1924). Together these three early books document in detail the advances made up to about 1923 regarding the methods of producing and indexing diffraction photographs and of using the structure factor for obtaining the atomic arrangement. Significantly neither Wyckoff's nor my book contains a main chapter on the intensity of diffraction; in spite of Lorentz', Darwin's, and Debye's work too little was known about it. Mauguin deals more fully with intensity.

MEANWHILE the x-ray crystallographers were becoming more ambitious. The first structures that had been determined, like NaCl, diamond, zinc blende and wurtzite, had been without a parameter; the atoms could not be moved away from the intersection of symmetry elements in the cell without admitting too many atoms to the cell. In pyrites and calcite, structures with one parameter had been solved by discussing the intensity sequence among the various orders of reflection of a face. It was still fairly easy to extend the methods of discussion to the case of two, and, in rare cases, of three parameters. But you could not set out to determine the structure of any given crystal, because in most cases it was likely to contain a large number of parameters. The purely optical principles of discussion then broke down. At this stage the idea of fixed atomic radii was introduced by W. L. Bragg and his school and eagerly expanded and modified by V. M. Goldschmidt and others. Nowadays it is a valid and much employed principle which is firmly based on a large body of experience. It appeared a risky principle in the mid-twenties and one would have liked first firmly to establish it on a large number of structures which had been determined without its use. This gave a special meaning to the collection of structure data which C. Hermann and I assembled as Vol. 1 of the *Strukturbericht*. In reporting the structure determinations we tried here to separate clearly the optical arguments, which seemed safe, from any doubtful additional hypotheses. Wyckoff followed the same line in his critical surveys of structures which were published in 1924, 1931, and 1935. This purist tendency has been dropped deliberately from the recent revival of the *Strukturbericht,* the Structure Reports.

In spite of auxiliary assumptions derived from atomic radii and structural chemistry the full structural analysis of crystals containing three or more parameters remained at best a hazardous undertaking. All problems seemed to end up in a sigh: if only we had a reliable means of measuring and evaluating intensities and of deriving from them the structure factors! It is true that in 1914 Darwin had given two expressions for the intensity of an x-ray reflected by the external face of a crystal, assuming either a perfect or a mosaic crystal. These expressions gave widely different values, and the measured intensities did not seem to fit either assumption too well. Besides, Darwin's papers were not well

written and were not properly understood. His restriction to the specular reflection from the net planes gave no indication as to what became of the cross-grating spectra which each of these planes would give owing to its own atomic periodicity. This was one of the reasons which prevented me from appreciating Darwin's work, and it was only after having set up my own dynamical theory of x-ray diffraction that I discovered that some of my results for a perfect crystal were identical with those of Darwin. Experimentally Bragg and James and Bosanquet showed in 1921 that the intensity of reflection depends largely on the treatment given to the reflecting surface of the crystal, such as grinding, polishing, etching.

By 1925 it had become apparent that the whole future of x-ray crystal analysis was at stake unless a solution to the intensity problem could be found. I learned that Wyckoff was coming to Europe and it occurred to me that this would offer an excellent occasion for having a joint discussion of all those who had worked on the intensity problem. After some correspondence I found a date in August, 1925, which was acceptable to nearly the whole group of experts and I arranged for five days of discussion at my mother's house in the country at Holzhausen on the Ammersee, some 25 miles from Munich. The little local inn was rented, a blackboard was borrowed from the nearest country school and a few boxes of cigars placed on the table in my mother's big studio (she was a painter). Then the exciting moments came of meeting my colleagues at the nearest railway station at which they, fortunately, all arrived on schedule. Remember that by 1925 the international relations had not yet been resumed on any large scale and that this was for most of us the first post-war meeting of an international character. Those present were, if I remember correctly, W. L. Bragg, Darwin and James from England; Wyckoff from U.S.A.; Brillouin from France; Fokker from Holland; Waller from Sweden; Laue, Mark, Ott, Herzfeld, and myself from Germany. Occasional visitors and participants were Debye and Fajans. Waller had just published his dissertation; the first part of this was a review and extension of the dynamical theory and the second an extension of Debye's work on the temperature factor. It was a very learned paper and required many years of development to be fully evaluated in its implications for the discussion of experimental results.

I think all of us enjoyed these full days of intense discussion in which Darwin finally got so entangled between his own papers of 1914 and 1922 that he promised to restate them and where Bragg declared emphatically at the end of one session: I will not be satisfied unless I can determine a structure with 19 parameters! This seemed utterly fantastic at that time, and yet, two or three years later, he was tackling the silicates and polytungstanates and was just about as far as he had wished. The Holzhausen conference was an important event in the history of x-ray crystallography. It coordinated at a critical time the various approaches, experimental and theoretical, British and Continental, i.e. reflection versus diffraction. It further made scientists meet after the war, many of them for the first time, and laid, I am happy to say, the foundations for a lasting personal friendship and respect. In doing so, it also paved the way for two of the future international enterprises in crystallography, the *Strukturberichte* and the *International Tables*. It stimulated experimental and theoretical work in the problems discussed at the meeting as is shown by a number of papers in the subsequent years. But the credit for overcoming the formidable deadlock of the intensity problem goes to W. L. Bragg who returned to Manchester to tackle it in a most systematic and realistic way. He first established a standard of intensity in the 400 reflections of a properly prepared rock salt face; together with James and Darwin he restated the results of the latter's theory in a *Phil. Mag.* paper of 1926; James, with Miss Firth, Brindley, and others, made a thorough experimental study of the temperature effect using high and low temperatures; Waller came to Manchester to help on the theoretical side. As a result of fundamental importance for all parts of physics the first direct confirmation of the zero point energy of an oscillator, here the crystal proper vibrations, was obtained. Hartree, then also at Manchester, tackled the remaining unknown intensity factor, the atomic factor, first on the Bohr orbit atomic model, and, after the advent of wave-mechanics in 1926, by the method of self-consistent field. Bragg, in 1927, reported on the atomic factor derived from the Thomas statistical model of the atom.

This may give an idea of the concerted effort which finally overcame the deadlock.

BY THIS TIME advantage was taken, also by the English workers, of the theory of space groups. Bernal came to see me in Stuttgart in 1925 carrying along a voluminous manuscript in which he had derived the 230 space groups in his own way. The problem was how to publish this work. As happened not infrequently to Bernal, the manuscript was interspersed with folding tables densely covered with symbols in order to accommodate all information on them. He had devised his own symbols for the space groups and it was all Greek to me. I well remember the three of us, Bernal, myself, and Carl Hermann sitting alongside on a sofa and the maps being unfolded on my knees. It took Hermann a split second to understand the tables, including the strange terminology, and to suggest various points of rearrangement of the tables in order better to bring out some of the subgroup relations which Bernal's arrangement did not show. The battle between them was fought out on my knees, and it took close to an hour to go through the tables.

Several new books had been published or were being written, such as Mark's book *Die Verwendung der Röntgenstrahlen in Chemie und Technik* (1926), Mark and Rosbaud's Space Group Tables, my own *Handbuch* article (1926), and a book by Schleede and Schneider which was being planned. Besides, the older books needed new editions by 1928. It was a matter of some

concern to me, and I am sure to the other authors also, how to get around the dilemma either of having to include detailed tables and illustrations of the 230 pages groups, or of writing a book that lacked practical value for the actual determination of crystal structures. Besides, there were proposals by Rinne and Schiebold, by Hermann, by Mauguin, and others for changing the nomenclature of the space groups so as to make it more descriptive than the Schoenflies symbols. A multiplicity of symbols for the same space group was going to create considerable confusion in an already sufficiently complex subject. The only way I saw out of this confused situation was the preparation of a set of impersonal tables containing a complete and standard description of each space group, a work to which each author could refer in his own textbook and from which he could pick examples on which to discuss the ideas of space group theory, without being obliged to bring a complete set of tables. I discussed this idea with Bernal and Mrs. Lonsdale at the occasion of a meeting of the Faraday Society in London, 1929, and together we laid it before Sir William Bragg who gave us every encouragement and promise of help and convened a meeting of a large group of crystallographers gathered for the Faraday Society, where this plan and others were discussed. Bernal and I undertook to prepare a detailed syllabus of such tables. We hit upon all kinds of difficulties, partly because decisions had to be taken concerning the symmetry axes of the second kind, the fixing of origins, the graphical representation of symmetry elements and of equivalent points, etc., and partly for reasons of a more personal nature because people in different laboratories and countries had become used to symbols and drawings which did not please those accustomed to others. A conference was the only way to thrash out these points, and, again taking advantage of a European trip of Wyckoff, Bernal and I prepared a meeting for July 1930. On the invitation of Niggli it was held at his institute. I took the chair at the four-day meeting, and we worked quite hard all the time. Those present were Wyckoff from the U.S.A., Bernal, Astbury and Mrs. Lonsdale from England, Mauguin from France, Niggli and Brandenberger from Switzerland, Kolkmeijer from Holland and myself, Hermann, Schiebold and Schneider from Germany. The questionnaire Bernal and I had circulated together with the invitation gave a lead to the discussions and some of the points were quickly settled. Hermann's notation was recognized to offer great advantages, and some simplifications which Mauguin proposed were accepted; Schiebold, somewhat reluctantly, refrained from pressing for the acceptance of his system. A rather lengthy skirmish took place over the graphical representations. The English were accustomed to the Astbury-Yardley diagrams, Niggli to those in his book to which most others were not partial. Mauguin circulated a neat set of cards which he used in his course showing the cell and a group of equivalent atoms for each space group but leaving out the indication of symmetry elements. It was finally agreed to give two figures for each space group, one showing the equivalent points in Mauguin's way, the other the symmetry elements in a modified form of Niggli's representation. Preference for taking the origin at a center of symmetry whenever possible, and of using rotation-inversion rather than rotation-reflection axes was soon agreed upon. Wyckoff's symbols for special positions were adopted and so was the product form for the structure factors, as given in Mrs. Lonsdale's Tables. It was further agreed to list the sub-groups for every group.

The discussion on the third day was on the second volume which deals with mathematical and physical tables. The details of the tables of trigonometrical functions were laid out in a form convenient for the calculation of structure factors; other trigonometric tables, useful for the calculation of absorption and other corrections were planned. The listing of absorption coefficients and of atomic factors, and many other details, were discussed along with the distribution of the work between the laboratories. It was also agreed to offer the Tables for publication to the publisher of Niggli's book.

In the preface of the *Internationale Tabellen* you will find details of the work supplied by the various groups, and a list of the Learned Societies whose generous subsidies made possible the publication of the work at a very reasonable price. The Tables appeared in 1935 and it has always made me happy that they were universally accepted and fulfilled the hopes in which they were conceived.

T HE NEXT international enterprise for which I worked was the foundation of an international organization of crystallographers. It began in 1944 when I was asked by the X-Ray Analysis Group to give an evening lecture at the March meeting in Oxford. The second part of this talk was the plea for the formation of an International Union of Crystallography. This plea was published in *Nature* (**154**, *628*, 1944). It sets out, as the main task of the Union:

(1) the publication of an international journal of crystallography;

(2) the establishment of archives for the storing of scientific results which would be too costly to publish in full;

(3) the abstracting, summarizing, and recording of crystallographic work, in particular in connection with the planned general scientific abstracting scheme;

(4) the preparation of a second edition of the *International Tables,* and their public ownership;

(5) the preparation or coordination of analytical tables for identifying crystals (Barker index, card index of powder lines);

(6) the representation of crystallography in the system of other international scientific unions.

The ball set rolling in Oxford was played in great style by W. L. Bragg who arranged an international congress of crystallographers in London in 1946. This was actually the second international congress, the first having been held in 1934 under the auspices of the

Union of Physics when Sir William Bragg was its president. It is unnecessary to recall the events in London which ended with the resolutions to found a Union, preferably an independent Union of Crystallography and, if this were not accepted by the International Council of Scientific Unions (ICSU), to form a group within the Union of Physics; further to start at once with the preparations for an international crystallographic journal, for the resumption of *Strukturberichte* in a new form, and for a new edition of *Internationale Tabellen*. The discussions of the committees nominated for these tasks began without delay while the foreign visitors were still about. In fact, the Russian delegation which arrived a day after the congress had closed, was just in time to take part in the discussions on the journal which took place a few days later in Cambridge. It is thanks to them that *Acta Crystallographica* carries its name.

The actual birthday of the International Union of Crystallography was the hot 3rd of August, 1948, at the Union's first Assembly at Harvard. It was the culmination of a long sustained effort of preparations, including ocean crossings and oceans of correspondence on the part of a great number of crystallographers. Everything was set for the Union to crystallize out at this meeting. Then an unforeseen inhibition occurred. The provisional executive committee had proposed to change the first draft of the Statutes of the Union in some points regarding the voting power of the delegates. The new formulation had to be accepted before the Statutes could be passed en bloc. So the changes had to be voted on, especially since there were some objections. Somebody raised the question: on what voting power is this going to be decided? Neither the first draft nor its amendment were binding, since neither had been accepted. Arguments for voting on the old or the new scheme went on in a freakish way. Finally the heat, I guess, must have concentrated the solute sufficiently, so that the inhibition was overcome and the Union at last crystallized out by the adoption of the revised statutes en bloc.

L OOKING BACK to 1946 and 1948 we may ask ourselves whether the foundation of the International Union of Crystallography was worth while. To answer this question we have not only to study what the Union has achieved, but also where we would be without it. Its main achievements are the journal *Acta Crystallographica,* the two, and soon four, volumes of *Structure Reports,* and the first volume of the new version of *International Tables;* besides, there are the two international Congresses and Assemblies—Harvard 1948 and Stockholm 1951—to which the third Assembly in Paris 1954 will be added next year. Furthermore, there is active work going on by correspondence in the commissions of the Union, as on Powder Data, on Apparatus and Standardization, and on Nomenclature. Within the system of International Scientific Unions the Union of Crystallography belongs to the small Unions, small by the number of adhering countries, by its financial demands, and by the limited importance of its international program which is not as vital in crystallography as it is in astronomy, geodesy and geophysics, or radio science, and not as extended as in chemistry and biological sciences. But as a small Union it has earned respect and acknowledgment by the determined effort to achieve internationally important results in the field of publication and standardization. Had the Union failed to materialize it is most likely that by now we would have three full-fledged crystallographic journals, in the States, in England, and in Germany. Each of these journals would be indispensable because each would contain important papers. There would be three editorial, and, worse, three publishers' policies regarding the scope and length of the papers, the yearly published volume, and the price. It is unlikely that private publishers would have received the generous subsidies on which *Acta Crystallographica* was started. In the first five years *Acta* has received altogether $17 400 from Unesco and from industry. These subsidies have helped over the first few years which are a critical time for a journal. Thanks to this help we have been able to accommodate an ever increasing influx of papers. The number of pages published has risen in the last three years by 78 percent, the production cost per page by 7.8 percent. The number of subscribers has been increasing steadily, at a rate of about 50 more subscriptions every year, and this is a healthy sign. Unfortunately, however, this rate is far too slow to offset the increase in cost of production. It means that at present *Acta* is adding to the "red" in the Union's books a deficit of about $10 000 per year. We are thus still in the midst of the teething troubles of our five-year-old baby.

It is not unnatural that this should happen. When the journal was started, it was done on the understanding that the price per volume be $10 and that the balance between production cost and sales be met by the subsidies. The $10 price was maintained for the first three volumes, then at the Stockholm Assembly the price was raised by 50 percent to $15, but meanwhile the volume had increased by 200 percent against the first volume. Now you might ask: is it necessary to publish that many papers? A moment's consideration will show you that it is a natural development. An increasing number of people trained in crystal analysis produces more and better papers every year; the advances in x-ray diffraction technique alone, and again in computational technique, allow an increased output of structure determinations, and the ever closer connection with chemical, biochemical, metallurgical, and physical investigation presents the diffractionist with problems of considerable interest in nearly overwhelming numbers. If the journal of the International Union of Crystallography is to tie together all this diverse diffraction work and be the forum for its adequate discussion, then we cannot afford to turn down good manuscripts because we are getting too many of them for a strictly limited volume. For the last few weeks I have felt very unhappy in following

this course after having received strict orders from the Executive Committee at our Paris meeting not to exceed last year's volume.

What then can be done with *Acta*? We have now some 1100 subscribers, that is double the highest figure ever obtained for the *Zeitschrift für Kristallographie*. This number is considerably below the saturation value, which I estimate at 2000. There are many university and industrial laboratories without *Acta*, in spite of their interest in the solid state. Many big establishments take only a single copy in spite of demand in different localities. There are also many among you who do not yet take advantage of the reduced price at which you can get your own copy to study at your leisure at home. *The Physical Review* is received by nearly every one of the 10 000 members of the APS and he pays for it in his membership fee. If we were similarly to arrange a general distribution of *Acta* to the 700 ACA members, of whom about 100 already take *Acta* at the reduced rate of $9, this would bring in one-half of the yearly deficit. But so far the ACA Council has not taken to this proposal.

There are two ways out of this rather desperate situation: One is to collect further subsidies, preferably guaranteed over a number of years, and to continue running the journal at a loss. The private subscriber may like this proposal because he is getting the benefit of the subsidy. But it puts the journal on an unsafe basis and endangers its financial independence. The other way is to increase the price of subscription so that the journal is self-sustaining. With the present volume and number of subscribers this point would be reached by raising the subscription rate from $15 ($9) to $24 ($15). Some income could also be gained by carrying advertisements, but this is not considerable. A page charge, while acceptable in the U.S.A., appears unacceptable to the European scientists. Further income will be necessary later for following the natural development, i.e. increasing the volume beyond the present 870 pages; this will have to be met by a further substantial increase in the number of subscribers.

I think it is important to explain this situation to a group such as is assembled here. We should not take the existence of scientific journals for granted. Each one of us should fight for their existence and make sacrifices, not only by saving space in his own publications by the utmost condensation, but also by subscribing and getting others to subscribe. Only by a deliberate and concerted effort can *Acta*, and also the two other publications of the International Union of Crystallography be brought over the inevitable difficulties of the first ten years. The gain these publications bring to the large fellowship of crystallographers all over the world seems to me to justify the existence of the Union and to make it worth while not to relax in sustaining its activities. We may, I think, be proud of what has been achieved so far and it seems unthinkable that the International Union of Crystallography should not be able to keep pace with the development of crystallography itself.

the founding
of the
AMERICAN INSTITUTE OF PHYSICS

The following talk was presented at the Banquet of the American Institute of Physics and the Member Societies in Chicago on October 25, 1951. Senator Brien McMahon, chairman of the Joint Congressional Committee on Atomic Energy, also addressed the gathering.

*By **Karl T. Compton***

PHYSICS TODAY / FEBRUARY 1952

FIRST may I add my greeting to Senator McMahon, and add my appreciation of his willingness to meet with us tonight. He and we have a strong bond in common. We physicists have been largely responsible for creating the activity for whose wise handling in the national interest he has so great a responsibility. And may I say, on the basis of several opportunities to see him in the discharge of these responsibilities, that we are very fortunate in having this aspect of our common interest in the hands of a man who has shown such real understanding of the basic conditions for scientific development and for advantageous application of the great potentialities of atomic energy.

Next let me try to give a bit of historical perspective to my reminiscences about the formation of the American Institute of Physics. This is its 20th anniversary, and 1931 was a milestone. There was another milestone, definite in character though not sharply defined as to date, about twenty years before that. This was the time when it was beginning to be respectable and effective for physicists to stay in the United States for their postgraduate study instead of going to Europe. During the ensuing two decades, physics grew rapidly, being part and parcel of the new development of postgraduate schools in this country, being stimulated by the teamwork of groups assembled for tackling some of the technical problems of World War I, and being greatly advanced by the program of National Research Fellowships supported by the Rockefeller Foundation shortly after the war.

But, in spite of this rapid development of physics during the "teens" and the "twenties", the general public was not very aware of this growing profession, soon destined to be of such earth-shaking significance, in both the figurative and the literal sense. For example, in the edition of Webster's New International Dictionary, published four years after the establishment of the American Institute of Physics, the preferred definition of a physicist was "One versed in medicine". The average citizen would associate the words physics and physical scientist with certain intestinal disorders or with gymnasium drill. In certain states, where some kind of registration of employees was required, the profession of physics was not recognized, and physicists had to register as either engineers or chemists, which some of them felt to be rather humiliating.

With this background of vigorous growth of this young, and then inadequately recognized, profession, let me proceed with the story of the organization of the American Institute of Physics.

WHEN Dr. Klopsteg asked me some months ago to give some reminiscences covering the establishment of the American Institute of Physics, I at first hesitated because of an impairment of my vocal apparatus. But I accepted because this Institute represents a momentous achievement in the development of organized physics in this country; and also because I owe a very great personal debt to the American Physical Society and the other societies associated with it in the Institute.

In 1909, just sixteen years after the establishment of *The Physical Review*, I submitted for publication my first piece of research, which was my master's thesis at the college of Wooster in Ohio. The college

Karl T. Compton, chairman of the corporation of the Massachusetts Institute of Technology, served as the AIP's first chairman from 1931 to 1936. A past president of the American Physical Society, Dr. Compton has held numerous positions of importance in industry, the government, educational institutions and foundations, and professional organizations.

at that time did not subscribe to *The Physical Review*, and I had no background of information regarding the proper form and length of a scientific article for such a journal. I consequently shipped the manuscript of my thesis on to Professor Merritt, who was then editor of *The Physical Review*—without realizing that its two hundred typewritten pages and numerous photographs would have constituted an article many times too long for publication, even in those days when editorial policy was far less strict than at present. In spite of the inappropriate length and character of this manuscript, I received from Professor Merritt a long letter giving detailed suggestions for rewriting the material. I tried a second time, and again Professor Merritt wrote back, saying that he felt the material had now been condensed to the point at which certain parts were not clear and again making suggestions for another revision; and this time the article was published.

I have often thought that this extraordinary help given by a great physicist to an unknown student in a small college, and involving on his part a great deal of work, was a splendid illustration of the helpful concern of the pioneers in scientific education in this country to encourage the development of their successors. Certainly, for me, it was both an inspiration and a lesson. Since that time I have always felt that any service which I could render to *The Physical Review* and to the profession of physics was an obligation as well as a pleasure.

During the decade following World War I, the rapid increase of research in the field of physics led to financial difficulties for *The Physical Review*.

To tackle the financial problem, the Council of the American Physical Society in the latter half of the 1920's appointed a Committee on the Financial Status of *The Physical Review*. The problem confronting this committee, of which I was a member, was not only financial, but also involved the great delay in publication caused by the accumulation of manuscripts, which the Society could not afford to publish promptly. To meet this situation, several steps were taken, including: a more rigid editorial policy, an increase in the annual dues of members, and introduction of the "per page charge to authors".

When this "per page charge" plan was put into effect, it was quickly accepted by some organizations but not by others. Very generously at this point our fellow member, Dr. Alfred L. Loomis, stepped into the breach and undertook for an introductory period personally to take care of the "per page charge" for institutions which reported themselves unable to meet the charge. Gradually, however, the plan gained general acceptance and is now a regular part of the financial basis of our physics publications, and has subsequently been adopted by other scientific organizations.

DURING that same period, in the late 1920's, another problem presented itself to the American Physical Society. This was the emergence of groups of physicists who felt that the main current of interest in the American Physical Society was not meeting their particular professional requirements. These groups undertook to establish new societies and new publications devoted to their important and special interests. Consequently, the American Physical Society was concerned over the centrifugal tendency to separate the basic science of physics into a number of independent groups. Very naturally, each of these groups had its own financial problems of publication.

My own attention was first drawn to the possibility of a better coordination of the various physics groups by a conversation which I had with Mr. William Buffum who was at that time the executive officer of the Chemical Foundation. I had gone to him for financial help for *The Physical Review*. He told me that he had also been approached by various other physics groups and it was his impression that the whole profession of physics was running away in different directions by independent groups without much coordination. He said that the Chemical Foundation did not feel that it would be a wise expenditure of its funds to support the separate groups, but that if some way could be found to bring them together in some sort of coordinated program, then he felt that the Foundation would be very much interested in helping to establish such a program.

From this point on, my recollection of events is very much amplified by excerpts from the records of the Council of the American Physical Society, which Karl K. Darrow very kindly dug out for me from the record books.

The first mention in the minutes of the Council of the American Physical Society of some official move toward coordination of the various activities in physics was taken on a motion of Professor G. W. Stewart at the Chicago meeting on 29 November 1929. On his motion the Council voted that a committee of three be appointed "with the President of the Society (H. G. Gale) as Chairman, which shall, after conference with the officers of the Optical Society of America, the Acoustical Society of America, and any other physics societies, recommend a plan of merger of these societies with the American Physical Society, and which shall present a preliminary report for discussion by the Council at the Des Moines meeting," in the following December. This committee consisted, in addition to President Gale, of G. W. Stewart, H. E. Ives, and D. C. Miller.

From this time on, until the actual establishment of the American Institute of Physics two years later, the problem of coordination of the various physics groups was a matter of discussion and report at every Council meeting.

The Council, at its April 1930 meeting in Washington, appointed a Committee on Applied Physics under the chairmanship of Dr. Paul D. Foote and comprising also L. A. Jones, A. W. Hull, H. E. Ives, L. O. Grondahl, K. T. Compton, George B. Pegram, and Henry G. Gale.

This committee made its first formal report to the Council of the Society at the meeting in November 1930, and I quote from its report, as follows:

"Dissatisfaction exists on the part of many physicists who feel that the activity of the American Physical Society is mainly confined to quantum physics and is not representative of physics in its broadest scope. This feeling is quite general, and whether justified or not, has been definitely evidenced by the formation of such organizations as the Optical Society, the Acoustical Society, the Rheology Society, and others. It is also evidenced by the contemplated formation of a society of Applied Physics and another society of Applied Mathematics, the latter being sponsored mainly by mathematical physicists. The feeling is still further evidenced by the fact that numerous papers dealing with pure and applied physics are not even submitted for the consideration of *The Physical Review* but are published in various chemical, engineering, photographic and geological journals. This state of affairs is a serious reflection upon the limited activity of the Physical Society in the general field of physics."

The Committee then went on to recommend a general organization, somewhat similar to that of the American Chemical Society, and that this organization should be started by the formation of two special divisions of the American Physical Society: one devoted to applied physics, and the other to mathematical physics. Each of these divisions should be, more or less, self-governing, somewhat according to the scheme of organization adopted by the various sections of the American Association for the Advancement of Science.

It was pointed out that this proposal would be in the nature of an experiment. The report went on to say: "If the developments under such action are successful, with a liberal policy of supervision and control, it is not improbable that the organization can be extended to include the groups which have already withdrawn from the Society." The report further suggested that such a federated organization would make it possible to establish a central business office and an administrative force which could serve all of the group. It also pointed out that funds for the advancement of physics would be more readily procurable because of better central and efficient business management.

THE FIRST MENTION of an Institute of Physics appears in the minutes of the Council of the American Physical Society on 29 December 1930, where several actions were taken to implement the preceding recommendations. One of these actions was to approve the establishment of a journal of applied physics. Another was to approve in principle the formation of sections within the Society and to encourage the affiliation of local physics clubs. Finally, and most importantly, the Council voted to propose the formation of an Institute of Physics for the purpose of coordinating various societies whose interests are primarily in the field of physics and for the purpose of supporting their publications.

As I recall it, the suggestion for an American Institute of Physics was first made by Dr. Foote, and the idea immediately took hold as a constructive method of dealing with the various complexities which I have just described. The proposal was submitted to the American Physical Society at its business meeting on the following day, and it was there approved.

The next steps were taken at the Council meeting of the American Physical Society in February, 1931, at which time a Joint Committee on the Proposed American Institute of Physics was established. This committee consisted of Messrs. Jones, Richtmyer, and Foote from the Optical Society of America; Fletcher, Arnold, and Saunders from the Acoustical Society; and Tate, Pegram, and Compton from the American Physical Society.

This joint committee promptly recommended several steps which were approved by the organizations concerned. These include the following:

That the American Physical Society the Optical Society of America, and the Acoustical Society of America cooperate in establishing the American Institute of Physics as an agency for studying the common problems of the organizations representing physics in America and for undertaking thereafter such functions as the cooperating societies may assign to it.

That each of the cooperating societies designate

three members to constitute with the others so designated the Governing Board of the American Institute of Physics.

That a full-time Executive Secretary be appointed by the Board.

That the Institute through its Board and its Executive Secretary undertake, in such order as may be deemed best by the Board, the study of the following subjects with a view to making proposals to the cooperating societies as to functions of the Institute:

(a) Publication problems and the possibility of benefits to be derived from cooperation or unification of effort in the business of publication.

(b) Possibilities and procedures for increasing income from subscriptions, advertising, and other sources of support.

(c) Suitable publicity for meetings and contacts with the press.

(d) Relations and contacts of the Institute with local groups interested in physics.

That the Board investigate the possibility of developing an international management for *Science Abstracts A,* with change of name to one more descriptive, and with improvement as to indexing and completeness.

That the Board consider the development of appropriate relations with other national societies which may or may not wish to become societies cooperating with this Institute, such as the Society of Rheology, the Meteorological Society, the Association of Physics Teachers, and others.

At the next Council meeting on the 10th of September, 1931, I reported, as Chairman of this Joint Committee, that Dr. Henry A. Barton had been elected Director of the Institute of Physics and Dr. John T. Tate had been appointed Advisor on Publications. I also reported the favorable action for affiliation by other physics groups and that the Chemical Foundation had given informal assurance that it was ready to spare no expense in furthering the interests of the Institute.

Thus was the American Institute of Physics established, and at a Council meeting on the 28th of December, 1931, I reported to the Council that there was no further need of this Joint Committee since its whole purpose had been achieved in the formation of the Institute. Dr. Darrow in his recent letter to me states very generously in this connection that "the committee was thereupon dissolved with honor. I think that no other committee in the history of the Society can have made so momentous an achievement."

From this point on, you all know the record of this new organization. It has served well during the past two decades in which the profession of physics has grown enormously both in numbers and in accomplishment. I think it has well solved the problem of coordination of the various important fields of physics, while at the same time giving free scope for initiative and freedom in the development of various aspects of the subject. It greatly alleviated the financial problem of publication, although I understand that this problem has again caught up with us because of the greatly increased amount of important material to be published and the increasing costs of publication.

IN CONCLUSION, I would like, for the record, to pay tribute to several individuals and organizations among the very large number who have contributed to the successful development of this enterprise.

I would pay a special tribute to our late colleague, Dr. John T. Tate, who, as Adviser on Publications, was principally responsible for the plan of uniform format and centralized editorial work which promoted economy and efficiency in publication. I would pay special tribute to Dr. Paul D. Foote, who so effectively guided the work of the Committee on Applied Physics, which was so largely responsible for the solution of this problem. The record would be notably deficient without recognition of the statesmanlike contributions of George Pegram in every stage of this program. His knowledge of organization, law, and finance, backed up by judgment and devoted interest, has been invaluable.

We owe a great deal to the Division of Natural Sciences of the Rockefeller Foundation, which helped us substantially to develop this program of scientific publication—a type of problem which was coming to the Rockefeller Foundation from many quarters—and I know that the Foundation took a great deal of satisfaction in having been able to assist in the development of this type of solution.

The Chemical Foundation helped out very substantially in providing the first quarters to be occupied by the Institute, and in underwriting a portion of the overhead in its early operations.

Special recognition also should be given to those physicists and friends of physics who contributed so generously to make possible the purchase of the fine headquarters building for the American Institute of Physics in New York. This building has not only provided operating facilities for editorial and other activities but has been a central gathering place for physicists of all categories, and it has also contributed space for some of the work of the United Nations and other good causes.

The Institute was especially fortunate in the selection of Dr. Henry A. Barton as its director, and we are all greatly indebted to him and to his loyal staff for the effective manner in which he has carried on the executive responsibilities of this organization and for the effective, yet very modest, way in which he has represented the interests of American physics in various national bodies.

I could go on to mention many others, but perhaps it can all be summed up by saying that each and all of those who have contributed to the development and operation of the American Institute of Physics have been but performing generously and effectively their professional duty.

The first fifty years of the AAPT

Melba Phillips

PHYSICS TODAY / DECEMBER 1980

Fifty years ago there was no way for physics teachers to communicate with each other, no way to share either their successes or their frustrations. Teachers had no professional standing as such, and teaching itself seemed to merit little if any recognition or reward. In December 1930 the American Association of Physics Teachers was organized as "an informal association of those interested in the teaching of physics." By the end of 1931 the Association had grown from an original 42 to more than 500. The AAPT now has more than 10 000 members and serves the entire physics community.

The growth of scientific societies

The first permanent scientific society of national scope in this country was the American Association for the Advancement of Science, organized in Philadelphia in 1848. In the beginning, it had two sections: "one to embrace General Physics, Mathematics, Chemistry, Civil Engineering, and the Applied Sciences generally, the other to include Natural History, Geology, Physiology and Medicine." More specialized interests were later represented by the establishment of separate sections; nine sections, including Section B, Physics, date from 1882.

As the country grew and science developed, the needs for communication among scientists increased. The journals were sometimes the first response to these needs. The American Physical Society dates from 1899, but Edward L. Nichols and Ernest Merritt of Cornell University founded *The Physical Review* six years earlier.

Melba Phillips, president of AAPT in 1966–67, is now an emeritus professor of physics of the University of Chicago.

Unlike the American Chemical Society, which embraced all aspects of chemistry, the recently-formed APS took a very narrow view of its role. Members might raise questions of applications and of pedagogy, but the decisions of the Council did not reflect these concerns. It is evident that much discussion took place that did not result in actions recorded in the formal Council minutes. A letter from Arthur G. Webster, the person most instrumental in founding the APS, to Elizabeth Laird of Mt. Holyoke College, dated 20 November 1905, states, "I have often tried to get the Physical Society to take up pedagogical questions, but without success." Applied physics and even fundamental physics related to applications suffered much the same neglect: the Optical Society of America came into being in 1916, partly because the Great War had cut off supplies of optical glass from Germany, but also because most of the influential physicists in APS took no interest in problems involving the principles of optics. The first article in the *Journal of the Optical Society of America* was written by Floyd K. Richtmyer, who was already an influential physicist; nearly twenty years later he was to write the first article in the new journal of the American Association of Physics Teachers.

The man who did the most to found the American Association of Physics Teachers, Paul E. Klopsteg grew interested in the problems of teaching physics at the University of Minnesota, where he became an instructor in 1913 with an MA and was promoted to assistant professor in 1916 on completing his PhD.[1] He did not return to Minnesota after serving in the US Army Ordnance Department (1917–18), but joined Leeds and Northrup Co, and moved on to Central Scientific Co (Cenco) in 1921. He made this last move largely because of the greater emphasis on

Three founders of the AAPT. At left, Homer Dodge, first president, canoeing on the White River in Vermont in 1948. Right, Paul E. Klopsteg in 1979, the man most responsible for founding the AAPT. Far right, a 1904 photograph of Floyd K. Richtmyer, who was instrumental in getting AAPT welcomed into the American Institute of Physics. Growing up in upstate New York, Dodge became expert at boating at an early age. Between 1953 and 1965, he retraced John Wesley Powell's journey of exploration of the Green—Colorado River Canyons, for the most part in an open canoe. Over the years he also ran all of the rapids of the St. Lawrence River, except for one stretch that was destroyed by a dam before he got to it.

As scientific societies proliferated in the 1920's and 1930's, physics teachers began to realize that their specific needs could best be served only by an association of their own.

instructional equipment at Cenco, and so remained in close contact with physics teaching.

It became evident that many were unhappy with the lack of attention to education in the American Physical Society. The sales manager for Cenco, S. L. Redman, had been a high-school science teacher himself, and was almost as concerned with physics teaching as Klopsteg. In travelling around the country he found William S. Webb and Marshall N. States of the University of Kentucky to be particularly sympathetic to the formation of a new society that would foster teaching and communication among teachers, being convinced that the APS would not offer the kind of forum they needed.

In April 1928 an article by John O. Frayne of Antioch College, entitled "The Plight of College Physics" appeared in *School Science and Mathematics*.[2] Frayne described the low level of physics teaching, especially in the universities, noted the negative attitude in APS and advocated forming a new organization devoted to the teaching of physics. Klopsteg got in touch with him, and they met in Chicago together with Glen W. Warner, editor of *School Science and Mathematics*. Between them they compiled a list of 115 people who might be interested in a society of physics teachers.

The association is born

But the AAPT as it finally emerged may be said to date from a conversation between Klopsteg, Redman and States at an APS/AAAS meeting in Des Moines in December 1929. The result was that 30 people, chosen from the "master list" prepared earlier, were invited to a luncheon on 29 December 1930 during the APS/AAAS meeting. Their avowed purpose was to launch a new organization concerned with physics teaching. The man they persuaded to

chair the luncheon meeting was Homer Dodge.[3] Dodge was known to have developed a particularly successful school of engineering physics at the University of Oklahoma.

Of the 30 invited, eight could not attend. Among those who vigorously supported the formation of a new society were Dodge, Klopsteg and Richtmyer. The decision was reached in unanimous passage of a motion made by Klopsteg "that there be organized an informal association of those interested in the teaching of physics; that officers be elected who shall remain in office for one year; that a committee be established for the purpose of preparing the plans for a formal organization; that these things be done without prejudice toward any possible approach from other organizations or societies looking toward affiliation." Officers were chosen: Dodge, president; Webb, secretary treasurer and Klopsteg, vice-president. It was also agreed that a meeting be scheduled at the time of the forthcoming Washington meeting of the APS, but there was more immediate work to be done, and it was decided to meet again on 31 December, and that those present invite others who might be interested. Forty-five people attended this second meeting, and a preliminary constitution was adopted. Karl T. Compton (who became a member of the first executive committee) was present and "discussed informally the plans for the formation of the Physics Institute of America (*sic*) to be constituted by an association of the several societies interested in various fields of physics. He advised that this society [AAPT] should take steps to cooperate with the APS in every way possible in the formation of the Physics Institute."

According to the minutes of the APS Council for 31 December, "The Council took notice of the organization on this day in the Case School Physics Laboratory of a new society to be known as the American Association of Teachers of Physics (*sic*)... The Society decided to have its first meeting in Washington at the time of the Physical Society meeting, at which time they invited Albert W. Hull to present an address on 'The needs of industry in the teaching of physics.' The Council instructed the Secretary of the Physical Society to make contacts with the new Society and to give them proper place on the first day of the Physical Society's Washington program." The address by Hull, who was director of research at General Electric Company, was actually entitled "Qualifications of a Research Physicist," and was later printed in *Science*.[4] It drew a large audience—other sessions were practically deserted—and Compton led a lively discussion.

Gaining the recognition of the AIP

Meanwhile the organization of the American Institute of Physics was proceeding. The first formal meeting was held 1 May 1931. Four societies participated: the Optical Society of America, the American Physical Society, the Acoustical Society of America, and the Society of Rheology, the last two having been organized in 1929. The AAPT was not invited; grave doubts by some as to the "eventual stability and success of AAPT" are reflected and refuted in a letter from Klopsteg to Compton, who was the first chairman of the AIP governing board. As a result of letters from both Klopsteg and Dodge and some intervention from Richtmyer, as well as a very successful first annual meeting of AAPT in December 1931 and the adoption of a more formal constitution, the AIP board, in February 1932, "expressed themselves unanimously as desiring your association to be included with the other founder societies of the AIP," and asked that three representatives be appointed to the board. Those chosen were Dodge, Klopsteg and Frederic Palmer of Haverford College. Klopsteg remained on the board until 1951 with a hiatus of only two years, and he was chairman of the board during 1940–47.

The AIP arose largely from the fragmentation of societies of physicists. According to Compton, "In one sense the American Institute of Physics is the child of the five parent national societies which have cooperated in forming it. In another sense, however, it has followed the more usual course of being born of two parents, the one financial distress and the other organizational disintegration."[5] Financial help was secured from the Chemical Foundation, a corporation formed by major chemical companies to take over German-owned patents after World War I. Its net free earnings were to be "used and devoted to the development and advancement of chemistry and allied sciences..." The impetus for the formation of AIP actually came from the Chemical Foundation, whose support was contingent on a "unified association of American physicists."

By late December 1931 a great deal of progress could be reported at the first annual meeting of the American Association of Physics Teachers, which was held in New Orleans with APS and AAAS. Of special significance was the appointment of a committee, headed by Webb, to develop ways and means of publishing a journal. The first issue of the *American Physics Teacher* (later to become the *American Journal of Physics*) appeared in February 1933 under the editorship of Duane Roller, then at the University of Oklahoma. Its lead article was entitled "Physics is Physics;"[6] in it Richtmyer pointed out that there are several aspects of physics—research and teaching, either at the high-school or college level—but they are still physics. But in his opinion "Teaching is an art and not a science." Although then only a quarterly the journal taxed the slender resources available; it was recommended that dues be raised from the original $2.00 to $3.00, and the change was later approved by a membership ballot.

Palmer had been something of a pioneer in the teaching of physics. His article, "Some properties of atoms and electrons as measured by students,"[7] a justification for and

description of an advanced undergraduate laboratory, had caught Klopsteg's attention and Palmer was invited to participate in the founding of AAPT. One particularly significant step taken in 1933 was to start the ball rolling to prepare an "encyclopedia" of lecture demonstrations; the idea was suggested by Claude J. Lapp of the University of Iowa. Palmer was instrumental in seeing that it was carried through: "I just went ahead and paid the bills to the extent of somewhere around $1500," he recalled. He also made available personnel and facilities at Haverford College; Richard M. Sutton of Haverford was the capable editor of *Demonstration Experiments in Physics*, published in 1938. The book was an immediate success; according to Palmer, "the 15% royalties amounted to enough so that I was paid back ... within three years. It's one of the best investments I ever made, I think."

At the December 1934 meeting in Pittsburgh an anonymous donor offered to finance for a period of three years an annual award (a medal and a certificate) for notable contributions to the teaching of physics. This form of recognition was to become the Oersted Medal, and the donor was later revealed to be Klopsteg. The first award, announced at the annual meeting in December 1936, was given posthumously to William S. Franklin (1863–1930). Franklin was described as a man of exuberant energy "who boasted that the teaching of physics was the greatest fun in the world." He was known for his frequent keen and clarifying comments on papers presented at Physical Society meetings, and he wrote prolifically—twenty-five volumes of textbooks, many contributions on "Recent Advances in Physics" in *School Science and Mathematics*, and a popular volume of educational essays dealing with the beauties of nature, in addition to his research papers. Much of his career had been spent at Lehigh University and MIT, and the Association placed bronze memorial tablets in the physics laboratories of both those institutions. If his death had not come in June 1930, the result of an automobile accident, he would have surely taken a prominent role in the organization of AAPT.

A **1928 summer institute** of the Society for the Promotion of Engineering Education (now the American Society for Engineering Education) at MIT. Here Paul Klopsteg spoke informally with people who taught physics to engineers. Posed in the front row are, from left to right: William S. Franklin, who was awarded, posthumously, the first Oersted Medal; A. Wilmer Duff, director of the institute and author of the physics text most widely used for many years, and O. M. Stewart. Behind Duff is Henry Crew and behind Stewart is Klopsteg. Klopsteg recalls a great unanimity of sentiment at that meeting in favor of an organization like AAPT.

Richtmyer's contribution to the first issue of the *American Physics Teacher* (later to become the *American Journal of Physics*), in which he argues that a successful physics teacher must have more than a thorough knowledge of physics—he must acquire the "art of teaching." ▶

The AMERICAN PHYSICS TEACHER

Volume 1   FEBRUARY, 1933   Number 1

Physics is Physics[1]
F.K. Richtmyer, *Department of Physics, Cornell University*

PERHAPS I can best elucidate the rather cryptic title of this paper by quoting a remark of the late Professor G.W. Jones, Professor of Mathematics at Cornell University from 1877 to 1907 and one of the best teachers who ever occupied a professorial chair. It is told that an embryo teacher, taking one of Professor Jones' courses, once asked him: "What must one do to become a successful teacher of mathematics?"; to which Jones replied: "To become a successful teacher of mathematics one must acquire a thorough knowledge of mathematics."

I am sure that every member of Section Q, and probably many educationists, would agree with Professor Jones' statement, *as far as it goes*. I am equally sure that these same persons would agree at once with the converse statement that no person can become a successful teacher of any subject unless he possesses an adequate knowledge of that subject, even though that person may have had all of the courses in education given in one of the larger universities — 79 of them at Cornell! May I point out, however, parenthetically, that the impression seems to be rather prevalent that there is another group of persons, composed r · 'cationists and educatic ' with this secc

from the fact that there are many excellent scholars who are poor teachers. (I hasten to add, however, that many such scholars who are seeming failures as teachers of the more elementary branches of a subject are most inspiring teachers of the more advanced courses.) Something else than a knowledge of the subject is necessary. That something is, I believe, the acquisition of the *art* of teaching. And it is primarily to this last statement that I wish to direct my remarks.

Teaching, I say, is an art, and not a science. In a recent address before Science Service[2] Dr. Robert A. Millikan characterized a science as comprising first of all "a body of factual knowledge accepted as correct by all workers in the field." Surrounding this body of knowledge is a fringe, narrow or wide as the case may be, which represents the controversial part of the science. And outside of this fringe is the great unknown. Investigators are constantly exploring this controversial region; making hypotheses and theories; devising experiments to test those theories; and gradually enlarging the boundaries of accepted facts. Without a reasonable foundation of accepted fact, no subject can lay claim to the appellation "science."

If **his ***inition* of a science be accept*** · * *

The Oersted presentation was not at first part of any joint ceremonial session as was the Richtmyer Memorial Lecture, but that has changed. For many years now, both events have been part of the ceremonial session, and both are regarded as prestigious honors.

Meetings, members, honors and awards

The pattern of AAPT meetings evolved gradually. After the AAPT was organized at an APS/AAAS meeting, AAPT meetings were held at those joint meetings until 1939, and at the APS meetings after that. In 1943 the annual meeting was shifted to January, and has remained so with only a few exceptions. The summer meetings were also joint at the beginning, but have been strictly AAPT affairs since the mid-1950's. These meetings are hosted by colleges or universities, and are on the whole less formal than the winter meetings.

At first, members of the AAPT were elected by the executive committee with a two-thirds majority needed for election. Those eligible were "(a) teachers in institutions of collegiate grade; and (b) those whose interest in education is primarily in physics of college and university grade." In December 1933 election of members was delegated to the officers, and there was much discussion in the executive committee of what was called "the secondary-school problem." The consensus of opinion was that requirements for admission be changed so that it would be possible for more secondary-school teachers to become AAPT members, but the constitution seemed to read otherwise. The solution arrived at was a new interpretation of eligibility requirement (b) above: "the executive board rules that all teachers of physics who have professional qualifications equivalent to those required of teachers of college physics are eligible for membership in the association." The quite unwarranted fear that the association might be taken over by the athletic coaches who taught physics in many of the small high schools of the day persisted for a number of years. Only in 1938 was eligibility requirement (b) changed to read "other persons whose election will, in the judgement of the

Council, promote the objectives of the Association." Also in 1938 the category of junior membership was established to admit college and university students with a major interest in physics and two years of college physics or the equivalent. The name of this category was changed from "junior" to "student" in 1975.

Despite the differences of opinion on the eligibility of many high-school teachers for membership, the AAPT paid a great deal of attention to the high-school teaching of physics from the beginning. As early as 1934, "support for work on the improvement of teaching in secondary schools" was listed as one of the major tasks of the Association. Prominent leaders in this area were Karl Lark-Horowitz of Purdue University and Robert J. Havighurst, the x-ray crystallographer well known for analysis of the structure of rock salt before he turned to social science and science education. Much of the emphasis was put on the problems of preparatory and continuing education for teachers. Teacher certification requirements in the various states merited much attention, particularly during the years that most students attended small schools, in which "one and the same teacher has to divide his attention among a great many unrelated tasks." Awards for high schools and for high-school teachers were set up later on; the exact nature of these awards for excellence in physics instruction has varied from time to time, but such programs have been continued and strengthened.

The Distinguished Service Citations "for important contributions to the teaching of physics" were initiated in 1952. The number of these awards per year has varied from two to ten; they are usually given to teachers but occasionally to other types of contributions to physics education.

It should be noted that none of the AAPT honors is restricted to members of the Association. The most recently established honor is the Millikan Lecture Award. It is used not only "in recognition of an individual for notable contributions to the teaching of physics" but also to serve as a highlight of the summer meeting. The first lecturer chosen by the committee (in 1964) was H. Victor Neher of

Caltech, a student and colleague of Millikan, but the lectureship had been made retroactive so that a lecture by Klopsteg in 1962 was designated as the first lecture. A medal accompanies this award.

Although not precisely an award it has been an honor since 1940 to be chosen to give the Richtmyer Memorial Lecture. Richtymer died unexpectedly in November 1939, and a proposal for the lectureship was approved the following year. The first Richtmyer Lecture was delivered by Arthur H. Compton on 30 December 1941. This was less than a month after Pearl Harbor, and Compton's title was very appropriately "War Problems of the Physics Teacher." This address has been reprinted in the volume *On Physics Teaching* (1979). The official description of the lectureship appears in a statement of policy approved by the AAPT Council on 30 January 1956: "It is not expected that the lecture should reflect any particular interest of Professor Richtmyer; the topics chosen for it are, rather, those in which he would have found interest were he still alive."

The war years and after

The Association was deeply involved in World War II, particularly in education and manpower. Many of its members, including several officers, went on leave from their teaching posts to work full time for the government directly or in war research laboratories, and other war-related activities were undertaken by the Association itself. Special committees prepared reports, and served to advise on training in physics both inside and outside the armed forces. The AAPT also worked with the War Policies Committee, which was established by the American Institute of Physics and chaired by Klopsteg. As the war progressed it became increasingly difficult to obtain equipment for teaching physics, and the Association, through the War Policies Committee, pressed for higher priorities for essential scientific teaching equipment.

It was clear almost from the start of the war that physics and physics teaching could never be the same again, that both would have new responsibilities in the world to come. Early in 1942 Edward U. Condon was already writing of "A Physicist's Peace."[8] Condon's concern for the social impact of physics was as great as his interest and enthusiasm for every facet of the subject itself.

The Oersted Response of George W. Stewart in January 1943, entitled "Teaching of Tomorrow," anticipated postwar changes, and stressed the necessity of making physics teaching even more useful to society. Vern O. Knudsen, who had been one of the founders of the Acoustical Society, charged in "The Physicist in the New World"[9] that we have trained too few students "to take important responsibilities in the practical world, and certainly too few to be independent scholars, thoroughly trained in fundamental and applied physics." The emphasis was dual: the education of professional scientists must be broadened, and science education must include the study of the relations of science to other human activities. Side by side with the strengthening of the curriculum within the discipline of physics there was an upsurge of interest in the role of physics in general education.

Interest in physics education increased markedly during the 1950's and so did the activities of the AAPT. The Apparatus Committee must be singled out for special attention, working on its own and also with the American Institute of Physics. An intensive study of apparatus used in physics teaching was carried out, and several valuable publications were prepared. In January 1959 the first Competition for New and Improved Apparatus was held at the annual meeting; this competition has been a popular feature of alternate meetings since that time. A new book on demonstration apparatus got under way. AAPT and AIP undertook a visiting scientist program for both colleges and high schools.

J. W. Buchta, first executive secretary of the AAPT and first editor of *The Physics Teacher*, as sketched by Fern Barber in the late 1950's.

The Association was much involved in the early efforts to support institutes for the continuing education of teachers, with J. W. Buchta of the University of Minnesota among the prime movers. The *American Journal of Physics* was able to expand; Thomas H. Osgood of Michigan State had taken over the editorship from Roller in 1948, and was succeeded by Walter C. Michels of Bryn Mawr ten years later, during a period of continuous growth. Michels, with his Falstaffian figure and red beard, was an influential figure in physics for many years. During his tenure as editor he often regretted the necessity for page charges, and would be delighted that the Association has now been able to dispense with them.

The second AAPT journal, *The Physics Teacher*, dates from the early 1960's. The national concern for high-school science teaching had grown during the late 1950's. The Physical Science Study Committee, initiated in 1956 under the leadership of Jerrold R. Zacharias and Francis L. Friedman of MIT, had produced PSSC Physics, and the National Science Foundation was supporting both Summer Institutes and Academic Year Institutes for the continuing education of science and mathematics teachers. It was clear that the AAPT should be of service to *all* physics teachers, including those in high schools, but broadening the *American Journal of Physics* to emphasize high-school concerns did not seem feasible. Under the leadership of Malcolm Correll, then AAPT president, a prospectus for a new journal was prepared in 1961, and a proposal was made to NSF for a grant to help it get under way. The first editor was Buchta, who was also the first executive secretary of the Association. He had served as editor of both the *Physical Review* and *Reviews of Modern Physics*, and had much first-hand acquaintance with American high schools. According to the masthead, "*The Physics Teacher* is dedicated to the enhancement of physics as a basic science in the secondary schools." Through the NSF grant all teachers of high-school physics received the journal without charge for the

Six former presidents of AAPT, caught here seated together in the second row at an AAPT meeting held earlier this year at Rensselaer Polytechnic Institute. They are, from left to right, Melba Phillips (president 1966–67), Robert N. Little (1970–71), James B. Gerhart (1978–79), Janet B. Guernsey (1975–76), Stanley S. Ballard (1968–69), and Robert Karplus (1977–78). Photograph by Reuben Alley.

first year. The first issue was that of April 1963. In 1968 it came into the capable hands of Clifford Swartz, SUNY, Stony Brook. Under the new editor and his associate editors (first Lester G. Paldy and then Thomas D. Minor) high-school physics is still central, but the journal embraces the teaching of introductory physics at all levels. It contains much of practical value, but its approach is by no means narrowly utilitarian. *The Physics Teacher* may be selected as the membership journal by AAPT members, or taken at a reduced rate in addition to the *American Journal of Physics*. (For a detailed history of the early years and an appreciation of Buchta, see the tenth anniversary issue of *The Physics Teacher*, April 1973.)

Establishment of an executive office

Until 1957 the Association had operated as an unincorporated body, but corporation papers were drawn up that year; the immediate reasons were to put the organization in a stronger legal position to deal with employees and to accept possible bequests. Not that there were many employees: the AAPT from the beginning was the product of volunteer labor except for secretarial help needed for the journal and to facilitate committee work as necessary. Even for that there was a great deal of institutional help, as there is now, from colleges and universities at which officers were resident, and in the early days Klopsteg made use of Cenco secretarial personnel. But the work of the AAPT expanded, with some funding assistance from NSF and other agencies, and the level of activities rose in the decade of the 1950's. In 1962 the Association was able to establish an Executive Office for the first time. Buchta was the first executive secretary; on retiring from the University of Minnesota he set up shop in Washington, initially with space rented from the National Science Teachers Association. A glutton for work, Buchta launched the new journal, *The Physics Teacher*, handled the collection of

AAPT dues and did what seems a thousand other things in addition to taking care of the Association business. It was nearly a one-man operation, except for typing and filing.

Buchta had always been a man of boundless enthusiasm and vitality; amid the inevitable respiratory infections of a Minnesota winter, which not even he could escape entirely, his undiminished cheerfulness and enterprise was almost exasperating. But after a brief illness his death came in October 1966, a blow to the Association since no backup had been provided. AAPT veterans and novices alike rallied to the aid of the officers in meeting the emergency. I was president at the time, and recall with pleasure the cooperation of many people in meeting the demands of the Executive Office. Since that time editing *The Physics Teacher* has been a separate operation, and dues collection, along with the maintenance of mailing lists, has been handled by the American Institute of Physics. The Association was very fortunate, at this critical time, in securing the services of Mark Zemansky as executive secretary. Zemansky, a former president of AAPT and retired from teaching, lived near New York, and it was possible to arrange for office space at the American Institute of Physics.

By 1968 it was evident that NSF would phase out the several national education commissions in scientific disciplines. The AAPT had from the start been very intimately involved with the Commission on College Physics, and it seemed to be the Association's responsibility to take over many of the activities and duties of that Commission.

The Commission on College Physics

To understand the existence and the role of the Commission on College Physics it is necessary to recall that the 1950's saw great intensification of interest in science education, particularly in physics. Physicists had contributed enormously to the winning of World War II, and people trained in physics had expanded their range of skills. New opportunities for employment arose in industry and in academic life; physicists had learned to extend their expertise to borderline and interdisciplinary fields as well as to many applications outside the demands of pure research.

Immediately after the war so many young people with experience in war laboratories or with sophisticated equipment in the field returned to study physics professionally that even the expanded market seemed to be satiated in the early 1950's, but things changed. It must be admitted that the Cold War played a not insignificant part in the renewed demand for physicists. The development of sophisticated weapons had not ended with the defeat of Hilter and Mussolini; this effort was if anything enhanced in the 1950's. The physicists active in education were not themselves motivated by the Cold War, but so many humanitarian reasons for improving education can always be found that they were glad to take advantage of the funds available for this purpose.

It is sometimes said that the Soviet launching of Sputnik in October 1957 led to all these efforts; that is not true, but there is no doubt the efforts were spurred by this event, and more Federal financing became available. That the USSR could surpass American technology in this fashion was an unexpected blow to American pride and causes were sought. President Eisenhower commented over a nationwide TV network: "According to my scientific friends, one of our greatest and most glaring inefficiencies is in the failure of us in this country to give high priority to scientific education."[10] Federal support of science education was forthcoming, and physics received the greatest amount, at least at first, partly because physics is basic to the technology required to build and launch missiles.

Until the very late 1950's NSF largely confined its support of science education to pre-college, predominantly high-school, study. The extension to college physics was promoted both within and outside the AAPT, particularly by the MIT contingent spearheaded by Zacharias, already engaged in the Physical Science Study Committee. The establishment of a separate commission was explicitly recommended in the report of three conferences held during 1959–60. The rationale was put cogently by the steering committee: "The development of physics teaching in the United States colleges and universities has largely been the result of individual efforts . . . The increasing role of physics in our scientific progress, in our technology, and in our society and culture, as well as the rapid advances taking place within physics itself, demands consideration of new approaches to the improvement of physics teaching. These should be broadly coordinated and national in scope."[11] The conference report described the basic aims of college physics courses and suggested activities to achieve them. There was a strong recommendation for the establishment of a "Commission for the Improvement of Instruction in College Physics."

A grant from NSF brought the Commission on College Physics into existence later in 1960. The Commission met four times a year, arranged and ran a large number of conferences, issued many publications and encouraged the development of a multitude of teaching aids.

By 1968 it had become evident that a surplus of professional physicists and physics teachers was in the making, and federal support of physics education began to diminish. In January 1969 the Commission was explicitly requested to plan an orderly phaseout, which was finally completed in August 1971. Many of the Commission's activities, duties and responsibilities had either to be taken over entirely by AAPT or abandoned. Zemansky wished to be relieved of his position as executive secretary in 1970, and it was at this time that the Executive Office was revamped to take on the larger role envisioned for AAPT. Wilbur V. Johnson, on leave from Central Washington State College, became Executive Officer in 1970, and opened an office in Washington D.C. He was succeeded by Arnold A. Strassenburg in September 1972, and the office was moved to Stony Brook, where it has remained. The Association's efforts to continue and to expand the Commission's services, coupled with retrenchment on the part of governmental and other sources of funds and the onset of double-digit inflation threatened the stability of the organization in 1973. The journals were particularly vulnerable to inflation: the price of paper rose by 30% in less than two years, and publication costs to the Association increased by 32% in the same period. Fortunately the leadership, notably the president, E. Leonard Jossem, was able to handle the situation. Rather stern measures were called for, and the person who put them into effect was Strassenburg, who managed to continue the expansion of services at the same time. There are Association members who still wince at what they consider his penny-pinching, however much they appreciate his capable and untiring efforts on behalf of the organization. Except for the two regular journals, most of the work of AAPT is carried out by or through the Executive Office.

Some of the activities begun during the life of the Commission were joint with AAPT from the beginning or were assumed entirely by AAPT almost from the start. The preparation of Resource Letters, annotated bibliographies on specific topics, had been introduced by Gerald Holton, then a Commission member. The Resource Letters appear first in the *American Journal of Physics*, and many of them were supplemented by Reprint Booklets containing some of the most useful papers cited in the parent letter.

Another ongoing activity fostered by the Commission is the Film Repository. This service was subsidized by the Commission during its first year, 1969, then taken over entirely by AAPT. Satisfactory film notes are mandatory, as is sound physics and technical excellence. The pricing is such as to cover only the cost of production; no remuneration goes to the maker of the film. The Film Competition has been a feature of the annual meetings in even-numbered years since 1968, alternating with the Apparatus Competition. Winning films in the competition are eligible for the Repository, provided they are accompanied by adequate explanatory notes for instructional use. Another part of the Film Repository is the distribution of sets of 35-mm slides that have been produced and developed by physics teachers. Each set of slides is accompanied by a Teacher's Guide.

Increasingly the Executive Office also makes available documents of other types. These can be categorized as follows:

▶ compilations of information useful to physics teachers, for example an annotated bibliography of films;
▶ reprint books of articles on specific topics from the AAPT journals, for example, *Apparatus for Physics Teaching*;
▶ instructional materials for students, such as a module on the bicycle; and
▶ conference reports, topical listings and journal reprints from a resource known as Information Pool, which was originally maintained by the American Institute of Physics. Typically the need for these products is identified by an AAPT member on committee. The production, marketing and order fulfillment are managed by the Executive Office.

Among the duties of the executive officer is the publication of the *AAPT Announcer*. The *Announcer* was started by Johnson in 1971, and is published four times a year and is sent free to all AAPT members; the May and December issues carry advance programs for the national meetings. The *Announcer* has grown steadily in coverage and importance.

The growth of local chapters

Local chapters of AAPT were authorized as early as April 1931, and the first chapter was recognized in 1932. The rationale for their existence has been primarily to provide meetings accessible to AAPT members and others interested in physics teaching. Individuals may be members of chapters (sections since 1947) without being AAPT members, and many AAPT members are not associated with any section. There are now 37 sections. The newest and one of the largest and most active is the Ontario section; most of the others have boundaries corresponding to states. All sections are represented on the Council of the Association, and the chairman of the section representatives is an influential member of the Executive Board. Sections form a vital and important part of the organization, but they do not help to meet one recurring difficulty: they make no direct contribution to the national treasury.

According to Dodge, who as first president was in a position to know, "finances were a serious problem right from the first minute. One reason was that physics teachers, then and now (1963), don't have much money." I have noted that Klopsteg financed the Oersted Medal in the beginning, and that Palmer advanced the costs of preparing the book on demonstrations; we find that somewhat later Marshall States contributed $500 to help initiate a volume of advanced undergraduate experiments as a memorial to Lloyd W. Taylor. To keep the *Journal* afloat, Richtmyer in 1937 obtained a grant from the Carnegie Corporation: $7500 to be spent over a five-year period.

The Association survived and remained active, but its funds were modest. Sears sometimes recalled in later years that when he became treasurer in 1952 the budget was prepared at a portable blackboard during annual executive-committee meetings. Records show that the first budget he proposed was $18 600. Now the gross budget amounts to more than three quarters of a million dollars and is the result of much advance preparation. It is true that the consumer price index has risen by a factor of four since 1952, but services to members and other physics teachers have multiplied by an order of magnitude. Most of these new services have been introduced within the past ten years, with the expansion of the Executive Office and wider committee activity.

There has been corresponding growth in meeting participation. Not only do meetings provide interesting papers and a forum for members but also tutorials on special topics and a multitude of workshops on many activities—most recently microcomputers, computers and programmable calculators, along with holography, have been especially popular.

It is impossible to put a period to a sketch of only fifty years of activity in behalf of physics teaching. There has never been a time when the variety and intensity of effort on the part of the Association has been greater. The AAPT is celebrating its anniversary by looking to the future—a time that will undoubtedly be more challenging than any before.

References

1. M. Phillips, Phys. Teach. **15**, 212 (1977).
2. J. Frayne, Sch. Sci. Math. **28**, 345 (1928).
3. J. Guernsey, Phys. Teach. **17**, 84 (1979).
4. A. Hull, Science **73**, 623 (1931).
5. K. Compton, Rev. Sci. Instrum. **4**, 57 (1933).
6. F. Richtmyer, Am. Phys. Teach. **1**, 1 (1933) (reprinted in Phys. Teach. **14**, 30 (1976)).
7. F. Palmer, J. Opt. Soc. Am. **7**, 873 (1923).
8. E. Condon, Am. J. Phys. **10**, 96 (1942).
9. V. Knudsen, Am. J. Phys. **11**, 78 (1943).
10. "Eisenhower Speaks on Science and Security," Bull. At. Sci. **13**, 359 (1957).
11. "Report of Conference on the Improvement of College Physics Courses," Am. J. Phys. **28**, 568 (1960).

The giant cancer tube and the Kellogg Radiation Laboratory

The early history of the world-famous nuclear physics laboratory involves a Nobel Prize winning physicist, a wealthy physician, the developer of a million-volt x-ray tube, and the cornflake king.

Charles H. Holbrow

PHYSICS TODAY / JULY 198-

Fifty years ago the W. K. Kellogg Laboratory of the California Institute of Technology was founded as a center of radiation therapy. Seven years later it abandoned medicine to pursue its development into what is today an internationally known center of nuclear physics. The behind-the-scenes negotiations surrounding the laboratory's founding, early history and abrupt change in direction give unusual insight into the administrative style of Robert A. Millikan, Caltech's chief executive. The early history of the laboratory was shaped in important ways by this Nobel Prize winning physicist's successes and failures in raising support money. His efforts, including a 13-year long attempt to take a horse farm away from the University of California, reveal why Millikan was so successful as head of Caltech, and show that it was just as difficult then to get support for pure research in a new field of physics as it is today.

The laboratory, established in 1931 to do research in the x-ray treatment of cancer, included a subordinate program of research in physics. What led

Charles H. Holbrow is Professor of Physics at Colgate University. In the summer of 1980 he was Visiting Associate at the Kellogg Radiation Laboratory of the California Institute of Technology, where this article was prepared.

to the events of 1938, when the physics program entirely supplanted the medicine? The direction of the laboratory, from the time of its founding through the time of this change and beyond, is tied to the interests and personalities of four men.

▶ Robert Millikan was so prominent and so completely identified with Caltech as a whole that we might overlook his special role in establishing Kellogg. His success as a fund raiser launched the lab; his vision fostered it; his influence protected its fragile program of pure research; and, finally, one of his rare failures as a fund raiser freed it from cancer research.

▶ Seeley G. Mudd, the physician son of a wealthy mine owner, played a role largely unknown until now. Yet he was a prime mover in creating the lab, in determining its purpose, and even in administering it. Only after he ended his participation did the lab change sharply in purpose and function.

▶ Charles C. Lauritsen, a Danish engineer, provided the innovation that was reason to create the lab—the million-volt x-ray tube. Also, it was Lauritsen who led the lab through that golden decade of nuclear physics before World War II, when Kellogg established its international reputation as a center of nuclear research.

▶ W. K. Kellogg of Battle Creek, Michigan, the eponym himself, provided the funds to build the Kellogg

Radiation Laboratory in 1931. Although he never fully understood what his money supported, especially as the work shifted more and more to nuclear physics, Kellogg believed in Millikan and contributed the operating expenses of the lab in certain critical years. He also drew Millikan into a bizarre fund-raising scheme to have the University of California shift millions of dollars of its endowment to Caltech for the Kellogg Radiation Laboratory.

Millikan and Lauritsen

In 1931, Caltech was quite a young institution. In 1891 Amos G. Throop, then 80 years old, founded Throop University, which two years later became Throop Polytechnic Institute. By 1901 the Institute consisted of over three hundred students distributed among a college, a normal school, an academy and a grammar school. In 1907, however, the Trustees decided to start over and create a first-class technical institution. They split off the elementary school, hired a new president, and a few years later changed the name to Throop College of Technology.

In 1911, the year the college moved to its present location in Pasedena, 31 students were enrolled. By 1921, the year after the name was changed to California Institute of Technology, enrollment had grown to almost four hundred, but only 15 were graduate students. The next decade, however, saw Caltech move toward eminence in research as the number of buildings increased from three to eight and graduate enrollment grew more than twelvefold. By 1931 the Institute was 11 years old—or 24 years old or 40 years old, but young and vigorous any way you counted its age.

Millikan's arrival in 1921 had much to do with the new Institute's rapid growth. Millikan was lured to Caltech from the University of Chicago by an exceptionally loyal, dedicated and generous group of Trustees who promised strong support for research in physics. Arthur Fleming, Chairman of the Trustees, promised $90 000 a year just for physics research. Also, if Millikan would come, Fleming agreed to pledge his fortune of $4.2 million to Caltech.

Although never officially "President," Millikan was the chief executive officer and was recognized as such. His extensive acquaintances, his personal charm, and his remarkable intuition as to what and who were important in physics enabled him to draw to Pasadena outstanding faculty and visitors: Tolman, Epstein, Lorentz, Einstein, Oppenheimer, to name a few. National Research Council Fellows came to Caltech in substantial numbers, and directly or indirectly Millikan also attracted good graduate students.

In 1926, Charles Lauritsen, a Danish engineer working for a manufacturer of radio sets, attended a lecture in St. Louis by Millikan. Inspired by Millikan, Lauritsen moved his family to Pasadena and at age 34 began work for a PhD. His thesis, completed in 1929, was a study of electron emission from metals in intense electric fields.

Work with high voltages and with field-emission electrons naturally involved Lauritsen with x rays. At that time no one had been able to design a single-stage x-ray tube that would exceed 350 kV. The principal problem was flashover due to the production of regions of high charge on insulators in the tube by backstreaming electrons. Lauritsen produced and patented a design with metal shields to protect the tube from backstreaming, and with a primitive ancestor of the equipotential rings of a Van de Graaff tube to smooth out the potential gradient. His tube could hold over one million volts.

There were obvious medical uses of such a tube. He wrote in his patent application of 1930

I have found that when [1 000 000 volt] potentials are employed, radiations substantially the frequency of the gamma radiation from radium may be obtained from the tube embodying my invention.

Such tubes can therefore be employed as the full equivalent of radium in the treatment of disease, or for therapeutic purposes.

Robert Andrew Millikan and Charles C. Lauritsen (left) stand atop a million-volt x-ray tube in the High Tension Laboratory at Caltech around 1930 or 1931. (Photograph courtesy of the Archives, California Institute of Technology). Figure 1

In fact, even before the patent was filed, Millikan and Lauritsen had begun to launch a program of research on the treatment of cancer with x-rays. Figure 1 shows Millikan and Lauritsen standing on the tube in the High Tension Laboratory where it was built and where the first patients were treated. From this work grew the Kellogg Radiation Laboratory.

Seeley G. Mudd

In 1930 Caltech began to suffer from the Great Depression settling on the United States. Within a year Fleming had lost his $4 million fortune pledged to Caltech. Trustees and donors more fortunate than Fleming rallied to help the Institute pay its debts. For example, Della M. Mudd agreed to delay the construction of the Seeley W. Mudd geology building she was donating in memory of her husband, and allowed the Institute to use the interest on that donation to meet expenses.

The generosity of the Mudds to Caltech is well known and evident. Funds from Seeley G. Mudd built the Seeley G. Mudd Geophysics and Planetary Science Building. The Millikan Memorial Library is also his gift. Both Harvey S. Mudd and Seeley G. Mudd served as Trustees for many years.

What is not known is that the Mudds also played a central role in the creation of the Kellogg Radiation Laboratory. It was natural for them to become involved, for although the family fortune was built on his father's investments in copper and sulfur mines, Seeley G. Mudd, after a brief try at mining engineering at Columbia, took up medicine at Harvard and received his MD in 1924. After three years of internship and residency at Massachusetts General Hospital, he returned home to Pasadena, a physician with experience in radiology as well as useful contacts in the medical profession.

In view of the close relation of the Mudd family to Caltech, it was as natural for Millikan to turn to Mudd as it was for Mudd to be interested in possible therapeutic uses of the 1 000 000 volt x-ray tube. Millikan enlisted Mudd's help in several ways. They decided to undertake a five-year study of the effectiveness of 750 kV x radiation in the treatment of cancer compared to the effectiveness of the 200 kV radiation then available from commercial tubes in hospitals. The plan called for expenditure of about $20 000 a year, roughly half of which would permit Mudd to sponsor and evaluate the treatment of several hundred patients over several years; the other half would fund a supportive program of physical research in which Lauritsen would continue his studies of high-voltage phenomena.

Again the Mudd family acted. Della Mudd, Seeley G. Mudd's mother, anonymously donated $75 000 for the project. (The 1931 dollar was worth six to seven times the 1981 dollar.) Seeley G. Mudd himself agreed to serve without salary as resident radiologist in the treatment center. As it turned out, in several years he also paid a large part of the operating costs of the project.

Mudd did the leg work of assembling a panel of local doctors to oversee the administration of the project, although Millikan wrote the coaxing letters to them. Millikan also saw to the publicity, which was vigorous. Figure 2 is a photograph that was described as showing the "World's Largest X-ray tube." Throughout 1931 and 1932 the headlines of the *Pasadena Star-News*, the *Los Angeles Times*, and the *Los Angeles Examiner* shout the same rhetoric as our own recent "war on cancer." Millikan was never shy about his accomplishments or those of Caltech, and he was very successful in raising money, facts which inspired the famous graffito shown in figure 3.

Kellogg and the laboratory

By late 1930 plans for construction of a building to specialize in cancer therapy were being drawn. Early in 1931 the estimates came in at $94 000, a sum uncomfortably greater than the amount promised by Mrs. Mudd.

Suddenly the fourth founder appears on the scene. In March 1931 the first correspondence between Millikan and W. K. Kellogg appears. The 71-year old Kellogg (figure 4) had come from Battle Creek, Michigan, to winter at his Arabian Horse Ranch in Pomona, California, where since 1925 he had bred

"The world's largest x-ray, a gigantic tube operating under the impulse of 1,000,000 volts of electricity, is being exhibited by scientists at the California Institute of Technology. . . . The ray, which is being used in complicated scientific experiments, is so powerful that it will penetrate two inches of lead, whereas a quarter of an inch of lead has heretofore stopped the most powerful X-Ray . . ." Photograph and text are from a Caltech press release. At the controls are Professors Ralph D. Bennett and C. C. Lauritsen. (Photograph courtesy of the Archives, California Institute of Technology).
            Figure 2

and raised horses for show. In the space of a bit more than two weeks there is an exchange of terse letters, which can be paraphrased as follows:

3 March 1931
Kellogg to Millikan

Meet me at my ranch on Wednesday, March 11.

15 March 1931
Kellogg to Millikan

I like your idea, but for $150 000 I should have some appropriate acknowledgment such as my name on the building.

18 March 1931
Millikan to Kellogg

We can work out something.

21 March 1931
Millikan to Kellogg

The structure will be built and equipped for about $120 000. An additional $30 000 will endow its maintenance.... the building to be known as the W. K. Kellogg Radiation Laboratory, and so inscribed in the tablet to be placed in the center of the east facade of the building.

22 March 1931
Kellogg to Millikan

I agree.

On 27 March 1931 the *Pasadena Star–News* carried the following:

Foundation work was started this morning at the California Institute of Technology on the Radiological Laboratory which is likely, according to experts, to put Pasadena in the front rank of the research centers in radioscopy and radiotherapy.

This must be one of the shortest lapses of time known between a donation for a building and the beginning of its construction. It gives the strong impression that Kellogg became a founder of the lab at the last minute. It also appears that Millikan underestimated Kellogg's potential as a donor. Evidence for this is the fact that the next year Kellogg donated his 700 acre horse ranch and $600 000, together worth about $3 000 000 to the University of California. As you will see, there is reason to think that for Millikan this was "the big one that got away," and it bothered him all the rest of his life.

But the laboratory was launched and construction was rapid, although full use of the x-ray equipment was held up because the two 750 000 V transformers were slow in coming from General Electric in New York. When they did come one had to be returned as faulty. Still, the board of trustees approved the budget of what Seeley G. Mudd called "The Seeley W. Mudd X-Ray Research Fund Program." Mudd was appointed Research Associate in Radiation to administer the biology half of the program. Figure 5 is a copy of his picture in the 1933 Caltech yearbook "Big T" which accompanied a description of his

work administering the treatment of cancer patients in Kellogg. The first patient was treated in the Kellogg Laboratory in September of 1932. While the laboratory was being readied, patients continued to be treated in High Volts, as the adjacent High Tension Laboratory was usually called.

Figure 6 shows the large interior room of Kellogg laboratory. You can see the thirty-foot long x-ray tube coming down through the treatment room. A rubber hose filled with water was used as an electrical conductor to carry the electric current from the high-voltage transformer to the x-ray tube. The potential drop was about 50 000 V. You can see the hose in the photograph; the pie tins are corona discs, which smooth the potential gradients. Often a spark would puncture

the hose and water would stream out. It was the job of the physicists to keep the tube in working condition, so a physics graduate student would get into a bosun's chair and swing out to replace the damaged section of hose.

In the treatment room (figure 7), vestiges of which remain today, four patients could be treated at a time, two sitting and two lying down.

From 1931 until 1936 Kellogg was devoted largely to radiation therapy, but there was another life that flourished among the "maintenance crew," the study of nuclear physics. Actually most of this work went on in the adjacent High Volts building and was not, strictly speaking, part of the Kellogg laboratory.

In High Volts, Lauritsen and Crane, reacting to the reports of Cockroft and

Graffito in recognition of Millikan's activities as a fundraiser for Caltech (circa 1937). Millikan was a controversial and very public figure. (Photograph courtesy of the Archives, California Institute of Technology.)
Figure 3

Walton's success, modified an x-ray tube to accelerate ions. They produced artificial radioactivity, and in January 1933 the *Pasadena Star–News* reported that Lauritsen and Crane had produced neutrons with an accelerator. They used the reaction

$$Be^9 + \alpha \rightarrow C^{12} + n$$

and detected the neutrons by putting paraffin linings into electroscopes invented by Lauritsen for his x-ray work. This was quite an accomplishment, considering that James Chadwick had only reported the discovery of the neutron a few months earlier. The excitement of the opening of an entire new field of research captured most of the interest the physicist might have had in x-rays for therapy. Millikan and Lauritsen in their minds had always kept the physics separate from the cancer treatment. In the correspondence eliciting support from physicians and money from W. K. Kellogg, Millikan seldom mentioned basic research, but all the publicity releases were clear on this point.
were clear on this point.

For example, the *Los Angeles Times* of 4 August 1931 says

> The largest and most powerful instrument ever devised for splitting the atom and combatting cancer was installed today in the California Institute of Technology new radiation laboratory. . . .
>
> The primary object of the institute's x-ray program, however, was to learn about the physics of high-speed electron particles.

And there was even a mention of the possibility of producing nuclear disintegration.

The doctors did cancer research; Caltech did physics. Nevertheless, the X-Ray Research Program supported Lauritsen with part of his salary, with equipment, and with postdocs and student assistants. The Caltech High Potential X-Ray Research expense sheets for 1934 show that Millikan understood overhead—he knew how to extract indirect costs from a grant before the term was invented. A half dozen students were budgeted for the year—among them one William A. Fowler—along with two post docs and equipment for "High Potential X-Ray Physics." Thus the "X-Ray Research Fund" paid tuition for graduate students and stipends for researchers who did nuclear physics most of the time and kept the x-ray tube running on the side.

Someone else's money

For the first five years the lab was supported by the money from Mrs. Mudd, but as 1936 approached Millikan had to search for more money for the lab. Naturally enough he went back to Kellogg, who aroused Millikan's hopes by holding out the prospect

Will Keith Kellogg of Battle Creek, Michigan (circa 1928). (Courtesy of the W. K. Kellogg Foundation).   Figure 4

of a $3 000 000 gift to endow the lab. These hopes launched Millikan and Kellogg on a bizarre fund-raising effort.

The trouble was that Kellogg had already given that money to someone else. Kellogg proposed to retrieve his $3 000 000 gift from the University of California and give it to Caltech! He was not happy with the way the University was running his horse farm; he felt they had not lived up to the spirit of the conditions he had placed on the gift and that they should return it and its endowment.

Millikan was realistic enough to foresee difficulties. He politely suggested that Kellogg not tell the California Regents that he was retrieving his property in order to give it to Caltech. Millikan also wanted his name left out of any negotiations to reconvey the property.

In December or 1936 Kellogg asserted his claim to the Regents. He wanted his gift back, he said, and went on

> I have in mind the gift of this property and endowment to California Institute of Technology, a California Corporation, primarily for the support of the Kellogg Radiological Laboratory established about five years ago and which is engaged in scientific and medical research in the field of radiology, particularly as applied to the treatment of cancer.

The suggestion was coldly received, and it provided meat for a wrangle that went on for thirteen years.

The first year was exciting. Kellogg marshalled his lawyers; Millikan arrayed his trustees. Together they probed the Regents. Millikan and University of California President Robert Gordon Sproul jousted politely, Sproul

holding out hopes but giving nothing away. When rumor came that only one particular Regent, a devout Catholic, was the focus of opposition, Millikan visited Archbishop Cantwell and asked him to intercede. The Archbishop sent back the message that the board was "unanimous in its opposition to the transfer." They tried to use Herbert Hoover's influence, but to no avail. Kellogg even suggested to Caltech trustee Harry Chandler, publisher of the *Los Angeles Times*, ". . . a short program which if carried on in the proper way through newspapers, might have some effect on the Regents. . ." but nothing came of it.

Kellogg donated $10 000 to keep the lab going for another six months, but after seven months of failing to get back his gift from the University of California, his spirits sank. Millikan tried to rally them in a long letter:

> . . . the transmutation of the elements, now an accomplished fact, plus artificial radioactivity, plus neutron beams—all three effects producible by Lauritsen tubes—plus the manifold uses of ultra-short wireless waves, therapeutic and otherwise, open up endless opportunities which with the addition of brains, persistence, energy and some financing should keep the Kellogg Radiation Laboratory for many years to come as it has been for the past six years an unexcelled center of physical and biological progress, exerting perhaps, as beneficent and as far-reaching an influence as any activity of the Kellogg Foundation.

This is one of the few times that Millikan is so explicit to Kellogg about the nuclear physics that went on in the lab in parallel with x-ray treatments. He concluded by suggesting that Kellogg simply endow the lab from his own money and work out the details of retrieving his property from the University of California later. Without additional operating funds, Millikan declared the lab would cease to operate on 1 November 1937.

Millikan's urgency is even more evident in his suggestion to the Caltech trustees that they negotiate some sort of compromise with the Regents:

> It isn't merely the cancer work that is at stake, it is the whole of Lauritsen's nuclear physics work which is as important as anything being done now in the country.

In fact, there was some reason to hope for compromise. The gift to the University of California required that the Arabians be bred and shown forever, but the endowment did not cover the costs. The ranch was a drain on the University budget, and it seems that Sproul was willing to return part of the gift in return for unrestricted use of the remainder. The Regents, however, foresaw the long-term value of land in Southern California; besides, they were not about to do Caltech any favors. After all, if Kellogg was so interested in radiation laboratories, they had one of their own at Berkeley; let him shift his gift there.

In October of 1937 Millikan admitted to Kellogg's lawyer that the prospect of reconveyance of the land looked hopeless. As he wrote a few months later to Kellogg

So far as my own activities are concerned, I have been virtually informed by one of the regents of the University of California that it would be wise for me personally to forget it.

The end of cancer research

It was time to open a new approach to funding the laboratory. Millikan went back to a traditional Caltech method. Find the best person available in the field, lure him to Pasadena for the winter, impress him with the quality of work at Caltech, and then use his reputation to garner further support. So with $5000 from Kellogg, Millikan and Lauritsen brought the eminent French radiologist Henri Coutard to Caltech.

Millikan now began to explore funding from the National Cancer Advisory Council of the National Academy of Sciences. He wrote to Dr. Ludvig Hektoen, the program's director, selling the need for studies of radiation therapy. Then, with some encouragement from Hektoen, Millikan drew up a proposal. Heavily using Coutard's name, Millikan asked for $62 500.

In the meantime Millikan went back to Kellogg. Almost 78 years old and in the throes of a series of operations for glaucoma, which left him blind for the remaining 13 years of his life, Kellogg was occasionally a bit testy at Millikan's importunings. He complained of the more than $2000 in legal fees in the fruitless effort to retrieve the gift from the University of California; he mentioned his painful and unsuccessful eye operations; he asked Millikan to leave him alone; and in January 1938 he wrote, "I am not prepared to commit myself for any further contribution toward the x-ray laboratory for the present year."

Millikan, imperturbable as ever, thanked him for his generosity, told of his efforts with Hektoen and promised to press on. But in February the proposal to the National Cancer Advisory Council was rejected. Some of the reasons for the rejection surely had to do with the growing recognition that x-ray therapy was not very effective.

The pathologist associated with the Kellogg Lab had written a report to Hektoen that seemed to show the treatments did little good and may have done harm:

Seeley G. Mudd in his office in the Kellogg Radiation Laboratory, about 1933. (Photograph courtesy of the Archives, California Institute of Technology).					Figure 5

I have no evidence in any material that any case of prostatic carcinoma has been destroyed by radiation therapy. There have been a few cases where no tumor was found at autopsy but there was profound necrosis in the prostate, apparently following radiation and trans-urethral resection, involving the entire organ so that no viable tissue remained.... Intense fibrosis has been produced in many of the cases and perhaps in some this has been very excessive—to the detriment of the patient. This particularly happened in the first years when the matter of dosage and length and number of exposures was not as well worked out as at present...

We have had much happier results in cervical carcinoma. We have had a few cases from whom we took repeated biopsies and have seen all the stages of cellular disintegration and destruction of cells as described by those who have used radium and 200 kV x-ray.... At the same time we have had plenty of cases where the tumor has progressed in spite of our attempts to destroy it.... However, I see only the dead cases representing the mistakes, failures and disappointments. The palliation that many patients get, the prolongation of life and other factors which our clinicians believe occur in a large number of patients, I do not see.

The tone of the pathologist's report of the results of seven years of research could hardly have been persuasive to Hektoen. But Millikan did not give up. He next went to the recently endowed Childs Foundation at Yale and prepared to approach the DuPont

family and to try Kellogg again.

But then came a change that spelled the end of the program of cancer treatment. Seeley G. Mudd, now Professor of Radiation Therapy, had served the project since it began. He saw the research as played out, and in April he wrote to Millikan, then in the East, to ask him to find a replacement,

... a highly competent and conservatively minded roentgen therapist to sit on the lid at Kellogg and keep the clinic running smoothly.

... You will recall that I am attracted to the biological and chemical angle in cancer research more than to attempt to make minor modifications in the existing therapeutic techniques using supervoltage irradiation.

After the rejection of his proposal by the Childs Fund in mid-May, Millikan had only one hope left. He went back to W. K. Kellogg. He wrote a long letter regretting the stubbornness of the Regents and sadly informing Kellogg that the lab would have to close on June 30. He recapitulated the accomplishments of the program but gave his description a twist that showed he had accepted the end of Seeley Mudd's support of the program.

Over seven years the program had cost about $1500 a month, Millikan told Kellogg. "Rather more than half of this had been required for the direct cancer treatment program,..." But for the future

Dr. Lauritsen's part, which has consisted in the development of the new physical techniques, must not be discontinued, and I am making vigorous efforts to find means of helping him at his present promising activities in the development of new radio-active elements and other nuclear physical problems, which have good prospects of extending still further the beneficent effects of radiation in the treatment of cancer and other human ills. He has just built a new tube which should be capable of working at two and a half million volts, and this tube in the hands of himself and his pupils, is pretty certain to open up new results in the field of nuclear disintegrations and atomic transformations, all of which are promising for the future of radiation therapy in the broad sense.

The "future of radiation therapy in the broad sense" is the precursor announcement of the change then under way in the Kellogg Radiation Laboratory. There is a certain lack of candor in referring to the 2-MV Van de Graaff as a "tube" as though it were just a further development of Lauritsen's x-ray work, but there is real ingenuity in presenting the study of the structure of light nuclei as promising for radiation therapy "in the broad sense," because strictly speaking it was true.

On 1 June Kellogg responded generously, asking Millikan if $10 000 would keep the lab going for another year. We can only speculate what would have happened if Kellogg had offered the $18 000–20 000 needed to maintain the full program. Or what the lab would be like today if the Childs Fund or the National Academcy of Sciences had supported Millikan's program. Kellogg's partial offer, however, allowed Millikan a graceful exit from the treatment program.

His 9 June 1938 reply to Kellogg is Millikan at his best. With $10 000 they could only support part of the program. Dr. Mudd would complete a statistical study of the nearly 800 cases already treated.

Dr. Lauritsen and his group, on the other hand, are eager to push up to higher potentials by new techniques with the aid of which there is the possibility that new radioactive substances may be artificially produced. If successful, this procedure may make the use of very penetrating rays much cheaper and vastly more convenient than it is when the patient must be brought to some point at which an expensive high potential tube exists. We propose then, on July 1st, to discontinue the use of the present tube for the time being and build, in the big room of the Kellogg Laboratory, a modified tube which will go to considerably higher potentials.

Depending on how things worked out, treatments might be resumed later with the new tube or the old. (And again there is a certain disingenuousness in Millikan's description of the Van de Graaff as a "modified tube.")

With unusual precision, then, we have dated the moment when the cancer treatment research was recognized as played out and the nuclear research that had been burgeoning on the periphery of Kellogg took over. The laboratory's destiny now lay firmly in the hands of the physicists.

Epilog

This would be the point to go back and trace the history of the nuclear physics that built the lab's worldwide reputation, but that fascinating story is for another chapter. Instead, we will tie up some loose ends by detailing the rest of Kellogg's contributions to the lab and telling what became of his project to give to Caltech what he had already given to the University of California.

In 1939 Millikan and Mudd visited Kellogg and reported the work of the lab. Kellogg offered $8000 that year, but announced to Millikan that the money was to be considered his last contribution to the program.

The central hall of the Kellogg Radiation Laboratory in 1933. Inside the cement balcony in the middle of the photograph is the treatment room, shown in figure 7. (Photograph courtesy of the Archives, California Institute of Technology).   Figure 6

But Millikan would never give up. A year later he sent a detailed report to Kellogg on the uses of the money. Acknowledging that the previous year's donation was to be Kellogg's last, Millikan asked him to suggest a new "friend" for the lab. Kellogg suggested the Kellogg Foundation, which, after a visit by Millikan to Battle Creek, gave $8000.

In 1941 there was another attempt to make a deal with the University of California. A fifty-fifty split was offered. President Sproul had a stormy session with Kellogg, who later wrote to Millikan

I regret the outcome of this contro-

versy, but hope that at some future time Dr. Sproul and his Regents will "see the light" and be willing to share this property with Cal Tech.

Emory Morris, the President of the Kellogg Foundation, washed his hands of the business.

Throughout 1940 and 1941 the nation was mobilizing for war. The changes at Kellogg Lab were dramatic as the lab became a major design and production center for rockets. Lauritsen headed a rocket project staff of more than 3000. How completely the lab had become a government facility was lost on Morris who in 1942, on behalf of the Kellogg Foundation, asked for a progress report on the use of the $8000 and offered more.

Millikan's reply is interesting. All the justifications for the support of research reduce to three: Support research for its useful applications; support research for the love of knowledge; support research because the people who do it are a valuable resource to be cultivated for a time of need. In all his dealings with Kellogg, Millikan had hammered home the theme of useful applications. Only rarely did he allude to the marvelous discoveries that were being made at the laboratory. Only by implication did he advocate support of Lauritsen for his unique qualities as a scientist.

Now, however, with the war on, the value of these physicists as a national resource was clear. In his response to Morris he developed the theme:

The outstanding place which that laboratory has taken during the past twelve years has been practically wholly due to the extraordinary effectiveness of Charles C. Lauritsen and his very able collaborator, William Fowler. My main concern in all my talks during the last half dozen years with Mr. Kellogg and the officers of the Foundation has been to arrange conditions such that this team could be kept working at maximum efficiency....

No matter what sort of jobs are assigned to them, a team of the Lauritsen–Fowler type is a great rarity, and it is a great credit to the laboratory that it has produced and maintained them.

The widespread acceptance of this approach after World War II led to the federal support of science on a scale never previously imagined. Millikan, with characteristic intuition, had grasped the central argument that would dominate postwar science.

Although the Foundation under Morris' leadership donated $15 000 to the laboratory in 1942, Millikan had no similar success after the war. Morris finally called a halt to Millikan's efforts in 1948 by writing

If the W. K. Kellogg name on the

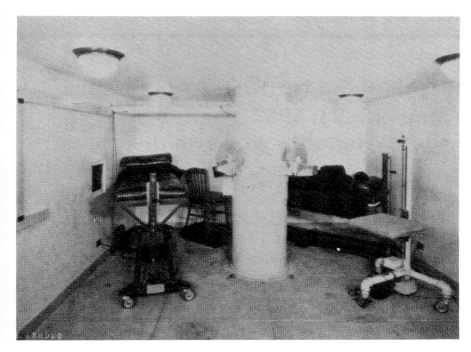

The x-ray treatment room in the Kellogg Radiation Laboratory around 1932. The x rays were generated when electrons from a filament in the base of the tube (below the treatment room) struck a target placed at the level of the tube portals (seen in the center of the photograph). Four patients could be treated at one time. (Photograph courtesy of the Archives, California Institute of Technology).

Figure 7

laboratory is going to jeopardize in any way the future support of the laboratory there should be no hesitancy on your part in removing the name on the laboratory in favor of any individual who would desire to create an endowment to perpetuate the work you are doing. I have conferred with Mr. Kellogg on this matter and he concurrs [sic] with this opinion.

Millikan made one last try. He kept up his annual visits to Kellogg. The two octogenarians would meet each year at Palm Springs where Kellogg now wintered. Millikan would report to him the activities of the laboratory and they would discuss more cosmic philosophical matters. It was through such contacts that Millikan learned that he had one more chance at the Kellogg Ranch.

In 1943 Kellogg had sent Millikan a copy of a letter from Henry L. Stimson, Secretary of War. The letter thanked Kellogg for arranging the donation of the Pomona ranch to the Army. "The gift will be a great asset to the Army horsebreeding plan." In return for donating the ranch to the Army, the University of California received clear title to the ranch's endowment. The University of California and Kellogg,

under the goad of patriotism, had finally come to an agreement.

After the war the government put the ranch up for sale. Kellogg was outraged at this violation of his generosity. His complaints and petitions from influential friends persuaded the government to give the ranch to the W. K. Kellogg Foundation.

Millikan learned of these events after a visit with Kellogg in early 1949. Writing to Morris of his visit with Kellogg, Millikan said

He asked me particularly to drop you a note to suggest that if the Kellogg farm, which is now reported as being turned back to the Foundation and has the possibility of being given by the Foundation to some philanthropic institution, the California Institute of Technology should, in view of past history, have first place in the picture.

Kellogg had deliberately minimized his influence with the Foundation from its start in 1930. By this time it was nil. The Foundation gave the land to be the Pomona campus of the California Polytechnic State University. Today Cal Poly breeds and shows the Arabian horses descended from Kellogg's original herd on the land that despite Millikan's Herculean efforts never became Caltech's.

* * *

With a few exceptions all the quoted materials in this article are from the Robert A. Millikan papers in the Archives of the California Institute of Technology. I appreciate the easy access to these papers and photos, which have been brought to a high state of organization under the leadership of the Institute's Archivist, Dr. Judith Goodstein. I am especially grateful for the help of Dr. Goodstein's assistants Susan Trauger and Carol Finermann.

the
evolution
of the
OFFICE of NAVAL RESEARCH

By **The Bird Dogs**

PHYSICS TODAY / AUGUST 1961

IT is not often that the birth of a Navy office, which certainly sounds like like a cold, administrative affair, makes history worth recording. But the birth of the Office of Naval Research was such an interesting one, participated in by so many famous and brilliant personalities, that a record of the events should serve a useful purpose. It might even bring inspiration to those who daily continue the struggle to evolve constructive changes in large government departments.

Soon the Office of Naval Research (ONR) will be celebrating its fifteenth anniversary. If the celebration is anything like the tenth reunion affair, a banquet will be held and a number of speeches will be made extolling the aims, purposes, and accomplishments of this remarkable office. Included will probably be a few remarks concerning the history of the formation of ONR.

Any reference to the history of ONR excites in the authors two responsive chords. The first is one of nostalgia brought on by fond and fascinating memories. The second is one of frustration caused by the realization that an authoritative history of the evolution of ONR has not, heretofore, been made public. Hence this attempt.

The campaign to sell the concept of establishing a central office to foster basic research and research coordination within the Navy Department was a lengthy, and sometimes bloody, struggle. The story of the evolution of ONR is really the tale of an educational process carried on over a five-year span (providing we are permitted to ignore pre-World-War-II struggles). This educational process required the concerted efforts of many people to create an atmosphere in the Navy Department, in the Executive Branch, and in Congress, which was favorable toward long-range research. Key people had to be convinced that future military strength depends to an increasing degree on the rapid and effective development of new weapons and weapons systems through a strong, balanced research effort.

It is recognized that history must be recorded from several points of view before all the facts are exposed. The story here presented was that as seen by a small group of Naval Reserve officers who were fortunate enough to have had a five-year worm's-eye view of the entire evolution of ONR from a vantage point within the Office of the Secretary of Navy. We were, in the parlance of the day, lowly skippers of LSD's (Large Steel Desks).

The Background

IT all began before the United States entered World War II, with the realization by such outstanding men of science as V. Bush, J. B. Conant, K. T. Compton, and F. B. Jewett, that this country was woefully weak in military research and development. Dr. Bush carried the idea of establishing a National Defense Research Committee "to coordinate, supervise, and conduct scientific research on the problems underlying the development, production, and use of mechanisms and devices of warfare (except problems of flight which were to remain under the NACA)" to President Franklin D. Roosevelt and Mr. Harry Hopkins early in June 1940. The White House acted rapidly and on June 15, 1940, the President signed letters appointing such a Committee with Dr. Bush as chairman. The Committee was to supplement rather than replace the activities of the military services so that links with the military were formed by the naming, as members, Brig. Gen. G. V. Strong of the Army and Rear Adm. H. G. Bowen of the Navy, in addition to K. T. Compton, J. B. Conant, F. B. Jewett, R. C. Tolman, I. Stewart, and C. P. Coe.

To further mobilize the scientific personnel and resources of the nation, President Roosevelt established by Executive Order on June 28, 1941, the Office of Scientific Research and Development. This group had as an Advisory Council Dr. Bush as Chairman, Dr. Conant, Chairman of the NDRC, Dr. J. C. Hunsaker, Chairman of the NACA, Dr. A. N. Richards, Chairman of the Committee on Medical Research, and one representative each from the Army and Navy appointed by the respective Secretaries.

The impact of this move led Secretary of Navy Frank

A recent photograph of several of those who took part in the early development of ONR. (The names of the original and second-wave "bird dogs" are italicized.) Front row, left to right: *Bruce S. Old, Ralph A. Krause,* R. Adm. Julius A. Furer, USN (Ret.), Jerome C. Hunsaker, *James H. Wakelin.* Back row: N. S. Bartow, Royal C. Bryant, *John T. Burwell, H. Gordon Dyke,* A. C. Body, *Thomas C. Wilson, James P. Parker.* The principal author of the present article is Dr. Old, who is now senior vice president of Arthur D. Little, Inc., Cambridge, Mass.

Knox to study what steps the Navy might take to increase its effectiveness in the prosecution and utilization of research and development.

There existed some controversy on this point. Rear Adm. H. G. Bowen, Director of the Naval Research Laboratory, had recommended on January 29, 1941, the centering of all research for the Navy in that Laboratory, giving it Bureau status; whereas the General Board in a rebuttal on March 22, 1941 had recommended that no change in Bureau cognizance for research be made and that the Chief of Naval Operations be made responsible for all research policies, including the operation of the Naval Research Laboratory.

Secretary Knox, at the suggestion of Rear Adm. J. H. Towers, therefore enlisted Prof. J. C. Hunsaker, the Chairman of the NACA as well as a member of the OSRD, and a graduate of the Naval Academy, to advise him. Out of this advice arose the first step in the long road to ONR. At the suggestion of Hunsaker, Knox issued General Order 150, July 12, 1941, which established the Office of the Coordinator of Research and Development in the Office of the Secretary of the Navy. This order provided that the Coordinator advise the Secretary broadly on matters of Naval research, and placed the Naval Research Laboratory under the cognizance of the Bureau of Ships.

The Office of the Coordinator of R & D

AT the urgent request of Secretary Knox, Dr. Hunsaker agreed to serve as the first Coordinator of Research and Development on an interim basis in order to get the Office organized and functioning. He was named Coordinator on July 15, 1941, and immediately selected a small staff consisting of two highly capable regular officers, Capt. Lybrand P. Smith and Comdr. E. W. Sylvester, and four young Naval Reserve officers having technical backgrounds. Hunsaker then proceeded to inspire these young men, whom he called "bird dogs," and train us in his effective manner in the basic elements of sound research program planning, administration, evaluation, and coordination. Another important facet of this training concerned the ways and means of getting things accomplished in wartime Washington in the face of odds, or even open opposition. With tongue in cheek, Hunsaker often asked the "bird dogs" to investigate situations and prepare brief memoranda. These he then waved around in the stratospheric secretarial or bureau-chief level to show that his position was obviously correct if even green reserve officers could quickly reach the same conclusion. (This procedure had a remarkable effect on the care with which memoranda were prepared, and on the morale of the staff through the display of confidence it represented.)

In order to carry out more efficiently his prior commitments to the NACA and OSRD, Hunsaker resigned the position of coordinator and turned it over to his carefully selected choice, Rear Adm. J. A. Furer, USN, on December 15, 1941. However, Hunsaker's superb advice and counsel always were, and at this writing still are, available to and continually utilized by the Navy Department. Other changes included the naming of Furer to the OSRD, Smith to the NDRC, the acquisition of Comdr. R. D. Conrad, USN, a truly brilliant technical man, as a replacement for Sylvester, and the addition of two more technical Naval Reserve officers.

During the first three years of World War II the work of this Office was aimed almost entirely at liaison between the NDRC and the Navy, assisting in the planning and establishment of research projects, following the progress thereof, and aiding in bringing about the utilization of the results by the Navy. (The effectiveness of this work under Adm. Furer's guidance is noted by J. P. Baxter in his history of the OSRD and NDRC entitled *Scientists Against Time*.) In addition, the Office coordinated Navy research efforts with the War Department, War Production Board, Coast Guard, National Advisory Committee for Aeronautics, National Research Council, and the United Kingdom and Canada.

However, from the very outset another important subject occupied the thoughts of the personnel of the Office of the Coordinator of Research and Development. All of us knew that the excellent OSRD-NDRC civilian research groups would probably evaporate as soon as the war ended. Therefore, at each step of the way, a gnawing thought occupied the minds of all: how could the Navy better organize and administer its *own* research? The Navy must be capable of developing the impressive and awful strength required to discourage any potential enemy to the end that the Navy could assist in avoiding further wars, or, at a minimum, avoid entering any future war without having all the advantages effective research could provide in modern weapons and weapons systems. Hunsaker sparked this thinking in 1941 within a matter of days after establishing the Office. One of the first tasks he assigned the group was a study of various Navy laboratories in order to determine wherein the Navy might be able to handle research work effectively so as to lessen the burden of the NDRC, and to determine wherein its research capabilities were lacking.

A searching analysis of Navy research strengths and weaknesses was actually a continuous task which came into consideration in practically all the work of the Coordinator's Office. In establishing liaison with the Bureaus, Offices, and Laboratories of the Navy Department, and in naming Navy liaison officers to the numerous NDRC projects, a rapid evaluation of the attitudes and capabilities of people was obtained—whether they were civil service, ensigns, or admirals. In a remarkably short time it was possible to categorize those persons who would do everything possible to stimulate better research programs, organization, and utilization of results, and those who stood firm upon the twin, and usually backward, defenses of cognizance and entangling red tape.

Fortunately within the Navy there arose almost immediately a solid core of highly intelligent people who welcomed and assisted the drive to push research on all frontiers. This was in part a tribute to the well-established Navy system of postgraduate study in various universities which had developed in many officers an understanding and appreciation of science.

This is not to say that all was peaches and cream. Well do we remember the time early in the war when we called in a top submarine officer and pointed out to him the magnitude of the US and UK antisubmarine

research effort. We postulated that the enemy was also doing work and out of it would come developments, like a homing torpedo, which would make life miserable for the submarines. Would he help us spark some pro-submarine research? Absolutely not—the subs in the Pacific were having a field day. This type of short-sighted thinking we paid for dearly later. We also remember probing into touchy areas, such as out-dated torpedo power plants, only to have officers rise in indignation based on rights of cognizance or secrecy. In fact, we had to develop a defense technique. Whenever we ran into a particularly salty, operational type who was bellowing in a manner destined to hold up the progress of research, we took out notebook and pencil and asked dutifully, "Would you mind repeating that statement so I could be certain to quote you correctly to the Coordinator of Research? He will be interested in your view, sir." This technique worked wonders.

We continually sought out and nurtured the progressive, intelligent core group. One of the first persons to be uncovered who showed vital interest in the postwar reorganization of research in the Navy was George B. Karelitz of Columbia University, a former Russian, who was working with the Bureau of Ships.

Two of the "bird dogs" began in 1942 to meet bimonthly at home in the evening with Karelitz. Tragically, Prof. Karelitz died in 1943, but fortunately not before he contributed immensely to the shape of things to come. Out of these sessions the initial pattern of ONR was almost completely conceived—the essential elements consisting of establishing a central research office in the Office of Secretary of the Navy, headed by an admiral, receiving funds from Congress for research projects, and having a powerful research advisory committee made up of top scientists, were actually all drawn up and recorded as early as November 1, 1943 by two "bird dogs." Of course much ground work involving numerous persons both within and outside of the Navy remained to be accomplished before any such plan could become a reality.

Since any research organization requires a sympathetic atmosphere in which to live, if it is to survive and be productive, the work of a large number of people during the first three years of World War II must be credited for setting the stage in the Navy Department for the later establishment of a central research office. Among the scientists who helped so materially in "selling" the importance of continued research to the Navy during this time were: Bush, Conant, Compton, Jewett, DuBridge, Adams, Rabi, Tuve, Tolman, Hunsaker, Terman, Loomis, Tate, Zacharias, Hunt, Kistiakowsky, Lauritsen, Morse, Stevenson, Suits, Ridenhour, Alvarez, Land, Kelley, Buckley, and Wilson.

And among those most receptive officers in the Navy Department who were sold and in turn helped to prepare the Navy for subsequent reorganization were: Briscoe, Bowen, Furer, L. Smith, Solberg, McDowell, Tyler, Schuyler, Entwisle, Bennett, Sylvester, Conrad, Lee, Hull, Dowd, Baker, Low, Cochrane, Mills, Ingram, Rickover, Piore, Bollay, Thach, de Florez, Strauss, Berkner, Teller, Lockwood, Kleinschmidt, Hatcher, Pryor, and Schade.

Certain official acts also occurred which spread the word on research in a fairly effective manner to the various Bureaus and Offices of the Navy Department. One mechanism was the establishment of the Naval Research and Development Board, headed by Furer, which consisted of the various Bureau research heads and the Readiness Division of Cominch, and had a "bird dog" as secretary. An interesting aspect of this work which soon arose was the need for better technical intelligence in order to aid in the rapid development of countermeasures to new enemy weapons. A new group was quickly established in the Coordinator's Office with two additional "bird dogs" to assist in this important work, and liaison was established with the Office of Naval Intelligence, G2 of the Army, and the Office of Strategic Services. The group performed outstanding work in piecing together data on German torpedo, ram jet, and guided missile work which resulted in the initiation of important new projects in the US. Out of this grew an awakening in the Navy to the important part scientists could play in intelligence. As a result the Navy sent one of the "bird dogs" on the first famous Alsos Mission, and later set up the highly successful Navy technical intelligence mission to Europe.

Plans for the Postwar Era

RECOGNIZING fully the almost total dependence of the Navy on NDRC research at this time, Rear Adm. Furer, as early as the fall of 1943, began to worry about what would happen to research in the Navy if the war suddenly ended. The first trial balloon was hoisted by him September 22, 1943, when he sent a memorandum to Vice Adm. F. J. Horne, Vice Chief of Naval Operations, suggesting a revision of General Order 150 giving a few expanded coordinating powers (but no money) to the Office of the Coordinator of Research and Development. Also he proposed that the Naval Research Laboratory be transferred back to the Secretary's Office. In this memo Furer invented the term "Chief of Naval Research" which was ultimately adopted as the title for the head of ONR. However, at this time the several Bureaus raised a large howl based on cognizance, and Admiral King was in no mood to favor any more power in the Secretary's Office, so the whole matter was dropped like a lead balloon.

Other factions were also beginning to awake. Mr. James Forrestal, Under Secretary of the Navy, on October 2, 1943 requested R. J. Dearborn, President of Texaco Development Corporation, to "make an extensive survey of Navy patent practices and the research situation of the Navy." Mr. Dearborn reported his findings on March 10, 1944 in which he recommended the "establishment of an Office of Patents and Research to be headed by a Coordinator of Patents and Research (which does not conflict with the present activities of the Coordinator of Research)." Mr. Forrestal was reluctant to take immediate action.

The Pot Begins to Boil

BUT, the lid could not be kept on the pot much longer. Dr. Bush warned the military members of the OSRD Advisory Council in the Spring of 1944 that, "The OSRD is a temporary war organization which automatically goes out of existence at the end of this war; so that in planning for peacetime research and development, we plan without that organization, which will presumably turn its affairs over just as soon as the war begins to end." This official warning note was a loud reminder that the scientists would in all probability flock back to their own laboratories as soon as the war appeared to be definitely won.

Furer immediately got busy and organized a large conference in the Navy Department on April 26, 1944, to discuss the problem. To this conference were invited all the top Army, Navy, and OSRD research personnel, some 43 in number. There was general agreement that the military would probably not be able to retain the interest of top scientists or obtain funds necessary for a vigorous research program. It was decided a committee should be established to study and recommend a proper organization for postwar military research. In retrospect, one of the most important points of the entire conference was one of omission. Not one person in the course of the meeting hinted in any way about the tremendous revolution soon to be thrust upon military research requirements by the advent of guided missiles, complex weapons systems, and the like.

As a consequence of this meeting Secretaries Stimson and Forrestal appointed a Committee on Post-War Research composed of Charles E. Wilson, chairman, four civilian scientists (Jewett, Hunsaker, Compton, and Tuve), four representatives of the War Department (Echols, Waldrin, Tompkins, and Osborne), four representatives of the Navy Department (Furer, Cochrane, Hussey, and Ramsey), and two secretaries, including one of the "bird dogs". At the first meeting on June 22, 1944, Chairman Wilson read to the press and news reels the following statement:

> The purpose of the Committee is to prepare a plan and organizational procedure which will insure the continued interest of civilian scientists after the war in scientific research for the Army and the Navy. The nation's scientists have been doing a splendid job since Pearl Harbor, and our task is to evolve a plan which will assure their continued interest in meeting the research needs of our Armed Forces after the War. In this way only can the United States keep ahead of all possible future aggressors in preparedness for National Defense.

The often heated deliberations of this committee finally resulted in the recommendation (drafted by Hunsaker) that an interim organization be established in view of the fact that Congress was considering several bills to create a new independent research agency

The first meeting (June 22, 1944) of the Committee on Post-War Research. Those present, left to right, were Col. R. M. Osborne, R. Adm. G. F. Hussey, Jr., K. T. Compton, Brig. Gen. W. F. Tompkins, R. Adm. J. A. Furer, F. B. Jewett, Charles E. Wilson, J. C. Hunsaker, Maj. Gen. O. P. Echols, R. Adm. E. L. Cochrane, M. A. Tuve, Maj. Gen. A. W. Waldon, and R. Adm. D. C. Ramsey. *US Navy photo*

to which the functions of the interim organization might better be transferred later. Accordingly, in a joint letter on November 9, 1944, Secretaries Stimson and Forrestal requested the National Academy of Sciences to establish the Research Board for National Security. F. B. Jewett, President of the Academy, proceeded immediately to organize the Board under the chairmanship of K. T. Compton. The Navy assisted materially in getting the RBNS organized and projects established with R. Adm. Furer taking the lead and the "bird dogs" helping in various secretarial and committee tasks. However, the RBNS was destined to enjoy but a brief existence, as President Roosevelt directed the Secretaries of War and Navy in March 1945 (with Bureau of the Budget urging) not to transfer funds for the use of RBNS pending a thorough review of the several bills before Congress for the organization of postwar research. The RBNS was finally killed by a joint letter from Secretaries Patterson and Forrestal dated October 18, 1945. Despite its short life the ill-fated RBNS served a very useful role in educating top people in the military services and Congress, thus preparing the way for more successful future actions on research organization.

In the period between the birth and death of the RBNS things were moving rapidly on other fronts.

In the summer of 1944 the "bird dogs", stimulated by the work of the Committee on Post-War Research, further developed their plan for a Navy office to key in with whatever outside agency Congress might establish. It was on September 6, 1944, that two of the "bird dogs" first set down a new organization chart for an Office of Naval Research which entailed the naming of

an Assistant Secretary of the Navy for Research with broad powers, an Advisory Committee, a Rear Adm. Chief of Naval Research, program emphasis on basic research work, and the transfer of the Naval Research Laboratory to the Office. This was a vitally important improvement over their earlier plan which had called for a rear admiral as head of the Office. This plan for an Assistant Secretary of the Navy for Research was discussed with Dr. Hunsaker and Dr. Bush who enthusiastically supported the idea. The whole scheme was then recorded by three of the "bird dogs" on September 23, 1944 as a beneficial suggestion to the Secretary of the Navy. But this mechanism was not needed, as Adm. Furer, Capt. Smith and Capt. Conrad all quickly endorsed the thought. Thus, Adm. Furer sent a memorandum to the Secretary of the Navy on October 11, 1944, recommending immediate implementation of such a move.

This was received coldly by Mr. Forrestal, who was considering only one new Assistant Secretary, and had him pegged in the field of supplies and logistics. Also he was about to spring a surprise which would soon lead to the replacement of Adm. Furer. In just eight days, on October 19, 1944, he established the Office of Patents and Inventions, with Vice Adm. Bowen in charge, as a first step in implementing the previously mentioned Dearborn report. This was followed by a series of moves which made it obvious that another change was coming.

This maneuvering became of concern to the "bird dogs" as we thought it might ruin our plans for an effective postwar organization. An incident which occurred caused us to take a rather desperate chance. By happenstance we came into possession of a comment made by President Roosevelt on a fat report by Adm. E. J. King concerning a suggested postwar reorganization of the Navy Department. The terse comment, handwritten on the cover, went something like this: "Ernie —I made you Cominch to fight the war, not to reorganize the Navy Department—FDR." This made it painfully clear that the President intended to control postwar departmental changes. We believed so strongly in our method of organizing research in the postwar Navy that we decided to take the risk of getting our USNR necks chopped off by putting our plan before the President. Evening meetings were held with some of his bright young men, who became most enthusiastic, and arrangements were set for a presentation upon the return of the President from his 1945 spring vacation. But, tragically, FDR died while still in Georgia on April 12, 1945.

The expected change in the Navy Department then happened, and on May 19, 1945, the Office of the Coordinator of Research and Development was swallowed up by the Office of Research and Inventions (ORI). Also the Naval Research Laboratory and the Special Devices Division of the Bureau of Aeronautics were also transferred to ORI. Adm. Furer was out and Admirals Harold G. Bowen and Luis de Florez took over with a bang. At the very outset it was a dreary time for Capt. Conrad and the "bird dogs" as we feared our dreams for the future would go down the drain. But we had miscalculated. In a very short while Admirals Bowen and de Florez took up the cudgels for an Office of Naval Research with great vigor. They solicited the powerful backing of men like Commodore Lewis Strauss, Under Secretary of the Navy W. John Kenney, and Assistant Secretary H. Struve Hensel. In June 1945 Dr. Bush's report to President Truman, entitled *Science, the Endless Frontier,* appeared and had great impact in Congressional and military circles. By September 1945 the "bird dogs" had a Congressional bill all drafted for the establishment of an Office of Naval Research to be headed, in deference to Mr. Forrestal, by a Rear Admiral. This draft, which included the establishment of a Naval Research Advisory Committee composed of eminent scientists, was to become known as the Vinson Bill.

There remained one serious hurdle, outside of Congressional action, before the establishment of ONR could become meaningful. This was to get the Universities, where the majority of basic research is performed, to be willing to accept Navy contracts. In this struggle Capt. Conrad became the recognized leader. Accompanied by various "bird dogs" Conrad visited many top universities in the winter of 1945. There was a definite feeling on the part of the scientists after four years of war to wish to forget the Navy and return to former pursuits. But Conrad was able to crumble all opposition by making superb speeches around the country, and by working with legal and contract people to pioneer an acceptable contract system. This would permit one over-all contract with a university with new

task orders to be attached as agreed upon, permit basic research to be contracted for, and permit the work to be unclassified and publishable. Once the legal eagles got this worked out, there was no holding the persuasive Conrad, and he was quickly able to get such institutions as Harvard, Chicago, University of California, California Institute of Technology, and MIT to agree to accept Navy work. Tragically, he contracted a lingering but fatal case of leukemia at his moment of triumph.

With Adm. Bowen and his influential partners and Capt. Conrad maneuvering effectively, the Vinson Bill passed with flying colors and became Public Law 588 on August 1, 1946. It turned out to agree almost verbatim with the 1945 draft by the "bird dogs". This was indeed a day of rejoicing, culminating some four years of effort entailing long hours of teaching, lots of perseverance, and even a little intrigue. As stated at the outset, the victory belonged not to a few, but to many scientists, naval officers, and political figures, some of whom are still unrecognized.

Unexpected Contribution

THE impact of this victory was destined to go far beyond the expectations of the authors. By dint of this far-sighted planning, coupled with favorable action by Congress, the Navy found itself the sole government agency with the power to move into the void created by the phasing out of the OSRD at the end of the War. While Congress still debated what to do about a national agency, Forrestal, Bowen, now Chief of Naval Research, and de Florez arranged for war-end money transfers, and ONR moved forward aggressively to bridge the gap. Sound policies set by the Naval Research Advisory Committee were admirably carried out under the guidance of Capt. Conrad and Dr. Alan Waterman. The world leadership of the United States in basic research in the decade following World War II has been largely credited by many experts to the timely and effective work of the Office of Naval Research.

Surprise Ending

IT was previously stated that the "bird dogs" suggested and worked for, even at some risk, the appointment of an Assistant Secretary of the Navy for Research as representing the ideal organizational solution to assure research the representation and emphasis it deserves in the development of a Navy second to none in this age of science. Continued education and pressure by many people finally brought about such a move in 1959. Perhaps it was a case of poetic justice, but, at any rate, the rest of the "bird dogs" are happy and proud to report that the very first man appointed to the office of Assistant Secretary of the Navy for Research was one of us.

If the recording of this brief history will but inspire continued constructive efforts by other lowly "bird dogs" in government, we will feel more than amply rewarded.

══ Chapter 3 ══════════════
Social Context

Like any human enterprise, physics is inextricably entangled in its social context, that is, in history at large. Articles in the previous section showed the internal workings of single, specific institutions; here we look at the whole community of physicists—as an entity of itself and as something profoundly affected by the rest of society. (Of course the physics community has had in return a tremendous impact on society, but that is another story.)

Surprisingly often, the history of social relations gets into topics that are of lively current interest. The physics community painted by Nagaoka's letters to Rutherford was a hundredth the size of our current enterprise, but the standard of courtesy and the method of learning by traveling remain valid whenever relations between "developed" and "developing" countries are as open as they were between Europe and Japan in 1911. Physics education in general, especially in its relations with engineering, is still more a topic where a look at history may help to save people from repeating, as if they were invented yesterday, arguments that have in fact been worked over vigorously for many decades. On the other hand, discrimination against women and certain other groups as scientists, although it is one of our oldest problems, was discussed all too little until recent years—the article in this section is the only extended historical treatment we have seen in any journal read by physicists.

Other issues of social relations have taken on grave significance only recently. "Recently" to historians means within the past few decades; there has been enough time to accumulate some experience. As the articles here suggest, the impact of economics on physics in the 1930s was repeated in some respects in the 1970s. As for questions of secrecy, and more generally of the role the federal government should play in science, questions which first became urgent around the time of the Second World War, these are still more urgent today. More generally still, the unprecedented and revolutionary reorganization of physics as a whole, both internally and in its relations to society, which took place within the past fifty years, is something that we can only come to grips with if we understand just what has changed and what has not.

Contents

Nagaoka to Rutherford, 22 February 1911

During 1910, the physicist Hantaro Nagaoka represented Japan at two international scientific congresses in Brussels and one in Vienna. This visit to Europe gave him an opportunity to observe the latest researches in the various centers of physics and to renew many acquaintances from his student days in Germany. He called at Manchester before continuing to the continent, and the letter he later wrote to Rutherford is both a description of the state of physics through the eyes of an acute observer and a "thank you" to Rutherford.

PHYSICS TODAY / APRIL 1967

by Lawrence Badash

WHAT WAS PHYSICS LIKE slightly more than half a century ago? One readily thinks of such famous names as J. J. Thomson, Ernest Rutherford, Marie Curie, Max Planck, Niels Bohr, H. A. Lorentz, Albert Einstein, et al, but these are the *highlights* of hindsight. For the *background* of perhaps lesser, but nevertheless significant and interesting efforts, we usually must look to the contemporary literature, since histories of science rarely have room for elaborate descriptions of a period.

The letter printed below contains the impressions of an eminent physicist who visited a good many physical laboratories in Europe during the last quarter of 1910. Years earlier, its author, Hantaro Nagaoka (1865–1950), had studied in Berlin, Munich, and Vienna, and was, therefore, renewing old acquaintances as well as familiarizing himself with the latest continental research activities. Since 1906, he had been professor of theoretical physics at the Imperial University of Tokyo; and many years later he was to become the president of the Imperial University of Osaka.

The recipient of this letter, Ernest Rutherford (1871–1937), needs no identification in PHYSICS TODAY, other than to indicate that at this time he was professor of physics at the University of Manchester. Nagaoka had visited Rutherford's laboratory in September 1910, and now, happily, thought to describe his trip in this letter of thanks for his host's hospitality.

Still classical physics

In this letter it is interesting to note the widespread activity in "classical" physics, which had by no means entirely been superseded by the increasing amount of research in "modern" physics. This is a point we too often overlook. One final note of interest is that, coincidentally, Nagaoka's best known scientific contribution derives its fame from the work of Rutherford. When the latter published his concept of the nuclear atom in 1911, it was seen that Nagaoka's "Saturnian" atom of 1903–1904 was something of a precursor. Though there was no direct influence of this earlier work upon Rutherford, and in fact their atoms bear many dissimilarities, these constructs of Nagaoka and Rutherford frequently have been associated in popular literature. It is not impossible, however, that the two discussed the Saturnian atom in September 1910 and that the concept remained subconsciously in Rutherford's mind, bearing fruit in the next year.

February 22nd, 1911
Physical Institute,
Tokyo University

Dear Professor Rutherford,

I have completed my "Studienreise" in Europe and returned home a few weeks ago, and have the pleasure of writing you some of my impressions during the journey. In the first place, I have to thank you for the great kindness, which you have shown me during my visit to Manchester. I have been struck with the simpleness of the

Lawrence Badash teaches history of science at the University of California, Santa Barbara. He did this work in Cambridge, England, supported by a NATO fellowship and an NSF grant.

apparatus you employ and the brilliant results you obtain. Everybody engaged with the investigations on radioactivity seems to be impressed with the same fact and expresses admiration of the splendid results, which you obtain with extremely simple means.

Lowest temperature yet

The "Kältekongress" in Vienna was too technical for me; it was in fact a congress for the industry of refrigeration. The only scientific paper of importance was a report by Kamerlingh-Onnes on the lowest temperature hitherto attained. By boiling liquid helium in vacuum, he claims to have reached the temperature of 2.5° from absolute zero. Later on I visited his laboratory in Leyden and saw his cascade process of reducing the temperature. He tells me that the greatest difficulty lies in the purification of gases; a millionth part of hydrogen mixed with helium would deteriorate the process of liquefaction. It will be quite interesting to experiment on the radioactivity at the temperature of −270°, if such cold can be maintained for a sufficient length of time. I met Planck in Berlin and asked his opinion as to the change which would be wrought on radioactivity. His conjecture on the change of λ in the neighborhood of absolute zero is in the affirmative, based on several considerations depending on the theory of radiation.

At the time I visited Vienna, the radium institute was not yet completely built, but I met St. [efan] Meyer in the old laboratory of Boltzmann and Exner. In Graz, I was happy to see my old friend Benndorf, who studied with me in Berlin and Vienna about 16 years ago. He was occupied with the registration of the atmospheric electricity and seemed much interested in seismology, which has special charm for Japanese on account of the volcanic character of the Japanese islands. It was very curious that most of my opinions respecting earthquakes were in accord with those of Benndorf, although I am quite at variance from Japanese seismologists.

Righi in Bologna was much interest-

Cast of Characters—People Mentioned by Nagaoka

Hans Benndorf (1870–1953). Physics professor, University of Graz, after 1910.

Ludwig Boltzmann (1844–1904). Physics professor, University of Vienna, from 1902 to 1906. Earlier at Munich. Famous for his part in the introduction of statistical mechanics.

Alfred Bucherer (1863–1927). Privatdozent, University of Bonn, after 1899; later professor.

Peter Debye (1884–1966). Privatdozent, University of Munich, 1910–1911. Later professorial positions at Zuerich, Utrecht, Goettingen, Berlin and Cornell. Nobel Prize in chemistry in 1936 for his studies of molecular structure.

Hermann Ebert (1861–1913). Mathematics professor, Technische Hochschule, Munich, after 1898.

Felix Ehrenhaft (1879–1952). Assistant in Physical Institute, University of Vienna, 1904–1910; professor after 1911.

Franz Exner (1849–1926). Physics professor, University of Vienna, after 1891. Interested in spectroscopy, particularly the lines of the ultraviolet region.

Carl Friedrich Gauss (1777–1855). The "Prince of Mathematicians" was mathematics professor and director of the Goettingen astronomical observatory, after 1807. Concerned also with terrestrial magnetism.

Ernst Gehrcke (1878–1960). Physicist at the Physikalische-Technische Reichsanstalt, Berlin; later director.

Charles Guye (1866–1942). Physics professor, University of Geneva, after 1900.

Friedrich Harms (1876–1946). Assistant in the Physical Institute, University of Wuerzburg, after 1901; later professor.

Johannes Hartmann (1865–1936). Astronomy professor, University of Gottingen, after 1909. Spectroscopist interested in continuous spectra due to atoms.

Hermann von Helmholtz (1821–1894). Physics professor, University of Berlin, 1871–1894; president of the Physikalische-Technische Reichsanstalt, Charlottenburg, 1888–1894; Famous for his work in physiology, sound and conservation of energy.

Heinrich Hertz (1857–1894). Physics professor, University of Bonn, from 1889 to 1894. Famous for his discovery of the electromagnetic waves predicted by Maxwell.

Ludwig Janicki (1879–????). Physicist at the Physikalische-Technische Reichsanstalt, Charlottenburg.

Heike Kamerlingh-Onnes (1853–1926). Physics professor, University of Leiden, after 1882. Nobel Prize in 1913 for low-temperature investigations.

Heinrich Kayser (1853–1940). Physics professor, University of Bonn, after 1894. With C. Runge, he determined that the distribution of spectral lines has a regularity.

Suekichi Kinoshita (1877–1933). Physics instructor, University of Tokyo, after 1909. Later professor.

Peter Paul Koch (1879–1945). Privatdozent, University of Hamburg; later professor.

Friedrich Kohlrausch (1840–1910). Physics professor, University of Wuerzburg, from 1875 to 1888; University of Strassburg, 1888 to 1895; then president of the Physikalische-Technische Reichsanstalt, Charlottenburg, 1895 to 1905. Explained electrolytic conductivity by dissociation hypothesis.

August Kundt (1838–1894). Physics professor, University of Berlin, from 1888 to 1894. Studied anomalous dispersion in liquids, vapors and solids; devised method of comparing sound velocities in gases and in solids.

Otto Lehmann (1855–1922). Physics professor, Technische Hochschule, Karlsruhe, after 1889. Discovered the unexpected existence of crystalline arrangement in some liquids.

Philipp Lenard (1862–1947). Physics professor, University of Heidelberg, after 1907. Nobel Prize in 1905 for his work on cathode rays.

Hendrik Antoon Lorentz (1853–1928). Physics professor, University of Leiden, after 1878. Shared 1902 Nobel Prize with Zeeman for his study of the influence of magnetism on radiation.

Otto Lummer (1860–1925). Physics professor, University of Breslau, after 1905. Noted for his experimental study of black-body radiation.

Stefan Meyer (1872–1949). Physics professor, University of Vienna, after 1908. In charge of the Radium Institute, and a leader in the field of radioactivity.

Alexander Pflueger (1869–1945).

ed with my model of Saturnian atom published in 1904. He showed me his different apparatus on electric waves and the so-called magnetic rays. O. Lehmann in Karlsruhe seems to have made similar experiment with a colossal tube of several meter length and arrived at results similar to Righi. In Geneve, I met Guye and Sarasin. The latter gentleman has been kind enough to show me all the notorieties and fine sceneries of Geneve. He told me of your visit there and how you laughed "von ganzem Herzen," if I may be permitted to use Sarasin's language.

80 000-gauss magnets

The speciality of the physical institute in Zurich seems to be electromag-

nets. Weiss showed me one of 1000 Kilogrm. weight, with which he can get a field strength of 80 000 gauss in space of 2 mm., a somewhat extraordinary figure.

In Munich, I saw Ebert's apparatus for registering the quantity of emanation coming out of the soil. What seemed to me new and interesting was the section for technical physics. There are various investigations going on in connection with the applications of physics to technical purposes. Unfortunately I could not see Röntgen, as he was away from the city. Koch tells me that he could measure Zeeman effect in field of 3 gauss by photographing the lines and comparing the intensity by means of Hartmann's pho-

NAGAOKA

Physics professor, University of Bonn, after 1905.

Max Planck (1858–1947). Physics professor, University of Berlin, after 1892. Nobel Prize in 1918 for discovery of energy quanta.

Erich Regener (1881–1955). Physics professor, agricultural Hochschule, Berlin, after 1914; later at Technische Hochschule, Stuttgart. Noted for method of counting alpha particles by scintillations.

Augusto Righi (1850–1920). Physics professor, University of Bologna, after 1889. Improved Hertz's vibrator, or wave-radiating apparatus.

Wilhelm C. Roentgen (1845–1923). Physics professor, University of Munich, after 1900. Nobel Prize in 1901, the first year it was awarded, for his discovery of x rays, made at the University of Wuerzburg.

Heinrich Rubens (1865–1922). Physics professor, University of Berlin, after 1906. Studied black-body radiation.

Jean Edouard Charles Sarasin (1870–1933). Geology and paleontology professor, University of Geneva, after 1896.

Clemens Schaefer (1878–????). Physics professor, University of Breslau, after 1910.

Arthur Schuster (1851–1934). Physics professor, University of Manchester, from 1881 until he retired in 1907 to allow Rutherford to succeed him.

Arnold Sommerfeld (1868–1951). Physics professor, University of Munich, after 1906. Noted for his refinement of Bohr's original atom

picture, by the introduction of orbital quantum numbers.

Johannes Stark (1874–1957). Physics professor, Technische Hochschule, Aachen, after 1909. Nobel Prize in 1919 for his discovery of the Doppler effect in canal rays and the splitting of spectral lines in electric fields.

Emil Take (1879–1925). Privatdozent, University of Marburg, after 1911; later professor.

Woldemar Voigt (1850–1919). Physics professor, University of Goettingen, after 1883. Explained Kerr effect using electron theory.

Wilhelm Weber (1804–1891). Physics professor, University of Goettingen, after 1849. Associated with Gauss in terrestrial magnetism, telegraphy, mathematical physics. Introduced absolute units in electricity.

Pierre Weiss (1865–1940). Physics professor, Polytechnicum, Zuerich, after 1903. Introduced the word "magneton" to represent an elementary magnet, in a theory of magnetism.

Johann Emil Wiechert (1861–1928). Geophysics professor, University of Goettingen, after 1898.

Wilhelm Wien (1864–1928). Physics professor, University of Wuerzburg, after 1900. Nobel Prize in 1911 for his study of black-body radiation.

Pieter Zeeman (1865–1943). Physics professor, University of Amsterdam, after 1900. Nobel Prize in 1902, shared with Lorentz, for discovery of magnetic broadening of spectral lines: Zeeman effect.

RUTHERFORD

DEBYE

tometer. When I studied in Munich in 1894 under Boltzmann, the institute was very poor, but it is now rebuilt and there is also an institute for theoretical physics under Sommerfeld, who is working on the principle of relativity, and Debye expounded mathematical formulae for the pressure of light acting on a dielectric or metallic sphere.

In Amsterdam I saw Zeeman investigating the effect bearing his name in various lines of research. In Leyden, Lorentz was discussing Ehrenhaft's curious result on the charge of electrons, but afterwards I learned in Berlin that the experiment was entirely wrong. Stark in Aix-la-Chapelle [Aachen] was propounding his "Lichtquantentheorie"; there is some doubt whether he will succeed in explaining the interference phenomena, or not. The Germans say that he is full of phantasies, which may be partly true. In Bonn I failed to see Bucherer, but his experiment on e/m is now repeated by C. Schäfer in Breslau, and I hope we shall be able to hear his result in the near future. Kayser's spectroscopic researches are worth seeing; instead of moving the grating and keeping the slit fixed, he uses the reversed method of turning the slit on a fixed circle while the grating is fixed on a stout pier. This will be sometimes advantageous in photographing the spectrum. Pflüger, with whom I worked in Kundt's laboratory in 1893, showed me a number of interesting apparatus. He has pasted thin quartz plates on a rocksalt prism and investigated the infrared as well as the ultraviolet rays with as much success as with fluorite or quartz prisms. A vacuum tube used by Hertz to demonstrate the passage of cathode rays through thin aluminum plate is one of the historical treasures of the physical institute of Bonn.

Discharges and x rays

The radiological institute in Heidelberg under Lenard is perhaps one of the most active in Germany. Professor Lenard and most of his pupils are working on the phosphorescence and photoelectric action. In Würzburg, I saw the room where X ray was discovered by Röntgen. Various researches on canal rays are going on under the direction of W. Wien. The famous

WIEN

ROENTGEN

SOMMERFELD

LENARD

LORENTZ

STARK

magnetic observatory without iron, built by Kohlrausch, is now rotten; more important works on vacuum discharge have absorbed the attention of Würzburg physicists. In the colloquium, Harms gave a report of your paper on the calculation of α particles, which was in progress when I visited Manchester. All the members present expressed great admiration at the splendid result obtained with such a simple device. It seems to me that it is only a genius, who can work with simple apparatus and glean rich harvest far surpassing that attained with the most delicate and complex arrangements.

[In his reply to Nagaoka, 20 March 1911, Rutherford noted: "I very much appreciate your kind references to myself and to my work. I did not know that the simplicity of my experiments was so unusual. As a matter of fact I have always been a strong believer in attacking scientific problems in the simplest possible way, for I think that a large amount of time is wasted in building up complicated apparatus when a little forethought might have saved much time and much expense."]

Liquid crystals

In Karlsruhe, the original apparatus of Hertz for demonstrating electromagnetic waves were most attracting. They are as simple as are most of your apparatus. Lehmann is busily occupied with the investigation of the so-called liquid crystals. The appearance of several substances in polarised light is quite phantastic and accords with the illustrations given by him. The only point of doubt is that these crystals appear only in the neighbourhood of the melting point, and may be closely connected with the changes in the aggregate condition. It is quite certain that in the present stage different views are entertained as to the nature of the liquid crystals.

Strassburg was interesting to me as a centre of seismological association; great changes are going on in the staff of the central bureau, and it is to be congratulated for the science of seismology that the reorganisation will produce good effect on the international investigation of earthquakes. We have to thank Prof. Schuster for the lively interest he takes for the as-

sociation, and the great effort he has made to strengthen the weak association, by recruiting it with personages, who can investigate earthquakes physically [better] than it has hitherto been examined statistically and with defective instruments.

Frankfurt, Leipzig, Breslau

Frankfurt has built a fine physical institute with rich equipments. It is curious that the city well known for its immense wealth has not yet established a university within its precinct. The magnetic properties of Heusler alloy is now being investigated by Take in Marburg; the artificial means of aging the alloy seems to effect interesting magnetic changes. Voigt's laboratory in Göttingen is justly celebrated for the numerous works, which are connected with the physics of crystals and the magneto- and electro-optics. There were more than 20 research students. The famous magnetic observatory of Gauss and Weber is now removed to the environs of Göttingen. Wiechert has installed an extremely sensitive seismometer, which records vibrations due to storm in the North Sea. He showed me traces of shocks due to dynamo engine in Göttingen.

The physical institute in Leipzig is perhaps the largest in Germany; but I find that the largest is not always the best. However poor the laboratory may be, it will flourish if it has earnest investigators and an able director. The size and the equipment of the laboratory seems to me to play a secondary part in the scientific investigations. The splendid institute in Breslau has been newly built by Lummer. The investigations are mostly optical; the different kinds of interferometers and the photometers are the essential equipments of the institute. Besides Lummer, C. Schäfer is working in electric waves and applications of integral equations to different problems of theoretical physics.

The works going on in the Physikalische Reichsanstalt in Berlin is somewhat akin to those in the National Physical Laboratory. Some measurements are nervously delicate that we can not help crying out *qui bono*. The [illegible]-rohr and Glimmlichtrohr of Gehrcke are very interesting, and the inventor claims to use the latter tube

as an oscillograph for high frequency up to 100,000 cycles per second. The investigations of spectral lines by Janicki will form a good contribution to our knowledge on the nature of atomic vibrations.

In the physical institute of Berlin, I saw Rubens who showed me his arrangement of "Reststrahlen" for isolating light waves of 96 μ. Regener was repeating Ehrenhaft's experiment and announced that the result was entirely wrong, so that there can not exist a charge, which is a fraction of that of an electron. While visiting the institute, I chanced to enter the rooms where I heard t h e l e c t u r e s b y Helmholtz and where I worked under Kundt in 1893. They made me deeply impressed how swiftly time is gliding; and while thus writing it reminds me that 5 months has passed away since I saw you in Manchester.

Cold in Siberia

I returned by way of Siberia and experienced the low temperature of $-44°C$ on the Chinese frontier. The car was comfortable, but the temperature difference of 60° in and out of the car was almost unbearable. The consequence was that I caught a severe cold and was confined to bed for about three weeks. I have as yet nothing to write you about the scientific investigations in Japan. Kinoshita is going to start radioactive works with the radium, which you have kindly procured for him.

Please remember me to Mrs. Rutherford and your daughter.

Wishing you much scientific success,

I remain

Yours faithfully
H. Nagaoka

* * *

For access to this letter and permission to print it and for permission to quote Rutherford's reply, I am indebted to: the family of Professor Nagaoka, Mr. T. Kimura and Dr. E. Yagi of the Committee for the Publication of Nagaoka's Biography, the grandchildren of Lord Rutherford, the authorities of the Cavendish Laboratory, and the Cambridge University Library. Nagaoka's letter is preserved at the Cambridge University Library; Rutherford's reply is in the possession of the Committee for the Publication of Nagaoka's biography. A few spelling errors in Nagaoka's 14-page longhand original have been corrected in the editing. □

American physics and the origins of electrical engineering

Pure physics applied: academic physics gave birth to a new practical discipline with its own priorities and its own departmental structure.

Robert Rosenberg

PHYSICS TODAY / OCTOBER 1983

At the same time that electricity was transforming American society in the last half of the 19th century, it was transforming the study of physics. During this period, electricity bridged the existing gap between pure science and useful applications, between thinkers and doers, scholars and tinkers, as no other technology had done before. It brought home to Americans the contributions of science to everyday life. It also quickened the pace of physics research in university classrooms and industrial laboratories.

Together, electricity and physics held immense promise for the future—a promise unnoticed at the Philadelphia centennial exhibition of 1876, with its small displays of telephones and dynamos, but visible to all at the opening in 1883 of the Brooklyn Bridge, illuminated by Edward Weston's arc lights. It happens that both dates are reference points for US physics. In 1876, Henry Augustus Rowland, educated as an engineer but dedicated to basic research, became the first professor of physics at the newly founded Johns Hopkins University in Baltimore, and, in 1883, Rowland proclaimed in his vice-presidential address to the American Association for the Advancement of Science that henceforth the word "science" should no longer be applied to the telegraph, telephone, electric light or electric motor. With the advent of electrical tech-

nology, American physicists could choose to be theoretical or practical—or both.

The connection and then disconnection of basic physics and electrical engineering had been made years earlier in Europe. In Britain, such theorists as James Clerk Maxwell and John William Strutt (Lord Rayleigh) at Cambridge University had a great impact on technology, but their immediate influence was indirect since few engineers could understand them. It took a creative effort almost equal to that of Maxwell and Rayleigh by Oliver Heaviside, a British engineer with no formal education past the elementary level, to translate their electromagnetic equations into a usable form, and even Heaviside's work was unintelligible to most engineers. Yet Maxwell and Rayleigh were among those physicists who consciously attempted to contribute to technology. Others include Heaviside's uncle, Sir Charles Wheatstone of King's College, London, who somewhat anticipated Samuel F. B. Morse in developing the telegraph, and William Thomson (Lord Kelvin) at Glasgow, who virtually single-handedly engineered the cables, galvanometers, and other electrical components for the first successful telegraph cable beneath the Atlantic Ocean in 1866.

By the 1880s, the need for rigorous training in electrical engineering was becoming clear to many. Werner Siemens, Germany's leading industrialist of the period, urged his country's technical schools to introduce courses in electrical engineering and, with a leading physicist, Hermann von Helmholtz, he persuaded the government to establish a national laboratory in 1882. Around that time, William Ayrton

attempted to organize in London the sort of laboratory instruction in electricity that he and John Perry had carried on in the late 1870s at Japan's Imperial College of Engineering.

Electrical innovations

Such examples did not go unnoticed in the US, though the order of events was somewhat different. By the late 1870s, the considerable body of knowledge produced by rapidly advancing research on electricity in Europe had crossed the Atlantic, and by the end of the century electric innovations in the US had provided an ineluctable justification for supporting physics teaching at universities and research work in companies. In the US it was not the physicist—such as J. Willard Gibbs at Yale or Henry Rowland at Johns Hopkins—who caught the public imagination, but the inventor—Edison, Charles Steinmetz, Nikola Tesla—working in commercial surroundings.

The success of electrical technology

Robert Rosenberg, a doctoral candidate in the History of Science Program at The Johns Hopkins University, has been recently appointed a research associate on the Edison papers project at Rutgers University.

had two effects on American physics. First, students eager to understand the new electrical technology and to contribute to it, as well as to profit from it, put an unceasing strain on the budgets and facilities of physics departments in universities, colleges, and technical schools. Of some 400 colleges and universities surveyed by T. C. Mendenhall for the US Bureau of Education in 1882, almost all offered some instruction in physics, but only 20 had even minimal laboratory facilities. In the many large physical laboratories built during the 1880s, the lion's share of space was devoted to the study of electricity and magnetism.

Second, the social impact of electrical technology confirmed the claim of physicists that their investigations led to material progress. In 19th-century America, this was an important point. Chemistry had already demonstrated its utility in agriculture and industry, and biology was linked with medicine, but until the growth of electrical tech-

nologies, physics held little claim to being utilitarian. The source of the new technologies was in research, both pure and not so pure.

Dynamo as symbol

As long as it emphasized power and light, electrical engineering needed a solid foundation of physics and mechanical engineering. It is somewhat surprising, then, that early electrical engineering education was under the direction of physics teachers. Mechanical engineers did not involve themselves because in the early 1880s mechanical engineers did not understand electricity. Thus, although a paper presented in 1882 to the American Society of Mechanical Engineers on the Edison Steam Dynamo—the combination steam engine and dynamo that was to power the Pearl Street Station in New York City—treated both components of the machine, the lengthy discussion that followed was entirely about the steam engine. One promi-

nent engineer, puzzled by the working of the dynamo, said, "There may be electrical reasons for this construction." What those reasons were, he had no idea.

Mechanical engineers recognized (and laughed about) the mechanical ignorance of many electrical engineers, and sometimes referred to Sir William Thomson's dictum that an electrical engineer should be 90% mechanical and 10% electrical. Until the end of the 1880s, however, when electric motors began to compete successfully with steam as a power source, mechanical engineering as a profession had little to do with electricity. By the time the mechanical engineers became concerned about the encroachment of electric power, the electrical engineers had their own discipline, their own professional image, and their own ideas about how to educate students.

In the early 1880s, the need for formal education in electrical engineering was becoming manifest. The editor of *The Electrician*, a New York trade journal, wrote in April 1882:

> There is now a rapidly growing want for men trained in the theory and practice of the science of electricity.... The demand is established, and it now behooves our foremost educators to devise a means of satisfying it.

An American just back from Europe wrote a letter to the student paper at Cornell in September 1882, urging undergraduates to consider the new profession of electrical engineer now being taught abroad.

> The enormous extension of the telegraph, telephone, electric light, etc., into all parts of the world will create a great demand for skilled electricians at no very distant day.

To which the editors added,

> We wish to recommend this specially to the students of Cornell University as a department well worthy of their careful investigation.

That fall, Edison wrote to the president of Columbia College suggesting that a course in electrical engineering should be given in the School of Mines and offering his electrical collection to the College as a museum.[1] Although Edison often publicly belittled academics and universities, he employed physicists, chemists and metallurgists and even consulted with college professors and read scientific journals. During the 1880s he contributed many thousands of dollars in equipment for electrical engineering programs at several schools. Columbia did not establish a course in electrical engineering until the end of the decade, by which time most universities were already actively teaching electrical science in their physics departments.

First course in EE

The first formally structured course in electrical engineering appeared in 1882. But the roots of that course were embedded in the 1870s, when such academic physicists as Charles Cross at MIT and William Anthony at Cornell began to shape their teaching around the new discoveries in electricity.

In 1869, Edward C. Pickering, professor of physics at MIT, established the first systematic laboratory instruction

in physics in the country.[2] In the 17 classes preceding the initiation of the electrical engineering course at MIT, only six of the 361 graduates took degrees in physics. The reason for the lack of interest in a physics degree is not hard to ascertain. It could be found in MIT's 1881–1882 catalog (and had been noted by Rowland at Johns Hopkins four years earlier): "Most of the students taking the course in Physics intend to make teaching their profession." Unfortunately, there were few openings for physics teachers in the 1870s and early 1880s.

Cross had graduated from MIT in 1870, one of a class of ten, the only student in the General Science and Literature course. He at once became an instructor in the physics department, a professor in 1874, and head of the department on Pickering's departure in 1877. Cross had an intense interest in electricity. In his 1873

report to the president of the Institute, he noted:

> The most defective portion of the apparatus designed for lecture-room use is that relating to electricity and magnetism, upon which a considerable sum must be spent in order to make it a fair representation of the present state of electrical science.

The next year some electrical apparatus, including an induction coil, was obtained by the department, and the electrical inventor Moses Farmer loaned the Institute one of his magneto-electric machines. In 1876, six electrical experiments were offered in the laboratory. The same year, Cross hired Silas Holman of the class of 1876 (in physics) as a laboratory assistant. Holman was an important part of the physics department for more than 20 years, contributing greatly to the electrical engineering program.

By the spring of 1878, electrical questions were appearing on examinations for second-year students of physics. Examples:

> What is a Thomson's galvanometer and what advantages has it over the ordinary form?
> What is a commutator?
> What is a shunt, and when used?

The next year, the first-term examination for the juniors had a question on Ohm's law. Four of seven questions on the same examination one year later (in January 1880) dealt with electrical subjects—the theory of the voltaic cell, Lenz's law, Ohm's law, and the operation of induction coils, telegraphy and dynamos.

In 1881, the MIT catalog announced:

> On alternate years a course of lectures will be given upon the scientific principles involved in the more recent applications of Electricity including the Telegraph, the Telephone, Electric Lighting, and the transmission of power by electricity.

The next year, with the addition of "an extended course of Laboratory instruction in electrical measurements," the lecture course became the senior-year instruction in the new "alternative course in Physics . . . for the benefit of students wishing to enter upon any of the branches of Electrical Engineering." Two years later the course would be formally called Electrical Engineering, but with no significant change in content. In fact, the establishment of the "alternative course in Physics" in 1882 involved little more than the shuffling of existing courses to effect a marriage of physics and mechanical

Henry Rowland of Johns Hopkins, one of the leading US physicists, in a portrait by one of the nation's greatest artists of the period, Thomas Eakins.

engineering. It was just the next step in a natural evolution, rather than a restructuring or redirecting of Cross's teaching.

At MIT, electrical engineering instruction kept the physics staff busy. Electrical engineering students had as much physics as the physics students and then some. In the first year of the course, 18 students were registered, and in the second year, 30. In successive years, it continued to grow, and in 1889 was the best-attended program at the Institute, with 105 students. Moreover, in 1891, some 23 students graduated in electrical engineering, while only three took physics degrees. In 1896, electrical engineering degrees were given to 48 students, while the number receiving degrees in physics was still three.

At Cornell, much the same evolution was taking place. Anthony had come to Cornell in 1872 with a high reputation in physics. When Anthony was hired away from the Iowa Agricultural College, Cornell's vice-president William C. Russel told the university's president, Andrew D. White, that the school had acquired a "tower of strength."[3] Anthony was an exceptional teacher and an adept experimentalist, and kept himself fully informed on current developments in his science. He possessed the idealism of a pure scientist and the practical bent of an engineer. The prospect of a position at Cornell was enticing. He wrote to Russel in 1872:

> I judge that your standard of scholarship is higher [than at Iowa], and that your aim is to make *scholars*, as well as impart "practical" knowledge. I want to get into an atmosphere where the grandeur and beauty of scientific truth are recognized and where science is valued for *itself*.

In 1873, after enumerating for White the many possible uses for physics in the modern world, he added:[5]

> But I should not consider the teaching of the practical application of physics to be the highest purpose of the physical laboratory. I should hope that young men would be found who would wish to pursue the science for its own sake. I should wish to furnish to such an opportunity to make investigations that would advance the interests of science.

To further this end, Anthony had made his acceptance of the job conditional on the university's purchase of at least $15 000 worth of apparatus in his first five years there.[6]

Funding problems

Had Cornell not fallen on hard times in the 1870s (as did MIT and many other institutions), the physics depart-

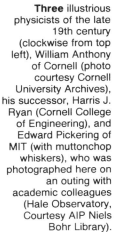

Three illustrious physicists of the late 19th century (clockwise from top left), William Anthony of Cornell (photo courtesy Cornell University Archives), his successor, Harris J. Ryan (Cornell College of Engineering), and Edward Pickering of MIT (with muttonchop whiskers), who was photographed here on an outing with academic colleagues (Hale Observatory, Courtesy AIP Niels Bohr Library).

ment might have achieved prominence earlier than it did. Certainly Anthony's career there would have been quite different. As it was, in the spring of 1873 Anthony had to give a course of popular lectures during vacation to raise money for apparatus, and when he resigned 14 years later it was partly out of frustration at being denied $1500 for instruments.

But although financial embarrassment was a hindrance to Anthony's department, the development of electrical science was tremendously stimulating. Anthony's interest in electricity was even more precocious than Cross's. In 1872, years before any commercial installations, Anthony already hoped to acquire an "electromagnetic machine for producing the electric light" to illuminate his lecture

room.[7] The next year, as part of a wish list of practical experiments for students to perform in the laboratory he did not have, Anthony included[8]

> Electrical measurements. Measurements of resistance and insulation, power of batteries, location of faults. Measurements of electromagnetic power, with reference to electromagnetic machines and motors.

The inclusion of motors was remarkably farsighted, for in 1873 the development of electric motors was barely under way—for the most part in Europe.

The next year, unable to get a Gramme dynamo from Europe, Anthony built one with the help of a student at Cornell and a machinist from Ithaca. The machine was a tribute to Anthony's talent, and became an early symbol of Cornell's eminence in electrical science. It was exhibited at the Centennial Exhibition of 1876, and on its return to the Ithaca campus it was used to power two arc lights, wired through underground cables of Anthony's design and manufacture. This was the first such permanent installation in America. The dynamo was used in the laboratory through the first decades of the 20th century, and is still in working condition today. By the early 1880s, electricity was occupying most of Anthony's time. Mechanical engineering undergraduates were writing theses on electrical topics under Anthony's supervision, and in early 1883 he was asked to draw up a curriculum for an electrical engineering course. Approved by the trustees and faculty, the course was offered that fall in the physics department.

Cornell's undergraduate degree program in physics had been no more popular than MIT's. In the ten years after 1876, only 13 students earned physics degrees out of a total of 678 undergraduate degrees awarded at Cornell. The student paper reported in 1876 that three-quarters of the undergraduates in scientific courses planned to be lawyers, physicians, ministers or journalists; the rest teachers, merchants or manufacturers, and "a very few, scientists." Although few students pursued a physics degree, some physics was required of nearly all students. This was also true at MIT.

By 1880, Anthony was irritated by crowding and the lack of laboratory apparatus. He told officials at Cornell that the department was "20 years behind the times."[9] That year the administration granted him his laboratory, and he requested a lecture room with 200 seats. Ten years later, the number of undergraduates in electrical engineering numbered 218—more than could fit in the lecture hall at one time.

In 1885, Anthony built an enormous tangent galvanometer, an instrument of extraordinary precision and utility. It represented the direction of the department: After 1882, almost all of Anthony's requests for appropriations concerned electrical apparatus. Defending one such a request in 1886, he protested:[10]

Is it to be supposed that, in 1872, I should have foreseen the demand that would be made by the extraordinary growth and the vast importance of the industrial applications of electricity? Is it to be wondered at that I should see possible ways of improvement now that I did not see then?

Unfortunately for Anthony, the sympathetic Andrew White had been succeeded as president in 1885 by the less scientifically inclined Charles K. Adams, who would only later learn to appreciate the place of technical studies in the university. Anthony's 1886 request was denied—repeatedly. Frustrated, he left Cornell in 1887 to take a position as consultant to an electrical manufacturer.

He suggested as his successor Edward L. Nichols, who would become a leader not only at Cornell, but in American physics as well. Anthony called him[11]

the best man I know to make a success of the Physical Department here in the directions both of pure science and its practical applications.

Nichols was a Cornell graduate who had spent four years in German laboratories, one year with Rowland, another year with Edison, two years teaching in Kentucky, and four years teaching at the University of Kansas. In his last year at Kansas, Nichols had prepared an electrical engineering course for the fall of 1887.

Nichols taught electrical engineering courses in his first year at Cornell. In the spring of 1888, however, an independent department was set up within the Sibley College of Engineering, with an associate professor of electrical engineering given responsibility for teaching "the construction of engineering work . . . peculiarly appertaining to electricity." By the end of the 1880s, the proper education of an electrical engineer was beyond a physics department. The new programs were run by electrical engineers with practical experience and scientific sophistication. Even so, physics departments were required to teach young electrical engineers the scientific fundamentals.

One of Anthony's prize students had just such training. Harris J. Ryan was a member of the first formally admitted class in electrical engineering and was Anthony's assistant. A year after his graduation in 1887, Ryan became an instructor in physics, and later the principal figure in the electrical engineering department.

Flourishing of EE

Although Cornell and MIT deserve special attention for establishing two of the earliest and most respected programs in electrical engineering, they did not have the field to themselves for long. In the same year that Cornell introduced its program, 1883, the Stevens Institute in Hoboken, New Jersey, began a course in Applied Electricity. A number of schools acknowledged the rise of electricity with subcourses in their physics departments—among them Lehigh in 1883 and Rose Poly-

Class of 1890 electrical engineering graduates, in frock coats and bowler hats, adorn stairs at MIT, then located in Boston's Back Bay, for a classic photograph of their halcyon days as students in a burgeoning field.   (Photo courtesy Archives, California Institute of Technology.)

Brooklyn Bridge, pictured just before its opening in 1883, became a symbol of American ingenuity, heralding the new era of electricity with its many lights.

technic and the Lawrence Scientific School at Harvard in 1884. The first two were well-attended, but the Harvard program was little more than a title in the catalog until the 1890s and even then was weak. By that time, electrical engineering programs existed in name, if not in fact, in schools throughout the country.

At the 1884 International Electrical Exhibition, Henry Rowland declared: "It is not telegraph operators but electrical engineers that the future demands." Accordingly, in 1886, he established a program in applied electricity at Johns Hopkins to train electrical engineers, and enrollment soon outgrew the new physics building. But when Hopkins's finances went sour in the 1890s and no outside sponsor could be found for the program, the subject was withdrawn.

Interest in the new technology reached into the Hopkins physics department itself. Rowland's first PhD recipient, William Jacques, given his degree in 1879, went to work immediately for American Bell telephone company as an "expert," a job that had not existed when Hopkins had opened its doors three years earlier. During the 1880s and 1890s, quite a few graduate students were admitted to Rowland's laboratory with the express purpose of gaining familiarity with electrical science. Many of them left to work in the industry. Rowland himself reigned for two decades in a dual role as America's foremost pure physicist and as America's foremost electrician. In

the language of the day, "an electrician . . . is a person thoroughly grounded in the theory of electricity and the laws by which it is governed, but it is not essential that he should have any special knowledge of its practical applications beyond laboratory work."[12] This definition was provided in 1884 by a trade journal in answer to a question about the difference between an electrician and an electrical engineer. In practice, the distinctions were unclear and largely semantic until the 1890s

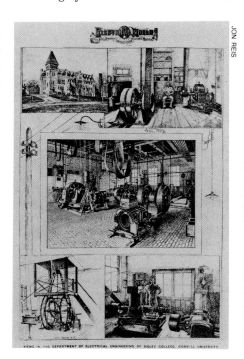

VIEWS IN THE DEPARTMENT OF ELECTRICAL ENGINEERING OF SIBLEY COLLEGE, CORNELL UNIVERSITY

when electrician began to assume its modern meaning—someone who can wire a house or fix an appliance—and electrical engineers became more particular about being called by their proper title. Rowland and other prominent physics professors—among them George Barker at the University of Pennsylvania, Henry Carhart at Michigan, and Cyrus Brackett at Princeton—had close ties to the commercial development of electricity as consultants and legal experts in patent squabbles.

Advancing truth and beauty

Besides stimulating departmental growth in the schools, electricity gave American physics research a utilitarian justification it had never before possessed. In 1876, at the time of the founding of Johns Hopkins, the champions of American physics numbered a mere handful. Besides those few physicists lucky enough to be in teaching positions or government service, the supporters were found primarily among the most educated in society. This group prided themselves in upholding high standards of culture. For them, those who pursued pure science were somehow ennobled as the vanguard of American civilization; they considered the study of physics the moral equivalent of the antebellum study of the classics. The discipline of the laboratory, enforced by Natural Law, they argued, would replace the discipline of conjugation and declension, enforced by the dusty pedant, and the beauty of Nature's Truth would excel the beauty of Homer and Horace. Although this group was also loud in proclaiming that disinterested, pure research was the basis of technological advance, their hearts were in the battle against the corruption and materialism of the Gilded Age. But the practical success of physics in the 1880s and 1890s was evident to all. Public and industrial reliance on electricity and the fortunes spawned by electrical products made the "physics as culture" argument unnecessary and obsolete.

The passion for practicality—and the concomitant lack of interest in the development of theory—had long been part of the American experience. Alexis de Tocqueville recognized this American trait in the 1830s and deplored it, maintaining that hardly anyone in the new nation was devoted to pursuing knowledge for its own sake. When John Tyndall lectured through the eastern states in 1872–73, he made a strong plea for the support of research and implored Americans to prove de Tocqueville wrong. In 1876, the astronomer Simon Newcomb bemoaned the nation's pitiful contributions to abstract science. Thus, when Henry Rowland stood before the physical science section of the AAAS in 1883 to deliver his celebrated "Plea for Pure Science," he was voicing frustrations of long standing.

But Rowland, speaking after the dawn of the Electrical Age, no longer represented the majority of his colleagues. Most contemporary physicists and their supporters welcomed the opportunity electricity offered to display the fruits of their labors. Few American physicists had the interest, ability, and opportunity that enabled Anthony and Cross to initiate electrical studies in the 1870s. Yet a decade or two later, virtually every physicist was celebrating the virtues of electricity and its applications. Maxwell, whom Rowland revered, had acclaimed the reversibility of the dynamo "the greatest scientific discovery of the last quarter of a century."[13] Within Rowland's immediate circle, Daniel Coit Gilman, the president of Johns Hopkins, found in electricity a justification for pure research. In an 1882 speech about the role of university research in the progress of civilization, Gilman claimed[14] that electricity had

> wrought greater changes in commerce than the discovery of the passage around the Cape; greater modifications in domestic life than any invention since the days of Gutenberg . . .

Indeed, through the 1880s, Rowland's successors as vice-president of the AAAS physical section either depicted the scientific mysteries of electricity or sang its praises as the gift of physics to the world—or both. In 1887, for example, William Anthony had rebuked Rowland by celebrating the patents taken by American physicists. All but two of the patents were electrical (and those two belonged to Rowland). A. A. Michelson began his 1888 "Plea for Light Waves" with a glowing description of the

wonderful achievements in the employment of electricity for almost every imaginable purpose. Hardly a problem suggests itself to the fertile mind of the inventor or investigator without suggesting or demanding the application of electricity to its solution.

And in 1889, Henry Carhart, in his "Review of Theories of Electrical Action," characterized for the decade the utility of physics:

> Of the practical applications of electricity it is not necessary to speak. They bear witness of themselves. A million electric lamps nightly make more splendid the lustrous name of Faraday; a million messages daily over land and under sea serve to emphasize the value of Joseph Henry's contribution to modern civilization. . . . The value of the purely scientific work of such men is attested by the resulting well-being, comfort and happiness of mankind.

Ironic turning point

The 1890s brought an ironic twist to the relationship of physics and electrical engineering in the US. By the end of the decade, electrical engineering educators complained that training in a course administered by a university physics department was bound to be inadequate. They questioned the value of abstract investigations in higher physics and argued that the curriculum should include only such physics as was fundamental to engineering.

As the electrical engineers parted company with the physicists, so did the public. The utility of the physicists had never been as clear to the general public as it had been to the educators and physicists themselves. In the schools, electrical engineering attract-

ed new laboratories and substantial funding. The research labs established by General Electric, Westinghouse and Bell Telephone were hailed by the press and public. Physics, by contrast, did not achieve significant academic or public recognition until after World War I, nor become preeminent among the sciences until World War II.

References

1. J. K. Finch, *A History of the School of Engineering, Columbia University*, Columbia U.P., New York (1954), page 68.
2. Background on MIT is in S. C. Prescott, *When MIT was Boston Tech*, Technology Press, Cambridge (1954); K. Wildes, "Electrical Engineering at the Massachusetts Institute of Technology," unpublished manuscript, MIT Institute Archives (1971). Student enrollment figures and course descriptions are in the annual *Catalogs*.
3. W. C. Russel to A. D. White, 8 August 1872, A.D. White Papers (Collection 1/2/2), Cornell University Archives (hereinafter ADW).
4. Anthony to Russel, 30 June 1872, ADW.
5. Anthony to White, September 1873, ADW.
6. Anthony to C. K. Adams, 11 December 1886, Executive Committee Minutes (Collection 2/5/5), Cornell University Archives (hereinafter EC).
7. Anthony to White, 5 August 1872, ADW.
8. Anthony to White, September 1873, ADW.
9. Anthony to Russel, 6 June 1880, ADW.
10. Anthony to Adams, 11 December 1886, EC.
11. Anthony to Board of Trustees, 19 June 1887, EC.
12. The Electrician and Electrical Engineer **3** (April 1884) page 93.
13. Quoted in H. Greer, Popular Science Monthly **24** (December 1883) page 254.
14. D. C. Gilman, *President Gilman's Address at the Euclid Avenue Church*, Fairbanks, Cleveland, Ohio (1883) page 23. □

Technological robot
~nevolently embracing
~an at the entrance to
~the Hall of Science at
the 1933 Chicago
Century of Progress
~position in the depths
of the depression.

Physics in the Great Depression

Hard times raised hard questions
that were not answered in the 1930's and
remain on the agenda now.

Charles Weiner

PHYSICS TODAY / OCTOBER 1970

In the spirit of the soul-searching seventies, physicists are now uneasily questioning the pace of physics and its proper place in society. They view with foreboding the changes in slope of the funding and employment curves that, along with assessments of changes in public attitudes, are the major social indicators of the health of the physics community. The immediate impact and long-range threat of reduced research funds, slackening employment opportunities and lower public esteem for physics are the apparent causes for concern. Threatened or imminent hard times are especially difficult to take on the heels of the high expectations that good times engender. This public statement by a distinguished physicist aptly characterizes the situation:

"Let us begin by facing the facts. Physics has enjoyed a place in the sun which it can not expect to hold permanently . . . Physicists would be more than human if they were not somewhat spoiled by the popularity they have enjoyed." [1]

Charles Weiner is professor of History of Science and Technology at MIT.

The need for analysis and planning was brought to the attention of the physics community by another leading physicist in his presidential address to The American Physical Society.

". . . this question of organized propaganda for physics and a thorough investigation of the sociological aspects of physics are the most important problems confronting our society. Physics in this country has simply grown like Topsy, and, unless some thought is given to these matters, we may have an autopsy on our hands." [2]

These assessments of the state of US physics, which certainly appear to fit today's scene, were made in the 1930's. The growth referred to took place in the 1920's, and the problems are those of the depression. It should prove informative to look back into that decade to see what gave rise to these statements and how the physics community responded to them. Glimpses of an earlier period can provide some perspective by showing the patterns of events and by identifying some of the issues and responses of the time. There is also value in questioning the assumptions so often made about the pre-

World War II development of US physics. These assumptions tend to minimize the achievements of that era as well as oversimplify its problems.

Coming of age in the twenties

The rapid growth of physics in the US, referred to by Paul Foote in his 1933 presidential address to APS, had occurred in the late 1920's, when physicists who were determined to build better departments at universities throughout the US received substantial financial support from private foundations. The major source of support was the Rockefeller-supported General Education Board, which between 1925 and 1932 provided 19-million dollars to help develop science departments in key US universities. At the same time that these efforts were being made, attention was being given to increasing the communication among US physics centers as well as between them and European centers. One of the most successful innovations was the establishment in 1919 of the National Research Fellowships, which enabled outstanding new US PhD's to pursue postdoctoral work at universities throughout the nation.

These fellowships were awarded by the National Research Council with Rockefeller Foundation funds.

Many physicists, as is often noted, went abroad to visit and to participate in seminars and research at the major European physics centers during the late 1920's, when the analytical force of quantum mechanics was being tried on a wide variety of physical problems. Now forgotten is that, at the same time, Europeans found the research facilities at US universities increasingly desirable. For example, of the elite group of 135 European physicists who, from 1924 to 1930, received international postdoctoral fellowships from the Rockefeller Foundation, one third chose to study at US institutions; more of them were attracted to the US than to any other country. In addition, some of the most distinguished European physicists accepted invitations to lecture at US universities in the late 1920's and early 1930's.

The annual University of Michigan summer school for theoretical physics was one of the special attractions for Europeans and Americans. Begun in 1927 by department chairman Harrison Randall, the school was famous for an informal atmosphere that encouraged lively discussion. The summer school staff consisted of Michigan faculty and invited lecturers, drawn from the ranks of the best physicists in Europe and the

US. The high level of the staff can be seen in this excerpt from a letter written 15 July 1930 to Gilbert N. Lewis by young Joseph Mayer, a participant in the 1930 summer school:

"[Paul] Ehrenfest, of course, rules the whole symposium like a somewhat childish Tsar, but it is a wonderful relief to hear quantum mechanics discussed with someone present who will not permit empty mathematical symbols and words to pose as explanations. For the first time since I left Berkeley I've again experienced some of the clarity and liveliness of the Monday evening colloquiums.

"[Enrico] Fermi is giving a course on [P. A. M.] Dirac's dispersion theory, Ehrenfest an unnamed course that so far has been the history of physics in the nineteenth century, and in addition there are two evening colloquiums in theoretical physics and one in experimental every week. [Philip] Morse is giving an introduction to quantum theory that I have not attended but that is said to be good.

"Fermi, by the way, is a very young and pleasant little Italian, with unending good humour, and a brilliant and clear method of presenting what he has to present in terrible English." [3]

Another innovation that demonstrates the growth of the US physics communi-

ty was the establishment in 1928 of *Reviews of Modern Physics*. John Tate, editor of *The Physical Review*, asked 45 leading US physicists whether they thought a review journal was needed in the US. Edward Condon, who had returned from his postdoctoral tour of Europe a year earlier, was one of the many who gave strong support to the idea. In a letter to Tate, dated 2 October 1928, Condon said:

"I have been thinking ever since I returned from Germany that the greatest handicap to physical research work here is the lack of an adequate literature in English . . . There is no question that our laboratories are better now than those abroad, but we lack the literature which brings the young men quickly into step with the research work in the various fields." [4]

The conscious effort to strengthen physics departments and to improve communication through personal interaction and professional journals produced a unique and vigorous physics enterprise in the US. US institutions were thus in close touch with contemporary work and were often in the forefront of many fields, as in the newly developing field of nuclear physics.

This new vigor was clearly in evidence during the 1933 APS meetings, which were held in Chicago to coincide with the Century of Progress Exposition. John Slater, who was then chair-

At the University of Michigan summer school in 1930. The informal group discussion includes Maria Mayer and Joseph Mayer, on the left, Lars Onsager on the right, and Paul Ehrenfest next to him. At the rear is Robert d' E. Atkinson. The lecturer in the other photograph is Enrico Fermi.

man of the Massachusetts Institute of Technology physics department, recalls that what impressed him most was "not so much the excellence of the invited speakers, as the fact that the younger American workers on the program gave talks of such high quality on research of such importance that, for the first time, the European physicists present were here to learn as much as to instruct." [5]

Impact of the depression

Physics in the US had grown rapidly during the 1920's and the physicists' expectations were high. Then the depression hit; its effects on the campuses were felt gradually and had greatest impact in the academic year beginning in the fall of 1933. Younger men were hurt most. In some departments junior faculty were dropped, but the main

brunt was borne by the new PhD's who found it extremely difficult to get jobs. Many subsisted on one small fellowship after the other; others were able to find assistantships that normally were given to graduate students; still others left physics. Faculty at the associate and professorial level were least affected, but they did receive salary cuts or "negative bonuses." These cuts were slightly offset by the decrease in the cost of living, but they still hurt. A comment from a letter written by Linus Pauling to Samuel Goudsmit in May 1933 characterizes the situation:

"I haven't the faintest idea as to where [your former student] can get a job. Caltech is filled with our own PhD's and former National Research Fellows hoping for a small stipend. It is a shame these able men should be without positions. We have had only a 10% [salary] cut, a year ago, but may well have another. I am hoping that conditions will improve soon."

Or, to put it in the terms used by Foote in December 1933:

"One does not require familiarity with the matrix mechanics to understand the principle of uncertainty as regards a physicist's employment during the past three years." [6]

Financial support for science was being severely reduced, and the outlook

NOVEMBER 11, 1931

OM NUCLEUS SEEN ELDING TO SCIENCE

A. H. Compton Visions New 'a in Physics When It Will at Last Be Smashed.

GENERATOR HINTS WAY

00,000 Volts Leap From $90 Device in Test Before the American Institute.

he atomic nucleus, the storehouse he vast energy of the atom, until practically impenetrable by ncies controllable by science, has ast begun to yield to experiments ch bid fair to disclose their in-t nature, it was said last night Dr. Arthur H. Compton of the versity of Chicago, Nobel Prize ner in physics, at a dinner given cientists and newspaper men by newly formed American Institute Physics at the New York Athletic b.

xperiments described by Dr. Comp-as "remarkable," and achieving t had hitherto been regarded by ntists as impossible, recently

SEPTEMBER 12, 1933

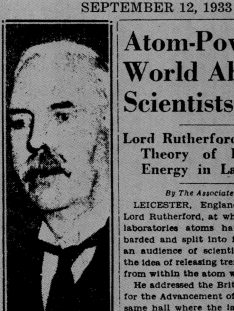

Lord Rutherford

Atom-Powered World Absurd, Scientists Told

Lord Rutherford Scoffs at Theory of Harnessing Energy in Laboratories

By The Associated Press

LEICESTER, England, Sept. 11.— Lord Rutherford, at whose Cambridge laboratories atoms have been bombarded and split into fragments, told an audience of scientists today that the idea of releasing tremendous power from within the atom was absurd.

He addressed the British Association for the Advancement of Science in the same hall where the late Lord Kelvin asserted twenty-six years ago that the atom was indestructible.

Describing the shattering of atoms by use of 5,000,000 volts of electricity, Lord Rutherford discounted hopes advanced by some scientists that profitable power could be thus extracted.

"The energy produced by the breaking down of the atom is a very poor kind of thing," he said. "Any one who expects a source of power from the transformation of these atoms is talking moonshine. . . . We hope in the next few years to get some idea of what these atoms are, how they are made and the way they are worked."

Sir Oliver Lodge, eminent physicist,

MARCH 30, 1934

USE OF THE ENERGY IN ATOM HELD NEAR

Dr. Compton Says New Experiments Show Its Practical Use May Be Possible.

CITES SUCCESSFUL TEST

Found Expenditure of 100,000 Volts on Atomic Bombardment Produced 3,000,000 Volts.

Science has obtained conclusive proof from recent experiments that the innermost citadel of matter, the nucleus of the atom, can be smashed, yielding tremendous amounts of energy and probably vast new stores of gold, radium and other valuable minerals, Dr. Karl T. Compton, president of the Massachusetts Institute of Technology, declared last night before a meeting of the Institute of Arts and Sciences of Columbia University at McMillin Academic Theatre, Broadway at 116th Street.

Although much energy must still be used to bombard matter in order to release atomic energy, the efficiency of the process is increasing and there are hopeful signs that eventual use of atomic energy on a practical basis may be possible. Dr.

was dim. A survey of the congressional appropriations bill by Science Service, published in July 1932, showed that funds for scientific research in the various government departments had been cut 12.5% for the 1932–33 fiscal year. Further cuts were made by President Herbert Hoover in the budget estimates he submitted to Congress in December 1932.[7] Operating funds of the National Bureau of Standards, the major government employer of physicists, were effectively cut 70% between 1932 and 1934.[8]

The impact of budget cuts at the universities can be seen in these telling excerpts from F. Wheeler Loomis's annual reports for the University of Illinois physics department, which, under his leadership, had been among the departments making rapid strides in the preceding years:

1931–1932 "The outstanding feature of this year in the physics department, as probably in all others, has been the curtailment of our activities made necessary by the financial emergency in the University. Since the time the economy orders were promulgated in January the department will have saved out of its appropriations about $3500, or 40 percent of the maintenance and operation budget for the year . . . most severely affected will be, of course, the research."

1933–1934 "The salient features of the past year in the physics department, have been the effects of the depression budget and the reduced enrollments in the courses. The department has had to get along with half the operating funds it had in the past and with no money at all for new equipment."

1934–1935 "The department, whose operating expenses have been reduced to a starvation point for over three years, suffered a financial crisis this winter and pretty nearly had to close up . . . It is almost impossible to convey an adequate idea of the extent to which our work, both in teaching and research, has been hampered and made inefficient and how all progress has been blocked by the inability to buy necessary articles. We should have had pretty nearly to cease activity in research if it hadn't been for the equipment which was bought in our three boom years 1929–32. . . ."[9]

The unfilled aspirations of budding physicists and the despair of department chairmen were amplified in public statements by spokesmen for the scientific community including William W. Campbell, astronomer and president of the National Academy of Sciences and Karl T. Compton, chairman of the board of the newly formed American Institute of Physics. In 1934 Campbell stressed that cutbacks in financial support of science had interrupted the careers of students and researchers and that many of them would be lost to science. The quality of research was still good, but, he warned, if the reduced scale of support continued for two or three more years, the results would be very bad indeed.[10] Compton and Henry Barton, director of AIP, called for an increase in government support of research to offset the decline in private support. But, if scientific research was to be supported by public funds, then the public had to be informed and convinced of the benefits of research. Determined efforts to do this were made by AIP.

Public image of physics

The public was not unaware of new discoveries in physics, especially in nuclear physics, which promised to yield new sources of energy. Newspapers and magazines described the exciting results of the "atom splitters," including artificial disintegration of the light elements and discovery of the neutron. Despite Ernest Rutherford's public ref-

JUNE 4, 1931.

PHYSICS INSTITUTE WILL BE ORGANIZED

Dr. K. T. Compton, Head of M. I. T., Announces Plan to Knit All Branches in Field.

SOCIETY WILL SERVE PUBLIC

Press Department to Explain New Laboratory Discoveries as They Occur.

CAMBRIDGE, Mass., June 3 (AP).— Plans for formation of a consolidated scientific organization to be known as the American Institute of Physics were made public today by Dr. Karl T. Compton, president of Massachusetts Institute of Technology.

Both science and the public are to be served. The institute will bring together several scientific organizations now separate but having common interests. It will also knit together a great group of men in industrial laboratories and manufacturing plants who, as physicists, play a most fundamental rôle in modern industry, but who have not heretofore constituted a well recognized unit. Also in schools and colleges, local or student branches of the institute may be founded.

ew York T

Copyright, 1934, by The New York Times Company.

NEW YORK, FRIDAY, FEBRUARY 23, 1934.

Leaders Deny Science Cuts Jobs; Warn Against 'Research Holiday'

Dr. Millikan Declares Technological Unemployment Is a Myth— Compton Proposes Federal Subsidies for Invention— Roosevelt Aids All-Day Symposium Here.

Science struck back at its critics yesterday, and with the aid of some of its inventions—the radio, sound cameras and loud-speakers—it told the world that science makes jobs and does not end them.

Fortified with statistics to confound the technocrats, armed with a message from the President, and bearing determined and bulky state- tries" in the near future and make many jobs.

The electron, he pointed out, had "gotten into industry" and had "created jobs" in enormous numbers, and research in pure and applied science undoubtedly would produce other brain children that, when harnessed, would make industries and provide jobs for thou-

NOVEMBER 28, 1

DR. CONANT OPPOS CURB ON RESEAR

'Planning' of Science W Check Intellectual Activ Harvard President Say

PRAISES CARNEGIE MET

Free Hand for Exceptional Urged—Dr. Keppel Sees Trusts Facing Change

Dr. James B. Conant, pres of Harvard University, took last night with the theory brakes should be put upon th crease of scientific research knowledge. He spoke at a di in the Waldorf-Astoria Hotel, ing the centenary celebration Andrew Carnegie's birthday.

The dinner was given by the negie Corporation of New Y which has had charge of the tennial celebration in the Ur States and the British dominio It was attended by members of

utation of the idea that tremendous amounts of energy could be released from the atom and harnessed, many scientists, science writers and laymen eagerly discussed the possibilities. Rutherford's statement has been frequently quoted in recent years:

"The energy produced by the atom is a very poor kind of thing. Anyone who expects a source of power from the transformation of these atoms is talking moonshine. . . ." [11]

Less well known now are the more optimistic views of other physicists. Attached to the *New York Herald Tribune* account of Rutherford's talk was another news item quoting Ernest Lawrence on the need to develop an efficient method of obtaining atomic power. Lawrence concluded: "I have no opinion as to whether it can ever be done, but we're going to keep on trying to do it." [12]

Despite the disagreement among scientists, public interest in atomic energy was high in the early 1930's. But although the public often associated the physicist with expectations of practical applications of atomic energy, it also increasingly linked him with unfamiliar complexities. Mingled with headlines such as "Use of the Energy in Atom Held Near," were such editorial comments as:

"It is not the electron that needs seven dimensions but the mathematicians. The world awaits another Newton to reveal simplicity. We are merely in the stage of experimenting with theories. Out of it clarity must issue if science is not to become irrational." [13]

In 1934, commenting on the complexity of the neutrino concept, the *Times* asked: "Can it be that nature needs eight particles in constructing the cosmos? Or is it the physicists who need them." [14]

Antiscience movement

More significant than the ambivalent public image of physics in the early 1930's was the changing public attitude towards science in general. As the depression wore on there was an increasing disenchantment with the "technological progress" that had long been associated with science. Science-based technology and the labor-saving devices it had produced were variously seen as uncontrolled, unplanned or misappropriated and, in all cases, as a major factor in the deepening economic crisis and the resulting despair. One proposed solution was to declare a moratorium on scientific research. The "Stop Science" movement dated back to the late 1920's in England and found increased social resonance in the US in the early 1930's. It is difficult to determine how widespread this attitude was, because it was only occasionally articu-

lated. It was, however, perceived as a major threat by leaders of the scientific community, because it occurred at precisely the time when scientists needed to make an effective case for increased public support.

Even before the economic crisis, criticism of science was gaining ground among those who saw it as a threat to humanistic values. Late in 1928, Robert A. Millikan commented on "the current opposition to the advance of science," and told his Chamber of Commerce audience that the *real* question was "how the forward march of pure science, and of applied science which necessarily follows upon its heels, can best be maintained and stimulated." [15]

By mid-1932, the realities of the depression sharpened the criticism of science and modified the response. In a speech dedicating the Hall of Science for the Chicago Century of Progress Exposition, Frank B. Jewett, head of the Bell Telephone Laboratories noted:

"In some quarters a senseless fear of science seems to have taken hold. We hear the cry that there should be a holiday in scientific research and in the new applications of science, or that there should be a forced stoppage in the extension of old usages by mandatory legislation." [16]

Jewett's response was a call for greater efforts to weave science into the social structure. The purpose of the Century of Progress Exposition, he said, was to increase understanding of the real place of science in the social structure and of those factors that have their roots in science and must influence the course of social controls in the years ahead.

Science was the theme of the exposition. Chicago was celebrating her centenary as a city, and the planners of the exposition wanted to show that the city's growth had been united with the growth of science and industry during the preceding century. The National Research Council enlisted the support of 400 scientists and businessmen to advise on exhibits. During the three years preceding the opening of the exposition, about 90 physics exhibits were devised and assembled by a group of physicists under the direction of Henry Crew. Similar exhibits represented the other sciences.

The exposition itself was opened in a dazzling manner to emphasize the accomplishments of science. Light that had ostensibly started its journey to earth from the star Arcturus 40 years earlier (at the time of the last great Chicago exposition) was relayed to Chicago from the 40-inch refracting telescope at Yerkes Observatory, Wisconsin, in the form of an electrical impulse to start the big show's night life. (The distance to Arcturus is now known to be about 36 light years.) The guidebook put it this way:

"A miracle, they would have said a hundred or even forty years ago. But today, the 'electric eye,' relays, vacuum tubes, amplifiers, microphones, which respond to the tiniest fluxes of energy, help to do the work of the world in almost routine manner. Progress!" [17]

The exuberant celebration of science and its applications took place in one of the worst depression years, and a major aim was to demonstrate that "the steady march of progress" could not be stopped by temporary "recessions." Considering the large number of unemployed in 1933, one wonders how the fair-going public reacted to the slogan boldly proclaimed in the official guidebook: *Science Finds—Industry Applies—Man Conforms!* [18] The use of science at the exposition may have been imaginative and entertaining, but it provided no real answer to critics who called for a research holiday. Instead it provided a dramatic reaffirmation of the relation of science to a technology that, in the eyes of some critics, had been misdirected and thus had contributed to existing social evils.

Response of the physicists

By late 1933 leaders of the physics community were alarmed about the criticism of science, because such criticism threatened to reduce public support of research even further. Their approach was to deny that science had caused unemployment. On the contrary, they asserted, science had created more jobs, and greater support of science could help to end the depression. Barton was among those who called for more flexible political and economic institutions and for methods to cope with "natural and unavoidable increases in human knowledge," [19] but the major emphasis of the scientists' response was simply that the answer was more rather than less science.

AIP and the New York Electrical Society (an engineer's group) responded to the crisis by conducting a well publicized symposium, "Science Makes More Jobs," in February 1934. Speakers included Karl T. Compton (who was president of MIT as well as chairman of the AIP Board of Governors), Millikan (president of the California Institute of Technology), Frank Jewett and William Coolidge (director of the General Electric Research Laboratory). Their talks, urging continued support of scientific research by government, universities and industry, were broadcast nationwide. Letters were read from President Franklin D. Roosevelt and from Albert Einstein, who pointed out:

". . . one cannot be sufficiently cautioned against the attempt to economize on scientific work. On the one hand, the progress of important branches of technology depends

Visitors in the Hall of Science at the 1933 Century of Progress Exposition. This photograph and the one on page 31 are used by permission of the Library, University of Illinois, Chicago Circle Campus.

on the results of experimental and even of theoretical science; and on the other, each disruption of scientific work causes lasting damage to the living body of research; that is to say, a partial forfeiture of previously expended labor and capital. . . . Hence, it is in the interest of this country to put on a secure footing the continuation of scientific investigations on the previous scale . . . ." [20]

The symposium was not unnoticed. A front page article in *The New York Times* the next day began "Science struck back at its critics yesterday . . .," and a full account of the meeting and the major points of the speeches followed. The editorial page, however, expressed disappointment:

"Neither the statistics nor the argument are new. Nor did any of the protagonists of the laboratory explain why there is poverty amid plenty, and idleness where we expect to hear the hum of the machine. We look to them for a way out of the slough, only to find them as helpless as the economists. As yet no one has devised the means of absorbing new technical developments with the least possible amount of distress. The question of pace is all important." [21]

The editorial went on to call for a government plan to apply science without neglecting "human aspirations" and "moral values."

The scientists simply had not addressed themselves to the immediately relevant questions of the social management of science and its applications. They had *assumed* that science and technology were the sources of progress that would lead to desirable improvements in the social condition. Although the tone was more defensive than the slogans of the Century of Progress Exposition, the message was the same. No wonder then, that even some friends of science tended to discount the scientists' statements as special pleading.

While science was attempting to answer its critics, efforts were also underway in Washington to improve coordination of government scientific work and to develop an emergency scientific program to combat the depression. Karl Compton was chairman of the Science Advisory Board appointed by President Roosevelt in 1933. Compton emphasized the need for major government support of applied science; some of the money could then be used to support basic science. His theme was expressed in the title of an article he wrote in 1935: "Put Science to Work: The Public Welfare Demands a National Scientific Program." [22] To bolster his point that science should have greater, rather than less, government support, he argued that other nations had more enlightened policies toward the support and organization of research. But the Board's activities were often marked by disagreements about the relative roles of the social and natural sciences. Another major roadblock was the fear of many scientists that government support would lead to government control and to the involvement of science in political and social issues. The Board went out of existence in 1935, and all concerned agreed that it had been a failure. [23]

In general, the arguments used during the early 1930's to encourage greater moral and financial support of science by the public went wide of the mark. It was not enough merely to reassert that the basic science–applied science–technology cycle would alleviate the economic and social crisis. In answering their critics, scientists did not respond adequately to the public's fear that uncontrolled and misapplied technology caused human misery. Karl Compton and others did urge individual scientists to analyze social, economic and political problems, and to ask at what points science could be usefully brought to bear on them. [24] But spokesmen for the US science community appeared reluctant to deal publicly with the social and political issues involved in revamping institutions and in discussing the rate and direction of the application of science. Justifying public support of science as a social good, however, implicitly involved assumptions about the social processes leading to eventual application of research. To ignore the growing concern with the need to analyze and improve these social processes was to weaken the argument for support of science.

The problem disappears

Things got better anyway. After 1935 the financial pinch eased and more academic teaching jobs became available; young physicists were needed to cope with the increasing enrollments in US colleges and universities. The improvement was only gradual, however, and in the middle and late 1930's the search for employment took many physicists into work they had not previously considered (for example, into oil fields as part of industrial geophysical research teams). A major effort was made by AIP during this period to call attention to the role of physics in industry, and symposia were held throughout the country to explore how this role could be increased for the benefit of industry, the nation and the physicists. These efforts raised occasional questions about whether the new physics PhD's were properly prepared and motivated for industrial positions, but apparently no major change in physics education resulted.[25]

By the end of the 1930's, growth curves looked good again. More than 1400 physics doctorates were awarded by US universities from 1931 to 1940, double the number awarded in the preceding decade.[26] The physics profession in the US had also been enriched by about 100 very talented physicists who had emigrated from Europe because they were unable or unwilling to continue their careers in Nazi-dominated countries.[27] By the spring of 1941 an estimated 4600 physicists were at work in the US, about half of them with doctorates,[28] and total expenditures for scientific research in the US had doubled during the decade.[29] Despite the depression crisis, physics had recovered and normal "progress" had returned.

During the dismal depression days questions about the internal dynamics of the physics community and about its relationship with the larger society were raised. These questions remained unanswered, to emerge again in other times of crisis. Although the vast changes that have occurred since the 1930's in physics, in the physics community and in the role of physicists in society have been accompanied by new questions and problems, the continuity of certain issues is clear. The perspective provided by the experiences of the depression emphasizes the pressing need and new opportunity to make (borrowing Foote's 1933 phrase) "a thorough investigation of the sociological aspects of physics." In the 1930's adequate answers were not provided to the challenges to the social relevance and human implications of physics, because significant changes in the economic and political situation permitted resumption of the growth of physics. But physicists now have another chance to respond, and they must if they are to cope with the present crisis and plan for the future.

References

1. A. W. Hull, "Putting Physics to Work," Rev. Sci. Instr. **6**, 377 (1935).
2. P. Foote, "Industrial Physics," Rev. Sci. Instr. **5**, 63 (1934).
3. Lewis Papers, University of California, Berkeley.
4. Niels Bohr Library, AIP.
5. J. C. Slater, "Quantum Physics in America Between the Wars," PHYSICS TODAY **21**, no. 1, 43 (1968).
6. P. Foote, Ref. 2, page 57.
7. Science **76**, 94 and 561 (1932).
8. R. C. Cochrane, *Measures for Progress: A History of the National Bureau of Standards*, US Dept of Commerce, Washington, D. C., (1966) page 322.
9. G. M. Almy, "Life with Wheeler [Loomis] in the Physics Department, 1929–40," manuscript, Niels Bohr Library, AIP.
10. W. W. Campbell, Science **79**, 391 (1934).
11. H. A. Barton, "Scientific Research in Need of Funds," Literary Digest **119**, 18 (1935).
12. *New York Herald Tribune*, 12 Sept. 1933.
13. *The New York Times*, 25 June 1933.
14. *The New York Times*, 11 March 1934.
15. R. A. Millikan, "Relation of Science to Industry," Science **69**, 30 (1929).
16. F. B. Jewett, "Social Effects of Modern Science," Science **76**, 24 (1932).
17. Chicago Century of Progress International Exposition, *Official Guide Book of the Fair*, page 20, Chicago (1933); L. Tozer, "A Century of Progress, 1833–1933: Technology's Triumph Over Man," American Quarterly **4**, 78 (1952).
18. *Official Guide Book of the Fair*, page 11.
19. H. A. Barton, "Shall We Stop Scientific Progress," Rev. Sci. Instr. **4**, 520 (1933).
20. A. Einstein to H. A. Barton, 21 Feb. 1934. Niels Bohr Library, AIP; talks published in Scientific Monthly **38**, 297 (1934).
21. *The New York Times*, 24 Feb. 1934.
22. K. T. Compton, The Technology Review **37**, 133, 152 (1935).
23. A. H. Dupree, *Science in the Federal Government: A History of Policies and Activities to 1940*, Harvard U. P., Cambridge (1957) page 350.
24. K. T. Compton, "Science and Prosperity" Science **80**, 387 (1934).
25. *Physics in Industry*, AIP, New York (1937).
26. National Research Council publications and *Dissertations in Physics* (M. L. Marckworth, ed.), Stanford (1961).
27. C. Weiner, "A New Site for the Seminar: The Refugees and American Physics in the Thirties," in *Intellectual Migration* (D. Fleming and B. Bailyn, eds.), Harvard U. P., Cambridge, Mass. (1969), page 190.
28. "Physicists in National Defense," mimeographed report, April 1942, Niels Bohr Library, AIP.
29. V. Bush, *Science the Endless Frontier*, Washington, D. C. (1945), page 86. (Reprinted by National Science Foundation, 1960). □

Scientists with a secret

While the Nazi war machine was gearing up, a few physicists
realized that a fission chain reaction was feasible—would they be able
to get all groups to agree to hold back publication?

Spencer R. Weart

PHYSICS TODAY / FEBRUARY 1976

What are physicists to do if they make a
discovery that promises to transform
industry but also threatens to revolu-
tionize warfare? Should they investigate the phenomenon within their traditions of free and open inquiry or keep
the deadly secret to themselves? This
is the dilemma that was faced by several groups of physicists who studied uranium fission in 1939 and 1940. In the
spring of 1939 one group, foreseeing the
unprecedented power of nuclear weapons, made a concerted attempt to restrict knowledge of chain reactions.
But it was not until over a year later
that censorship—imposed by the community of physicists on itself—became
fairly complete.

Any attempt to keep a secret must by
its very nature follow a course that is
difficult to observe, creating confusion
and misunderstanding. But this
course, which the participants could not
see clearly at the time, can now be
pieced together from collections of papers made available to researchers, supplemented by oral history interviews
conducted by the Center for History of
Physics of the American Institute of
Physics.

Fears of disaster

The first arguments over nuclear secrecy revolved around the unlikely figure of Leo Szilard. A short, round, exuberant Hungarian, Szilard in 1939 had

Spencer R. Weart is the director of the Center
for History of Physics, American Institute of
Physics, New York.

neither a job nor a home. But he was
uniquely qualified to face the issues of
nuclear energy and secrecy because for
over five years he—and he alone—had
been concentrating on these problems.

Since 1933 Szilard, then recently arrived in England to escape the Nazi
persecution of Jews, had wondered if
there was a way to release the energy
that physicists knew to be bound up in
nuclei.[1] The answer came with his realization that if one could bombard
some element with a particle (say, a
neutron) and make it radioactive in
such a way that it emitted two particles,
a chain reaction of awesome power
might be induced. The possibility
seemed much closer the next year, when
Frédéric Joliot and his wife Irène Curie,
working at the Radium Institute in
Paris, discovered that, with alpha particles, one could indeed make nuclei radioactive artificially. Szilard decided
to devote himself to nuclear physics and
set out to search for some type of nucleus in which a chain reaction might be
sustained.

From the start Szilard feared the consequences of his work. He attempted
to gain some control by the only means
then available to a scientist who wanted
to restrict the use made of his work:
He took out a patent on his ideas. Furthermore, he persuaded the British government to declare the patent secret;
there was a small but real possibility, he
warned them, of constructing "explosive bodies ... very many thousand
times more powerful than ordinary
bombs."[2] Meanwhile Szilard brashly

tried to alert his colleagues in Britain.
His ideas, he told one professor in 1935,
could cause an industrial revolution but
might cause a disaster first. It would
be necessary to bring about something
like a conspiracy of the scientists working in the general field. In a letter to F.
A. Lindemann, the head of physics at
Oxford, he offered a mechanism to ensure secrecy—an agreement to make experimental results in the dangerous
zone available only to those working in
nuclear physics in England, America
and perhaps one or two other countries,
while otherwise keeping quiet.[3]

Szilard foresaw only too well the likely reaction to his efforts: "Unfortunately it will appear to many people
premature to take some action until it
will be too late to take any action."[3]
And indeed the leading physicists in
Britain were cool to Szilard's obstreper-
ous advice. They thought his proposed
chain reaction entirely unworkable (as
was in fact the case for the mechanisms
Szilard was then considering). They
were suspicious when he sought to patent his ideas, suspecting that he was
seeking pecuniary return, a motive incompatible with British traditions of
disinterested science. Finally, they
found the idea of scientific secrecy entirely alien. Even those scientists who
felt most keenly the responsibility of
scientists for the consequences of their
discoveries traditionally felt that secrecy is abhorrent and that interference
with the normal process of open criticism would not only impede scientific
progress but pervert it.[4,5]

Szilard went on to study various elements for a possible chain-reaction mechanism; he had not quite reached uranium when he learned that Otto Hahn, Fritz Strassmann, Otto Frisch and Lise Meitner had discovered uranium fission. When Szilard heard of this in January 1939 in New York, where he had moved to escape the war that appeared ever more imminent in Europe, he discussed his concern with scientists at Columbia University.

Private messages

The leading nuclear physicist there was Enrico Fermi, who had fled Italy because Fascist race laws affected his Jewish wife, and who had arrived in New York scarcely three weeks ahead of the news of the discovery of fission. Like Szilard and other physicists, Fermi quickly recognized the possibilities this discovery opened. According to one account, he made a rough calculation of the size of the hole a kilogram of uranium would make in Manhattan Island if it underwent an explosive chain reaction.[6] However, he soon concluded that this would never happen: When a uranium nucleus was struck by a neutron and split in two, it seemed unlikely that it would release enough neutrons to sustain a chain reaction. When Szilard approached Fermi about the need to keep fission work secret, Fermi's response was direct: "Nuts!"

> From the very beginning [*Szilard recalled*] the line was drawn; the difference between Fermi's position throughout this and mine was marked on the first day we talked about it. We both wanted to be conservative, but Fermi thought that the conservative thing was to play down the possibility that this [chain reaction] might happen, and I thought the conservative thing was to assume that it would happen and take all the necessary precautions.[1]

Rebuffed by Fermi, Szilard remained alert for a way to control events. At about this time, late January, a telegram arrived at Columbia, addressed from Hans Halban, a physicist in Paris, to his colleague George Placzek. As Szilard recalled it long after, the telegram was opened by a secretary by mistake, and Szilard learned the contents: "JOLIOT'S EXPERIMENTS SECRET." Placzek had just come from a visit in Paris, and Szilard assumed that Placzek had learned of an experiment Joliot was doing; apparently Joliot had now decided to keep the experiment quiet for the time being. Szilard had little doubt what experiment would be so important as to require secrecy.

What Szilard felt was involved here was the sort of secrecy that had been traditional in science for centuries—the caution of the scientist who holds back his results until he is ready to publish

them, so they will not be broadcast in a distorted form and so that others will not take advantage of a hint to beat him to the next result. This was quite different from the sort of secrecy Szilard had in mind. There was some misunderstanding here, for Joliot did not actually begin fission experiments until late January, after Placzek had left Paris, and it is not clear what Halban and Placzek were corresponding about. But Szilard now believed (correctly as it happened) that Joliot's group was working on fission, and decided to send him a letter.

The only reason he was writing, Szilard said, was that there was a remote possibility that he would be sending a cable after some weeks, and the letter was to explain what his cable would be about. Some scientists in New York were concerned about the possibility that neutrons would be liberated in fission. Obviously, if more than one neutron would be liberated, a sort of chain reaction would be possible. In certain circumstances this might then lead to the construction of bombs which would be extremely dangerous in general and particularly in the hands of certain governments. Perhaps steps should be taken to prevent anything on this subject from being published. No definite conclusions had been reached, but if and when any steps were agreed on, Szilard would cable Joliot. Meanwhile Fermi was doing experiments to see whether the danger was real, and these would perhaps be the first to give reliable results. But if Joliot got definite results sooner, Szilard would be glad to have the uncertainty ended. Also, if Joliot felt that secrecy should be imposed, his opinions would be given very serious consideration.[3]

Neither Joliot nor his close collaborators Halban and Lew Kowarski responded. The letter was obviously a purely personal venture, and this impression must have been reinforced by a letter Fermi sent Joliot two days later. On 4 February 1939 Fermi wrote that he was then engaged in trying to understand what was going on in uranium fission—as was, he thought, every nuclear physics laboratory. After thus informing Joliot's team that they had competition, Fermi went on to ask help for another Italian refugee scientist and closed without saying a word about keeping secrets.[7] There was every reason to believe that Fermi would publish first if the French held back their own results.

Even as a personal request Szilard's letter made little impression on the French, for it stated that it was only meant to help them understand a cable that might follow. Weeks passed, no cable appeared, and the French, as Kowarski recalled, "considered that probably the whole idea was abandoned. We simply published."[8]

This publication, the first result of the joint efforts of Halban, Joliot and Kowarski, contained important news: Neutrons were indeed liberated when a uranium nucleus fissioned.[9] The experiment was of a kind that would only have been done in a few places, requiring ingenuity, a powerful source of radioactivity and an interest in chain reactions. It had not been easy to detect the few neutrons produced in fission amidst the flood of neutrons that had been required to provoke some fissions in the first place, nor had it been obvious that these neutrons were important. Although the French, like Fermi, believed scientists everywhere were

---

The many "secrets" of the atomic bomb

There was no single discovery that showed how atomic bombs could be built, but a combination of discoveries made at various times. Here is a partial list:

Published discoveries

1934 Artificial radioactivity can be produced with alpha particles (Joliot and Curie, *France*) or neutrons (Fermi, *Italy*).

December 1938 Neutrons can cause uranium to fission (Hahn and Strassmann, *Germany*, Frisch, *Denmark* and Meitner, *Sweden*).

March 1939
▶ Neutrons are produced during fission (Anderson, Fermi and Hanstein, *US*; Szilard and Zinn, *US*; Halban, Joliot and Kowarski, *France*).
▶ Two or three neutrons are emitted per fission (same groups).
▶ U^{235} is the fissionable isotope of uranium (Bohr and Wheeler, *US*).

Unpublished discoveries

June 1939–February 1940 A self-sustaining nuclear reactor can be built if a suitable moderator can be found (Szilard, *US*; Halban, Joliot, Kowarski and Perrin, *France*; Heisenberg, *Germany*; various groups, *USSR*).

Spring 1940
▶ Carbon is a suitable moderator for a nuclear reactor (Anderson and Fermi, *US*).
▶ Nuclear reactors can be used to produce a fissionable element, plutonium (Turner, *US*)—from this resulted the bomb that devastated Nagasaki.
▶ It is possible to isolate sufficient U^{235} to make an explosive critical mass (Frisch and Peierls, *UK*)—from this resulted the bomb that devastated Hiroshima.

hard at work on the question, there was in fact only one other group then carrying on a similar experiment—the group at Columbia.

Chain reaction—and invasion

By mid-March Fermi and Szilard, working with Herbert Anderson, Walter Zinn and others, had done their own experiments and independently learned the distressing news that neutrons were produced in fission. This was still far from proving that a chain reaction was possible, for that would depend on the precise number of neutrons emitted in each fission, a thing still more difficult to measure. The group estimated that there were about two neutrons per fission, which made it appear only barely possible that a chain reaction could be sustained (the true value is about 2.5 neutrons per fission).

On 15 March, as the Columbia physicists finished writing up their experiments for publication, German troops invaded the remnant of Czechoslovakia that had survived the Munich agreement. With this action, many felt, Hitler crossed his Rubicon, subjecting for the first time a non-German people and giving a clear signal that world war was inevitable. Despite their concern over this, the physicists sent their papers to the *Physical Review* the next day.

Szilard was not satisfied, and three days later he met with Fermi and with another Hungarian refugee physicist, Edward Teller. As Szilard recalled the meeting, he and Teller pressed for keeping their work secret, but Fermi was repelled by this idea, holding that publication was basic to scientific morality. "But after a long discussion, Fermi took the position that after all this is a democracy; if the majority was against publication he would abide by the wish of the majority ..."[1] Fermi therefore arranged to ask the *Physical Review* to delay the publication indefinitely.

Szilard was now on the point of cabling Joliot, but before he did so he heard of the French team's note, just published in *Nature,* which revealed that some neutrons are emitted in fission. Fermi felt that there was now no secret to keep, so that there was no longer any sense in refusing to publish. Szilard denied this (the crucial number of neutrons emitted per fission was not yet published), and argued that "If we persisted in not publishing, Joliot would have to come around; otherwise, he would be at a disadvantage, because we could know his results and he would not know our results." Fermi was not convinced but, determined to be fair, he reluctantly agreed to put the matter before George Pegram, administrative patron of the Columbia group and a respected physicist. Pegram delayed his

SZILARD

KOWARSKI, HALBAN AND JOLIOT

decision for some time. Szilard's arguments were forceful, but others at Columbia replied that an attempt to restrict publication was both futile and an undesirable breach of scientific custom.[1,3]

Warnings

While Pegram deliberated, Szilard and his friends were determined to waste no time. Several of them talked the matter over, among them Victor Weisskopf, an emigré Austrian physicist. "We were very much afraid of the Nazis," Weisskopf recalled. "We knew this was a hopeless thing but we thought we had to try ... And then the question was ... how do we get to Joliot." As Weisskopf said in a recent interview, he had met Joliot's collaborator Halban years earlier and the two had become close personal friends, so Szilard and Weisskopf drafted a telegram to Halban, which Weisskopf signed. The telegram asked Halban to advise Joliot that papers on neutron emission had already been sent to *the Physical Review*, but that the authors had agreed to delay publication for the reasons indicated in Szilard's letter to Joliot of 2 February. The telegram continued:

NEWS FROM JOLIOT WHETHER HE IS WILLING SIMILARLY TO DELAY PUBLICATION OF RESULTS UNTIL FURTHER NOTICE WOULD BE WELCOME STOP IT IS SUGGESTED THAT PAPERS BE SENT TO PERIODICALS AS USUAL BUT PRINTING BE DELAYED UNTIL IT IS CERTAIN THAT NO HARMFUL CONSEQUENCES TO BE FEARED STOP RESULTS WOULD BE COMMUNICATED IN MANUSCRIPTS TO COOPERATING LABORATORIES IN AMERICA ENGLAND FRANCE AND DENMARK ...[7]

The proposed scheme was similar to the one Szilard had conceived in 1935, with the additional idea that papers should be sent to journals, not for publication but to certify priority of discovery.

At the same time Weisskopf also cabled P.M.S. Blackett, a leading British physicist, asking whether it would be possible for *Nature* and the Royal Society's *Proceedings* to cooperate in delaying publication of fission research. Meanwhile another of Szilard's Hungarian physicist friends, Eugene Wigner, wrote P.A.M. Dirac and asked him to support Blackett. The matter was rather urgent, Wigner said; although American scientists were willing to cooperate, they realized that their interests might be prejudiced if scientists in other nations published results and they did not.[3,10] Blackett and another prominent physicist, John Cockcroft, promptly replied that they would support the secrecy plan. *Nature* and the Royal Society were expected to cooperate.[3]

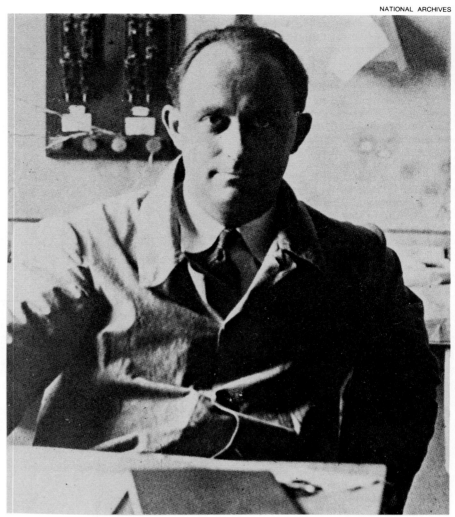

FERMI

Szilard, Teller, Weisskopf and Wigner also talked the problem over with Niels Bohr, who was visiting the United States. Bohr doubted very much that fission could be used to cause a devastating explosion. And he thought that at any rate it would be difficult if not impossible to keep truly important results secret from military experts—the matter was already public. Nevertheless he agreed to go along with the attempt and drafted a letter to his Institute in Denmark (which apparently he did not immediately mail):

The Columbia group is busy organizing cooperation among all the physics laboratories outside the dictatorship countries, to keep possible results from being used in a catastrophic way in a war situation, and I must therefore ask you, if work along these lines is going on in Copenhagen, to wait before you publish anything until you have cabled me about the results and received an answer.[11]

But the conspirators still had to win the agreement of other American laboratories.

The most immediate problem was a group headed by Richard Roberts working under Merle Tuve at the Carnegie Institution in Washington, DC. They too had recently seen some neutrons released from uranium. But the neutrons they saw were emitted over a period of some seconds after fission occurred: These were not the true fission neutrons, but occasional neutrons produced as a side effect of the radioactivity of the fission fragments.[12,13] The development was announced in a news release of Science Service dated 24 February, written by Robert D. Potter, a science writer who kept in touch with the Columbia physicists and was infected with their excitement over chain reactions. Potter headlined the possibility of an explosive chain reaction propagated by neutrons. He carefully noted that Roberts's delayed neutrons might not be enough to sustain a chain reaction—in fact they are not—but he quoted Fermi as saying that the possibility of a chain reaction was certainly present.[14]

Szilard and his friends quickly approached the Washington group, who promised to cooperate in withholding future publications. The proposal was spread further within the United States by word of mouth and letter. Maurice Goldhaber of the University of Illinois was included and Ernest Lawrence of Berkeley was probably informed of the matter when he visited New York on 3 April.[15] John Tate, editor of the *Physical Review,* was brought in, for nearly all important physics papers in the United States passed through Tate's office; anyone else who showed an interest in fission neutrons could thus be put in touch with the conspirators. The attempt to restrict the circulation of information to physicists outside the dictatorships was well begun. It lacked chiefly the acquiescence of the French.

The French reply

The French knew what Weisskopf's telegram implied, for they were as alarmed as he by Hitler's march towards world war. However, like Bohr and Fermi, the French believed an atomic bomb was not likely to be built for many years, if ever. In this they were entirely correct, so far as atomic bombs were then conceived—masses of tons of natural uranium. Nobody had yet seriously considered the likelihood of isolating a substantial quantity of the rare fissionable isotope U^{235}, still less of the undiscovered element plutonium; and these two substances are the only ones that could in fact be used for a nuclear weapon. Unaware of these possibilities, Joliot and his collaborators thought that industrial nuclear power from nuclear reactors was a much more immediate prospect than weaponry.

It was up to Joliot, as head of the team, to answer Weisskopf's telegram, but he discussed it at length with his colleagues. Thinking back, they recalled a number of factors that entered their decision.[8,16] For one thing, Joliot believed strongly in the international fellowship of scientists, and in principle had little sympathy with secrecy.[17] For another, if he and his colleagues failed to publish, they might well be eclipsed by those who did. For they could scarcely believe that everyone would adhere to an unprecedented pact, a pact pushed forward, so far as they knew, only by two Central European refugees on the outskirts of the Columbia scientific community. (Had Fermi, Bohr or a leading American scientist written them about the scheme, the French might have found it more plausible.) And if they failed to be first to publish discoveries, the French might have trouble getting the money they would need to pursue the development of industrial nuclear energy. Finally, even if all the laboratories joined and stuck by the agreement, there would remain a powerful objection, the same one noted by Fermi and Bohr. It was scarcely likely that copies of papers circulated privately around America, France, Britain and Denmark could be kept out of

BREIT

Germany and the Soviet Union; moreover, German and Soviet scientists were surely aware of the importance of fission chain reactions.

Ideas of fission power and weapons had begun to show up in the popular press. The French were aware of at least some of the sensational news stories that emanated from the United States. The French were not in close touch with what was happening there, but it is very likely that they had seen a copy of a Science Service news release of 16 March, which summarized their own report, published in *Nature* on that date, of neutrons resulting from fission. Presumably they were not pleased to read that they had apparently been beaten to the discovery: Their result, the release said, "is comparable with, and a confirmation of, the announcement (Science Service, 24 February 1939) that scientists at the Carnegie Institution . . . had been able to observe the same important reaction in atomic transmutation."[18] This was an error, but it made it seem that the most important facts were already leaking out in America.

For all these reasons, the team cabled Weisskopf a discouraging reply around 5 April.

SZILARD LETTER RECEIVED BUT NOT PROMISED CABLE STOP PROPOSITION OF MARCH 31 VERY REASONABLE BUT COMES TOO LATE STOP LEARNED LAST WEEK THAT SCIENCE SERVICE HAD INFORMED AMERICAN PRESS FEBRUARY 24 ABOUT ROBERTS WORK STOP LETTER FOLLOWS

JOLIOT HALBAN KOWARSKI[3]

Szilard was well informed on the work of Roberts's group through their publications and through letters from Teller, who had visited them various times, and on the next day, Weisskopf having left New York, Szilard answered on his behalf. Roberts's paper, he noted, concerned delayed neutron emission, which was harmless. But the group had been approached and had promised to cooperate. The American group had delayed publishing papers; were the French inclined to delay their papers too, or did they think everything should be published?

That same day the French sent their final answer:

QUESTION STUDIED MY OPINION IS TO PUBLISH NOW REGARDS JOLIOT.[3]

The scheme fails

This reply, along with the preceding French publication of the fact that fission does produce some neutrons, doomed the attempt to restrict publication. Pegram, who was not aware how much progress Szilard and his friends had made aside from the French, after some days of deliberation decided that any attempt to impose secrecy was hopeless. Szilard was forced to give in. The Columbia scientists asked the *Physical Review* to print their papers.[19]

On April 7, the day of the final exchange of cables with Szilard, the French sent *Nature* the results of experiments and calculations that estimated the number of neutrons emitted per fission at between three and four. The report was duly published on 22 April 1939. This note convinced many physicists that uranium chain reactions were a real possibility. In Britain, George P. Thomson decided to warn his government of the dangerous prospects and meanwhile to begin experimenting with uranium.[20] In Germany, Georg Joos wrote a letter to the Reich Ministry of Education; independently and simultaneously, Paul Harteck and Wilhelm Groth wrote a joint letter to the War Office.[21] News of the French work may also have played a role in the startup of Soviet nuclear energy research, perhaps provoking the letters on uranium which I.V. Kurchatov and others sent the Soviet Academy of Sciences about this time.[22] Thus in Britain, Germany and perhaps the Soviet Union, publication of the French results precipitated offically-supported programs of research into nuclear energy. The effort of Szilard and his friends, after coming within an inch of success, had failed disastrously.

Nevertheless, by the end of 1939 a blanket of secrecy had settled over fission research in certain countries. After war broke out in September, scientists in France, Germany and Britain withheld publication on fission and any other subject remotely of military interest. But in the United States, the Soviet Union and other neutral countries, publication was scarcely impeded.

US government: Do it yourself

Szilard continued to work on the problem. With Albert Einstein he set in motion a chain of events that led to the formation of an official government committee, under Lyman J. Briggs, which was supposed to support and coordinate fission work.[23] From the beginning Szilard hoped that the committee would also do something about secrecy. When he took up the matter with Briggs he added another element to his by now increasingly well developed scheme. Presumably to counter objections he had faced from younger men at Columbia, he wrote:

> For a physicist, who has not yet made a name for himself, refraining from publication means a sacrifice which he should not be asked to make without being offered some compensation. Some addition to the salary which he is normally drawing from the university might therefore be desirable and might require the creation of some special fund.[3]

But the Briggs committee remained all but inactive, leaving everything up to the physicists. As late as 27 April 1940, when the committee held one of its rare meetings, the only response Szilard could get was a suggestion from Admiral Harold Bowen, present as an observer, that the scientists working on uranium might get together and impose upon themselves whatever censorship they felt necessary. The government itself would do nothing.[3]

Szilard had already taken the single step that was entirely within his power: He withheld from publication a paper of his own. This paper, completed in February 1940, contained elaborate calculations of the characteristics of a nuclear reactor and concluded that there was a strong possibility of making one work. Had the article been published, it surely would have been a great stimulus to nuclear reactor work in various countries. But when Szilard sent it to the *Physical Review* he requested that printing be delayed until further notice.[2] For a second specimen of a withheld paper, in late April Szilard persuaded Herbert Anderson, a graduate student who had worked closely with Fermi on fission from the beginning, to hold back his doctoral thesis on neutron absorption in uranium, which was then already in proof.[24,25]

Anderson and Fermi had meanwhile been measuring the neutron-absorption cross section of carbon. This difficult-to-determine quantity was central to the question of whether or not a nuclear reactor could be built, for carbon seemed the only feasible moderator, and even carbon could be used only if it absorbed virtually no neutrons. This turned out to be the case: The cross section was extremely small. Szilard now approached Fermi and suggested that the value for the cross section should not be published. "At this point," Szilard recalled, "Fermi really lost his temper; he really thought that this was absurd." But while Fermi stuck by his principles, Pegram had second thoughts and finally asked Fermi to keep his work secret.[1]

This decision came late, but still in time: If the value for the carbon cross section had been published, the course of World War II might conceivably have been changed. For German scientists, using experiments they carried out later in 1940, wrongly concluded that carbon had a substantial neutron-absorption cross section. From that point on they abandoned carbon as a moderator and attempted to use the extremely rare isotope deuterium, which they never managed to get enough of.[21,26] Soviet scientists too at first did not seriously consider carbon as a moderator.[27] The French scientists were also committed to deuterium. They escaped to England when France fell to the Germans, and thereafter the British followed their lead in matters of reactors, regarding carbon as an unlikely choice. Anderson and Fermi's work could have put all these groups on a different track.

Prescription for a bomb

This was not the only hole in the dike that had to be plugged. In late May, Louis Turner at Princeton sent Szilard a copy of a paper on "Atomic Energy from U^{238}." In this paper Turner pointed out that if U^{238} were bombarded by neutrons, as would happen in a nuclear reactor, a series of steps would give rise to a new element. This he predicted to be fissionable—it was the element later named plutonium. Although Turner had not realized it, he had written the prescription for the easiest route to building an atomic bomb.

Szilard wrote back at once to say that his own paper was secret, implying that there was an official move underway to withhold papers. He persuaded Turner to write the *Physical Review* and delay publication.[3] It was well he did so: Turner's paper could have been an essential clue for the Germans and others. Meanwhile Szilard approached Harold Urey and asked him to try to set up a committee to regulate fission publications.

Before much progress had been made, the 15 June issue of the *Physical Review* appeared, containing a letter from Edwin McMillan and Philip Abelson at Berkeley. They had observed the production of neptunium when ura-

Two history-making releases from Science Service, as reprinted in *Science News Letter*. After reading an erroneous statement in the later (lower) article, which said that their results had already been published in America, the French team rejected Szilard's request for secrecy.

nium was bombarded with neutrons. This was the first and most essential step of the process that Turner had predicted should lead to plutonium. But Abelson and McMillan had simply failed to see the connection between their work on transuranic elements and the fission problem.[15,28]

This publication brought down a flurry of protest, which helped to settle the secrecy issue. From as far as Britain, scientists interested in fission protested the publication of such revealing information. But the most important news came from Gregory Breit at the University of Wisconsin. Breit had known Szilard and Wigner for years, and was awakened to the secrecy problem through long conversations with them. Around the beginning of June Breit found a way to circumvent the problems Szilard and others were running into. Recently named to the National Academy of Sciences, he had been put in the Division of Physical Sciences of the Academy's National Research Council. At a committee meeting he spoke up in favor of censorship. There was some skepticism, Breit later recalled, but a committee on publications was appointed to consider the problem. Breit was made chairman of a subcommittee concerned specifically with uranium. Acting on his own initiative, he immediately began writing letters to journal editors, proposing a voluntary plan under which papers relating to fission would be submitted to his committee before publication. Sensitive papers would be circulated only to a limited number of workers. Breit added that he expected ultimately to publish the papers in book form or otherwise, with a statement of the original date of the paper and with a suitable acknowledgment of the public spirit of the authors.[15]

There were some raised eyebrows, but the editors of scientific journals and other leading scientists agreed to the plan. "As recently as six months ago," Lawrence wrote Breit, "I should have been opposed to any such procedure, but I feel now that we are in many respects essentially on a war basis."[15] German troops were pursuing the remnants of the defeated French army, and none could doubt that the international situation was desperate.

Better than never

Within a few weeks Breit, who swiftly set up close communications with Fermi, Urey, Wigner and others involved in parallel efforts at secrecy, had imposed total censorship on American fission research. After passing the papers around by mail for comment, Breit's committee let some through as innocuous; other they withheld from publication.[25] Because of this procedure, carried out entirely by physicists

with no government participation, long before the United States went to war it was keeping vital scientific information within its own borders.

The extraordinary coincidence that history's most dangerous scientific secret appeared at the moment history's greatest war began made possible this unique case of scientific self-censorship. It was imposed against the grain—even some of the conspirators, like Szilard and Teller, would later argue strongly for the advantages of open publication. But it is worth noting that if self-censorship is difficult, under sufficiently deadly circumstances it can be achieved, and that if it may seem to come late, late may be far better than never.

* * *

I wish to thank first of all Gertrud Weiss Szilard, who kindly gave me permission to use the Szilard Papers and to publish the excerpts above. Thanks are also due to Hélène Langevin, who kindly made available the Joliot-Curie Papers; to Monique Bordry, who gave invaluable assistance in their use; to Gregory Breit, Otto Frisch, Victor Weisskopf, and particularly Lew Kowarski, who answered the questions I posed them, and to Charles Weiner, who assembled interviews and other materials at the Center for History of Physics of the American Institute of Physics. For further details see Weart and G. W. Szilard, eds., Leo Szilard: His Version of the Facts, *MIT Press, Cambridge (1978); Weart,* Scientists in Power, *Harvard University Press, Cambridge, (1979). All translations are my own except for the Bohr letter, for assistance with which (and for much else) I thank John Heilbron.*

References

1. L. Szilard, "Reminiscences," *The Intellectual Migration, Europe and America, 1930–1960* (D. Fleming and B. Bailyn, eds.) Harvard U.P., Cambridge, Mass. A revised and expanded version is in *Leo Szilard: His Version of The Facts* (see note above).

2. *The Collected Works of Leo Szilard, Volume 1, Scientific Papers* (B. T. Feld, G. W. Szilard, eds.), MIT Press, Cambridge, Mass. (1972).

3. Szilard papers, La Jolla, Calif.

4. Bainbridge collection, American Institute of Physics, New York.

5. J. D. Bernal, *The Social Function of Science*, Routledge & Kegan Paul, London (1939), pages 150, 182.

6. Pegram collection, Columbia Univ. Library.

7. Joliot-Curie papers, Radium Institute, Paris.

8. Testimony of L. Kowarski before the US Atomic Energy Commission's Patent Compensation Board, Docket 18, 16 March 1967, Energy Research and Development Administration, Germantown, Md.

9. H. von Halban, F. Joliot, L. Kowarski, Nature **143**, 470 (1939); *The Discovery of Nuclear Fission: A Documentary History* (H. Graetzer, L. Anderson, eds.), Van Nostrand Reinhold, N. Y. (1971).

10. Copies are in ref. 3; the original is in Dirac papers, Churchill College, Cambridge, UK.

11. Bohr Scientific Correspondence (copies are held at the American Institute of Physics, New York; American Philosophical Society Library, Philadelphia; Bancroft Library, Berkeley, and Niels Bohr Institute, Copenhagen).

12. R. B. Roberts, R. C. Meyer, P. Wang, Phys. Rev. **55**, 510 (1939).

13. R. Roberts, L. R. Hafsted, R. C. Meyer, P. Wang, Phys. Rev. **55**, 664 (1939).

14. Science Service, 24 Feb. 1939; reprinted in Science News Letter, 11 March 1939, page 140.

15. Lawrence papers, Bancroft Library, Berkeley, Calif.

16. R. Clark, *The Birth of the Bomb: The Untold Story of Britain's Part in the Weapon that Changed the World*, Horizon, New York (1961); B. Goldschmidt, *Les Rivalités Atomiques 1939–1966*, Fayard, Paris (1967), page 27; interview of Kowarski by Weiner, American Institute of Physics.

17. F. Joliot-Curie, *Textes Choisis*, Editions sociales, Paris (1959), page 154.

18. Science Service, 16 March 1939; reprinted in Science News Letter, 1 April 1939, page 196.

19. R. B. Anderson, E. Fermi, H. B. Hanstein, Phys. Rev. **55**, 797 (1939); L. Szilard, W. H. Zinn, Phys. Rev. **55**, 799 (1939).

20. M. Gowing, *Britain and Atomic Energy, 1939–1945*, St. Martin's Press, New York (1964), page 34.

21. D. Irving, *The Virus House: Germany's Atomic Research and Allied Counter-Measures*, William Kimber, London (1967), page 32.

22. I. N. Golovin, *I. V. Kurchatov: A Socialist-Realist Biography of the Soviet Nuclear Scientist* (H. Dougherty, transl.), Selbstverlag Press, Bloomington, Ind. (1968), page 31.

23. R. G. Hewlett, O. E. Anderson Jr, *The New World: A History of the United States Atomic Energy Commission, volume 1: 1939–1946*, Pennsylvania State U. P., University Park, Pa. (1962), page 16; Briggs Committee correspondence, Atomic Energy Papers, Office of Scientific Research and Development, National Archives, Washington, DC.

24. E. Fermi, *Collected Papers, volume 2: United States 1939–1954*, (E. Segrè et al, eds.) University of Chicago Press, Chicago (1965), page 31.

25. Samuel A. Goudsmit collection, Library of Congress, Washington, DC.

26. W. Bothe, P. Jensen, "Die Absorption thermischer Neutronen in Elektrographit," 20 Jan. 1941, captured German report G-71, Technical Information Service.

27. Bulletin de l'Académie des Sciences de l'URSS, Ser. Phys. **5**, 555 (1941); a translation by E. Rabinowitch, Report CP-3021, is available from Technical Information Service, Oak Ridge, Tenn.

28. E. McMillan, P. H. Abelson, Phys. Rev. **57**, 1185 (1940). □

Some thoughts on Science in the

By E. U. Condon

PHYSICS TODAY / APRIL 1952

The following is an address given by Dr. Condon on September 25, 1951, less than one week before his resignation as director of the National Bureau of Standards became effective. His talk was prepared for delivery at the National Academy of Sciences in Washington.

AS MY NEARLY SIX YEARS of service as Director of the National Bureau of Standards draw to a close, it seems that an important final part of that service would be to set down some over-all views concerning the scientific work of the Federal Government growing out of that experience. Our governmental institutions are so close to us that I had some experience with them before entering Federal service full-time, especially during World War II, and likewise I expect to have association with such matters in the future while in private employment.

It seems to me that the scientific research activities of the Government are on the whole good but nevertheless, like all things human, capable of improvement, and it is to some suggestions for improvement that I will principally turn my attention.

The first general point I wish to make is the very obvious fact that the whole complex of modern material civilization arises from application of scientific knowledge. All modern engineering and industry, agriculture and medicine is based on the results obtained by consciously planned laboratory experimentation within the last three centuries, and largely within the last century. It is this new type of activity which has in the last century made greater changes in our material ways of life than have occurred in thousands of years before. The improvement of the conditions of life through the

lightening of burdens by the development of mechanical power from flowing water and from fuels, the improvement of our homes and clothing by modern products of applied science, the more effective production of foods and the use of refrigeration for their large-scale preservation and wide distribution, the increased knowledge of nutritional principles, the improvement in all kinds of techniques of medicine and surgery—all these may be counted as great blessings to mankind resulting from the cultivation of science and its application to our material needs.

Even greater perhaps than all those material benefits, however, is the benefit that comes from the freeing of men's minds and spirits from the oppressiveness of superstitious belief and the growing realization that we live in a world of law and order that is intelligible to us if we will but discipline ourselves to the effort necessary to understand its structure and workings. Certainly this spiritual blessing, in common with the material blessings already mentioned, should combine to produce in all of us the recognition that scientific study is one of the most rewarding fields of human endeavor possible in the world today.

Science is a method by which we learn to know in ever wider ways and with ever greater precision about the world in which we live. The study of science can make genuine and wholesome contributions to char-

Federal Government...

Ewing Galloway photo

acter development not the least of which is an uncompromising demand for truth and honesty in all the affairs of life and a proper humility before all the many wonders which surround us. But great as I think are the values which science has brought and will bring to humanity, I would not wish to leave you with the impression that man can live by science alone, for science does not provide him with the ethical guidance nor the spiritual insights which are needed to realize our ideal of the good life.

Not all of the consequences of this enormous increase in man's knowledge of the world have been beneficial nor can it be said that we are effective in making the fullest use of the knowledge we already have. We have been slow to bring about a widespread distribution of these benefits to all of the population of even a wealthy and favored nation like the United States. While steady progress is being made—at a lamentably slow pace—the fact is that we have done very little toward slum clearance in our major cities or toward providing adequate schools and hospital service for all of the population. We are doing very little to assist the underdeveloped countries to bring the benefits of modern applied science to improve the welfare of the hundreds of millions of their population.

We talk of bold new programs in this direction, and we feel uneasily that much more needs doing than we

have undertaken so far, and still we do essentially nothing about it. Our carelessness here is storing up great trouble for us in the future. We in America and in Western Europe are a small minority among the world's peoples. Other hundreds of millions of persons, chiefly in Asia, have caught a glimpse of what modern science can do for them and they are determined to have it. If we help them we can have their friendship as equals. If we do not, they will get these benefits for themselves anyway in the course of time, and on terms which will involve a great deal of strife and difficulty for us. It is true we have done much to assist in the reconstruction of Western Europe, but we have done practically nothing to assist the development of Asia and Africa. We have not even made effective plans for relief and reconstruction in the devastated areas of Korea.

The effort in this direction that I feel is necessary will be very great but it is my sincere conviction that effort of this kind is the most important thing we can do to preserve and extend the kind of Christian democratic civilization which we believe in. I believe that this kind of constructive effort to assist in bringing the

E. U. Condon is director of research and development for the Corning Glass Works, Corning, N. Y. He was director of the National Bureau of Standards from 1945 to 1951. He has been scientific advisor to the Special Committee on Atomic Energy of the U. S. Senate since 1945.

benefits of modern science to the whole world is the only kind of effort which will accomplish the construction of the kind of world in which peace and goodwill can reign. I do not regard this required effort as a burden but as a great opportunity which has been presented to us which we should be grasping with eagerness and enthusiasm.

While it may be necessary, under present conditions, to use our scientific knowledge and our industrial productive capacity largely for building up our military strength, I am convinced that we are, perhaps unconsciously, placing too great an emphasis on this, as if it would give us the means of solving the difficult social problems with which we are confronted. All that we can hope for from military strength is that it will enable us to preserve a situation in which Western civilization will have an opportunity to share its wealth-producing techniques with the other peoples of the world, instead of having them snatched from us by angry hordes of men who outnumber us ten to one and who will have come to resent bitterly the seeming hypocrisy of our attitudes toward them. I will not therefore go so far as to say that under present conditions the building up of military power on which we are again engaged is now avoidable. But this course of action by itself may prove fruitless unless it is accompanied by a very great program—one whose scale of effort is at least as great as that we are putting into building up our armaments— that is designed to help all peoples of the world who are willing to work with us, to achieve the benefits of modern science which we enjoy. If we do this we shall derive great spiritual benefit from the increased happiness of these millions of God's people and material benefits from our participation in the contributions which their intelligence can bring to our unsolved problems.

THERE IS ANOTHER ASPECT of recent tendencies in development of military armament which we need to consider very carefully. War at best is an evil thing in which peoples resort to force to impose their will on each other instead of using love and compassionate efforts at mutual understanding to arrive at a solution of their difficulties. The opening years of this century were marked by all kinds of efforts in the way of agreements for the humanitarian treatment of prisoners, in agreements to confine the fighting to organized military forces, and even in agreements to avoid the use of certain particularly horrible weapons such as dum-dum bullets. In the two world wars of recent years, and in the military rearmament which is now going on, such ideas have become quaintly old-fashioned.

No longer do we give the slightest consideration to the distinction between military and civilian populations. In World War II both sides gave very little regard to avoiding destruction of the civilian population of their enemies, and enormous damage was done to other than strictly military objectives. A minute part of this terrible destruction was made by the use of the bombs based on the fast neutron fission of uranium and plutonium. The loss of life in Japan **alone** due to fire raids using napalm was much greater than that due to atomic bombs.

A large part of our organized effort in modern science goes today into putting enormous teams of men to work on developing even more deadly and destructive weapons than the world has ever seen before. We talk openly of germ warfare and nerve gases and we almost never hear of these being criticized as inhumane and revolting to the consciences of Christian men and women. No, we hear them criticized because it is difficult to produce germ cultures or gases in sufficient quantity or concentration to wipe out the whole population of a city as their proponents would say is possible and therefore we should devote our attention to the creation of some other fiendish thing like the hydrogen bomb. This, in turn, we hear criticized, not in terms of revulsion that men would use such things against each other, but that maybe its destructiveness is too concentrated and that the same effort put on more conventional types of atomic bombs would enable destruction to be carried out over an even greater area.

At San Francisco a few weeks ago the President spoke unspecifically of fantastic new weapons too horrible even to describe. The press was thereby filled with all kinds of science fiction speculations about what these horrible new wonders might be. Within a few days Congress increased the already enormous appropriations to the Air Force by five billions. In a matter of hours the Congress gave five billion for fantastic new weapons of which it knows next to nothing—the same Congress which after long debates finally cut one billion dollars out of the foreign aid program, the same Congress which by its long delays did much to nullify the effects in promoting goodwill of our finally supplying a credit (not a gift) for $190 million for grain to alleviate severe famine in India; the same Congress which refuses to provide $300 million in Federal aid to our overcrowded and inadequate school system, the same Congress which has lopped off the paltry appropriation of $14 million for the National Science Foundation which was intended to give some slight nourishment and sustenance to the fundamental scientific research on which rests the whole structure of modern industry, agriculture, and medicine.

Some may think that in referring to $14 million for the National Science Foundation as a paltry sum I speak like one of those terrible bureaucrats who has no regard for the burdens which the taxpayer must bear. I am concerned about taxes but I also want us to show some sense of proportion. Congress is this year spending $60 billion of new money or a total of about $100 billion of available funds on the Department of Defense. It has just increased this by another $5 billion for "fantastic" new weapons which the newspapers say can "conquer the atmosphere," whatever that means. It is spending $6 billion on foreign aid much of which is for rearmament rather than economic development.

Included in the military appropriations is about $1.5

billion for military research and development, a staggering sum of money which, if invested at 6% interest, would produce annually as much money as Congress has appropriated to the National Bureau of Standards in the entire fifty years of its existence. But it cannot spare $14 million next year for strengthening basic scientific research.

Today every activity of Government is being adjudged solely on the basis of its contribution to defense. I doubt whether such vast sums can be spent wisely for the purpose intended, and whether it is wise to put so much of our reliance on military strength while thinking so little about peaceful and constructive solutions of the difficult domestic and international problems before us.

If so much of our scientific effort is directed toward military weapon development, it must necessarily mean neglect of the basic science on which future progress must be built and neglect of the application of modern science to improving human well-being in our own and other parts of the world. There is another reason why we might be disturbed at the extent to which science is devoted to military purposes today. Although it seems to be very little in evidence at the moment I believe that deep in the consciences of men and women there is a horror and revulsion at the terrible means and methods of modern warfare which will some day find expression in a new and powerful and constructive determination to live together peacefully, and effectively to renounce war as an instrument of national policy. If in the years to come science and the scientists are closely identified in the public minds as the wizardry and the wizards who have made all the fantastic new weapons of mass destruction that Governments are now so eagerly urging them to produce, this horror and revulsion of war may, in that illogical and irrational way that so many things go in politics, be extended to science and the scientists. If this were to happen it would be bad not only for the scientists, but it would be bad for society, for a rejection by society of the method and power of scientific inquiry will stop progress in understanding and tend to retard the extension to all mankind of its beneficial applications. If men's consciences reawaken to the absolute necessity of abolishing warfare, then there may be serious danger that science may be the baby which is thrown out with the bloody bath which is War.

This situation poses very difficult problems for scientists in general and especially for those in official positions in our Government. Speaking personally, all of my friends know with what strong conviction I hold the general views which I have tried here, rather inadequately I am afraid, to outline. When I came to Government service at the close of World War II, I hoped and believed that there was to be an era of peace in which fundamental research in science would flourish and be supported by society as a whole as a worthy intellectual activity and for the constructive benefits to man's well-being which it can bring. At that time, only six years ago, the United Nations had just been born and many of us believed that the experience of wartime alliance had taught the lessons which would gradually enable the growth of a mutual confidence and trust between Russia and the United States and other principal nations of the world which would remove any basis for future war of major proportions. In such a setting one could hope for a steady reduction of national armaments, with the enormous economic waste which they imply, and their replacement by an international police force. In such a setting we hoped that all efforts in the field of atomic energy would go to peaceful purposes in chemical and medical research and in making available new sources of power.

At this time it seemed that Congress and the people of the United States, impressed by the contributions which applied science had made during the war, were prepared to support a National Science Foundation in a really adequate way—by this I mean to the extent of several hundred million dollars a year—and that science in other countries would be aided by a major program of the United Nations Educational Scientific and Cultural Organization as well as by local efforts in those countries.

During my first year in Washington, 1945–46, my attention was largely taken up with assisting the Special Senate Committee on Atomic Energy of the 79th Congress, as scientific adviser, when it was developing the Atomic Energy Act of 1946 by which the present Atomic Energy Commission was established.

During that first year the Senate held extensive hearings on proposed legislation for the National Science Foundation and passed a bill, but this was allowed to die in the House when the situation became confused by behind-the-scenes lobbying of those who insisted on a large part-time board for the Foundation. Otherwise the National Science Foundation probably would have started operating five years ago with an annual appropriation of about two hundred million dollars. If this had been allowed to happen we would have been incomparably better off than we are today from every point of view. Fortunately, the vacuum thus left was quite well filled by the enlightened scientific research program of the Office of Naval Research. This was conducted as liberally and as intelligently as any purely civilian program could possibly have been conducted and has made a wonderful contribution to the development of basic science in America during the post-war period.

SOON AFTER THAT FIRST post-war year it became clear that expenditures for scientific research for military purposes would be maintained at a high level and expanded above the minimum reached in the demobilization period. Work in this field has always been an important part of the program of the laboratories of the National Bureau of Standards. The Bureau has a long history in meeting such military needs, having first developed the optical glass industry in World War I, having initiated the atomic bomb project in World War II, and also having played a large part in the development of proximity fuzes, having developed the only fully automatic guided missile to be used in warfare thus far, and having done much to improve knowledge of long-distance radio transmission on which the continuity of military communications depends. This latter service was initiated during World War II and organized as a permanent service in the Bureau during the first post-war year. Congress has been willing to support this work reasonably well and has made provision for splendid new laboratories for the radio work of the Bureau to be built in Boulder, Colorado. This radio work is, however, essentially the only new activity of the Bureau for which it has been possible to get direct financial support from the Congress during the post-war years.

In this period, to be sure, and particularly during the last year, there has been a great expansion in the level of activity of the Bureau. But this has not been by direct Congressional support, but rather by doing project work in Bureau laboratories for the armed services and with funds provided by them from their own appropriations. For example, this fiscal year the Bureau will operate on a total budget of some 60 million dollars, less than $10 million of which is directly appropriated by the Congress, nearly all of the rest being paid by the military for work done for them. To get some idea of the disparity of figures involved it is interesting to note that this year the Bureau will spend on electronic ordnance developments alone about 50% more money than the $14 million which the House has refused to give the National Science Foundation for Federal support of basic science.

The amount of military work done by the National Bureau of Standards has increased almost by a factor of seven during the time that I have been Director. Provision has been made for expanded facilities for such work in Washington, in Boulder, Colorado, where large new Bureau-operated laboratories are being built for work of the Atomic Energy Commission, and also in Corona, California where some unused former Naval hospital facilities have been converted into a splendidly equipped development laboratory for guided missile work for the Navy. Another Bureau development of military importance has been the establishment of a department of applied mathematics with facilities both in Washington and Los Angeles, and the development of an important electronic digital computer, the SEAC, which has been in service for more than a year on military problems. These are just highlights of a program which involves dozens of research projects of specifically military interest some of which relate directly to fantastic new weapons which cannot even be mentioned. I think therefore that the National Bureau of Standards is in a stronger position than ever before to make important contributions to military needs.

Turning to the fundamental support of the civilian program of the Bureau the situation is far from satisfactory. The National Bureau of Standards is a Cinderella whose Prince Charming has yet to come along. In spite of its long record of splendid accomplishments, its scientific program was crippled by severe budget cuts in 1933 as one of the economy acts of the Roosevelt administration. Valiant efforts were made by my predecessor, Dr. Briggs, to hold an effective staff together in spite of this short-sighted action but the Bureau is still suffering from the effects of that decision.

Except for the expanded radio work the direct support available for the Bureau in the post-war years has remained nearly constant, as expressed in dollars, and therefore has declined steadily in real purchasing power for goods and materials. This is a most serious situation, for it has occurred at a time when there has been a steady growth in the amount and complexity of the needs for standards of precise physical measurement.

Every kind of physical quantity is being measured, in connection both with scientific research and with more accurate control of industrial processes, with greater precision than before, and over a wider range of extreme conditions, and the need for exact calibrations of measuring instruments arises from a much greater number of research laboratories and industries than ever before. This has put a burden of work on the National Bureau of Standards with which it is quite unable to cope within the framework of its present appropriations. Try as we will we have not been able to keep up with the demands for such services. The result is, of course, that much money is wasted by others in duplications of calibrating set-ups which the Bureau should have and that many scientific jobs are done with a lower grade of accuracy than desirable and than would be possible if the National Bureau of Standards were allowed to render an adequate service.

I confess that I do not know how to do a better job of bringing this need to the attention of the Government. It has received a great deal of my attention in the last five years but with essentially no results. I hope that my successor in office will be able to do better on this than I have. Here it is important for him to realize that not all of the difficulty is with Congress. The budget of the National Bureau of Standards has to pass four hurdles before it is approved. It must first be approved by the budget officers of the Department of Commerce. It comes before them as a peculiarly difficult-to-understand technical item which amounts to less than two per cent of the total budget requirements of that Department. Since it is such a small part of the Departmental budget it is only natural that these budget officers have no scientific and technical background. At this stage efforts to get even what increase is necessary to keep abreast of the declining purchasing power of the dollar are pretty well nullified because these men are working under a general over-all limitation as to what the Department of Commerce itself may have.

After the Department of Commerce has finished its consideration, the Director and his staff must write up the whole thing again in great and specific fiscal detail for the Bureau of the Budget. This is supposed to show that the whole program of proposed work has been very thoughtfully considered. Having filed all this data with the Bureau of the Budget, several hours are spent explaining the needs to staff officers of the Bureau of the Budget. Here again because scientific research is diffused over the whole structure of the government one is dealing with individuals who have very little background either in the over-all needs of the Government for scientific research, or in the accomplishments of the National Bureau of Standards in particular, or for the methods and aims of physical science in general.

This process goes on intermittently during the first half of the fiscal year preceding the one for which the budget is being prepared. Out of it comes an official determination by the Bureau of the Budget of what each governmental agency will be allowed to ask for in going before the Appropriations Committees of the House and Senate. The end result of this process when carried out for all the agencies of the Government appears in a large document which is printed and transmitted to Congress as the President's Budget. This is now official, and sometime in the spring the Director and his staff are summoned down to present the Bureau's part in the President's Budget to his subcommittee of the House Appropriations Committee and then to the Senate Appropriations Committee. Before doing this, however, his own staff of budget officers have had to rework completely the elaborate document by which the plans for the coming fiscal year were submitted to the Bureau of the Budget.

It is hard to convey any idea to persons outside of the Government of the extent to which the working agencies are called on to supply over and over again statistical reports about their work which cover essentially the same ground in slightly different forms.

Each agency sends up a large budget document to the Congress for the use of the Appropriations Committee in advance of the hearings. At the hearings the questioning often indicates that the Congressmen have very little understanding of the particular scientific needs of a technical agency and that perhaps they have not had time to look at the contents of the elaborate budget document which was prepared for them.

SOME OF MY most treasured memories of Government service are connected with incidents which occurred in these appropriations hearings. One feels rather nervous and tense on these occasions for on their outcome hinges the whole fate of the Bureau's work.

One time while waiting our turn outside the committee room, the budget officer of the Patent Office came out of the door looking pale and fell on the floor of the hall in a dead faint. We bustled around administering first-aid and when he came to partially he muttered deliriously, "It's awfully hot in there." Later, when it was my turn to go in, I found that he was right. That was during the Eightieth Congress at a time when the Alsops referred to the House Appropriations committee as a bunch of blind men pruning a jungle.

I remember one time one Congressman had me quite upset because he was scowling through the whole of my presentation. When it came his turn to ask questions he asked me, "Doctor, where *is* the National Bureau of Standards?" I told him it was out on Connecticut Avenue and he said excitedly, "Is *that* what that place is?" and became quite friendly.

On another occasion a Congressman was questioning the chief of the Bureau's radio division, who had been talking about the scarcity of space in the radio frequency spectrum for the many needs of communication services. He said: "Doctor, I understand that among you scientists there are two theories: some say space is finite, others say it is infinite. I want to know, where do you stand?" The witness started to explain the limitations of using very low and very high frequencies but the Congressman interrupted him to say, "No, I mean space, you know, *space*," making a large and globular gesture toward the part of the three-dimensional continuum in front of him.

The witness squirmed and looked at me for guidance, quite willing to make it finite or infinite for the sake of the budget, but I could only indicate with a gesture that I did not know which was the preference of that particular Congressman. So he gulped hard and said, "I think it's infinite." "Thank you very much, Doctor, that's all I wanted to know", replied the Congressman and passed on to another topic.

When these hearings deal with science they are apt to be rich in non sequiturs. For example, only yesterday I was reading the Senate defense appropriation hearings (p. 1177) where an Army colonel is asking for funds for an electronic computer for logistic planning. A Senator asks him: "Now, is there any relationship between the number of equations that have to be developed and the time the machine is in operation?" And the colonel replies: "Electricity travels 186,000 miles per second, sir, so it is an infinitesimal difference."

There would be no point in describing this procedure in such detail unless I had a suggestion to offer. I do have.

I am convinced that the over-all importance of scientific research in Government has become so great that it requires careful attention and study by a new standing committee of the Congress. It is at least as important as atomic energy which has a permanent Joint Committee, affording an organized means by which Congress can study these problems. A similar means is needed for scientific research broadly if we are to get intelligent action and focus attention on unwise actions or inactions. Such a committee would study and deal with legislative problems affecting scientific research.

In addition it would be very desirable if the Appropriations Committee of the Congress would find a formal way to give some unified over-all attention to the scientific research requirements of the Government. A legislative committee on science in the Congress would not be enough unless the Appropriations Committees were also prepared to have a look at the whole program of all the large variety of specialized agencies in the government which are doing scientific work.

The most natural way for the Congress to deal with

science in a unified way would be for the scientific agencies of Government to be gathered up into what would be in effect a single Department of Government. I believe that the general importance of scientific research in the Federal Government has become so great that this should be done. If it were not considered desirable to establish a new Department of Scientific Research then I would recommend that the Smithsonian Institution be used for this purpose. I believe that the new Department or enlarged Smithsonian Institution should include all of the scientific agencies of Government including the major military research laboratories, the research laboratories of the Atomic Energy Commission, the National Bureau of Standards, the National Institutes of Health, the laboratories of the Department of Agriculture, the Weather Bureau, and the Bureau of Mines, the Geological Survey, the National Advisory Committee for Aeronautics, and, of course, the National Science Foundation.

Whether a new Department of Scientific Research in the Executive Branch or an enlarged assignment of responsibilities for the Smithsonian Institution represents the better proposal I am not prepared to say. But it seems clear to me that a unified administration of the scientific affairs of the Government, including unified treatment of them by the Bureau of the Budget and by appropriations and legislative committees of Congress, would surely be an improvement over the present situation. I am inclined to favor the adaptation of the Smithsonian for this purpose over the creation of a new Department, for the reason that each cabinet member is on the board of the Smithsonian and thereby the relation of science activities to the other government activities they support would be preserved while giving scientific research as a whole a coordinated administration.

The suggestion that the Bureau of the Budget should have a special staff for study of the needs of scientific research is not a new one, having been made as a recommendation in the Steelman report, entitled "Science and Public Policy." But it has never been acted upon, I suppose, because of the difficulty of finding properly qualified individuals to do the job and the Budget Bureau may feel that it is better to do it not at all than to do it badly.

If there is any merit in the general suggestions I am making I would like to see the Bureau of the Budget call on the National Academy of Sciences for a study and recommendations and also to ask the Academy for its help in reviewing the budget of the existing agencies. The Congress too should recognize the many ways in which it could get help on scientific problems from the Academy and call on it for help more often on large broad issues than it has in the past.

As part of such a plan, the National Academy of Sciences, the National Research Council, the American Association for the Advancement of Science, and the specialized scientific societies would retain the independent status which they have now but would work in close cooperation with the new science administra-

tion to make sure that the Government's research program is effectively carried out in a way best suited to serve the national interest in relation to the professional needs of the scientific work in the universities and in industry.

ONE OF THE MOST REMARKABLE omissions in the report of the Hoover Commission on the reorganization of the Government was its almost complete lack of any recommendations for improving scientific research in the Government. This is hard to understand for surely the men who developed that report appreciate the importance of science today in Government, and cannot have felt that the present diffusion of responsibility over many separate agencies is a form of organization which cannot be improved upon.

It seems to me all the more important that a unified central body for science in Government be set up because research is a very fashionable thing these days and every new agency feels it must do research in order to have status in the world of bureaucracy. While it is very difficult to get adequate support for the established research agencies, it is always possible to set up a research program as a small part of a general need to which assent has been given and by indirection to obtain vaster sums of money than the established agencies can get for research. For example, it would be extremely difficult the way things are now to get a modest increase in the funds available to the National Bureau of Standards for research on synthetic rubber in spite of a splendid record of past achievement, whereas a quite substantial amount of support is carried along by the Government as an incidental to the operation of the Government-owned synthetic rubber plants. But I am convinced that when the work is carried on in this way, with uncertainty as to its continuance, and therefore an unusually high personnel turnover, it is not nearly so effective as if it were part of an over-all coordinated scientific program supported on a more stable basis.

Another example of an agency of Government which has recently entered the field of science is the Department of State. It has established a science liaison office and looks forward to having scientific attaches serving in various of our embassies in leading capitals of the world. I believe that there is an important service to be rendered in fostering international relations in the field of science. But I do not believe this can be done effectively under circumstances where it is just one minuscule activity under the supervision of men who are so busy with so many other matters that they are unable to give it their attention. All such activities of the United States Government could probably be better handled by a general science agency, of the kind suggested.

Another recent venture in organization of Government science that many feel could be improved is the Research and Development Board of the Department of Defense. This was established by one section of the National Security Act of 1947, the law which established a single Department of Defense and was intended to be the means for bringing about a close coordination of the scientific research and development work sponsored by the Army, the Navy, and the Air Force. Experience has shown that it has not been a very effective tool for doing this. I think that this outcome might have been foreseen from the outset and for the reason that the Research and Development Board was set up as a purely advisory body, without operating responsibilities. Operating responsibilities for research continue to belong to the three services and their individual bureaus. Because the RDB lacks direct responsibility it is not an attractive place for scientists of real ability to work, so it has been unable to attract staff of sufficient competence to cope with the very difficult problems presented by an extremely complicated situation. I am convinced that the RDB cannot perform a useful function as long as it functions in a purely advisory way, and that the situation could be greatly improved by putting all of the military research laboratories completely under civilian management of a Department of Scientific Research or a new Smithsonian Institution.

Next I would like to make a few observations on the Federal Government as an employer. Uncle Sam is a reasonably good employer so far as salaries, retirement plan, vacations, and the like are concerned. But the salaries paid for positions of major responsibility are in no way commensurate with the rewards which can be obtained in private life for similar services. Some curious inequities develop in this way. The tax position of many corporations is such that it costs the Government more in decreased tax revenues paid by the private employers to have a man work in private industry than the salary which the Government will pay that man to work full-time for the Government.

The curious thing about the low salary scale which the Government pays to scientists is that one way and another the Government is finding it impossible to compete with itself in securing the services of the men it needs. Many private employers of scientists use them on Government contract work on a cost-plus basis so the Government pays the man's full salary at higher than Civil Service rates as part of the cost of the contract. This possibility has, in the post-war years, led to a new development which is having devastating effects on the ability of Government operated laboratories to recruit qualified staff. More and more there is a tendency to assign Government research programs to *ad hoc* groups organized as private corporations solely for the purpose of taking a Government contract and even in some cases for the purpose of staffing and operating a Government-owned facility. In this way Government money is used to pay salaries in excess of Civil Service rates and all manner of operational red-tape is avoided, but the Government finds itself paying much more for the same services than it would pay if the work were done in its own laboratories This is not good for the morale of loyal Government workers. The proper remedy would be to improve the rules affecting the Civil Service instead of inventing ways to evade them.

Aside from questions of salary alone, some members of Congress so often give expression to attitudes of contempt and distrust toward the thousands of dedicated, conscientious, and intelligent citizens in the Government service that they have to be quite thick-skinned indeed not to have their morale in some measure impaired by such treatment. No private employer would think of saying the kinds of things about his employees which are often said about Government employees and expect to retain any of their loyalty or devotion. This situation has been greatly aggravated in recent years by the use of dishonest smear tactics in Congress giving rise to an artificial hysteria which has led to widespread injustices toward Government employees carried out in the name of the Loyalty Program. Everyone in Washington knows dozens of stories of great suffering caused by silly and trivial accusations in this connection. For example only recently I heard of a labor relations expert who was employed by our Government in Japan to work to diminish communist influence in labor unions there, who was officially commended for his work, and then later had to defend himself before a loyalty board against the charge "that when you were in Japan you evinced a great interest in Communism". I know of a case of a woman who was accused of disloyalty on the grounds of sympathetic association with her own husband. I know of another who was charged with acquaintance with a scientist who is in fact the man who is entrusted with a major role in the hydrogen bomb development.

Not the least of the evils associated with the actual functioning of the Loyalty Boards is the slowness with which they operate. Often a person is kept in a state of nervous suspense for months after a hearing is held before he gets word of a decision. In general the processes are carried out in an altogether too formal and unsympathetic manner. No man can become a psychoanalyst until he himself has been analyzed. I think the situation would be improved if no one served on a Loyalty Board until he had a laboratory course in the Golden Rule by having himself been given a protracted experience with a Loyalty Board.

In conclusion let me thank you for your courtesy in listening to this rather too dogmatically expressed recital of opinions of one who sincerely believes in the importance of Government service and of science in the modern world of affairs and who only hopes that some of these thoughts may make a slight contribution toward working out some improvements. No one who has ever been entrusted with major governmental responsibility can fail to be impressed with the importance of the American government and of strengthening the American contribution to the welfare of all the peoples of the world. I am not an old soldier and I hope not to fade away. I leave the Government service happy in the friendships and experience it has given to me and hoping that I may still in private life be responsive to the duties of citizenship in this our beloved America.

Fifty years of physics education

PHYSICS TODAY / NOVEMBER 1981

A.P. French

A.P. French is a professor of physics at MIT. He was consulting editor of *The American Journal of Physics* from 1973 to 1975 and chairman of The International Commission on Physics Education from 1975 to 1981.

Physics education isn't what it was 50 years ago. The most obvious changes have been quantitative: In 1930 in all of the US about 1000 students received bachelors' degrees in physics; about 200, masters' degrees; and about 100, PhDs. In 1980 the corresponding numbers were 4500, 1500 and 1000; and in the peak years around 1970 the figures were about 6000, 2500 and 1500 (see page 52). These large increases, corresponding to a factor of ten or more for graduate degrees, also point to the extent to which physicists have become essential to the technical development of our society and to the increased acceptance (at least in principle) of the idea that every citizen, in today's technological society, ought to have some knowledge of science in general and of physics in particular. Few would go so far as to embrace C.P. Snow's view that no person can be considered truly educated without knowing the sec-

ond law of thermodynamics. ("My partner doesn't even know the *first* law," said Michael Flanders, in mock disgust, in the show "At the Drop of a Hat.") But many would give vigorous assent to what Henry Perkins[1] said in 1969: "As we are living in a scientific age, physics is just as much of a cultural subject as the older humanities. Not to know something about the basic principles of mechanics, electricity and, at least, to have a smattering of atomic structure stamps the modern man as only half-educated, just as ignorance of Latin and Greek indicated relative illiteracy to our forefathers."

The task of the physics teaching profession is to address both the professional and the cultural needs of students; the purpose of this article is to provide a brief survey of progress in both areas. (One place this was done was in the masterful survey of physics education in the US as of 1970, published in 1973 by the National

"*I'm trained, yes, but not highly trained.*"

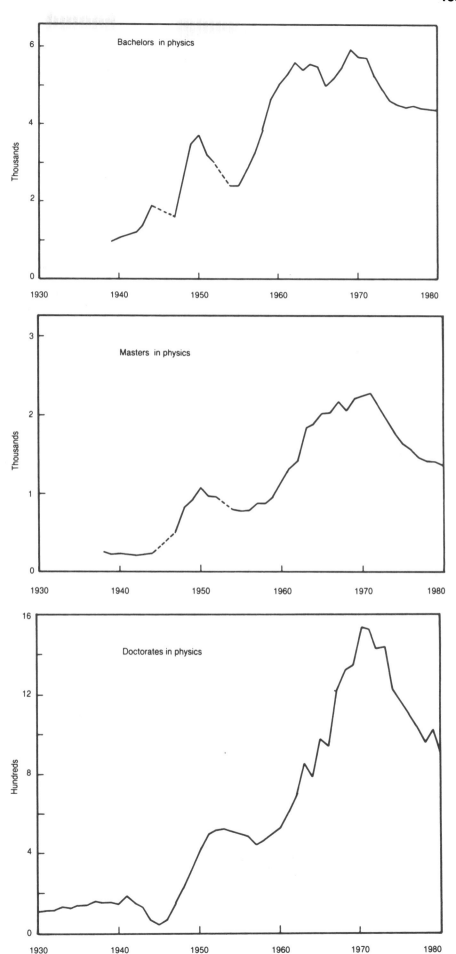

Academy of Sciences in *Physics in Perspective*, Volume 2, Part B.) I will, however, look only briefly at graduate physics education. I will deal mainly with pregraduate courses in high school and college. This area, with which I am most familiar, has changed more than the relatively stable apprenticeship system of graduate training.

Physics in the high schools

The most dramatic changes in physics education in the US during the past few decades have occurred not in the colleges and universities but in the high schools, even though the high schools present a numerically far more formidable challenge. During the last year of secondary school, before which most US students do not have access to a physics course, just over 20%, about 600 000 students, take a physics course of some kind. (My statistics, here and elsewhere, are based mostly on data published by the AIP Manpower Division.) These courses are taught by about 15 000 teachers (of whom, however, it has been estimated that only about 4000 ± 1000 are "adequately trained for the purpose," according to Clifford Swartz[2]). Although these numbers are large, the possibility of change and reform at this level is eased by the organized structure of secondary education systems, contrasted with the near anarchy represented in 2000 independent and largely autonomous college physics departments.

Two ambitious programs have transformed the teaching of physics in high schools; the first was provided by the Physical Science Study Committee. Formed in 1956 (*before* Sputnik I, as its chief instigator, Jerrold R. Zacharias, likes to point out), it brought the methods of big physics research

The Physics Community—A Retrospective

into the educational domain. Its goal was to develop a course that would present physics not as a catalog of facts and formulas to be learned, but as an intellectual adventure concerned with exploring and understanding the real world. Bringing together several hundred high-school and college teachers and millions of dollars supplied chiefly by the National Science Foundation, the PSSC created a completely novel physics course from scratch, with a full panoply of teaching aids—textbooks, teachers' guides, tests, laboratory experiments and apparatus, films, and supplementary monographs. It accomplished most of the job in about five years. Nothing like it had ever hit the educational scene before, but other sciences were quick to follow this lead.

The PSSC course, the work of highly respected physicists, offered the US high-school student (and teacher) a presentation of basic physics of a sophistication never before available at that level. Use of the PSSC course grew rapidly at the expense of the old, traditional courses, until it was being taken by at least 30% of high-school physics students.

The PSSC did not, however, realize one of its aims—to increase the numbers of high-school students choosing to take physics at all. This was admittedly optimistic, since the PSSC course was acknowledged to be a difficult one. Could there be another new course of high scholarly caliber that would appeal more to the student whose interest in physics was more general? In 1962 Gerald Holton and his chief collaborators, F. James Rutherford and Fletcher G. Watson, at Harvard took up this challenge. They created Harvard Project Physics through a major developmental program very similar to that used for PSSC. Like the PSSC, Harvard Project Physics captured a substantial fraction of the high-school physics clientele. In fact, it seems to have come to

be the more popular of the two. Although it is hard to obtain reliable numbers, rough estimates suggest that these two courses together reach about 30% of present high-school students who take physics. Their influence, however, reaches further, for various features of their approach have been incorporated in other widely used high-school textbooks.

Introductory physics at college and university

According to AIP Manpower Division statistics, about 350 000 out of 2 000 000 first-year college students take introductory physics in some form or other. The numbers may be substantially higher, because many students take physics courses in institutions that for various reasons (lack of a separate physics department, lack of a complete bachelor's degree program) are not represented in the AIP data.

Of the total, something like 5% take physics courses suited to future physicists, chemists and engineers. Another 15% or thereabouts, including premedical students and biology majors, enroll in courses that present physics in a fairly analytical but mathematically less demanding way. The very large remainder—about 80%—engage in a wide variety of courses that require minimal mathematical skills.

With a few exceptions, such as the Physical Science for Nonscientists Project, the physics teaching profession has given organized attention to curriculum reform and innovation more or less inversely to the size of the various constituencies identified above. Admittedly there have been a few conferences concerned with the improvement of physical science courses for liberal arts majors, at one of which H. R. Crane[3] presented his famous caricature (see below) of the kind of traditional approach that repels such students. Since the early 1960s, however, progress has consisted mostly in the development of a great variety

SYLLABUS FOR DETECTIVE STORY
AS WRITTEN BY A PHYSICS PROFESSOR

CHAPTER I   ORIGINS OF LAW IN BABYLON

II   CONSTITUTION OF UNITED STATES

III   BASIC ORGANIZATION OF POLICE DEPARTMENT

IV   ELEMENTS OF COURTROOM PRACTICE

V   THEORY OF FINGERPRINTS

XXX   (LAST PAGE)   THE CORPSE
(SOLUTION LEFT TO THE STUDENT)

SALES: ZERO

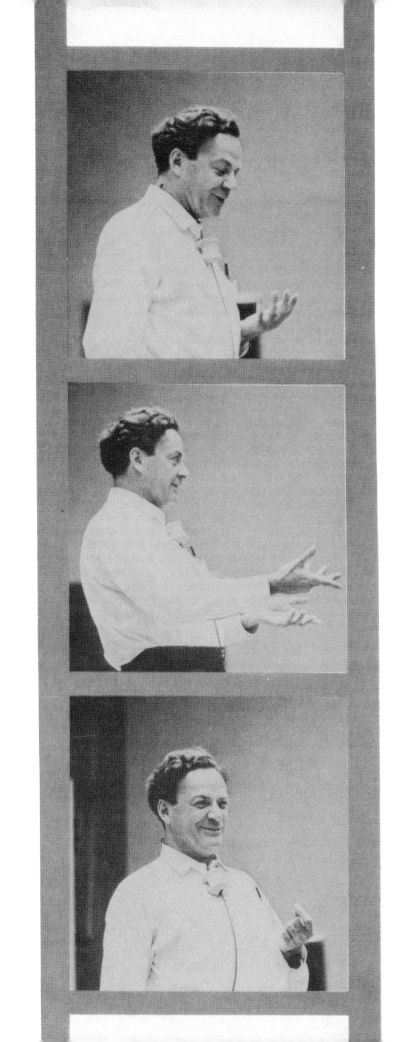

of new courses by individual instructors or departments, here described by Arnold A. Strassenburg[4]

"Since 1964 the proliferation of elementary college physics courses has been phenomenal. The need for more diversity to meet the demands of an increasingly large and varied student body was loudly proclaimed by several leaders of the physics education community, and these exhortations, without doubt, set the stage for major change.

In the late 1960s, however, a sharp decline in the attractiveness of physics as a major provided a much more powerful stimulus to seek a new clientele among students committed to other disciplines. Offerings such as "Physics for Poets," "The Physics of Music," and "The Physics of the Environment" popped up at virtually every college and university in the US."

The Commission on College Physics sponsored conferences and a summer workshop in 1963 to address the needs of the middle group of students, those going on to study life sciences. An initial conference tried to design curricula suited to these students; subsequent workshops aimed at producing materials—monographs, experiments, computer programs and films—that could be used in these and other curricula. However, fifteen years later it is hard to identify any direct influence of such efforts on the shape of intermediate-level physics courses.

When it comes to introductory physics courses for prospective engineering and, above all, physics majors, university physics teachers feel far more confident about what needs to be done and are more willing to commit their time and effort toward the creation of new courses. There was a consciousness by the early 1960s that with the renovation of high-school physics, led or inspired by the PSSC program, colleges and universities needed to upgrade their offerings. It would have been irresponsible—and frustrating to incoming freshmen with improved backgrounds—not to rejuvenate the college curricula. In fact, some of those who had been involved in the PSSC—in particular, Francis L. Friedman, Philip Morrison and Zacharias—were among the leaders in pressing for similar efforts at a higher level.

Among a great wave of curriculum improvement projects during the 1960s, two projects call for special mention because of their sheer size. In 1960, MIT established its Science Teaching Center (later to be called the Education Research Center) under the direction of Friedman; less than two years later, with Charles Kittel as chairman, the committee formed that guided the creation of the Berkeley Physics Course. With massive backing of the National Science Foundation, both groups embarked on large-scale projects: the Berkeley group, to produce a set of texts and accompanying laboratory materials; the MIT group to devise a more diversified program, which came to include

physics texts as well as films, demonstrations and student experiments. Notable features of these projects were an increase in conceptual and mathematical sophistication in the presentation of physical ideas, an injection of substantial amounts of "modern physics" (relativity and quantum ideas) into the first-year curricula and an organized use of many different kinds of learning aids in the presentation of the subject matter.

In the midst of all these collective enterprises, a virtuoso performance by a single person, Richard P. Feynman, extended the horizons of every teacher of introductory physics for prospective majors. At Caltech from 1961 to 1963 he gave the lectures that have become a standard part of every instructor's library (see the photographs on page 54). In his characteristic fashion, Feynman thought through everything from scratch, enriching each topic with his marvelous insight and originality. The result of Feynman's venture was a two-year course covering all the basic ideas of classical and modern physics. While it is too strong meat for all but the very best students, it provides a precious resource for teachers of physics at all levels.

The physics major

The physics major is a multiple entity. The latest available AIP statistics show that in 1978 about 4500 students graduated with bachelors' degrees in physics. Of these, about a third went on to graduate work in physics, and another quarter into graduate school in other fields (such as engineering, mathematics, computer science, medicine and law). The re-

mainder, just over 40% of the graduates, either went directly into jobs or were planning to do so. The bulk of these did in fact find immediate employment; if past experience is any guide, more than half made direct use of their physics training in industry, government, teaching, and so on, but a substantial fraction (perhaps 15% or more) did not. Therefore, to regard a major in physics as being primarily a basis for further academic training in the subject ignores some of the facts.

Recognizing this diversity of interests and career paths on the part of physics majors, the Commission on College Physics, founded in 1960 with support by the NSF (which also funded similar commissions in other sciences) helped organize several conferences on undergraduate programs for physics majors and recommended the creation of two basic curricula. One curriculum, for students going on to graduate study or work in physics, corresponded closely to existing programs for physics majors in most institutions; the other was for students who would become secondary-school teachers of physics or would enter other fields. This latter curriculum would have been a novel development had it been completed, but it did not materialize in any definite or unique form, although many individual courses and teaching aids were developed. With the present decline in funds for curriculum innovation activity in this area has markedly decreased.

On top of the retrenchment forced by shrinkage of resources, there is a trend to return to older and more conservative programs. "Back to Ganot" is a phrase that has been jocularly applied to this development, Ganot being the author of an extraordinarily successful nineteenth-century physics text that went through many editions in its original French and then gained a comparable success in an English translation. Still, one has only to make a close comparison of Ganot's book with a typical text for physics majors of today to realize that the growth of sophistication, the shift from descriptive to analytical and conceptual, the sheer elevation of levels, has been immense. As Feynman remarked a number of years ago: "When I was a student, they didn't even have a course in quantum mechanics in the *graduate* school; it was considered too difficult a subject. . . . Now we teach it to undergraduates."

Tools of the trade

Some of the most notable developments in physics education over the past 50 years—and particularly during the last 25—have been in instructional techniques and aids. I comment on some of these developments briefly below, referring especially to the introductory college level, for which a large fraction of the innovations have been designed.

Films. The use of films in physics instruction has grown tremendously over the last 30 years. The PSSC gave this trend a great boost by producing more than 50 films devised and executed under its control. The films were typically about 25 minutes in length, each expounding and demonstrating a particular topic and designed to be an integral part of the teaching program.

A number of similar films were made for use at the college level. However, because college teachers are in general far less willing than high-school teachers to surrender substantial amounts of classroom time and their own initiative as instructors to films of this type there has been a major shift to short filmed demonstrations to be used within the instructor's own presentation of a topic. The development of 8-mm film-loop projection has further encouraged the use of such films, both inside and outside the classroom, with the result that there has been a great increase in the variety of physics demonstrations available to teachers (see figure on page 58).

Some of the obvious advantages of film are its ability to capture events that occur on too short or too long a time scale to be demonstrated directly, are on too small or too large a scale of size, or are just too complicated or expensive to be performed by the average instructor. Theoretical problems solved by computer with a visual display are another 'natural' for presentation on film. The figures on this page show examples of filmed demonstrations of these various kinds.

Television. Although television is an obvious resource of enormous potential for education, so far in physics that potential has gone largely undeveloped. While this may represent a failure of imagination on the part of the profession, there are obstacles to the effective use of the medium outside of such straightforward applications as the use of closed-circuit television to feed a lecture to overflow audiences or to make visible a small-scale demonstration. There are also objections to a form of instruction that is so frozen in format and allows for no reaction from the viewer. Perhaps the biggest obstacle to the expanded use of television is its sheer cost.

Laboratories. Many years ago a colleague of mine, at the end of a freshman lab class that he had been supervising, remarked to me: "Well, Ohm's law has survived another onslaught!" Such cynicism about the efficacy of laboratory instruction, at least in introductory physics courses, is widespread among both teachers and students. "There is little empirical evidence," said Leo Nedelsky at the beginning of an article about introductory physics laboratory,[5] "that laboratory instruction in elementary courses contributes significantly to the student's understanding of physics. In the absence of such evidence, physics teachers argue *a priori* that contact with phenomena is a necessary constituent of adequate training in physics." Most physics teachers, probably all, accept that last proposition; the problem is how to make the laboratory an instructive experience. If this problem has not been solved, it is not through lack of effort; more, and more varied, attention has gone to laboratory work than to almost any other component of physics education. (See, for example, the report of a 1978 international conference on the role of the laboratory in physics education.[6])

Technological developments have certainly brought about dramatic refinements in student apparatus. Electronic instrumentation has improved beyond recognition, of course. In some cases there have been reductions in cost. (The current catalog of one equipment manufacturer shows two versions of a free-fall apparatus. One of

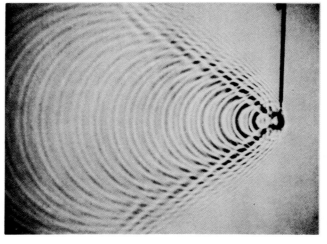

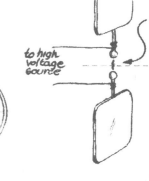

New texts. *Above: the PSSC prepared many films, such as this one showing ripple tank phenomena (courtesy Kalmia Co.); at right: the Harvard Project Physics Course characteristically used free-hand sketches. This one illustrates electromagnetic radiation*

them, a traditional version with a spark timer and paper tape, costs well over $1000; a modernized version with a digital electronic timer costs less than $200!) The availability of inexpensive microcomputer modules has made sophisticated instrumentation a possibility in all kinds of elementary-laboratory experiments.

For more advanced students, particularly for physics majors, there has long been a tradition of high-quality laboratory work whose value has been recognized by students and teachers alike. For the most part this has taken the form of "cookbook" experiments, albeit of a high order. There has, however, been a substantial move in the direction of involving undergraduates in ongoing research, which is a healthy antidote to the abstractness and formality of the classroom teaching that makes up the bulk of a student's training.

Computers. In his AAPT Millikan Award Lecture in 1978, Alfred Bork began by saying

"We are at the onset of a major revolution in education, a revolution unparalleled since the invention of the printing press. The computer will be the instrument of this revolution. . . . By the year 2000 the *major* way of learning at all levels, and in almost all subject areas, will be through the interactive use of computers."

Many physics teachers—including some of those for whom the computer is an essential research tool—feel almost embarrassed at the difficulty they have in incorporating its use into their teaching of physics. Bork's belief, implied in the quotation above, is that the computer offers an unparalleled opportunity for students to learn through interactive dialog. They can do this at their own paces, without inhibitions, helped by an imperturbable instructor of infinite patience (unless it is programmed to be otherwise!).

It is Bork's contention that full exploitation of the computer as teacher will serve to humanize, rather than dehumanize, instruction by relieving teachers of many avoidable chores and freeing them to deal directly with individual students. Perhaps this is indeed the wave of the future, but its acceptance in practice will call for a massive change in the views of most instructors who, outside certain centers of enthusiasm, view the computer's role much more conservatively.

One of the most ambitious efforts has been the development of the PLATO system at the University of Illinois.[7] Beginning in 1970, this project has developed instructional programs for a number of topics in physics. In addition, it has provided tutorial materials for a complete introductory mechanics course, with work by students at the terminals replacing half of the lecture time. The system clearly "works" inasmuch as students taking the course in the PLATO mode perform at the same level as those following the conventional route. It does not, however, seem to demonstrate any marked superiority.

Some have made optimistic claims for the use of the computer in a less dominant role as tutor, helping students with the solution of homework problems. For the present, at least, it seems that this method, allowing students to determine their pacing, may be the most effective use of the computer as teacher.

It has been suggested, on the basis of the time distribution of published papers about the use of computers in teaching physics, that interest in this topic peaked in about 1972 and has since decreased.[8] However, this conclusion can be questioned, even with respect to the restricted use of the computer as tutor. The more conservative use of computers for calculations and data analysis, and as a laboratory instrument, has undoubtedly been growing steadily.

Keller Plan. In 1968 Fred S. Keller, a professor of psychology, published a paper with the title "Goodbye, Teacher . . . ."[9] It described a scheme by which students would proceed at their own rates through a course, its subject matter being divided into a number

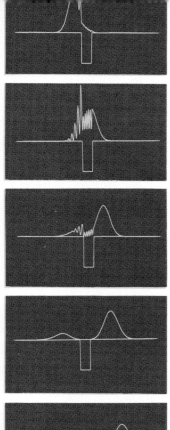

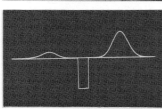

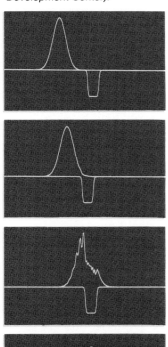

Stills from films *showing probability distributions for particle packets on potential wells. Above: zero surface thickness; below: surface thickness of about one-eigth well width (courtesy Education Development Center).*

of separate learning units. In this scheme, which later acquired the title PSI (personalized system of instruction) lectures became secondary or nonexistent. Students work from a textbook with guidance from supplementary notes and tutors and take unit tests when they feel ready. Testing is for mastery of the material, and students do not proceed to a later unit until they have shown sufficient knowledge of the previous one. This scheme, enthusiastically taken up by a number of physics teachers, has had substantial success (as judged by student performance) in classes of widely different size, all the way from freshman physics to graduate courses. It provides the benefits of immediate feedback (and Skinnerian reinforcement) but, as one of its proponents (Robert G. Fuller) admits, its influence to date has been relatively small, perhaps in part because it involves such a change of role for the instructor, who ceases to occupy center stage, and in part because it may entail an increased investment of instructor time to work well. Also, if a self-paced course in this mode is competing with other courses that have strict deadlines and scheduled examinations, students are likely to let it slide down their lists of priorities.

Examinations and testing

Fragment of a physics oral examination[10] at Oxford University, about 1890:

 Examiner: What is Electricity?
 Student: Oh, Sir, I'm sure I have learnt what it is—I'm sure I *did* know—but I've forgotten.
 Examiner: How very unfortunate. Only two persons have ever known what Electricity is, the Author of Nature and yourself. Now one of the two has forgotten.

Devising satisfactory testing and meaningful evaluation of students continues to plague physics instructors everywhere. It is, however, a matter that is absolutely central to the educational process. Everyone knows that for most students the content and character of examinations come close to defining their whole attitude to a course. As Eric Rogers put it in his AAPT Oersted Medal

lecture on examinations in 1969:

"Examinations tell students our real aims, at least so they believe. If we stress clear understanding and aim at a growing knowledge of physics, we may completely sabotage our teaching by a final examination that asks for numbers to be put in memorized formulas. However loud our sermons, however intriguing the experiments, students will judge by that exam—and so will next year's students who hear about it."

Students have many pressures on their time besides the physics courses we are teaching, and they know that grades are important in this harsh world. Few students (even among those who will become professional physicists) are so unworldly or so temperamentally secure as to let their interests wander far afield from what they can expect to be asked in their examinations.

Much thought and effort has, of course, been devoted to these matters. The PSSC program, for example, devised a complete set of tests that emphasized methodology of physics and aimed to assess a student's ability to deal with novel or unfamiliar situations, to draw conclusions from given information, to make logical predictions or to suggest new lines of investigation. Another development, the unit-test system of PSI, uses examinations as far as possible as entrance, rather than exit, tests, such that satisfactory performance entitles a student to proceed to the next stage.

The physics teacher

Some essential differences exist between the advancement of knowledge in an area of research and progress in education—in physics or in any other field. The basic and obvious point, of course, is that research is concerned with the cumulative growth of objective knowledge, whereas education means nothing unless it focuses on the development of the human individual. As the body of objective knowledge expands, the problem of reconciling the needs of the learner with what is to be learned becomes ever more acute. Although the nature of the problem is unchanging, each new generation has to tackle it afresh. Progress in education is to be measured primarily in terms of the vigor and imagination with which this challenge is met. Research needs to be done on the process of education itself—a field to which most physics teachers, at least at college and university level, have been far too indifferent. While physics teachers have their own subjective impressions of successful or unsuccessful teaching experiences and many institutions solicit student feedback and evaluation, the matter has received remarkably little methodical study. Paul Kirkpatrick, in his AAPT Oersted Medal lecture in 1959, commented on this in a provocative fashion:

"Across the campus from where you and I work [he was at Stanford] there is a school of education, or maybe a department. We *hear* that it is there, but we don't visit it very often. We are trying to teach the young and realizing that the job is too hard for us. Probably there isn't a teacher here who feels that he is a complete success. But while we go on with these feelings of partial success there is nearby this company of specialists in the philosophy and technique of the teaching process, and we make no use of them. When we are stuck on a chemistry problem we go to the chemists; if we get beyond our depth in mathematics we know mathematicians who throw us a rope. . . . But when we are wondering how to educate we do not go to the educationists. . . . I cannot dismiss this old boycott, as some apparently can, by declaring that ninety percent of the educationists are fools led by the ten percent who are smart rogues. . . ."

Now don't imagine that this pure company is being infiltrated by a disguised educationist. I have muddled along for forty years without the benefit of a course in education, but I am not proud of my ignorance as most science teachers are. Whether or not the educationists know the answers, they deal with great questions. What is the real nature of the process of learning? How can one train the mind? Can people be taught to think, and if so how do you do it? To what teaching operations are the different types of students susceptible, and how are the various types to be recognized?"

In the past two decades we have seen physicists, in partnership with educationists and psychologists, beginning to explore these fundamental questions. Some are dedicated followers of Jean Piaget; others, less committed to particular models of learning, are simply watching and listening to what goes on in the classroom or seeking to understand how students learn by themselves through reading, laboratory experience, and thought.

Experimentation with new teaching strategies has been a prominent feature during the past two decades. Teachers have been challenging the complacency of relying on traditional modes of instruction, based primarily on lectures and textbooks. However, innovations need to be seen not as ends in themselves, but as means to the end of effective teaching and learning of the subject matter of physics. John Rigden, editor of *The American Journal of Physics*, commented forcefully on this in a recent editorial[11]:

"We [in the AAPT] have allowed ourselves to become overly enamored with one educational novelty after another. There was programmed instruction, the Keller plan, self-paced instruction, computer-assisted instruction, Piagetian psychology, and education psychology more generally. I have been through them all and in each case the disciples of the new pedagogical savior led me up the mountain to view their promised land—a promised land that has never materialized in the form proclaimed. Lest I appear curmudgeonly, I believe all of these educational strategies have merit. The good teachers I have seen use features of all of them, but they do so without making a big deal over it. . . . Of course, a good teacher cares about students, but that concern must focus first and foremost on how the student is responding to the discipline. A good physics teacher loves the discipline, keeps up with the discipline, knows the discipline, and contributes to the discipline."

Rigden's remarks concern physics teaching and teachers in colleges and universities, where improvement is needed but calamities aren't widespread. At the high schools, however, there are causes for major concern. I cited earlier the estimate that only about a quarter of the approximately 15 000 teachers who teach physics in high school are "adequately" trained for the job. Far too often the teaching of physics is a minor assignment for a teacher whose main responsibilities and whose own training are in a different field. Furthermore, the greater material rewards of industry are drawing away many of the most able teachers: Only about 5% of new recipients of masters' degrees in physics now go into high-school teaching, compared with about 20% a decade ago. While it always has been a rarity for a person with a good degree in physics to go into teaching at the precollege level, the recent trend is alarming.

The role of the NSF

No account of progress in physics education in the US would be complete without reference to the

Physics labs of the past at Brown (far left) and at Wellesley (near left) and of the present at Carleton courtesy the AIP Niels Bohr Library.

role played by the National Science Foundation.

Almost immediately after its creation in 1950, the NSF launched its graduate fellowship program, its first major contribution to science education. In the first full year of operation (FY 1952) this program accounted for $1.5 million out of a total appropriation of $3.5 million for the Foundation's work; the Foundation funded over 500 predoctoral fellowships (in all branches of science). The program grew rapidly, until by the mid-1960s it was supporting well over 3000 predoctoral students through fellowships or traineeships; about a third of them at this point were in the physical sciences (about 10% in physics in particular). From 1967, when a maximum of $45 million per year was reached, student support declined rapidly. By 1975 the allotment of funds was down to about $13 million (equivalent to only about $8 million in 1967 dollars). The decline for physics was even more rapid, since the *fraction* of the fellowships awarded in the physical sciences was also sharply reduced. The total demise of such support is now imminent. The impetus the NSF gave to graduate education through its graduate fellowship programs has certainly been of enormous value; its elimination is to be deplored.

A similar story describes the NSF's support of other aspects of science education, which began in a major way in 1956–57, with the PSSC program and the summer institutes for science teachers. The latter, supplemented by academic-year institutes, fellowships for science teachers, and so on, rapidly became a major component of NSF's science education budget; by the early 1960s the NSF was supporting the institute programs at a level of over $40 million per year, that is, about two-thirds of the total funds then being allocated to science education (excluding the graduate fellowship programs). These institutes and various related programs brought large numbers of secondary-school teachers into contact with teachers from colleges and universities, which was very beneficial to both groups and to the health of science education generally.

As time went on, the Foundation expanded and diversified its programs of aid to science education. In 1965, the total support for such programs had risen to about $80 million per year; of this, nearly

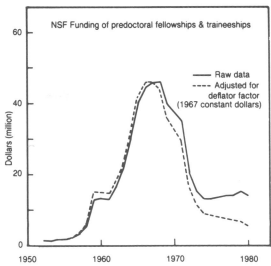

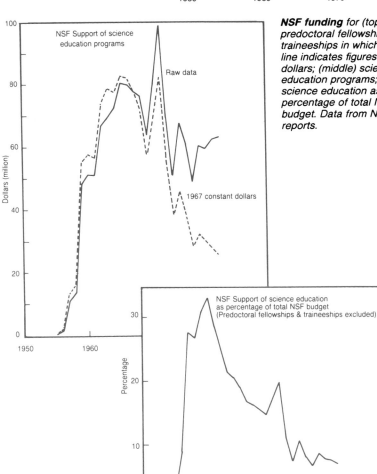

NSF funding for (top) predoctoral fellowships and traineeships in which the dotted line indicates figures in 1967 dollars; (middle) science education programs; (bottom) science education as a percentage of total NSF budget. Data from NSF annual reports.

60% went into institutes and other teacher-improvement programs, about 20% into course content and curriculum development, and about 20% into direct aid to colleges and schools (in a ratio of roughly 4:1) for scientific equipment.

The middle graph on page 61 shows the total of NSF support for science education (excluding predoctoral fellowships and traineeships) as a function of time. It may be seen that, corrected for inflation, the NSF did not maintain support at the level of the early 1960s. By 1980 support was down to about one-third of that peak. Even more striking is the bottom graph on page 61, which

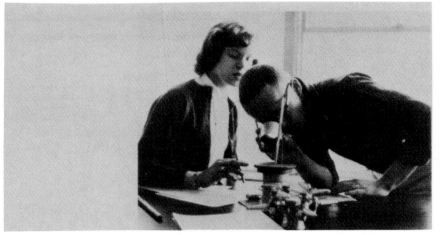

Students at Garden City High School recording observations in a physics lab (courtesy AIP Niels Bohr Library).

shows NSF support of science education expressed as a percentage of the total NSF budget; clearly the support of education has lost a great deal of ground relative to research.

It is true, of course, that even after the cutbacks, the level of NSF support for science education in FY 1980, at about $60 million per year, was by no means negligible. There were summer courses and academic-year workshops reaching over 15 000 teachers for grades 5–12 and short courses for 3000 faculty members at two-year and four-year colleges. About 18% of the total program funds went to direct support of minorities. "Science and society" programs received $7.3 million (more than 10% of the total), and nearly $14 million went into science education and research, much of it for the development of instructional techniques and materials and for research into the learning process. The emphasis has thus shifted markedly from the earlier days, when scientific subject matter and curriculum development loomed large and broader social concerns were hardly considered. These statistics for NSF support of science education in general apply with an appropriate scaling factor to physics in particular.

The importance of NSF contributions cannot be overestimated; they have been used, as we have seen, for major curriculum improvement projects, for upgrading of physics facilities at the secondary and college levels, for the activities of the Commission on College Physics, for extensive teacher training and for many other things. As with the graduate fellowships, the erosion of this support must be regarded as a serious blow to the state of physics education in the US. The prospect that the present Administration might phase out all the funds for such activities is grim indeed

and suggests a strange view of priorities in a society whose health is so directly based on scientific achievement.

Problems and prospects

This article will, I hope, have justified the belief that there is indeed progress in physics education—and, more, that it is at present a field of great activity and perhaps of more diversity than ever before. If the problems confronting us seem much the same as those of 50 years ago, that is no cause for despair—it is in the nature of the subject. However, it would be wrong to pretend that all is rosy; among the areas in special difficulty are these:

▶ The health of the teaching profession itself. Except for the college and university level, the educational system is not producing an adequate supply of new, competent teachers; in the past year alone, the membership of the Association of High School Science Teachers decreased by 10%. There is also a serious lack of contact between teachers at different levels, in particular between teachers at secondary schools and universities. The wave of activity in this direction during the 1960s, sponsored mainly by the National Science Foundation, has subsided with the fading of that support.

▶ The lack of success teachers have had in communicating knowledge or appreciation of physics to students who are not going to become professional scientists. This difficulty has its origins at the earliest levels of education, well before students have the chance to elect (or, more often, reject) physics as a subject of study toward the end of high school. Physicists (and other scientists) are not succeeding in spreading scientific literacy; they are best at producing their

own successors. This must be seen as a most urgent problem in our society because, until it is solved, there will be a continuing gulf preventing understanding between scientists and the public, in which latter category are many in positions of power and responsibility.

* * *

I wish to thank Susanne D. Ellis for sending me a quantity of statistical data from previous and current AIP Manpower Surveys and E. L. Jossem for communicating a number of materials and suggestions pertaining to this article.

[Note added, February 1985:

Since this article was written, one major development has been the reestablishment (in October, 1983) of the Science and Engineering Education Directorate within the NSF, with particular concern for pre-college programs, in response to wide recognition of a crisis in science and mathematics education in the U.S. (In this connection, see the special issue of Physics Today, September, 1983.)

At the graduate level, the feared elimination of NSF fellowships did not in fact take place, and this program has been maintained since 1981 at a fairly constant level.]

References

1. Henry A. Perkins, Am. J. Phys. **17**, 376 (1949).

2. Clifford Swartz, The Physics Teacher **17**, 422 (1979).

3. H. R. Crane, Commission on College Physics, Newsletter No. 7, April 1965.

4. Arnold A. Strassenburg, Change **10**, (1) 50 (I978).

5. Leo Nedelsky, Am. J. Phys. **26**, 51 (1958).

6. *The Role of the Laboratory in Physics Education.* J. G. Jones, J. L. Lewis, eds., Association for Science Education, Hatfield, UK, 1980.

7. S. G. Smith and B. H. Sherwood, Science **192**, 344 (1976).

8. A. Douglas Davis, Am. J. Phys. **49**, 391 (1981).

9. Fred S. Keller, J. Appl. Behavior Anal. **1**, 79 (1968).

10. Falconer Madan, *Oxford Outside the Guide-Books*, in Jan Morris, ed., *Oxford Book of Oxford*, Oxford U. P., 1978.

11. John Rigden, Am. J. Phys. **49**, 809 (1981).

Women in physics: unnecessary, injurious and out of place?

Despite eight years of affirmative action
more changes are necessary to create an atmosphere where
women are equally accepted in the field of physics.

Vera Kistiakowsky

PHYSICS TODAY / FEBRUARY 1980

The subtitle for this article is taken from a Strindberg essay written at the end of the 19th century opposing the appointment of the mathematician, Sonia Kovalevsky, to a professorship at the University of Stockholm, in which he attempts to prove "as decidedly as that two and two make four, what a monstrosity is a woman who is a professor of mathematics, and how unnecessary, injurious and out of place she is".[1] It is certainly a much more extreme statement than anything likely to be voiced publicly today but it does vividly and tersely encapsulate many of the opinions that have been expressed to me in much more veiled and discursive form over the last ten years. Largely because of these continuing though muted attitudes I have accepted an invitation to write this article for PHYSICS TODAY. I will very briefly sketch the history of women's participation in physics as a background to the current situation and then discuss some statistical information about women physicists in the recent past and present in the United States. It will come as no surprise that the percentage of physicists who are women is small and that their employment patterns are different from those of men. I will discuss the possible reasons for this situation. Finally, I will comment briefly on recent changes and what expectations one may have for the future.

History from Arate to Whiting

Since physics as we know it today only emerged at the beginning of the seventeenth century, I should perhaps start my mention of women's participation with this period. However, having grown up

Vera Kistiakowski is a professor of physics at MIT and does research in experimental high-energy particle physics.

with a pre-history of science, that of the Greek natural philosophers, in which women were conspicuous by their absence, I can not resist remarking that there is evidence that women natural philosophers did exist. Arate of Cyrene was supposedly a contemporary of Socrates (5th century BC) who taught and wrote on natural philosophy in Attica.[1] She was, however, not the first; women were equal members of the Pythagorean school in the 6th century BC[2] and Theano, the wife of Pythagoras, assumed the leadership of the school after his death.[3] Moving forward a millennium we find Hypatia, a neo-Platonic philosopher and mathematician who spent the last part of her brief life teaching at the university in Alexandria at the beginning of the 5th century AD.[3] In the middle ages the physical sciences languished; and, although the convents produced a number of notable women scholars, their writings were mainly in the areas of the biological sciences and medicine. However, one of these women, St Hildegard, the Benedictine Abbess of Bingen-on-the-Rhine in the 12th century AD, wrote on a heliocentric universe in which "the sun attracts the heavenly bodies as the earth attracts its inhabitants," an early intimation of gravitation.[1]

Unfortunately, the beginning of the scientific age coincided with a wave of opposition to the education of women in Europe and Great Britain. The few women who contributed to physics were either of high enough social status that they could follow their inclinations despite the general prejudices of the times, like Emilie de Breteuil, Marquise du Chatelet and Laura Bassi, of the early 18th century[1,3], or like Mary Somerville (early 19th century), who was known principally as a mathematical astronomer,

self-educated over the opposition of their families. This situation remained about the same until the end of the 19th century.

In the US the situation of women improved somewhat more rapidly than it did elsewhere. The Boston public schools were started in 1642, and although they did not admit girls until 1789, this occurred considerably earlier than was the case in Europe and Great Britain. Many secondary schools in the US were opened to women at the beginning of the 19th century, apparently because more school teachers were needed. Finally, in 1837, two hundred and one years after the founding of Harvard College, Oberlin College admitted the first three women to the bachelor's degree program.[5] Due to both the economic and feminist pressures for women's education, a few more male institutions became coeducational, and several women's colleges were established. However, the number of these institutions remained small until after the Civil War, and many of the women's colleges were of inferior quality. The lack of greater change in opportunities for women could be considered part of a general pattern where educational reforms which included the establishment of scientific, technical and graduate education remained blocked until after the war ended in 1865. Then both academic science and women's education blossomed and the numbers of women scientists increased. We know of no woman recognized as a physicist prior to this period; the earliest woman scientists in the US of whom there is a record were a botanist, Jane Colden (1724–66), and an astronomer, Maria Mitchell (1818–89). Two of the first women to achieve recognition as physicists were Margaret E. Maltby (1845–1926) and Sarah F. Whiting (1847–1927),

*"I want you to know, gentlemen, that at this moment I feel
I have realized my full potential as a woman."*

Drawing by Franscino; © 1973 The New Yorker Magazine, inc.

who taught at Barnard and Wellesley College respectively.[4]

Women physicists in the USA

Margaret Rossiter[6] has given us a very detailed picture of the situation of women scientists in the US at the beginning of the 20th century using the information given for individual men and women in the 1906, 1910 and 1921 editions of "American Men of Science." Among the physicists included in her sample are 23 women, a number that corresponds to 2.6% of the total number of physicists listed. It is not surprising that 11 of these women received their undergraduate education at women's colleges and that 21 of them were employed at women's colleges at some point in their career. These colleges were both an important source and the employer of a majority of academic women at the beginning of the century. Three of the women also spent extended periods of time as secondary school teachers, whereas this was true of none of the men, another difference common in fields other than physics. None of the women physicists had married. It was generally accepted before 1920 that the pursuit of a scientific career required a single-minded determination, which was incompatible with marriage for a woman. A wife was expected to be totally dedicated to that role and to subordinate her interests and activities to the aspirations of her husband.

By the end of the 19th century the PhD had become the scientific union card, and one may begin to trace the participation of women in physics through the percentage of doctorates awarded to women. In Rossiter's sample, 65% of the women and 71% of the men physicists had PhD's. The percentage of physics doctorates awarded to women increased until 1920,

a year in which four women received physics PhD's, 19% of a total of 21.[7] The figure on page 34 gives the number and percentage of physics doctorates awarded to women from 1920 to 1978. The corresponding numbers and percentages for astronomy doctorates are also shown because some of the statistical information I will discuss later in this paper is available only for physicists and astronomers lumped together. It can be seen in the figure that the percentage dropped steadily to a low of 1.8% in the 1950's. The numbers of women physicsts increased in this period, but less rapidly than was the case for men. The reasons for this pattern, which is also seen in most other fields, include the subsiding of the first wave of feminism, which exhausted itself on the achievement of suffrage and universal education in the early 1920's. The improvement of women's role in marriage, which also occurred, was not far-reaching enough to make marriage and career generally compatible. The depression that followed was a further deterrent to the aspirations of women; any money available in a family was usually dedicated to the education of the men, who were still considered the primary breadwinners. And in World War II, although women went to work by the millions, graduate study did not seem an appropriately patriotic endeavor. After the war the massive return-to-the-home propaganda campaign presented the women of my generation with a clear and explicit message—husband and family came first and this should be the exclusive concern of women. The decline continued, reaching a low point in the 1950's. In the 1960's, when physics was mushrooming in post-Sputnik euphoria, the percentage of doctorates awarded to women began to increase, probably due to the

many changes of that decade which affected social attitudes, and marital and economic patterns. These include the resurgence of the feminist movement which became increasingly vigorous in the later 1960's, leading to further changes reflected in the continuing increased percentages for women in the 1970's.

The 1973 New Yorker cartoon in the figure on this page very accurately portrays this change. The phraseology of its caption is that used to describe the womanly woman who was the paragon in the previous three decades, which sounds wildly inappropriate when applied to success in a mostly masculine field.

This renaissance of feminism was felt by professional women and led, among other things, to studies of the situation of women in the various professional societies in the early 1970's. In The American Physical Society, Brian Schwartz started the ball rolling. Through the Forum on Physics and Society, he organized a session on Women in Physics, chaired by Fay Ajzenberg-Selove, at the 1971 Annual Meeting of the APS. This was a most thought-provoking occasion, not only because of the presentations by the speakers but also because of the less than informed comments from some members of the audience. The most memorable was the statement, "If I had been married to Pierre Curie, I would have been Madame Curie," by a well-known male physicist. This session inspired a letter cosigned by 20 women physicists requesting that the APS Council establish a committee on women in physics to study their situation and make recommendations for appropriate actions by the Society. At the 1971 spring meeting in Washington the Council did establish such a committee and with the help of Jerome B. Wiesner, president of Massachusetts Institute of Technology, it obtained the Sloan Foundation grant that made the study possible. A report and a roster containing the names of women physicists were prepared and submitted to the APS Council at the 1972 annual meeting.[7] Seven years of affirmative action later we are all, perhaps, accustomed to the statistics, but at the time it was novel information. For example, an eminent physicist whom I encountered at an information-gathering session of the Committee on the Future of the APS asked me why the Committee on Women in Physics was wasting its time on a study when there were only two women physicists in the United States and both of them were happy. Obviously, he was aware that there were more than two.

However, most physicists would have numbered their women colleagues in the tens and not in the hundreds, which was the outcome of the study. He was also misinformed on the question of happiness. One of the women he had in mind was a member of the Committee, the other was actively supporting it, and neither was happy with the status quo. Only two of the 451 doctoral women physicists who responded to the survey indicated any lack of enthusiasm for the work of the Committee, and a majority of the respondents were strongly supportive. This interest of many women physicists in the issues raised has continued to be active and the Committee has therefore continued with a changing membership carrying out a variety of projects.

Let me briefly summarize the findings of the 1971 study. It described a situation that was little changed from that described by Rossiter for the period before 1920. Women physicists in both studies were employed mainly in academia, were found more frequently in the lower faculty ranks and non-faculty positions, and worked at the less prestigious institutions. In both studies a larger percentage of women than of men were found to suffer from involuntary unemployment and under-employment, and the average salaries of employed women were lower. An interesting difference between the situation in 1971 and that before 1920 is that 60% of women physicists in 1971 were married, compared with none in 1920. The APS study drew the conclusion that overt discrimination, prevalent societal attitudes and the practical problems of combining career and marriage had played important roles in causing the differences observed between the women and men who had chosen physics as a career.

The situation in the 1970's

Let us look briefly at the statistics for the participation of women in physics during the last eight years. The figure on page 34 shows that the number and percentage of doctorates awarded to women have continued to increase since 1971 but the percentage increase is much more dramatic. This has been partly due to the continuing increases in the number of bachelor's degrees in physics awarded to women (see the figure on page 36) and also because the fraction of women students leaving graduate study with only a master's degree has decreased. Thirty-three percent of the women receiving physics baccalaureates in the 1950's went on to a master's degree within an average period of two years and 37% did so in the 1960's. The ratio of the percentages is 1.12, indicating only a small (12%) increase. However, the comparable figures for those completing a doctorate an average of seven years later were 10% for those receiving baccalaureates in the 1950's and 17% for the 1960's, a 70% increase.

The percentage of the doctorates awarded to women in the various subfields of physics in the periods 1960–69 and 1970–76 were not significantly different from those for all subfields combined in those periods, respectively 1.9% and 3.5%. This percentage includes astrophysics in the later period (4.9 ± 1.0%). The percentage of the doctorates in astronomy awarded to women in 1970–76 was significantly higher (8.4 ± 1.4%), as was the percentage of doctorates in astronomy and astrophysics combined in 1960–69 (6.4 ± 1.2%). However, since the astronomy doctorates were only about 5% of the number awarded in astronomy and physics combined in both periods the statistical information for these combined

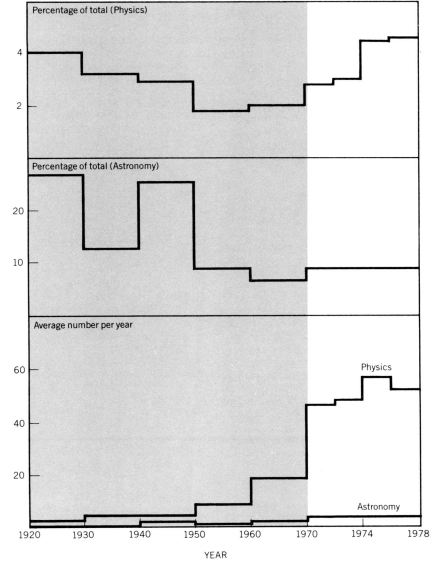

Doctoral degrees awarded to women. From 1920 to 1970 the numbers are averaged over and the percentages are calculated for each decade. From 1970 to 1978 the physics numbers are averaged over and the percentages are calculated for each two year period. The astronomy number is averaged over and the percent is calculated for the eight year period. These data are taken from "Doctorates Awarded from 1920 to 1971 by Subfield of Doctorate, Sex and Decade," National Research Council (1973) and "Summary Report (Year). Doctorate Recipients from United States Universities," National Academy of Sciences, for the Years 1972 through 1977.

fields will not be significantly different from that for physics alone.

Some further comments are possible concerning the physics and astronomy doctorates of recent years. For example, 63.3% of the women and 63.5% of the men receiving doctorates in 1974 through 1977 were married, reflecting the very major change in attitudes toward the possibility of combining careers and marriage since the beginning of this century.[8] In the years 1973 through 1975 11% of the black and American Indian doctoral recipients were women, a percentage which is based on very small numbers and is, therefore, not significantly different from the corresponding percentage for whites, of whom 3.4% were women. However, the percentage of foreign citizens awarded doctorates in the years 1974 through 1977 who were women is 7.7%, which is significantly higher than 4.2%, the corresponding percentage for US citizens.[8] In this period both the median age when receiving the doctorate and the median length of time between baccalaureate and doctorate were the same for men and women.[8]

The percentage of doctorates in the physics/astronomy labor force (those employed or seeking employment) who were women rose from 2.0% in 1971[7] to 2.5% in 1975.[9] The percentage of women who were foreign-born US citizens or foreign citizens in the labor force in 1975 was 21.8%, which is not different within the uncertainties from the percentage, 20.6%, for men.[9] The table at the top of this page indicates that the percentage of women employed part time or full time was 89% in 1973, whereas the similar percentage for men was 97%. The percentage of those unemployed and seeking employment was about four times greater for women than for men. Approximately eight times more women worked part time, but in 1973 about half of them were seeking full-time employment. In 1977 the percentage of women doctorates in physics and astronomy in the labor force who were seeking employment was 5.7%, still much higher than that for men.[10] However, between 1973 and 1975 the percentage of women doctorates in physics and astronomy who were working part time and seeking full-time employment dropped from 8.4% to 2.7%, although it was still more than three times greater than the corresponding percentage for men.[11]

The table at the bottom of this page gives the distribution of men and women physicists and astronomers with respect to type of employer. The percentage of women in educational institutions in 1973 was greater than that for men, but decreased from its 1971 value of 77%, with corresponding increases in the percentages in government and nonprofit employment.[7] The percentage of men in industry decreased from 26% in 1971, whereas that of women increased very

slightly.[7] It should be noted that the percentages of doctoral women who taught in junior colleges and secondary schools in 1973 are larger than those for men. However, a study of women high-school physics teachers showed that these women are a small minority and, in fact, most women high-school physics teachers do not have any physics degree.[12]

The median salaries for men and women for the various types of employers were consistently lower for women by 5 to 20% in 1971, 1973 and 1977.[7,10] In any number of studies it has been found that further subdivisions of the sample does not remove the differences. For example, in 1977 the median salaries for all age groups of women doctorates including the youngest were significantly less than those for men.[10]

Because the major employer of physicists is the educational institution, it is interesting to examine the situation there more closely. The table on page 37 presents the number and percentage of women in various types of physics departments in 1971–72 and in 1978–79. It is seen that the percentages for the total of all types of departments have decreased except for assistant professors and "other." In the PhD-granting departments the changes are not significant except for assistant professors. The increases in the percentages in the "Top Ten" physics departments are particularly striking but should be interpreted with caution since seven of the eleven women are at MIT. Similarly, although 7.3% (ten women) of all the assistant professors appointed between 1972 and 1979 in these ten departments were women, the figure drops to 4.4% (five women) for the nine departments excluding MIT. It should be noted that except for the "Top Ten" category, the institutions in the various categories are not exactly the same in the two years studied, and thus the changes in percentage and number are a composite of changes in degree-granting type and changes in the employment of women. Eight years of affirmative action can hardly be said to have caused major changes in the presence of women on physics–department faculties. Nonetheless, there has indisputably been an improvement for women at the assistant-professor appointment level.

In summary, the predominant impression gained from looking at the statistics is that there has not been very much change since the beginning of the century or since the 1971 APS study. The exceptions are the continuing increase in the percentage of PhD's awarded to women and presence of a few more women on the faculties of departments in research universities.

Reasons and remedies

If one wishes to speculate on the future it is important to consider the reasons for

the low participation of women in physics and for the differences between the careers of men and women. I will discuss various reasons that have been suggested, grouping them into five categories: innate ability, environment, discrimination, career conflicts, and the Matthew effect. I will also comment on remedies.

The question of an insurmountable difference in innate ability between the sexes has become somewhat of an unmentionable topic these days, thanks to the raised level of public sensitivity. There are few Lionel Tigers who will argue in the public press that since males dominate the baboon society, females must be subordinate in human society.[13] However, there are many studies investigating sex differences in various attributes, and it is necessary to deal with this topic by taking a close look at the situation concerning innate and unalterable sex differences. It has been difficult for a non-specialist to get a clear picture of the cumulative outcome of such studies due to the prolixity of the experimental situation, but there is now an encyclopedic compilation and discussion of this research by Eleanor Maccoby and Carol Jacklin.[14] Although there is not universal agreement with all of the conclusions drawn by the authors, their overall picture is generally accepted and disagreement is focused on interpretation of experiments in certain areas. The tabular arrays of experimental results presented in Mac-

Employment status of Men and Women PhD Physicists and Astronomers in 1973

Employment Status	Men	Women
Full-time	94%	66%
Part-time	1.7%	16%
Part-time seeking full-time	0.8%	7%
Unemployed seeking employment	1.7%	7%
Unemployed not seeking, retired, other	3.0%	11%
Total number in sample	17 481	471

Data from 1973 Survey of Doctoral Scientists and Engineers, National Research Council.

Employers of Men and Women PhD Physicists and Astronomers in 1973

Employer	Men	Women
Educational Institution	56%	67%
PhD Granting	41%	44%
MA Granting	5%	4%
BA Granting	9%	15%
Jr College	1%	3%
Secondary School	0.3%	1%
Government	15%	16%
Industry	21%	10%
Nonprofit	5%	4%
Other	3%	3%
Total Number	16 689	387

Data from 1973 Survey of Doctoral Scientists and Engineers, National Research Council.

coby and Jacklin's book clearly make the point that the result of a single experiment, or those of a small group of experiments, are never adequate to yield a definitive answer to any general question in this field. The sample choice, the experimental technique and the interpretation of what is measured permit contradictory results for any attribute studied. However, certain patterns do emerge

and they are relevant to aptitude for scientific work. First of all, there are eight attributes for which sex differences are commonly believed to exist but for which the evidence is conclusive that this is not the case. These include rote-learning ability, higher-level cognitive processing, analytic ability and achievement motivation. For all of these no sex differences of any origin have been found. For seven

other attributes, including competitiveness, dominance and compliance, Maccoby and Jacklin conclude that there is not sufficient evidence to decide the question. They also conclude that there are four areas where sex differences are well established. For two of these, verbal ability and mathematical ability, available evidence does not indicate a sex-linked genetic component, and the sex differences can be attributed completely to environmental effects. The magnitude of the sex differences in mathematical ability varies widely, depending on the age group studied, from none for young children to significant differences for adults. The differences between medians of the relevant test scores for men and women are at most 0.4 of the standard deviations, and the test score distributions extend over the whole range of values for both sexes.

Finally, there are two attributes for which Maccoby and Jacklin believe evidence exists for a sex-linked genetic component. The first is aggression, which is probably not positively correlated with scientific competence since, as it is defined, it does not include achievement motivation, competitiveness or dominance. Furthermore, since the learned component of this attribute is important and aggression is negatively correlated with intellectual ability in boys, the greater male biological priming for learning aggressive behavior appears to be a negative indicator for a male scientific career. It is interesting to note that the correlation is positive in girls and that aggressiveness could be a positive indicator in their case.

The second attribute for which the authors believe there is evidence for a sex-linked genetic difference is one type of visual-spatial ability. There is some disagreement with this assessment, but, even if it is correct, it only means that one of a number of genes contributing to high spatial ability is sex-linked. Furthermore, there is also an equally important learned component to the exercise of these abilities. The differences observed between the medians of relevant test scores for males and females vary widely between various cultures and are at most 1.4 of the standard deviations. Since there is no information concerning the correlation of spatial ability with scientific achievement it is hard to assess the effect of this attribute. However, it is clear that the one sex-linked genetic component is not a major factor and that the differences could be substantially reduced by an educational process which stresses development of visual-spatial abilities equally for both boys and girls.

Thus, it is extremely unlikely that sex-linked genetic differences are an important factor in the observed differences in scientific participation. There remain, however, the differences that are environmental in origin, and their importance

Bachelor's and master's degrees awarded to women. The numbers were averaged over and the percentages calculated for the periods 1948 to 1960 and 1960 to 1970. Annual numbers and percentages are given for the period 1970 to 1976. These data are taken from Table PS-P-2, "Professional Women and Minorities," B. M. Vetter, E. L. Babco and J. E. McIntire, Scientific Manpower Commission, Washington, D.C. (1978).

is evident. It is impossible to establish cause and effect, but I would suggest that the same environmental pressures that have led to the differences on mathematical ability test scores are also responsible for the sharp decrease in the participation of girls in mathematics and physical-science courses in secondary school with the level of the course, rather than mathematical ability itself. The difference in participation is much too great to be plausibly accounted for by the small differences in the medians of the test score distribution. What are these pressures? They start in early childhood when girls are rewarded for "feminine behavior" and given "girl's" toys. They escalate in adolescence when conformity to a particular feminine role is considered necessary to attract boys. To be good at science and math has been considered to be inappropriate for a girl, a threat to her popularity and unnecessary for her future role in society. Alison Kelly has pointed out in a paper describing the substantial differences in participation in secondary school physics in Great Britain, that girls' schools have a significantly better record than coeducational schools, presumably because in that environment there is more faculty encouragement and peer support for achievement in physics.[15]

These effects are also felt at the undergraduate college level, where women's participation in physics continues to be low in spite of the academic selection that has taken place. In general, a lower percentage of women than men prepare themselves for graduate school in any discipline. The seven women's colleges that are linked with the Ivy League men's colleges (The Seven Sisters: Barnard, Bryn Mawr, Mount Holyoke, Radcliffe, Smith, Vassar and Wellesley) have a uniquely excellent record for both the number and percentage of their graduates who have continued to a doctorate and to professional recognition.[16] This record includes the fields of mathematics and the physical sciences, and one can again speculate that a supportive environment is a cause.

That self-selection also plays a role is evident from the excellent record of a few coeducational colleges (Oberlin, Reed and Swarthmore) and from the fact that a greater percentage of women with baccalaureate degrees from MIT later received a doctorate degree than was the case for any other academic institution with a significant number of women (11% versus 9.7% for the next highest).[17] This could hardly be attributed to a reputation for a supportive environment, since, although MIT granted its first degree to a woman in 1867, it was not until nearly one hundred years later that women were recognized as an important part of the undergraduate community. However, my own experience and that of many other women has been that the supportive environment of a woman's college made

it much easier to study mathematics and science with the expectation of pursuing careers in these diciplines.

The question of math and science avoidance has been discussed by many authors, notably Shiela Tobias,[18] and a number of programs to counteract this situation have been established. One of these is an informal network of women scientists and mathematicians working in San Francisco area schools and colleges to encourage girls to take science and math courses and to tell them about career options in the various fields. The program, originated primarily by Lenore Blum and Nancy Kreinberg, presently involves more than 400 women scientists and mathematicians.[19]

Societal views of appropriate roles for women are changing. Admittedly, the progress is uneven, but I do not think that there can be a pre-teenage girl whose family owns a television set who views marriage and motherhood as the only option for a woman, even though this may be the only option of interest to her. She

knows that there are women in many "men's" fields, including the physical sciences, and gradually this should result in increases in the numbers of girls who take physical sciences and advanced math in high school and who can therefore consider such majors in college. Again, the changes are slow, but since our society is now one in which the majority of women are employed outside of the home for a major part of their adult lives, they should lead to much more substantial numbers of young women laying the foundation in high school and college for graduate work in physics.

In the past, there has also been substantial attrition in graduate school, with twice as many women graduate students in physics terminating with a master's degree than was the case for men.[7] Again, anecdotal evidence indicates that negative peer attitudes concerning the appropriateness of scientific careers for women were an important factor, together with the perception that job opportunities were limited for doctorate-level women

Women Faculty in Physics Departments

Department Type and Rank	1971–72[a]	1978–79[b]
"Top Ten"[c]	*Percent (number)*	
All Professors	0.8 (4)	2.7 (11)
Full Professor	0.6 (2)	1.0 (3)
Associate Professor	1.1 (1)	5.8 (3)
Assistant Professor	0.9 (1)	7.7 (5)
Other[d]	2.8 (1)	5.0 (1)
PhD Granting		
Number of Departments	(158)	(212)
All Professors	1.5 (74)	1.7 (88)
Full Professor	1.0 (23)	1.2 (38)
Associate Professor	1.8 (24)	1.5 (22)
Assistant Professor	2.0 (27)	4.5 (28)
Other[d]	5.9 (17)	4.5 (15)
MA Granting		
Number of Departments	(133)	(123)
All Professors	2.3 (28)	2.5 (27)
Full Professors	1.9 (7)	1.5 (8)
Associate Professor	2.2 (9)	2.9 (12)
Assistant Professor	2.6 (12)	4.4 (7)
Other[d]	4.6 (6)	16.2 (11)
BA Granting		
Number of Departments	(743)	(606)
All Professors	5.4(144)	3.9 (93,
Full Professor	5.8 (55)	3.2 (29)
Associate Professor	4.9 (33)	3.9 (35)
Assistant Professor	5.2 (56)	6.9 (39)
Other[d]	9.5 (44)	11.4 (27)
All Three Types		
Number of Departments	(1034)	(941)
All Professors	2.8(246)	2.5(218)
Full Professor	2.4 (85)	1.7 (75)
Associate Professor	2.7 (66)	2.5 (69)
Assistant Professor	3.3 (95)	5.5 (74)
Other[d]	7.6 (67)	8.3 (53)

a. 1971–72 data from "Women in Physics", report of the Committee on Women in Physics of the American Physical Society, *Bull. Am. Phys. Soc.* **17**, 740 (1972).
b. 1978–79 data compiled from the "1978–79 Directory of Physics and Astronomy Faculties," American Institute of Physics (1978). Astronomy Faculty are not included.
c. The top ten in 1970 according to the American Council on Education: Berkeley, Caltech, Chicago, Columbia, Cornell, Harvard, Illinois, MIT, Princeton and Stanford. The same institutions were included for 1978–79 sample. There were two additional women in the Division of Physics and Astronomy at Caltech who were designated astronomy faculty.
d. Lecturer, instructor, research professor, etc.

physicists. Bluntly, why get a PhD in physics when you can't get an interesting job and it makes it harder to be married? Other contributing factors that have been mentioned are isolation, not being included in the collegial interactions of the peer group, and "invisibility"—not being perceived as a serious student by professors. Here again, changing attitudes concerning appropriate roles for women and the changing views of marriage must also have improved the general situation in the last ten years. Furthermore, affirmative action, ineffective as it has been on the whole, has created the impression that doctoral women scientists can get jobs. It comes then as no surprise that more women now continue to a doctorate.

The third category of reasons for the difference between the statistical patterns for women and those for men listed at the beginning of this section is discrimination. Although it is generally hard to document, there is considerable direct evidence that discrimination has been an important factor. Universities have had overt policies of not accepting women graduate students, of not hiring women faculty even though they educated women students, and of favoring men for promotions and pay increases because they "needed it more, they had a family to support." There is also considerable anecdotal evidence of discriminatory attitudes. For example, there is the thesis supervisor who advised a woman student to look for a job as a scientific editor, since such a job would be more compatible with marriage and a family than a position requiring her to do research. Or the numerous professors who said that they did not want women graduate students because they once had a very good one who quit to raise a family as soon as she got her degree. It is interesting that, although I have heard this from so many individuals that it

should be a significant statistical effect, the evidence is quite to the contrary. Approximately 95% of the women who received a PhD have remained professionally active, although a substantial number took time off or worked part time when their children were small.[7]

There have also been regulations that were *de facto* discriminatory, such as nepotism rules invoked mainly against wives. The classic example is Maria Goeppert Mayer, who was denied a paid scientific position for a major part of her scientific career and did not receive a full-time professorship until after the publication of her Nobel prize–winning work on the nuclear shell model.[20]

And finally there has been an inability to recognize women as plausible scientists, which certainly must have colored the reactions of those men so affected toward the hiring or promotion of a woman scientist. An experimentalist recently commented to me that physics departments were obviously "leaning over backward to appoint women as assistant professors" because in the last five years the percentage of these appointments that have gone to women has been about the same as the percentage of recent physics doctorates earned by women. The phase "leaning over backward" clearly reflects an attitude about the qualifications of women in general which can not help but influence decisions on individuals. This perception of women physicists is still quite widespread and is not only held by older scientists. The person who made this remark is a generation younger than I. Discriminatory attitudes also frequently manifest themselves in an unwillingness to admit that a woman could succeed. I have heard a number of people say that Enrico Fermi "gave" Maria Mayer the nuclear shell model, or that Pierre Curie was mainly responsible for the Nobel prize shared

with Marie Curie. The evidence supports neither assertion. It is, of course, difficult to assign credit when work is done jointly by husband and wife. However, in numerous articles mentioning Marie Curie as a scientist who won a Nobel prize in 1903 jointly with her husband and Antoine Becquerel, there is no mention that she received a second unshared Nobel prize in 1911 for the discovery of radium and polonium after her husband's death and no mention of the fact that she was the only person to receive two Nobel prizes until 1962. These stories are not as trivial as they may seem, because they translate to "Oh, her husband (professor, coworker, and so on) did the important part of the work" when such attitudes are encountered by less famous women scientists. Only time can cure such attitudes, as the men who hold them retire and are replaced by others who have had women physicists as professors and peers, and are at ease with them.

The fourth category of reasons for the differences between the participation of men and women in physics stems from conflicts between the demands of a career and those of personal life, particularly if these involve marriage and children, because these conflicts have in the past generally been seen as a problem that the wife must resolve. An interesting consequence of this was observed by Lindsey Harmon in a study of early performance indicators, such as high-school grade point averages and college entrance tests, of individuals who subsequently received doctorates. Almost without exception in all fields the married women ranked highest on all indicators, with single women ranking next, followed by single men and finally by married men. This was a totally unexpected result for which Harmon suggested the following explanation: "When the superiority of women over men doctorate-holders was noted in

Women Nobelists. Opposite page, Maria Goeppert Mayer, her husband Joseph E. Mayer, Robert d'Escourt Atkinson, Paul Ehrenfest and Lars Onsager lounging on the lawn of the University of Michigan summer school in 1930. Left, Marie Curie. Rarely is it mentioned that she received a second Nobel Prize in 1911 for the discovery of radium and polonium after the death of her husband. She remained the only person to receive two Nobel Prizes until 1962. (Photos courtesy of APS Niels Bohr Library.)

the study of 1958 graduates, the hypothesis was advanced that this was due primarily to the greater hurdles the women had to overcome to attain the doctorate degree . . . It is assumed . . . that marriage and its attendant responsibilities is a handicap rather than a help in further academic attainment for the women".[21]

This is true not only in the US. In the USSR women participate in substantial percentages in all branches of science and technology through the first level of the universities, but there is a steady decrease in the percentages for higher levels of achievement. For example, in 1970, 50% of the junior scientific assistants, 24% of the senior scientific assistants, 21% of the docents (roughly postdoctoral level) and 10% of the professors were women. Twenty-seven percent of the candidate degrees in science (roughly PhD level), but only 13% of the doctorate degrees in science (a higher level) were awarded to women.[22] A number of sociologists, both Soviet and non-Soviet, have suggested that this is due to the fact that Soviet women are mainly responsible for the care of the household and children.[23] Although it is obvious to even the occasional visitor that other factors such as discrimination also contribute to the differences in the Soviet Union, it is clear that the much greater difficulty of maintaining a family in the USSR would be a crushing burden to a research career.

In theory, evidence that marriage adversely affects women's careers could be observed in terms of differences in rates of publication. Experimentally, different studies give different answers to this question.[24,25] I am personally aware of a substantial number of women scientists who have combined an active research career with raising a family. However, in numerous surveys it has been found that in the past women scientists have frequently accepted less demanding careers

because of their roles in their marriages. They have been willing to put their husband's career first, to move to areas where there were no or inferior job opportunities for the wife, to assume the major share of household labor and the responsibility for children, and to choose teaching over research because it meshed better with their family duties. In recent years, however, there has been a change in the attitudes toward marriage and roles in marriage. Many young couples are considering having no children or, at most, one, and many marry with an explicit understanding that their careers have equal priority and that they share equal responsibility for all facets of their married lives. It will be interesting to see the effects of these changes in the next decade.

Returning to my list of possible causes for the differences between men and women physicists there remains the Matthew effect, first so identified by Robert Merton. In the words of the apostle:

"For unto everyone that hath shall be given and he shall have abundance; but from him that hath not shall be taken away even what he hath" [Matthew 13: 12].

In Merton's words, the Matthew effect in science "consists in the accruing of greater increments of recognition for particular scientific contributions to scientists of considerable repute and the withholding of such recognition from scientists who have not yet made their mark".[26] The existence of a scientific elite has been discussed by sociologists, notably Johnathan and Stephen Cole, Merton, and Harriet Zuckerman, and the pattern is clear.[25–27] Those scientists who work in leadership positions at the research universities accrue grants and students that result in publications which are in turn rewarded by more grants, students, and

prizes in a spiral of success. On the other hand, those who are in secondary positions or at less prestigious institutions (categories in which women have been heavily represented) do not receive this type of support and are unlikely to join the elite. Even women with tenure at major research universities may be outside this circle, whose members are known to each other and who are proposed by one another for leadership or advisory positions, prizes and other forms of recognition. If the women scientists are perceived as outsiders, it is unlikely that they will develop the contacts to become members of the scientific old boys' club. I was quite distressed when an eminent theoretical physicist said to me about five years ago that it would take two generations before there were good women theorists. I was unhappy at the possible impact of this point of view, and appalled at the apparent callous disregard of existing women theorists. But in terms of the Matthew effect, he was correct. These women were not part of the inner circle, and given the small numbers at the top universities and the slow change in the attitudes toward woman physicists held by people like himself it will take time for women theorists to attain significant representation among the elite, but hopefully not two generations.

Unnecessary, injurious, out of place?

It must be fairly clear by now that the adjectives in the subtitle of this article are not as extreme as they may have seemed initially. They have all been used many times with respect to women physicists. Therefore let me use them as a framework for some comments on what the future may hold.

Is it unnecessary that women have equal opportunity and encouragement to become physicists? It is both as necessary and as unnecessary as is the case for men. Depending on how you look at it, the job outlook for the future is bleak (prestige academic positions), or better than most fields (physics-related positions). I think that it is safe to say that a reasonable employment situation would continue if the number of doctorates remains approximately constant; and because the percentage of physicists who are women is so small their participation could increase by a factor of five to six without increasing the number of doctorates if there is a corresponding continuation in the decrease in the number of men receiving doctorates in physics. But why encourage this to occur? The answer is simple. Women in this country face a future in which most of them will work during most of their adult lives. They therefore deserve a society in which they can choose employment according to their interests and abilities, and for which they will receive the same rewards as men. And it can only benefit the profession to move closer to a situation where rewards are based on a perception of scientific merit that concerns itself with the substance of performance, not with the externals of sex or race.

The question of whether increased participation by women in physics would be injurious has two aspects. One is the indubitable fact that, if a field or job category has become identified as a woman's field, it has in the past been accorded lower prestige and a lower salary. Since women are reaching out into almost all careers these days their entry into various fields is unlikely to continue to have this effect. The other aspect of the question is that it has required and will continue to require external pressures such as affirmative action to effect equality of opportunity; this is viewed by some as an infringement of personal or institutional prerogatives by the government and a dilution of quality. In view of the small increases that have been achieved by eight years of affirmative action it is not possible to tell what the effect on quality has been. As for the question of infringement of prerogative, I would argue that no one should enjoy the prerogative to choose faculty in a manner biased by preconceptions and misconceptions of women. Affirmative action is still necessary to prime the pump, to increase the visibility of successful women physicists in order to create an atmosphere where women are accepted and rewarded for their contributions in all aspects of the profession. If

some appointments are made which are not successful, it will not be a new phenomenon. Many men hired by academic institutions have been denied promotion and tenure in the past without any discussion of injury to the profession.

Finally, there is the question of whether women are out of place in physics. There is no compelling evidence that girls are not equally endowed with the abilities necessary to become successful physicists. There is overwhelming evidence that the attitudes of society and the pressures of marriage and family have made this much more improbable for women than for men. A prominent physicist once remarked to me, "It is too bad that you were not born a man." And indeed, there are very few women physicists for whom there has not arisen some career obstacle, whether internal or external, directly attributable to their sex. But, if we are indeed to take seriously the ideal that participation in physics should be based on interest, aspiration and ability, then certainly no individual should be discouraged on any grounds other than these.

References

1. H. J. Mozans, *Women in Science*, Appleton (1913); reissued by MIT, Cambridge, Mass. (1974).

2. G. Sarton, *A History of Science*, Harvard U. Cambridge, Mass. (1952).

3. L. M. Osen, *Women in Mathematics*, MIT, Cambridge, Mass. (1974).

4. E. T. James, *Notable American Women 1607–1950*, Harvard U., Cambridge, Mass. (1974).

5. F. Rudolph, *The American College and University*, Knopf, New York (1962).

6. M. W. Rossiter, *American Scientist* **62**, 312 (1974).

7. APS Comm. Women in Physics, Bull. Am. Phys. Soc. II, **17**, 740 (1972).

8. D. M. Gilford, P. D. Syverson, "Summary Report [*Year*] Doctorate Recipients from US Universities," Nat. Acad. of Sci. (1972 through 1978).

9. D. M. Gilford, J. Snyder, *Women and Minority PhD's in the late 1970's: A Data Book*, Nat. Acad. of Sci. (1977).

10. B. D. Maxfield, N. C. Ahern, A. W. Spisak, *Science, Engineering, and Humanities Doctorates in the United States. 1977 profile*, Nat. Acad. of Sci. (1978).

11. B. D. Maxfield, N. C. Ahern, A. W. Spisak, *Employment Status of PhD Scientists and Engineers. 1973 and 1975*, Nat. Acad. of Sci. (1976).

12. M. E. Law, J. Wittels, R. Clark, P. Jorgenson, Bull. Am. Phys. Soc. **21**, 888 (1976).

13. L. Tiger, *New York Times Magazine*, 25 October 1970, page 35.

14. E. E. Maccoby, C. N. Jacklin, *The Psychology of Sex Differences*, Stanford U., Stanford, Cal. (1974).

15. A. Kelly, *Phys. Bull.* **30**, 108 (1979).

16. M. J. Oates, S. Williamson, *Signs*, 795 (Summer, 1978).

17. M. E. Tidball, V. Kistiakowsky, *Science* **193**, 646 (1976).

18. S. Tobias, *Overcoming Math Anxiety*, Norton, New York (1978).

19. C. E. Max, "Opportunities for Women in Physics," U. California Rad. Lab. Report UCRL-80943 (1978).

20. J. Dash, *A Life of One's Own*, Harper and Row, New York (1973).

21. L. R. Harmon, *High School Ability Patterns*, Nat. Acad. of Sci. (1965).

22. G. F. Schilling, M. K. Hunt, "Women in Science and Technology: US/USSR Comparisons," Rand Paper Series P-239, Santa Monica, Cal. (1974).

23. W. M. Mandel, *Soviet Women*, Doubleday, Garden City, N.Y. (1975).

24. J. A. Centra, *Women, Men and the Doctorate*, Educ. Testing Serv., Princeton, N.J. (1974).

25. J. R. Cole, S. Cole, *Social Stratification in Science*, U. Chicago, Chicago, Ill. (1973).

26. R. K. Merton, *Science* **159**, 56 (1968).

27. H. Zuckerman, *Scientific Elite*, Free Press, New York (1978).   ▫

The last fifty years – a revolution?

PHYSICS TODAY / NOVEMBER 1981

Spencer R. Weart

Spencer R. Weart is director of the Center for History of Physics, American Institute of Physics, New York.

Laser-assisted machining.
Lasers are based on old theory; what is new is their uses, which range from experiments in fundamental physics to the machining operation shown here. Uncovering the subtle complexities of Nature and making use of the results is the hallmark of modern physics. (Courtesy General Electric Company, Research and Development Center)

In some periods great conceptual revolutions shake the world of physics; at other times research seems to plod ahead within the confines of an established framework. And the structure of the physics community must change in a way that somehow matches the changing style of research. What, then, has been the form of physics during our own lifetime, and how has it changed? This is a difficult, but not impossible question. Only history can give us an inkling of the answer.

To place ourselves here in 1981, on the fiftieth anniversary of the American Institute of Physics, we need to imagine how physicists fifty years ago saw their own place. Suppose there had been a fiftieth anniversary of something back in 1931—what would those physicists have said about their position in time? In fact we have a good idea of that, because people back then wanted to orient themselves in time just as much as we do now, and they often recorded what they thought of their situation.

Physicists in 1931 saw themselves at the crest of a great, spreading wave of new knowledge. They were right to think so, considering what physicists had done in the half-century up to 1931. Most striking, perhaps, was the development of electromagnetic theory and practice. Only in 1888 did Hertz detect electromagnetic waves, sealing the process

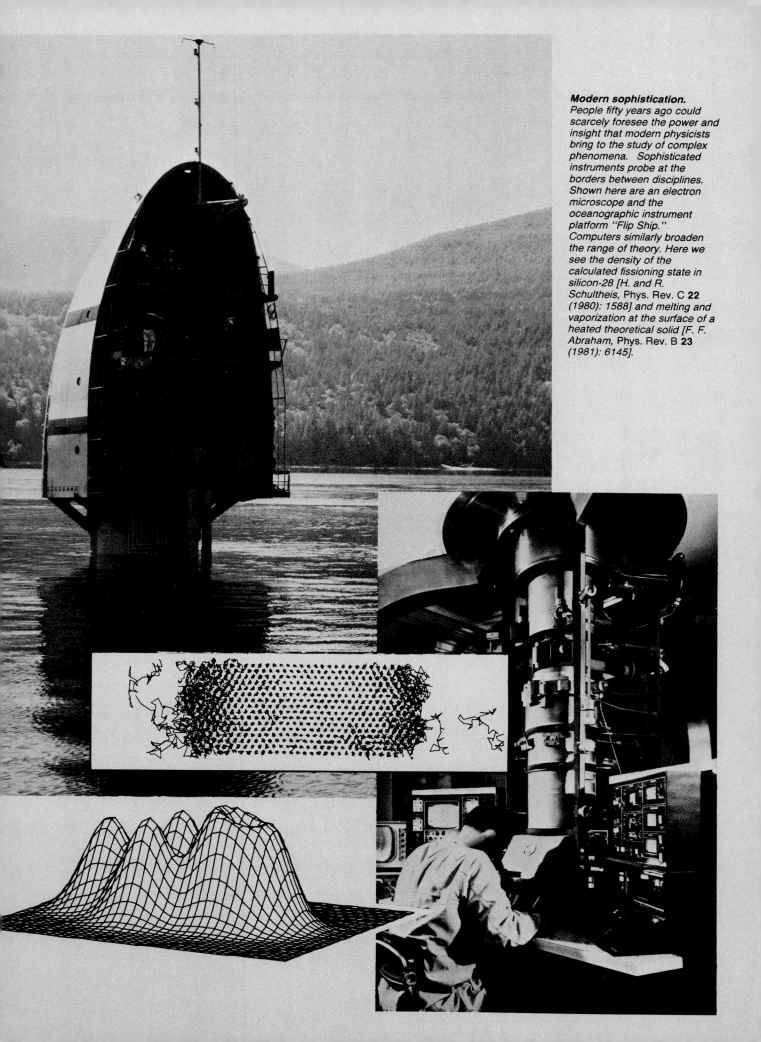

Modern sophistication.
People fifty years ago could scarcely foresee the power and insight that modern physicists bring to the study of complex phenomena. Sophisticated instruments probe at the borders between disciplines. Shown here are an electron microscope and the oceanographic instrument platform "Flip Ship." Computers similarly broaden the range of theory. Here we see the density of the calculated fissioning state in silicon-28 [H. and R. Schultheis, Phys. Rev. C **22** (1980): 1588] and melting and vaporization at the surface of a heated theoretical solid [F. F. Abraham, Phys. Rev. B **23** (1981): 6145].

The Physics Community—A Retrospective

by which Maxwell's equations came to be accepted as definitive. It was between then and 1931 that most homes got their dozens of electric lights and their dozen or so little electric motors. In 1931 the silver-haired dean of American physicists, Robert Millikan, told the *New York Times* that this was the greatest change of the previous couple of generations: the substitution of electrical power, driven by fossil fuels, for human muscle power. No past time had known such a great change, he said, and he could not imagine that physics could bring any change so great in the next couple of generations.[1] The revolution in communications, symbolized by radio and telephone, was also largely completed.

In the more abstract kingdom of theory, the physicists of 1931 could look back on equally great changes. Fifty years before, statistical mechanics had barely started, and some leading scientists even refused to agree that atoms existed. Then the work of Boltzmann, Gibbs, and many others established the statistical atomic theory beyond question, solving one of the oldest problems of science. This atomic view had then been pressed forward to the discovery of that most fundamental particle, the electron. The discovery of x rays and radioactivity added to the excitement: At last the structure of matter was becoming known.

But that was only an appetizer. During their own careers the physicists of 1931 had overthrown the commonsense view of how atoms must behave, creating the new quantum mechanics. To many the quantum seemed incredible, bizarre. Yet by 1931 the quantum view had been capped with the Dirac theory. And the positron, just discovered, confirmed Dirac in a most surprising way.

And even that was not all. Einstein had replaced Newtonian mechanics with his special theory of relativity, and had gone on to build a new general theory that explained gravity in a far deeper way than before. Just as the discovery of the positron had unexpectedly underwritten Dirac's theory, so the discovery of the expansion of the universe gave an astounding demonstration of the usefulness of Einstein's equations.

The only word for all of this is revolution. The physicists of 1931 were keenly aware that their generation had upset previous ways of seeing the universe, as thoroughly as Lenin's generation had upset the social structure of Russia. Said Millikan: "The discoveries which I myself have seen since my graduation transcend in fundamental importance all those which the preceding 200 years brought forth."[2] A revolution is a complete turnabout; that well describes what had happened to the world-view of physics in the fifty years up to 1931.

A social revolution

Physics had known a social revolution too, almost as radical as the new theories. This was particularly true in America. Back when Millikan and the other senior physicists of 1931 began their education, American physics scarcely existed. Then came the foundation of The American Physical Society and the *Physical Review*, which together defined the existence of an American physics community. These were joined by other institutions, such as the American Association of Physics Teachers, brand-new in 1931. There were something like 3000 physicists in the United States in 1931, where fifty years earlier there had been at best a couple of dozen. In that year 1931 the *Physical Review*, for the first time, was cited more often in the physics literature than its chief rival, the German *Zeitschrift für Physik*.

This rise of American physics was a world-historical change, more significant in the long run than the bloody useless battles of the First World War. The leaders of physics, when they thought about the effects of that war, thought particularly of how physicists had proved themselves in industry. As recently as 1900 there had scarcely been such a thing as an industrial physicist, but in 1931 about one-fifth of the members of the Physical Society were in industry. There was talk of forming a society of industrial physicists— splitting up the Physical Society.[3]

Along with this growth had come an important change in the public attitude toward physics. Our science had always been respected for its deep understanding of nature, and also for its promise of making life on earth easier. But in the fifty years up to 1931 the admiration had redoubled. The discoveries of quantum mechanics and relativity put physics, for the first time, beyond the reach of the intelligent layman. Einstein was the first physicist ever to be regarded, even by intellectuals, with the uncomprehending awe once reserved for sorcerers. Meanwhile, the practical value of science had been proven in the war and in the industrial laboratories. The physicists of 1931 could look in any magazine and find advertisements declaring that some particular toothpaste or refrigerator was made better by laboratory scientists. That was something new, something revolutionary.

How confident was the public in 1931 about science? Nothing shows it better than peoples' desire to make themselves radioactive. Radium could help cancer patients, of course, but that was not all; many people thought that a little radioactivity could be a healthful stimulant. Spas in many countries were proud to advertise the natural radioactivity of their waters. A 1929 pharmacopoeia listed no less than eighty patent medicines based on radioactivity.

You could take radium by capsule, tablet, compress, bath salts, liniment, cream, inhalation, injection or suppository. You could eat mildly radioactive chocolate candies, then brush your teeth with radioactive toothpaste. The manufacturers claimed that their nostrums would give relief from tuberculosis, tumors, rickets, baldness and flagging sexual powers.[4]

Despite all the enthusiasm for science, the public had some doubts. The increasing application of physics to industry in war and peace, and the increasing failure of most people to understand physics, led to criticism. Physics was said to stifle the spontaneous, unthinking wholeness of life, to destroy moral values, to reduce workers to robots or throw them out of work entirely.

New problems require new solutions. Many leaders of physics in 1931 saw a serious need to keep the public trust in physics. They also needed to deal with the rise of industrial physics, which threatened to cause a schism between industrial and academic physicists. But most pressing, they had to reorganize the finances of physics journals, for the journals were losing money as the physics community became increasingly specialized. It was to meet all these needs that the American Institute of Physics was founded. The newborn AIP had the practical task of making publication more economical by consolidating the production of the various physics journals. But it had even more important tasks, as the people who founded it saw things: To cement a bond between industrial and academic physics, and to serve up reliable information on physics to the press and the public. With the founding of the AIP, the structure of American physics was put in order.

When physicists of 1931 looked ahead to our own time, they were sure that physics would continue to grow and spread into every field of human activity. They said that power would become still cheaper and more widely used. They expected that by the 1980s we would have widespread use of television, transportation much better than steamships and some kind of industrial robots. In all this they were quite accurate.

Physicists were less accurate when they looked at their own science. In 1931 the problem of the nucleus had grown so pressing that it seemed to tremble on the edge of a grand solution, and many hoped for a new revelation, as exciting and surprising as quantum mechanics or relativity. Niels Bohr even suggested in 1931 that the law of conservation of energy might have to be junked. In George Gamow's draft for a textbook, wherever he talked about the internal constitution of the nucleus he drew a little skull and crossbones in the margin, to warn the reader how uncertain the current speculations were.[5]

While nuclear physicists awaited their revelation, others worked in another direction to undermine the recently won achievements of quantum mechanics. These others were few but they were led by the greatest of all, Einstein. The union of electromagnetic and gravitational theory seemed not far off. And would that not put both general relativity and quantum mechanics to rest, as mere shadows of some far grander and deeper unified field theory? Such were the dreams of 1931.

Uncovering new subtleties

Where are we now in 1981? Are our times much like the times of the people fifty years ago? Is the history of physics in our lifetimes much like the history that they lived through, or are there qualitative differences? There can be no simple answer, for physics is an uncommonly diverse subject, but I can mention a few particularly salient features.

The study of nuclear forces and particles has not brought the new revelation, the overthrow of quantum mechanics. Instead of wholly new laws, we have been uncovering ever more layers of complexity. Quite a lot has been said recently, in connection with the 1980 Nobel prizes,[6] about theoretical developments, so I will touch on some of the experimental results. First there were the mesons, and then a number of other particles, particularly the strange particles—as the name implies, quite a surprise for physicists. This particle zoo, as it was called, gradually sorted out; the discovery of the omega minus gave an encouraging confirmation that the zoo had a comprehensible layout. And lately there came the experiments that most of us would call the confirmation of quarks: another layer of complexity.

Did all this add up to the sort of revelation that physicists of 1931 had seen in their past or hoped to achieve with nuclear physics? I think the answer must be no—not exactly. Certainly the appearance of new particles has been surprising and exciting. People in 1931 really thought that with the electron and proton (and conceivably the neutrino) they had counted up all the fundamental particles. That sort of thinking has been overthrown. Yet the overthrow was in no way comparable in scope to the discovery of the electron or to quantum mechanics. The idea of fundamental particles is as useful now as it was then;

ey just turn out to be more nu-
erous and subtle than people ex-
ected.

I am not speaking now of the
ew unified force theories. After
ll, these still involve a pile of em-
rical parameters. Historians of
e future may well say that the
ear 1981 marks the center of a
ow revolution in our view of
rces and particles, but it is too
arly for a historian to write about
at yet.

What has been most striking is
ot the revolutionary nature of
ese fundamental theories but
eir continuity, their tortuous step-
ise development out of earlier
eories. In some ways it has
een less a process of inventing a
ew theory, in the sense of a new
orld-view, than a process of sep-
rating out the valid theories from
e almost infinite variety spread
rth by quantum mechanics.
ontrary to what many scientists
xpected in 1931—above all Ein-
tein—quantum mechanics re-
ains and is more solid than ever.

But it has not remained un-
hanged. There was, of course,
onfirmation that it really worked,
ight down to the incredibly fine
etail revealed by the Lamb shift.
hat was more than some expect-
d. But there was also a new ap-
roach to quantum electrodynam-
cs associated with the names
Feynman, Tomonaga, Schwinger
and Dyson. In one sense this is
only a better means of calculation,
but we must beware of saying
"only" a means of calculation,
when that is all any theory comes
down to when you go into the laboratory. It was not easy for many
people to swallow renormalization,
and it was not easy to swallow the
notion, so clearly pointed out by
the diagrams of particles interact-
ng, that a positron can be represented as an electron going backwards in time.

*Albert Einstein, Hendrik
Antoon Lorentz and Arthur
Stanley Eddington* in 1923.
These theorists pursued an
underlying unity and simplicity.
Similar work over the last half-
century has advanced less
rapidly than work in their day.
(Photograph courtesy of AIP
Niels Bohr Library.)

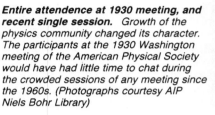

Entire attendence at 1930 meeting, and recent single session. *Growth of the physics community changed its character. The participants at the 1930 Washington meeting of the American Physical Society would have had little time to chat during the crowded sessions of any meeting since the 1960s. (Photographs courtesy AIP Niels Bohr Library)*

On top of these subtleties came the discovery of parity violation. Even more fundamental was the violation of CP, which is to say, the experimental proof of time asymmetry, an even more fundamental reversal than the reversal of an electron into a positron. These discoveries surely did upset old preconceptions. Yet again, I will not call them revolutionary in the sense that quantum mechanics had been. No old system of ideas was turned on its head. Rather, people were set free to consider a greater range of ways the universe can behave—this freeing-up was indeed necessary to open the way to the new unified force theories.

As with quantum mechanics, so with the general theory of relativity, the years have seen not an overthrow but a strengthening. The unification of electromagnetism with gravitation—the project on which Einstein spent half his life—is not yet done, nor does quantum mechanics seem any easier to reconcile with a theory of curved spacetime. Yet this does not mean there has been no progress. To begin with, general relativity has passed exacting experimental tests, much as quantum mechanics did, and that is great progress. And there have been wonderful developments in the theory—not new equations but new theorems spinning out from the old equations, each more astonishing than the last.

Consider, for example, the relations among cosmology, thermodynamics and relativity. It has gradually dawned on physicists that the direction of entropy, the direction of time, is somehow embedded in the general relativity solution for the expanding universe, in the elementary sense that time's arrow points the direction away from the Big Bang. Moreover, we have learned how even that "singular" solution of the equations, a black hole, can have its own time scale, a lifetime determined by statistical, indeed quantum emissions.

A hundred years ago, in 1881, there was simply Newtonian time, a concept scarcely different from that handed down from Aristotle, a concept of crystalline simplicity. By 1931 this was done away with, replaced by relativistic time—a new way of putting time into our equations, an astounding revolution. Yet Einstein's idea of time was as easy as Newton's, once you got used to it, and even simpler; that was why physicists liked it. But what has happened since then? Relativistic time is still basic. But the concept has been wonderfully enriched. Time is reversible; time symmetry is not even conserved; time plays fantastic tricks around spacetime singularities; time is tied up with all the majestic expansion of the universe. The physicist's conception of time is today far more complex than in 1931, much richer and more subtle.

So when I say that there has been no revolution in the last fifty years comparable to those of the fifty earlier years, I'm not heaping scorn on recent progress. Physics does not always have to advance in a revolutionary way.

Sometimes it advances precisely by coming to more complexity, more layers, more calculations and models, more subtlety. No doubt the universe is characterized by great simplicities, not all of them known; but the universe is also characterized by an intricate physical texture, which it is also the task of the physicist to understand.

Freeman Dyson makes a similar point when he divides scientists into "unifiers" and "diversifiers."[7] As an example of a unifier he suggests Einstein, always searching for underlying unities; a diversifier would be someone like the great majority of our colleagues in biology, always studying the marvelous diversity of specific creatures.

This is in fact a fundamental division in the way humans can approach the universe. Many years ago, in the classic study of mysticism, Underhill pointed out that mystics may approach God in two ways. They may see God as transcendent, wholly other; or they may see God as present in all things, diverse and evolving.[8] In a similar way, the search for some transcendent unity beneath the surface of things was an important root of modern science, but the love of diversity, of particular things in the world, was no less important.

The two feelings could be combined in one person. Galileo was certainly a unifier, and he found fundamental laws beneath the motions of things. But he was also a diversifier. The old unitary theory of his day saw the sun as a perfect sphere, and the planets car-

ried around the sky on perfect crystal spheres. Galileo, peering hour after hour through his telescope, discovered the moons of Jupiter and the sunspots, and messed up that beautiful, clean theory. Galileo loved change and diversity; dirt, he said, was better than diamonds. If the whole earth were a perfect crystal sphere, said Galileo, he would consider it just "a wretched lump . . . in a word, superfluous."[9]

A Broader Scope

Both unifiers and diversifies are important in science. But there may be times when one type of thinking can make swifter progress than another. And in physics of the last fifty years, while much attention has gone to the efforts of unifiers, I think much of the finest work has been in the direction of diversity.

Take, for example, astrophysics. Compared with what we know now, the people in 1931 knew almost nothing. They did not even know whether red giants are an early stage of the evolution of stars, or (as is the case) come later. Today, the evolution of stars is better understood than the transformations of a tadpole into a frog. Then, more recently, there was all the development of radio astronomy. A whole new universe, the so-called violent universe, is now open to us.

Yet none of this is what I would call revolutionary. Some of what astronomers guessed in 1931 turned out to be wrong, but no

strongly held astronomical world-picture was overturned. It was not that astronomers had a wrong idea of the radio universe or of stellar evolution, so much as that they admittedly did not understand these things at all. Modern astrophysics has not been like a revolution overturning an established government; it has been more like a wave of colonization that sets up new nations in an uninhabited territory.[10]

This colonization was made possible because of the alliance of astronomy and physics. Nuclear physics and spectroscopy and electronics and optics have all been essential to the advance of modern astronomy. Indeed, for some time now about half of all new astronomers have brought their PhDs from physics. In return, physics has been enriched beyond measure by what the astronomers know.

This kind of cross-fertilization is another aspect of what I have been talking about: the increase in richness and complexity that has been the main feature of physics for the past half century. Astrophysics is not the only example. Another would be geophysics. The 1930s saw a massive invasion of oilfields by physicists with gravimeters and the like. Since then there has been a true scientific revolution among our colleagues in geology, the development of plate tectonics—the view, stoutly resisted by many old-timers, that the continents slip about like so many cakes of ice on a churning ocean. While many lines of evidence converged on this rev-

elation, not least in importance were techniques brought in from physics, such as measurements of the radioactive ages and magnetic orientations of rocks.

It was not by a fluke that physics became an indispensable part of the tool chest of many other sciences. The great discoveries preceding 1931, statistical mechanics, radioactivity, the electron and all, laid a firm conceptual foundation not just for physics itself, but for all the sciences. It remained only to apply these tools to the thousands of old problems that awaited them. And who could do this better than physicists?

The most exciting example of all was molecular biology. In 1931 physicists and biologists had little to do with one another. Then came the discovery of artificial radioactivity. By the end of the 1930s, in laboratories around the world—Berkeley, Paris, Copenhagen, Tokyo—cyclotrons or other particle accelerators were being built. But most of them were not built primarily to explore the nucleus. These devices were funded above all to provide artificially radioactive isotopes for biological and medical research.

The new coalition between physics and biology spread after the Second World War. Erwin Schrödinger went so far as to suggest that if physicists went into biology they might discover, in those huge complex molecules, revolutionary new laws of physics. That was a fantasy, but physicists, inspired by Schrödinger, gave biologists important help in deciphering the genetic code. More important, the analytical ways of thought pioneered by physicists conquered certain fields of biology. And most important of all were the physical techniques, especially radioactive tracers. It is hard to say where molecular biology would be today without all that—certainly far behind where it is now.

When physicists back in 1931 looked ahead they foresaw something of this. "Questions of life and health," said Arthur Holly Compton, "will probably be in the forefront." And Millikan said, "It is rather in the field of biology than of physics that I myself look for the big changes in the coming century." They predicted this because they foresaw that physics was bound to enter and inspire biology.[11]

The last fifty years, then, have revealed an ability of physics, a surprisingly powerful ability, to help make sense out of the most complicated phenomena, even in fields far from home. But most striking has been the way that physics has done this in its own central area, the understanding of everyday matter.

The physicists of 1931 would certainly be gratified to see the advances that have been made in understanding collective phenomena, not only in inaccessible places like the nucleus or a neutron star, but even in ordinary matter. For example, behavior near the critical point is understood now in a far more satisfying way than formerly; the unifiers have done well here. But no field exemplifies so clearly as solid-state physics the urge to look into diversity and understand it.

The band theory, the study of point defects and their consequences, the theory of superconductivity and the study of lattice vibrations are just part of a list that could go on for pages. I wish I could tell in a few words the story of this field, because in many ways the history of solid-state physics, its growth into condensed-matter physics, has been at the heart of the history of physics over the last fifty years. We all know of the great applications, not only the long-predicted televisions and robots, but also the computers, with their little-anticipated power to help along every field of science. But I think many people do not realize the fundamental interest of this

Brookhaven National Laboratory *in 1962, as seen from the air, looking south. Only national governments could support science on such a scale. (Photograph courtesy of AIP Niels Bohr Library.)*

field. The condensed-matter physicists are the ones who provide an explanation of the physical characteristics of everyday matter: they can literally tell us why the things we see and handle look and feel as they do. This is primary among the ancient, homely tasks of physics, and it is a task that has been largely accomplished in our time.

I wish, I say, that I could tell the story of this development, but I can't. The story has not yet been put together by historians. Why has fundamental solid-state physics gotten less public attention than many other fields? I suspect it is because the field is obviously not revolutionary. Once again, it has been more a matter of people colonizing unknown territory, through steadfast continuous work, rather than overturning what was known. Cyril Stanley Smith has written about this.[12] Solid-state physics, he feels, was held back because of an overemphasis on "good, clean" Newtonian methods. Only when people accepted a messier, more approximate way of dealing with things could solid-state physics be done. "I rather suspect," Smith writes, "that solid-state physics has in it some of the future of science in dealing realistically, not purely statistically, with complicated systems, and not being purely

reductionist as almost all physics was until 1940 or so." He calls the history of solid-state physics "the history of an emerging science of buildings, not of bricks." I think something like that could be said for much of the history of physics over the past fifty years.

Certainly there are times when revolutionary ideas are adopted— and noboby would dare say such times may not be here today. However, there are also times for diversity, for the advance of a science of buildings, not of bricks, and those can be exciting and important times too.

An institution transformed

Turning now from physics as an intellectual field to physics as a community of people, what has happened in the past fifty years? Again I do not see revolutionary changes. There has been nothing comparable to the preceding burst of activity that took American physics from a nonentity to a field with its own journals, societies and Institute. Today as in 1931, physicists are organized in the Physical Society and others, with the *Physical Review* and some other journals. Today there are still a fifth of the Physical Society members

Research teams and equipment grew vastly in fifty years. Carl Anderson (above) designed by himself the device he used to discover the positron in 1931; Samuel Ting's team (right) fitted easily into one corner of the apparatus they used to discover the J/ψ particle in 1974.

in industry, with most of the rest employed by academic institutions. Yet underneath this there have been changes. And just as in physics itself, the changes were no less important for being complex and subtle rather than revolutionary.

For example, those people employed by academic institutions today are in large measure paid by the federal government. This is particularly obvious for the quarter of them who work at government-contract laboratories, perhaps less so for professors who indirectly draw part of their pay from the government's tuition subsidies. This dependence on federal money would have horrified Millikan, a sturdy free-enterpriser. Yet he should have foreseen it, for even in his day the United States was a holdout, a country of privately employed physicists in a world where the salaries of most physicists were paid by national governments. This great change in American physics does not seem so revolutionary, then, when seen in the perspective of world history. Physics tends to be strongly supported by the state, a fact that has been clear in most countries for many years.

Another change is also not surprising, except in its scope: the rise of American physics to world dominance. In 1931 Millikan predicted that by our time, "the United States and Germany will probably be the world leaders in science." Only two years later Hitler came to power, and the cream of Central European scientists began to make their way to American shores. Since then, the United States has been the location for more important theories, experiments and instruments than all the rest of the world put together. This dominance of a field of science by one country is without precedent in modern history.

It was government funds as much as anything that allowed this, promoting a great increase in the number of physicists. Where there were some 3000 physicists in America in 1931, there are over 30 000 now. Any physicist in 1931 could have predicted some such increase, simply by extrapolating the exponential growth that had already been going on for generations. In fact, an extrapolation would have indicated close to 100 000 physicists in the 1980s. However, around 1968 the growth reached saturation—the maximum number of physicists that society was willing to support. The end of exponential growth demanded a number of painful readjustments,

which are still underway.

It would have been harder for the physicists of 1931 to understand what the increase in numbers would mean for their way of life. The break came sometime in the 1950s when American physicists could no longer all know one another as the people in a small town know one another. Relationships shifted. Some obvious indicators are the innumerable parallel sessions at meetings, the insuperable thickness of the *Physical Review*, and the need for weighty grant applications rather than a simple chat with your patron.

Another indicator is the rise of team research, and the clustering around great instruments, a way of working that would have been wholly alien to the physicists of 1931. Yet that is no revolution, really, for the old-style physics still goes on where it can. It is again a matter of increased complexity, of diversity, of deeper levels of understanding and organization making it possible to break into new territory. (Perhaps in some way the nature of the social organization parallels the nature of the knowledge it makes available; that deep question cannot be answered here.)

As one indicator of the in-

Twenty-three-year-old cartoon. *Close new connections of physics with the military since World War II altered the physics community and increased the public's ambivalence toward science. (Drawing by Modell; © 1958 The New Yorker Magazine, Inc.)*

*"From the cyclotron of Berkeley to the labs of M.I.T.,
We're the lads that you can trust to keep our country strong and free."*

creased complexity of the physics community, look at the American Institute of Physics itself. There has been no revolution, for it is still the old AIP established fifty years ago. But what a difference! Instead of a director and one secretary operating in a free-wheeling way out of a tiny office, AIP is today an organization as diverse as a large bank, with 400 employees clustered around computer terminals and the like. Besides its old task of publishing journals (now using physics-based electronics technology, of course), it addresses the problems jointly faced by the various physics societies through an array of sophisticated mechanisms: a public affairs committee, representation in copyright clearance organizations, a marketing apparatus, manpower studies, a public information service and even a history center. In short, AIP and the physics community, along with the subject of physics

itself, have become far more complex and more intricately interconnected with the rest of the world in the past half century.

In this matter of connections with the rest of the world, there is one more thing I must talk about. Robert Millikan and Arthur Compton would have been most surprised to see it. That is the fact that today, as through the past forty years, something like a quarter of all American physicists are employed rather directly in military research. Beyond this, the armed forces have given generous support even to research that seems quite pure—for generals too have grown sophisticated, and understand the long-term sway of science. Most physicists, like myself, have benefited at one time or another from Department of Defense contracts. I think that the physicists of 1931 would find it overpoweringly strange that so many scientists now work on

weapons, and I think that some, for example Compton, would be distressed. They believed that the physics research of today is the main factor in determining the world of tomorrow. So they would want to know what sort of a world we have it in mind to create.

It may well be that this revolution, this infection of physics by military problems, also has something to do with changes in public attitudes toward our science. I do not think there has been a complete revolution here; the public is still mostly enthusiastic about our enterprise. Yet consider how people think of radiation. Nobody wants to become radioactive any more. On the contrary, people have become as unreasonable in their fear of radioactivity, as they were unreasonable in their hopes for it fifty years ago. The change can be dated very precisely: it was caused by Hiroshima. Since then, any public support for physics has had within it a certain fearful reservation, and rightly so, if you consider our situation. This is another of those subtle changes, a new complexity, a new wisdom I suppose, that we must live with.

Fortunately, many physicists themselves have responded by taking their social responsibilities more seriously, and dealing with them in a more understanding and sophisticated way. The growing recognition that even in the abstract acts we perform in our research, physicists are human beings living in society, is one of the most subtle and most hopeful of all the changes we have seen.

To answer, finally, the question I began with, I do not think that physics in our times has been like the physics of fifty years ago. Our times have been less revolutionary, but more diverse and penetrating; less welcoming to dreams of vast revelations, but no less exciting and rewarding. Of course, this history shows us no way to predict whether our fifty years of development have built a platform for another revolutionary leap, or whether the steady process of extending and reinforcing the structure of our science will continue for many years. But the history does show that if we are to keep physics vigorous, we must always be ready for changes in our social arrangements and even in our approach to knowledge.

References

1. *New York Times*, 30 September 1931, X:3. See also Millikan, *Science and the New Civilization*, Scribner's, Boston (1930), page 73.

2. Millikan, *Science and Life*, Pilgrim, Boston (1924), page 68.

3. S. Weart, "The Physics Business in America, 1919–1940: A Statistical Reconnaissance," pages 295–358 in Nathan Reingold, ed., *The Sciences in The American Context: New Perspectives*, Smithsonian Institution, Washington, D. C. (1979).

4. Public attitudes to radioactivity will be discussed in my book, *Nuclear Fear*, now in preparation.

5. Gamow, *Constitution of Atomic Nuclei and Radioactivity*, Oxford U. P. (1931).

6. Nobel Prize lectures, published in *Rev. Mod. Phys.* **30** (1980) and elsewhere.

7. F. Dyson, "Infinite In All Directions," address to American Association for Advancement of Science, Toronto, January 1981.

8. Evelyn Underhill, *Mysticism*, 12th edition, Dutton (1961), pages 40–41, 99, 128, 291. See also Arthur O. Lovejoy, *The Great Chain of Being*, Harvard U. P., Cambridge, Mass. (1964), pages 83–84 and *passim*.

9. Quoted in Arthur Koestler, *The Sleepwalkers*, Grosset & Dunlap, New York (1963), page 474.

10. David O. Edge and Michael J. Mulkay, *Astronomy Transformed: The Emergence of Radio Astronomy in Britain*, Wiley, New York (1976), pages 386–94.

11. *New York Times*, 30 September 1931, X: 3.

12. C. S. Smith to author, 31 October 1980.   ☐

—Chapter 4—
Biography

Biography is one of the favorite modes of historical writing, so widespread that it has almost become a genre of its own, as shown by the separate racks given to "Biography" in paperback bookstores. Some of the articles in this section, particularly the ones on Oppenheimer and Urey, follow the normal historical mode of looking into just what happened during a particular period of the protagonist's life, but other articles are more impressionistic. All of them deal with something that goes beyond history of physics into other forms of literature: the investigation of character.

The people described in this section were at the very top of their profession (and so are many of the authors). Biography is not interested in ordinary lives but in those that touch greatness. We read biography to participate vicariously in such extraordinary lives, and also to try to understand how some people manage to rise above the common. In science the matter is not only of personal interest but of practical importance, for studies have shown that a large fraction of the most important research has been done by a very small fraction of all researchers.

What makes some people great scientists? The articles here, and many other studies, show that raw intelligence is not the only answer. Some of the people discussed, such as Rutherford, themselves insisted that their minds were not remarkably brilliant or subtle. What seems to have been more important for all successful scientists is what the Victorians called moral energy—a drive to work hard and take chances, an indefatigable boldness, combined with honesty at the most basic level, a common sense that does not hesitate to identify the scientist's own errors.

Note too that every one of the scientists memorialized in this section showed high ability at getting on with other people. As students they worked well under the direction of their seniors, and when they became leaders in turn they inspired not only hard work but admiration and even love among their own students and their colleagues. They had less pleasant characteristics too, sharp edges which many of our authors have hesitated to discuss in public; none of these scientists succeeded without being ready to fight for what they wanted. In any case these articles imply that the comic-book image of the scientist—an inhumane genius cogitating solitary thoughts in a sterile white laboratory—portrays only the kind of person who makes little impression on history.

Contents

The Two Ernests—I

Some personal recollections of Ernest Rutherford and Ernest Lawrence in the period 1927-1939. Rutherford, who dominated the Cavendish Laboratory, gave his physicists a minimum of equipment but a maximum of personal interest in their research. Lawrence developed the Radiation Laboratory into a prototype facility for research with large, expensive equipment. Both inspired others to produce and interpret nuclear reactions.

PHYSICS TODAY / SEPTEMBER–OCTOBER 1966   *by Mark L. Oliphant*

ON 11 JANUARY 1939 after a visit to Berkeley, I wrote a letter to Ernest Lawrence that contained the following paragraph:

"I find it very difficult to thank you for the magnificent and instructive time which I had in Berkeley. It was truly fine of you to be so liberal of time and of thought on my behalf. I know of no laboratory in the world at the present time which has so fine a spirit or so grand a tradition of hard work. While there I seemed to feel again the spirit of the old Cavendish, and to find in you those qualities of a combined camaraderie and leadership which endeared Rutherford to all who worked with him. The essence of the Cavendish is now in Berkeley. I am sincere in this, and for these reasons I shall return again some day, and I hope very soon."

Now, in 1965, after many subsequent visits to the Radiation Laboratory, which Lawrence created and which is now named after him, I remain intrigued by both the many similarities, and the differences, between Rutherford and Lawrence.

John Cockcroft and Ernest Walton first observed nuclear transformations produced by artificially accelerated particles, and James Chadwick discovered the neutron, in the Cavendish Labora-

tory, Cambridge, in 1932. Lawrence conceived the cyclotron principle in 1929, in the University of California, Berkeley. By 1932, with his colleagues Niels Edlefsen and M. Stanley Livingston, he had made the cyclotron a successful instrument with which he was able to confirm the results of Cockcroft and Walton, and carry them to much higher bombarding energies. The period between these great discoveries and that of the fission process by Otto Hahn and Fritz Strassmann in 1938, was of the greatest importance in the development of modern physics. In this article, I endeavor to set down some recollections of that period and of two individuals who gave it such momentum that it changed the whole course of physics and led, inexorably, to the development of nuclear weapons and nuclear energy. No pretense is made that this account is complete, or that the facts presented are in accordance with the recollections of others who lived through those stirring days. The study of the effects produced in the atomic nucleus by bombarding it with nuclear projectiles had transformed knowledge of matter and its properties. The parts played by Rutherford and Lawrence, directly and indirectly, will remain outstanding contributions to that work.

Ernest Rutherford and Ernest Law-

rence, in two succeeding generations, built around them great schools of investigation that laid the foundations of physics as it is practiced today. These two men, so much alike, and yet so strangely different, were parts of totally different worlds. Together, their lives spanned the period of the greatest revolution in knowledge of the physical universe since Newton's time. Each was a pioneer, and each was the descendant of pioneering parents who chose to build a new life in a land far removed from the home of their ancestors. It is revealing to review the early life of each.

Rutherford, early years

Rutherford's grandfather, George Rutherford, migrated from Scotland to New Zealand in 1842. His son James, then three years of age, grew up in

Sir Mark Oliphant, K.B.E., F.R.S., worked with Ernest Rutherford in the Cavendish until 1937, when he went to the Univ. of Birmingham. In 1950 he became professor of particle physics and director of the Research School of Physical Sciences at the Australian National University.

HOUSE, in South Island, New Zealand where Rutherford lived as a child.

the colony and followed his father's trade as a wheelwright. James met and married a widow, Caroline Thompson, who had left England for New Zealand with her parents, in 1855. They settled near Nelson, in the South Island, where James Rutherford had a small farm and worked as a contractor building the railways. Ernest Rutherford was born on 30 Aug. 1871, the second son in a large family of twelve children. When Ernest was eleven years of age, the family moved a short distance to Havelock, where his father established a mill to treat the native flax of the area, and a small sawmill. At the primary school there, Ernest was influenced by his teacher, J. H. Reynolds, who taught so well that Ernest won a scholarship to Nelson College, with almost full marks in the examination. He entered the College at 15 years of age, and was much helped by one of the masters, W. S. Littlejohn, a classicist who taught also mathematics and science. Ernest had a broad education, excelling in mathematics, but winning distinctions in Latin, French, English literature, history, physics and chemistry, and becoming head of the school. He was a scholar of distinction, but played games reasonably well and entered fully into the life of the school. A. S. Eve, in his biography of Rutherford, quotes a fellow student as saying, "Rutherford was a boyish, frank, simple and very likable youth, with no precocious genius, but when once he saw his goal, he went straight to the central point." He took photographs with a home-made camera, dismembered clocks, made model water wheels such as his father used to obtain power for his mills. Under the influence

of his mother and his fine teachers, Rutherford developed a wide taste for literature and read avidly all his life. He became especially interested in biographies.

In 1889 he won a scholarship to Canterbury College, Christchurch, a component college of the University of New Zealand. There, as one of 150 pupils in the small institution, he enjoyed five very full years, obtaining successively his B.A. and M.A. degrees, the first in Latin, English, French, mathematics, mechanics and physical science, and the second, at the end of his fourth year, with a double first in mathematics and physical science. During his fifth year, Rutherford concentrated on his science, carrying out many experiments on the electromagnetic waves discovered by Heinrich Hertz, and investigating the effects of the damped oscillations of the Hertzian oscillator upon the magnetization of steel needles and iron wires. He showed that the magnetization was confined to a thin, outer layer of the metal, by dissolving away the surface in acid.

Rutherford was able to use these magnetic effects to detect the wireless waves from his oscillator, and demonstrated that these waves travelled for considerable distances, passing through walls on the way. He reproduced Nikola Tesla's experiments on the high voltages that could be produced with a resonant transformer, and developed techniques for measuring intervals of time as small as 10 microsec. He spoke to meetings of the Science Society on his work and on the evolution of the chemical elements, and he published two papers in the *Transactions* of the New Zealand Institute. He found it necessary to supplement his scholarship by coaching students, and went to live with a widow, Mrs. de Renzy Newton, whose daughter Mary he later married.

In 1895 Rutherford applied for an 1851 Scholarship, which was awarded to a New Zealand student in alternate years. The examiners of the 1851 Royal Commission, in London, awarded this to a chemist, J. C. Maclaurin, but were impressed enough by Rutherford to urge the award of a second scholarship, which was not given. However, Maclaurin gave up the scholarship to

accept an appointment in the civil service; so Rutherford was offered the award. He elected to go to the Cavendish Laboratory, in Cambridge, to work under J. J. Thomson, and had to borrow the money to pay for his passage to England. He and John S. Townsend, of gas-discharge fame, arrived at the Cavendish Laboratory almost simultaneously, to become the first of the new category of research student recently established in the University of Cambridge. There he joined Trinity College and began fresh experiments on the detection of electromagnetic waves by use of the effects of high-frequency currents upon the magnetization of iron wires. He soon established himself as a research worker of great promise, of whom Andrew Balfour wrote, "We've got a rabbit here from the antipodes and he's burrowing mighty deep." Rutherford was ambitious and anxious to qualify for a post that would enable him to marry Mary Newton. He thought that the detector using very fine magnetized steel wires surrounded by a solenoid in which high-frequency currents reduced the magnetization might make his fortune. Before Guglielmo Marconi, he was able to detect radio waves at a distance of half a mile.

Rutherford developed early an extraordinary ability to recognize, and concentrate upon, the puzzling problems of frontier knowledge in physics. He was never content to follow pedestrian paths of measurement or rounding off of investigations initiated by others. George P. Thomson, in his Rutherford Memorial Lecture, pointed out that Rutherford was working in the Cavendish Laboratory when two completely new physical phenomena were discovered. These were the discoveries of x rays, by Wilhelm Roentgen, and of radioactivity, by Henri Becquerel and each opened up hitherto unsuspected areas of investigation destined to change the course of physics. It is not surprising, therefore, that when J. J. Thomson invited Rutherford to join him in the investigation of the ionization produced in gases by x rays, Rutherford seized the opportunity to move into more exciting fundamental studies.

Rutherford showed that the ionizing effect of x rays was due to the production of positive and negative ions in equal numbers and devised ingenious methods for measuring the velocity of drift of these ions in an electric field. Then in 1898 he investigated the ions produced when ultraviolet light fell on a metal plate, showing that they were all negative ions and that their properties were identical with the ions produced in the gas by x rays. Upon hearing that the radiations discovered by Becquerel to be spontaneously emitted by uranium and thorium were able to ionize gases, Rutherford made observations of the properties of the ions produced, and found them identical with those that he had investigated previously. He showed that two kinds of radiation were present, an easily absorbable and strongly ionizing component which he called "alpha rays," and a much more penetrating radiation to which he gave the name "beta rays." He had found the field of physics in which he was to spend his life.

In August 1898 Rutherford was appointed to a professorship of physics at McGill University. He had applied for the post reluctantly, after assessing his prospects in Cambridge, mostly because of his desire to get married, but, having made the decision he accepted enthusiastically. Upon arrival in Montreal he rapidly established himself, and was soon at work on the further studies of radioactivity that were to establish him as the greatest experimental physicist of his day. In the summer of 1900, he went to New Zealand to collect his bride, returning to McGill in the autumn. In 1901 their only child, a daughter, was born.

Rutherford's subsequent work in Montreal, Manchester, and Cambridge, is part of the history of science, in every textbook.

Lawrence, early years

Lawrence's grandfather, Ole Lawrence, left his home in Norway to settle in Madison, Wisconsin, in 1840. There he became a school teacher in a primitive, pioneering community. He sent his son, Carl, to the University of Wisconsin, from which he graduated in 1894. Carl followed his father's profession as a teacher and showed that

he inherited the pioneering spirit, for he moved farther west to South Dakota as a Latin and history master. He became superintendent of public schools in the small community of Canton, and while there, married Gunda Jacobsen, the good-looking daughter of Norwegian immigrants, in 1900. Ernest Lawrence was born to them on 8 Aug. 1901.

Ernest's parents were good people, in the old-fashioned sense of these words. Although his father had a degree in arts, and had taught the humanities, he was not a scholar. The mother, a teacher of mathematics before her marriage, became an excellent wife and mother. She was a strict Lutheran, mingling high principles and loving care in the upbringing of her two sons, Ernest and John. From his parents Ernest acquired a strict moral code and a belief in the inherent decency of most human beings. Carl's ability as an administrator, combined with his integrity, led to his becoming in turn head of the Southern State Teachers' College in Springfield, and then of Northern State Teachers' College in Aberdeen, South Dakota. So, the family enjoyed modest means, but not sufficient to enable the boys to indulge in extravagances without earning money for themselves.

Ernest grew to be a tall, gangling youth. Unlike Rutherford, he did not enjoy the rough and tumble of team games like football but enjoyed tennis, which he played well, if not brilliantly, throughout his life. His career at high school was not outstanding, and though he showed promise in science, he performed indifferently in English. He read very little, and in later life was sarcastic about and impatient of his humanist colleagues, seeing little practical good in their work. He was never a cultured man and had few of the social graces so that he made few friends among girls and did not shine in extracurricular activities of the school. However, he was by no means antisocial, these traits arising from indifference towards any activity that did not fire his interest. He was ambitious and worked hard and consistently, so that he graduated from high school at 16 years of age after three, instead of the usual four years.

During the long summer vacations,

RUTHERFORD AT 21, while a student at Canterbury College, University of New Zealand. Photo from A. S. Eve, *Rutherford*, Cambridge University Press.

LECTURING AT McGILL University, 1907, after Rutherford left Cambridge.

JESSE BEAMS shares a laboratory with Lawrence at Yale University, 1927, where they developed a technique to observe the lifetimes of excited atomic states.

Lawrence worked on farms in the district, as a salesman for aluminum ware and in other ways earned the money required to buy the necessities of an American boy with a mechanical turn of mind—motor cars of various vintages, radio receiving equipment, tools and electrical gadgets, and so on. No doubt under the influence of the concern for others of his parents, he decided upon a career in medicine, and he was sent to a small private college, St. Olaf's in Minnesota, to begin his preliminary studies. He was too young and unsettled to do well there. After a year he moved to the University of South Dakota. He soon applied to the dean, Lewis E. Akely, for permission to build and operate a radio transmitting equipment. Akely was much impressed with the knowledge and ambition of the youth, and persuaded him to turn to physics, providing him with individual tuition in the subject in order to give him a start. After graduation in chemistry— he had not abandoned his ambition to do medicine—Lawrence was persuaded by his close friend, Merle Tuve, and by the offer of a fellowship, to move to Minneapolis. There he worked with W. F. G. Swann, an English immigrant who had been working in geophysics in Washington, but who had joined the University of Minnesota in order to work in more basic physics. Leonard Loeb recalls that Swann was not popular with his colleagues but that he got on extremely well with young graduate students,

inspiring them to do research of quality and encouraging them with help and discussion. Under his influence, Lawrence abandoned his desire for a medical career. Swann introduced him to the exciting field of experiment arising from development of the quantum theory. His early interest in electromagnetism was stimulated and developed. He took his master's degree early in 1923, and later that year moved with Swann to Chicago.

In Chicago Lawrence found himself in a very different environment where research was vigorously pursued by an outstanding group of physicists. He was stimulated greatly by contact with Arthur Compton, at the time completing his work on the Compton effect. But he found himself also in a department run on strictly European lines, where the professor was all-powerful and status determined the relationships among members of the laboratory. Neither Swann nor Lawrence was at ease in this atmosphere, and when Swann accepted a post at Yale, a year later, Lawrence went with him. In Chicago Lawrence had learned the real meaning of research, and he threw himself into it with complete devotion. But it was at Yale that his gifts as an experimenter, aided by his energy and enthusiasm, really flowered. For his PhD he worked on the photoelectric effect in potassium vapor, carrying out beautiful experiments that demonstrated clearly that he was a physicist of high quality. Under a National Research Council Scholar-

ship, and after appointment to an assistant professorship, Lawrence continued with his researches. He made precise observations of the ionization potential of mercury vapor, of importance in the determination of the value of Planck's constant h and devised an elegant method of measuring the ratio of charge to mass of the electron. With Jesse Beams, who became his firm friend, he developed a beautiful technique for measuring very short time intervals, which was applied to observations of the lifetimes of excited states of atoms.

In 1928 Lawrence was offered an associate professorship at the University of California, in Berkeley, having turned down an earlier offer of an assistant professorship. A lengthy correspondence with Elmer Hall, the chairman of the physics department, and with Raymond Birge, who had called on Lawrence in Yale and was much attracted by him, has been faithfully recorded by Birge in the history of the department that he is writing. It seems that Lawrence was attracted to California by the opportunity to teach an advanced course and to direct the work of research students, activities reserved in Yale for more senior members of staff. Birge pointed out the good opportunities for rapid advancement of a good man in Berkeley, contrasting this with the policies at Yale, Harvard and Princeton, where it was almost impossible to "get anywhere, after one was there, except under very special circumstances. . . ." Lawrence wrote to Birge saying that some men in Yale were very "sore" that he should even consider a position in California to be comparable with one in Yale. "The Yale ego is really amusing. The idea is too prevalent that Yale brings honor to a man and that a man cannot bring honor to Yale."

Lawrence accepted the offer from Berkeley, and arrived there in August 1928. He set to work at once to continue his work on the photoionization of cesium vapor, used the techniques which he had developed with Beams for the measurement of short time intervals in observations of the early stages of the spark discharge, and one of his research students, Frank Dunnington, developed his method for measuring the charge-to-mass ratio of the electron. He was not committed to this type of investigation, however. He felt that the current challenge in physics was the investigation of the atomic nucleus, rather than of the atom as a whole. He was impressed by the limitations of the methods of investigation developed by Rutherford, who bombarded nuclei with alpha particles emitted by naturally occurring radioactive substances. Like Cockroft, he appreciated Rutherford's desire to be provided with much more intense beams of even more energetic particles with which to probe the internal structure of nuclei.

Lawrence has recorded how, early in 1929, he read a paper by Rolf Wideröe on the use of high-frequency voltages for accelerating charged particles. He recognized that it should be possible to use a magnetic field to curl the paths of such particles into a spiral, and that because the Larmor time-of-revolution in the field was independent of the energy, they could remain in resonance with the voltage across an accelerating gap. Robert Brode has told me of a visit to him by Lawrence the day after seeing the article, enquiring whether the mean free paths of ions could be made long enough for them to suffer negligible scattering by residual gas in their very long spiral paths. Lawrence's colleagues agreed that his calculations were correct, but they were dubious whether the method could be applied in practice.

In 1930, Edlefsen, who had completed his PhD thesis, constructed crude models of the system and observed some resonance effects. Livingston joined Lawrence, after Edlefsen left that summer, and built an improved model that showed resonances corresponding with the rotation times of molecular and atomic ions of hydrogen. By Christmas 1930, a 6-in model surprisingly like a modern cyclotron, was in operation, producing hydrogen ions with energies of 80 000 eV.

The "magnetic-resonance accelerator," as the cyclotron was first named, had become a reality. Lawrence had found his life's work.

In 1932 Lawrence married Molly Blumer, daughter of a distinguished medical man, whom he had met while at Yale and whom he had courted for some years. They had six children, two boys and four girls. He was happy with his family, and the children enriched the life of both. Lawrence appears to have been a normal scientist-father, much preoccupied with his work, alternatively indulgent and too strict, with his serene and capable wife holding the balance and creating the home.

The two compared

The similarity between the early careers of the two men is apparent. The earliest interest of each was in radio. However, while Rutherford abandoned that field completely when he turned to the study of radioactivity, the radio-frequency problems of the cyclotron kept alive the interest of Lawrence. With David Sloan and Livingston he built his own oscillators, and after the war he developed a picture tube for color television that is now manufactured by the Japanese firm, Sony. Each moved from radio into atomic physics, and then to the study of the atomic nucleus. Each was single-minded, working indefatigably towards a goal once it was chosen. Each showed tremendous enthusiasm, which he was able to convey to others.

In his early work, Lawrence showed an insight into physics very like that of Rutherford. Whereas Rutherford continued throughout his life to explore in the frontiers of knowledge, however, Lawrence chose to contribute to physics less directly. After the discovery and successful development of the cyclotron, Lawrence's flair for organization and his business ability enabled him to build the first of the very large laboratories in which massive and expensive equipment was designed, built and used by the able teams of men he attracted to work with him for investigations into basic problems in physics in which he played little part, personally. This pattern of research has become the modern approach all over the world. Rutherford, on the other hand disliked large and expensive equipment. He preferred to remain involved, personally, in almost all the work going on in his laboratory. His interest and ability in administration and finance were rudimentary. He dominated the laboratory by his sheer greatness as a physicist and provided for his colleagues and students only the very minimum of equipment required for an investigation. Rutherford, with his roots in the soil and the hard, practical life of New Zealand, bucolic in appearance, became the deep thinker and the originator of new physical concepts. Lawrence, brought up in an academic atmosphere, impressive and scholarly in appearance, became the originator of new techniques and of the large-scale engineering and team-work approach to discovery.

Both men were extroverts and good "mixers" in company. Donald Cooksey recalls that when Lawrence entered a room filled with great industrialists or successful politicians, his presence was at once noticed, and his impact upon them was profound. Rutherford, however, could be taken for a farmer or shopkeeper, and it was not till he spoke that he was noticed by those who did not know him. Neither was a good speaker or lecturer; yet each influenced and inspired more colleagues and students than any other of his generation. Both built great schools of physics that became peopled with other great men, and Nobel prizes went naturally to members of their laboratories. Each was most generous in giving credit to his junior colleagues, creating thereby extraordinary loyalties.

Rutherford and Lawrence were self-confident, assertive, and at times overbearing, but their stature was such that they could behave in this way with justice, and each was quick to express contrition if he was shown to be wrong.

Neither Rutherford nor Lawrence could tolerate laziness or indifference in those who worked with them. Rutherford said to a research student from one of the dominions, at tea before a meeting of the Cavendish Physical Society, "You know, X, I do not believe that you are in and at work because your hat is hanging behind your door!" Such a remark was far more effective than any reprimand. During the hectic days of the Manhattan Project in the war years, Lawrence spoke to me several times of individuals whom he felt did not share his sense of urgency and complete dedication to the task in hand. "I don't know what has gone wrong with Y. He's lazy and his attitude is affecting those round him. I think we'd better get rid of him."

Rutherford had a great and affectionate regard for Niels Bohr, who had worked with him in Manchester. Lawrence could not understand the attitude of the gentle theoretician, who had been smuggled out of Denmark by the British and brought to Los Alamos, where it was thought that his genius could aid the design of a nuclear weapon. While the task was not completed, Lawrence could see no sense in Bohr's worries about how it should be used, or his concern about the part the devastating new weapon could play in the creation of a world without war. Great as was his admiration for the man who had made a living reality of Rutherford's nuclear atom, he felt that Bohr was actually holding back progress and would be better away from the project. On his part, Bohr found it difficult to understand the complete objectivity of Lawrence over an undertaking which created a crisis in human affairs to which men of science could not be indifferent.

Although wholly dedicated to the pursuit of scientific knowledge, both Rutherford and Lawrence delighted in the company of men who had achieved greatness in other spheres. Because of their positions and reputations, they made many contacts and a multitude of friends among industrialists, politicians, lawyers, medical men and the higher echelons of the civil service. They were at home in such company and enjoyed the good living which many such men accepted as part of their existence. But there was one great difference. Rutherford enjoyed what has been called smoking-room humor. Although his own memory for such stories was not good, his great roar of booming laughter was to be heard after dinner as he savored the subtlety of some lewd tale. I never heard Lawrence swear, under any circumstances, and his reaction to off-color humor was not encouraging.

Both Lawrence and Rutherford could be devastatingly blunt and uncompromising when faced with evidence of lack of integrity, or of gullibility,

RUTHERFORD, IN 1926, visits New Zealand as Cawthron Lecturer.

LAWRENCE AT CONROLS of the 37-in. Berkeley cyclotron, about 1938.

in scientific work. I recollect an occasion when Rutherford was asked to advise whether the inventor of a diagnostic machine, which had been reported upon favorably by one of the Royal physicians, should be paid a large sum of money for rights to use his equipment. Diseases were alleged to be diagnosed by connecting electrodes to the patient and observing the deflections of meters indicating excess or defect of various elements in the patient's body. The inventor explained that the "black box" contained radioactive varieties of each of the elements, where-

upon Rutherford became very angry, pouring scorn on both the fraudulent inventor and the gullible physicians who believed in the efficacy of his machine. I am told that Lawrence was invited to examine the claims of a chemist in Berkeley who maintained that isotopes of the chemical elements could be detected, and their proportions measured, in incredibly small concentrations, by observation of certain optical resonances in polarized light, which were characteristic for each individual isotopic mass. Looking through the eyepiece, he could find no evidence whatever of the maxima and minima which were said to exist. He burst into laughter, in a cruelly embarrassing manner, at the self-delusion of the young observer, who had been persuaded by the senior perpetrator of the hoax that there was something to observe.

Politics

In politics, Rutherford was what would be called nowadays, a woolly liberal. My wife and I spent many periods with the Rutherfords at their country cottage, "Celyn", in the beautiful Gwynant Valley of North Wales, and later at "Chantry Cottage" in Wiltshire, where the walking was less arduous. He and I often had political arguments, which were particularly hot at the time of the abdication of Edward VIII. I thought that no harm would come if Edward were allowed to marry Mrs Simpson, whereas Rutherford argued that it would do irreparable harm to the monarchy. His main concern was that science should be used properly in the development of the economy, and on one of his rare appearances in the House of Lords, he advocated the establishment of a ministry of prevision to keep the government informed about the advance of science and technology and the probable impact upon industrial development. He was most generous and openhearted, and did all that he could to aid the victims of Nazi persecution. He was as suspicious of communism as he was of extreme conservatism, but he liked Stanley Baldwin, one of the most conservative prime ministers Britain ever had. At heart, he was apolitical, but when pressed, declared that he was a liberal.

Ernest Lawrence was both an idealist, who cared intensely about the future of his children and all mankind, and a pragmatist, who saw little good in the obsession of some of his colleagues with the examination of social and political schemes for alleviating the lot of humanity. Sometimes during the war, he and I walked up or down the hill between the Radiation Laboratory and the campus of the university. The downward trip usually began by his drinking a carton of cold milk, which I loathed, the liquid portion of which often fertilized one of the stately euca-

lyptus trees planted on the hillside. We would pause on the way to gaze down over the unforgettable beauty of San Francisco Bay. Then, and while walking, he would tell me of his deep concern that science be used fully to aid the development of the human race, and of his admiration for the practical steps that Franklin Roosevelt was taking to enable this to happen in the United States. He would outline what he could see ahead in the application of physical knowledge in communications, and the productivity of industry and agriculture. He would express his conviction that knowledge of matter and radiation would transform the biological sciences and provide tools for medicine that would alleviate, cure and prevent disease. He felt that this was a task for mankind, and not only for America, and he was anxious to help create a world situation in which all knowledge could be shared by all men. In a practical way he did this whole-heartedly, helping us all, wherever we were, to build cyclotrons, by providing freely drawings, full details, and even his thoughts about improvements upon what had been built in Berkeley. Of course he could not escape entirely the atmosphere of the times, and after the end of the war, he veered somewhat towards a more restricted and less generous view of the part that his great country should play in maintaining the peace and assisting other nations. But this was true only of his politics, and his deep commitment to the defense of America. In his science, he remained the same open-hearted believer in openness and in the value of exchange of knowledge and of information in the removal of international misunderstandings.

However, Lawrence was genuinely apolitical. He had inherited liberal democratic leanings from his parents, but he could not become excited about political issues. For instance, he was quite unaffected by the "loyalty oath," which the university imposed upon members of its staff, and which caused great dissension among some of them. Although unable to appreciate the strong objections of many of his colleagues to what he regarded as a trivial obligation imposed by those who generously supported his laboratory, never-

theless, he fought hard for them as individuals.

Advice on cyclotrons

It is interesting here to recall that the first inquiry Lawrence received from anyone about the possibility of construction of a cyclotron elsewhere, was from Frédéric Joliot, of Paris. On 14 June 1932, he wrote from the Laboratoire Curie, saying that he had read with great interest Lawrence's publication on the production of ions with high velocity. "Votre travail me paraît remarquable, et les études que l'on peut faire avec de tels rayons sont d'un grand intérêt." [Your work seems remarkable to me, and the studies that can be made with such rays are very interesting.] He would like to build an apparatus of a similar type, and to do it rapidly. To this end, he requested two reprints of the article, and any details of construction of the "points les plus délicats" [the most delicate points]. On 20 Aug. Lawrence replied, apologizing for the delay, and told Joliot that he might be able to obtain a magnet made for a Poulson arc radio transmitter, similar to one that Lawrence had obtained in the United States, which he understood was being dismantled at Bordeaux.

The generous attitude of Lawrence towards others desiring to build cyclotrons of their own is well illustrated by the following extract from a letter to Kenneth Bainbridge, dated 6 Feb. 1935:

"I have just received a letter from Professor [George] Pegram at Columbia, saying that they want to embark upon the construction of a cyclotron provided that I have no objections. I am writing him that, rather than having objections I am more than delighted that they are planning to build a cyclotron. The cyclotron to my mind is by far the best ion accelerator for nearly all nuclear work, and it would give me a great deal of pleasure if many laboratories would build them."

On 27 Nov. 1935 Lawrence wrote to Chadwick, congratulating him on the award of a Nobel Prize, and offering to give him every help in building a magnetic-resonance accelerator in Liverpool. He said that the Cavendish must miss Chadwick greatly, but that this was compensated by the fact that he would build in Liverpool another great center of nuclear physics. Chadwick replied that he felt rather lucky to get a Nobel Prize and thanked Lawrence for his offer to help to build "your magnetic-resonance accelerator, which ranks with the expansion chamber as the most beautiful piece of apparatus I know." In letters about the construction of cyclotrons by others, Lawrence always emphasized that, contrary to the ideas of many, the cyclotron was not a difficult piece of equipment to get into operation.

The word "cyclotron" did not appear in any publication from the Radiation Laboratory till 1935, in a paper by Lawrence, Edwin M. McMillan and Robert Thornton,[1] where the following footnote is inserted:

> "Since we shall have many occasions in the future to refer to this apparatus, we feel that it should have a name. The term 'magnetic-resonance accelerator' is suggested. . . . The word 'cyclotron,' of obvious derivation, has come to be used as a sort of laboratory slang for the magnetic device."

Running their laboratories

The Cavendish Laboratory, under Rutherford and his predecessors, was always short of money. Rutherford had no flair and no inclination for raising funds. Only under extreme pressure, first from the ebullient Peter Kapitza, and later from Cockcroft and me, was he prepared to fight hard for money for large or complex equipment. He never sought riches and died a comparatively poor man. Lawrence, on the other hand, had shrewd business sense and was adept at raising funds for the work of his laboratory. Apart from his early interest in medicine, he realized early the medical possibilities of the radiations produced by the cyclotron, and did not hesitate to use these in his search for funds. In 1935 he wrote to Bohr:

> "In addition to the nuclear investigations, we are carrying on investigations of the biological effects of the neutrons and various radioactive substances and are finding interesting things in this direction. I must confess that one reason we

have undertaken this biological work is that we thereby have been able to get financial support for all of the work in the laboratory. As you well know, it is so much easier to get funds for medical research."

Similarly, after the war, he made full use of the wartime achievements of the Radiation Laboratory in raising the support required for the very large expansion of its activities. However, it was his concern for the defense of his country and his belief that it was unwise to confine the development of nuclear weapons to Los Alamos, which led him to establish a branch of the laboratory devoted to this work at Livermore.

Lawrence's phenomenal success in raising money for his laboratory was undoubtedly due to his able handling of executives in both industry and government instrumentalities. His direct approach, his self-confidence, the quality and high achievement of his colleagues, and the great momentum of the researchers under his direction bred confidence in those from whom the money came. His judgment was good, both of men and of the projects they wished to undertake, and he showed a rare ability to utilize to the full the diverse skills and experience of the various members of his staff. He became the prototype of the director of the large modern laboratory, the costs of which rose to undreamt of magnitude, his managerial skill resulting in dividends of important scientific knowledge fully justifying the expenditure. But in achieving this, he had to give up personal participation in research. His influence on the laboratory programs remained profound, and his enthusiasm radiated into every corner of the institution. William Brobeck, who joined the Radiation Laboratory in 1936 as an engineer, recalls that Lawrence took an animated part in all discussions of technique and showed an extraordinary ability to see a piece of equipment as a whole, avoiding becoming bogged down in detail. Lawrence was a regular visitor to each section of the laboratory until illness caused him to appear very seldom outside his office.

Rutherford's method of running a laboratory was in striking contrast to that of Lawrence. He was not much

interested in the apparatus for its own sake, believing that techniques grew from the demands of the experiment. Like Lawrence, he advocated a simple, preliminary approach, a sort of skirmish into the territory to be explored, followed by refinement if the reconnoiter showed promise. He would roam round the laboratory, discussing results and the physical knowledge they revealed, rather than apparatus. His stimulus was enormous, and his influence direct. A glance at any list of publications from the Cavendish Laboratory, or from the laboratories in McGill or Manchester in his periods there, reveals how deep was his influence on the researches carried out. Lawrence worked to give others the opportunity to achieve important results; Rutherford was so great a physicist that almost every member of his laboratory found himself working upon some problem that Rutherford had suggested, or that arose directly from Rutherford's own work. This dominance was not imposed upon his colleagues and students. They often began work along lines of their own choosing, but rapidly found that the instinct of Rutherford's genius was a surer guide to interesting and important results.

Both Rutherford and Lawrence gave coherence to laboratories inhabited by workers of differing temperaments and varying abilities. Under their influence, each gave of his best; all rejoiced in the outstanding achievement of one of their number, and each felt himself to be part of the whole, sharing its triumphs and its vicissitudes.

Seventh Solvay Congress

Although Lawrence had made a very rapid tour of Europe with his friend Beams in the summer of 1927, he and Rutherford did not meet till 1933. In that year, the Seventh Solvay Conference, held in Brussels from 22 to 29 Oct., was devoted to nuclear physics, and, naturally, Lawrence was invited to attend. He was eager to go, since this would give him the opportunity to meet the principal workers in his field. Those taking part included:

From Cavendish Laboratory:

Ernest Rutherford
James Chadwick

John Cockcroft
Patrick Blackett
Paul Dirac
Cecil Ellis
Rudolf Peierls
Ernest Walton
From Institut du Radium, Paris:
Marie Curie
Irene Joliot-Curie
Frédéric Joliot
M. S. Rosenblum
From the Physical Institute, Leipzig:
Werner Heisenberg
Peter Debye
From elsewhere:
Neils Bohr (Institute of Theoretical Physics, Copenhagen)
Albert Einstein (then living in Belgium)
Erwin Schrödinger (Physical Institute, University of Berlin)
Wolfgang Pauli (Physical Institute, Zurich)
Louis de Broglie (France)
Marcel de Broglie (France)
Enrico Fermi (Physical Institute, University of Rome)
George Gamow (Institute of Mathematical Physics, Leningrad)
Abraham Joffe (University of Physics and Mechanics, Leningrad)
Walther Bothe (Physical Institute, University of Heidelberg)
Lise Meitner (Kaiser Wilhelm Institute, Berlin)
Francis Perrin (Institute of Chemistry and Physics, Paris)
Léon Rosenfeld (Institute of Physics, University of Liège)
H. A. Kramers (Institute of Physics, University of Utrecht)
Nevill Mott (University of Bristol)

Ernest Lawrence, the only American invited, naturally was greatly pleased to find himself among this group of eminent physicists who, together, represented almost all that was then known, from experimental and theoretical investigation, of the atomic nucleus. His invitation from the President, Paul Langevin, asked him to participate in "l'examen de questions relatives à la constitution de la matière" [the examination of questions relative to the constitution of matter], and reports were to be read by Rutherford, Chadwick, Bohr, Heisenberg, Gamow, Cockcroft, and M and Mme Joliot. It was clearly to be an exciting meeting, as it was only a year earlier

that the neutron had been discovered, and transmutation of nuclei by artificially accelerated beams of charged particles had been achieved.

In a letter to Langevin, dated 4 Oct. 1933, written after he had read the papers that had been circulated to those invited, Lawrence stated that he wanted particularly to make some rather extensive observations on Cockcroft's report, and that he might wish to comment on papers by Chadwick, Joliot, and possibly Gamow. He was able to obtain funds to meet the costs of his trip, but owing to his commitments in Berkeley, he could stay in Europe for only a very limited period.

At this time, Lawrence and his co-workers had used the cyclotron to confirm the results of Cockcroft and Walton on the disintegration of lithium by proton bombardment, and had extended their observations on this and other transformations to higher energies. Lawrence had eagerly availed himself of the opportunity offered by the success of Gilbert N. Lewis, at Berkeley, in producing almost pure samples of heavy water, and had accelerated the nuclei of the new hydrogen isotope in the cyclotron. His team observed an enormous emission of protons and neutrons from every target that was bombarded, and this similarity of results, irrespective of target material, had led Lawrence to put forward the hypothesis that the nucleus of heavy hydrogen, called the "deuton" by Lewis, was unstable, breaking up in nuclear collisions into a proton and neutron. Meanwhile, Lewis had presented samples of heavy water to many investigators, including Rutherford, and we had been making observations in the Cavendish Laboratory that were not in accord with Lawrence's view that the deuton was unstable.

Lawrence went to the Solvay Conference prepared to defend his hypothesis and to back the cyclotron as the type of accelerator most versatile for experimental work in nuclear physics. The marginal notes made by him on the copies of the reports presented, give interesting information about his attitudes. Some of these are vigorous, as the large cross over Cockcroft's assertions that "only small currents are possible" from the cyclotron, and when Cockcroft restated this

WATSON DAVIS, SCIENCE SERVICE

CYCLOTRON MODEL is held by Lawrence in 1930, year after conception.

later, he wrote, "Not true," boldly in the margin. In several places he complained that the deuton-breakup hypothesis received no mention, and it becomes clear that he did not appreciate fully the calculations of neutron mass given by Chadwick, or the observations of Cockcroft, and of Rutherford and me, which were not in accord with his idea. He showed particular interest in those observations reported by the Joliots on gamma rays produced from atoms bombarded by alpha particles, both those collisions that result in capture of the alpha particle, and those in which a nucleus is excited, without actual capture.

Lawrence's meticulous care to give credit to his colleagues for their part in the work in his laboratory is evident from his insistence upon the addition of their names—Malcolm Henderson, Milton White, Sloan, Lewis and Livingston—wherever Cockcroft's paper mentioned only Lawrence.

Chadwick recalls, in a letter to me, that Rutherford was much impressed by the vigorous young Lawrence, and remarked to Chadwick, "He is just like I was at his age."

Lawrence paid a brief visit to the

Cavendish Laboratory after the Solvay Conference, and it was then that I met him. We had a vigorous discussion, with Lawrence sticking firmly to his concept of an unstable deuton. When he had gone, Rutherford, said, "He's a brash young man, but he'll learn!"

Cooksey tells me that he met Lawrence at the boat in New York on his return to America. Lawrence was bubbling over with enthusiasm for all that he had seen and learned. He was particularly enthusiastic about the great power of the neutron as an agent for disintegrating nuclei, and expressed the view that, before long, these would make possible a self-propagating reaction, and hence the practical release of energy from nuclei. A truly prophetic remark.

Deuton instability

After his return from the Solvay Conference, Lawrence wrote to Cockcroft informing him that, with Livingston and Henderson, he would concentrate upon the origin of the protons, with a range in air of about 18 cm, which were emitted from all targets bombarded with deutons. Firstly, they would try to clear up the uncertainty about contamination of the targets, and if this did not turn out to be the source of the particles, they would "continue the experiments to shed further light on the origin of the 18 cm protons." He reported also that, on his way back, he had visited Washington, where Tuve had a beam of protons with an energy of 1.5 MeV from his Van de Graaff accelerator.

"I persuaded Tuve to investigate the origin of the 18 cm protons and the hypothesis of the disintegration of the deuton right away. I want to get the matter cleared up as soon as possible and it will be a great help if Tuve, with his independent set-up, will investigate the problem."

He wrote also to Gamow on 4 Dec. 1933, saying that he had been paying particular attention to the hypothesis of the disintegration of the deuton, using clean targets and carefully purified materials. "However, we find that the yield of protons and neutrons produced by the bombarding deutons is quite independent of our endeavors

to clean the targets." They found that 2.8-MeV deutons produced disintegration protons in the same proportions as observed at 1.2 MeV. On 28 Dec. 1933 he wrote again to Gamow:

"The experimental evidence that the deuton disintegrates is growing. Lately, we have observed the emission of long range protons (up to about 20 cms) resulting from the bombardment by protons of targets containing heavy hydrogen. Though perhaps the matter cannot be regarded as entirely settled yet . . . certainly it must be admitted that the evidence is preponderantly in favor of the hypothesis of the energetic instability of the deuton."

Cockcroft, in a letter to Lawrence of 21 Dec. 1933, reported further work on the long range protons produced by bombardment with deutons from lithium, carbon and boron, and noted that while iron gave a small yield of protons, none were observed from copper, gold or copper oxide.

"We have so far not worked beyond 600 kV, and it may well be that some groups appear at higher voltages. I feel myself, however, that the evidence so far is against your interpretation of the break up of H^2."

Lawrence replied on 12 Jan. 1934:
"It seems to me that you are hardly justified in feeling that the evidence obtained by you so far is against the interpretation of the break-up of the deuton, since you have not worked at voltages above 600 kV . . . it seemed pretty evident from our first preliminary observations that the yield of the group of protons which we ascribe to deuton disintegration is in all cases very small below eight or nine hundred thousand volts. Despite your greater intensities, on the basis of our observations we would hardly expect that you would observe the disintegration of the deuton at the voltage you have been using. . . . I hope that you will soon raise your voltage to eight or nine hundred thousand. Meanwhile I have written Tuve your results and asked him to look into the matter, as I understand he is able to work now above a million volts. I am anxious that the hypothesis of deuton disintegration will be

settled to everyone's satisfaction, and to that end it seems essential that independent experiments be carried out in another laboratory."

Cockcroft wrote again on 28 Feb. 1934:

"We have been working steadily on the question of disintegrations by heavy hydrogen. In addition to the results on lithium I reported to you in my last letter, we find three groups of protons from boron. . . . We have been investigating copper, copper oxide, iron, iron oxide, tungsten and silver, with stronger heavy hydrogen, and we find from all of these we get three groups of particles of identically the same range. The first is an alpha particle group having a maximum range of 3.5 cm, the second is a proton group of about 7 cm, and the third is a proton group of about 13 cm. This latter group is the one which you ascribe to the break up of the deuton. It seems in the first place clear that these three groups cannot all be due to this break up, and we therefore feel strongly that the alpha particle group and the 7 cm proton group are at any rate due to an impurity which is probably oxygen. We are not yet certain about the 13 cm group, but are carrying out experiments with white hot tungsten targets which I hope may finally dispose of this possibility. We can observe all these groups at voltages as low as 200,000, and the voltage variation shows the standard Gamow tail to the curve. . . .

"I feel, however, that we have still very good justification for refusing to commit ourselves to your hypothesis of the deuton break up until further experimental work has been carried out."

To this typewritten letter, Cockcroft added the following handwritten postscript:

"We have now found that on boiling in caustic and cleaning thoroughly the 13 cm group is reduced by a factor 10; on heating to 2,600 by a further factor. The 2.5 and 7 cm groups disappear on heating and reappear on oxidation and seem due to oxygen. . . . Oliphant is getting queer results with $H^2 + H^2$."

Lawrence replied on 14 March 1934,

agreeing that Cockcroft's observation that boiling tungsten in caustic reduces the 13-cm group by a factor 10 showed clearly that this is due to a contamination.

"I think it is quite possible that the effects we observed when bombarding targets of heavy hydrogen with hydrogen molecular beams were due, as [C. C.] Lauritsen suggested, to an increase in deuton contamination resulting from partial decomposition of the targets. I cannot understand my stupidity in not recognizing this possibility when the experiments were in progress. Needless to say, I feel there is now little evidence in support of the hypothesis of deuton instability. . . .

"Rather than continuing with preliminary and exploratory experiments at higher voltages, we have decided to embark on careful investigations of the nuclear effects brought to light and we shall make as precise and trustworthy measurements as we can. These recent experiences have impressed upon us forcibly the fact that much of our work has been of too preliminary character to be of value. I regret very much that the question of deuton instability involved you in so much work, and I want to thank you very much for stepping in and clearing the matter up so effectively and so promptly."

Lawrence and his colleagues were relatively new to nuclear physics, and it is not at all surprising that they made mistakes in interpretation of a complex phenomenon. It was characteristic of the young Lawrence that he held tenaciously to his concept of deuton instability, but that when presented with definite evidence that it was wrong, he immediately set to work to change the approach of his team to its experiments in such a way as to avoid similar pitfalls in the future.

Deuton stable after all

Meanwhile, the explanation of the origin of the proton group that had led Lawrence astray had been found in the Cavendish Laboratory. On 13 March 1934, Rutherford wrote to Lawrence:

"I have to thank you for the very interesting letter you sent me some time ago giving an account of your work. The whole subject is certainly in an interesting stage of development and reminds me very much of my early 'radioactivity' days before the theory of transformations cleared things up.

"I think you have heard from Cockcroft about some of our observations the last few months. Oliphant and I have been particularly interested in the bombardment of D with D ions, and I am enclosing a note from Oliphant giving an account of our results. I personally believe that there can be little doubt of the reaction in which the hydrogen isotope of mass 3 is produced, for the evidence from all sides is in accord with it. The evidence for the helium isotope of mass 3 is of course at present somewhat uncertain but it looks to me not unlikely.

"You will see that Oliphant like myself is inclined to believe that the proton group which you observe for so many elements arises from the reaction I have mentioned. We have made a large number of observations with beryllium and other elements but the results are not easy of interpretation. We think the information we have found about the D-D reaction will be helpful in disentangling the data. As you no doubt appreciate, it takes a lot of work to make a reasonably complete analysis of the groups of particles from any element and then it has to be done all over again with the other compounds to try and fix the origin of the groups. There is an enormous amount of work that will have to be done with the lighter elements to be sure we are on firm ground.

"You will have seen about Cockcroft's results due to the bombardment of carbon by protons. This no doubt produces the radio-nitrogen of the Joliots but we can obtain quite strong sources of positrons by this method. I heard that Lauritsen or yourself had observed similar effects with D bombardment. The whole subject is opening up in fine style. You will also have seen that Oliphant and Co have separated the lithium isotopes and confirmed the tentative conclusions we put forward before." My note went as follows:

"You may have heard of the experiments which we have carried out during the last week or two on the effects observed when heavy hydrogen is used to bombard heavy hydrogen. As I believe these are intimately related to your own work, I should like to tell you what we have found."

The letter went on to give details of the results, and of their interpretation as due to two competing reactions, the first leading to the production of hydrogen of mass 3 and a proton, with ranges of 1.6 cm and 14.3 cm respectively, and the second to helium of mass 3 and a neutron.

"We suggest, very tentatively, that your results may be explained as due to the bombardment of films of D and of D compounds. Our results with C, Be, etc., could all be accounted for by the presence of less than one monomolecular layer of D. . . ."

On 4 June 1934 Lawrence replied to my note, saying that the late answer was due to his desire to be able to send some news of interest.

"Your experiments on diplons, together with Cockcroft and Walton's recent work, have certainly cleared things up in beautiful fashion. There can no longer be any doubt that our observations which we ascribed to diplon break-up, are in fact the results of reactions of diplons with each other."

He ended his letter with a reference to Cockcroft's contention, in his Solvay Conference paper, that the cyclotron gave only small currents:

"Dr. Cockcroft might be interested to know also that we are gradually increasing our currents of high velocity ions, and that now we are working regularly with more than a microampere of either 3 MV diplons or 1.6 MV protons and several microamperes of 3 MV hydrogen molecule ions."

Lawrence had already replied to Rutherford's letter on 10 May 1934, saying:

"I want to thank you for your very much appreciated letter. Everyone here was delighted to learn of the extraordinarily interesting experiments you have been doing on the reactions of D-ions with each other (perhaps I should say diplons. I do appreciate the force of your arguments in support of diplon,* but all of us here have become quite accustomed to deuton and it would be some effort to change).

"It is difficult for me to understand how we could have failed to detect the effect of diplons on each other. We did notice about twice as many long range protons from the heavy hydrogen target under bombardment by diplons, but the difference between the targets was much greater under proton bombardment. The fact that the calcium hydroxide targets decompose readily may in some way account for our observations. Professor Lewis has prepared some ammonium chloride targets and we shall investigate the matter soon.

"The manuscript of Cockcroft and Walton's admirable paper has just arrived. There can hardly be any doubt any longer that most of the effects which we ascribe to disintegration of diplons are in fact due largely to a general contamination of heavy hydrogen in our apparatus. I certainly appreciate the manner in which this complexity of nuclear phenomena already brought to light makes it clear that it is easy to fall into error, and that a good deal of cautious work must be done for trustworthy conclusions.

"Fermi's observation of radio-activity induced by neutron bombardment is a case in point. When we bombard various targets with three million volt deutons, large numbers of neutrons are always produced, which among other things produce the types of radio-activity discovered by Fermi. On receiving Fermi's reprint announcing the effect, we looked for it and found that it was no small effect at all. For

example, we found that a piece of silver placed outside of the vacuum chamber about three centimeters from a beryllium target bombarded by a half micro-ampere of three million volt deutons became in the course of several minutes radio-active enough to give more than a thousand counts per minute when the silver piece was placed near a Geiger counter. We are now studying this type of radio-activity induced in various substances and will not return to the effects produced by diplon and proton bombardment until we understand pretty well the neutron effects.

"Dr. [Franz] Kurie has been photographing with the Wilson chamber the recoil nuclei and disintegrations in oxygen produced by neutrons from beryllium bombarded by deutons. Although the Wilson chamber is about twenty inches from the neutron source and therefore subtends a rather small solid angle, the neutron intensity is sufficiently great to give him something like five or ten recoil oxygen nuclei in each picture and about one disintegration fork per ten pictures. Most of the disintegrations appear to result in C^{13} and an alpha-particle; but Kurie has a dozen or so which seem to involve the emission of a proton and therefore the formation of N^{16}. But these conclusions are highly tentative. At the moment Kurie is busy making measurements on his photographs.

"We have sent off for publication a manuscript on the transmutation of fluorine by proton bombardment and I am enclosing the essential curves of the experimental results. As far as we can determine, the alpha-particles from fluorine have a range of between six and seven centimeters, depending on the energy of the bombarding proton. These results support the possibility suggested in your paper that the 4.1 cm alpha-particles observed by you are due to boron.

"Dr. McMillan has been studying gamma radiation from various substances and finds among other things that fluorine emits under proton bombardment, a five million volt monochromatic gamma radiation of

considerable intensity. Some day perhaps a short range group of alpha particles from fluorine will be found to account for this gamma radiation.

"But possibly the most interesting result that McMillan has found about this radiation is its absorption coefficient. He finds that the absorption per electron of the five million volt gamma radiation varies approximately linearly with atomic number, reaching a value for lead double that for oxygen. In other words, nuclear absorption (pair production presumably) is so great that in going from two and a half to five million volts the absorption coefficient in lead does not decrease a great deal.

"I am glad to hear that you are very well. You need not have told me that you are kept very busy in the laboratory, but I was very glad to hear that the government has given you a substantial grant of money for research and that you are responsible for its disbursement. Also your comparison of your early radio-activity days with the present is very much appreciated. I remember in the course of my graduate studies what a 'kick' I got out of reading of the early work on radio-activity, but I did not even hope at that time that I would have the opportunity to work in a similarly interesting new field of investigation. . . .

"Please tell Dr. Oliphant that I appreciated his letter very much and that I will be writing him directly before long."

Rutherford's brash young man learned very quickly, as Rutherford predicted he would. From that time onward, the contributions made to nuclear physics in the Radiation Laboratory were above reproach and of rapidly increasing importance, as the energy and intensity of beams available from the cyclotron increased. □

(*This is the first of two articles on Ernest Rutherford and Ernest Lawrence. The second will appear in the next issue.*)

Reference

1. E. L. Lawrence, E. M. McMillan, R. L. Thornton, Phys. Rev. **48**, 493 (1935).

---

* The evident confusion in nomenclature arose in this way. G. N. Lewis had proposed the name "deuton" for the nucleus of the atom of heavy hydrogen. Rutherford objected strongly to this, feeling that it would inevitably lead to confusion with neutron, especially in the spoken word. After discussion with his classical colleagues, he proposed the name "diplon," for the nucleus, and 'diplogen' for the atom, terms derived from Greek, and analogous to proton and hydrogen. The dual nomenclature was given up eventually, and the compromise "deuteron" and "deuterium" was accepted. It was said by one cynic that Ernest Rutherford was happy when his initials were inserted into deuton!

The Two Ernests—II

Sir Mark continues his personal recollections of Ernest Rutherford and Ernest Lawrence. By 1935 precise mass determinations with nuclear reactions were being made at Cavendish. In the following years Rutherford was arranging for new facilities at the laboratory. Meanwhile Lawrence began to use the cyclotron for medical research, learned to extract a beam from the accelerator and found a lot of unexpected radiation. Two years after Rutherford's death, the discovery of fission opened a new era.

PHYSICS TODAY / SEPTEMBER–OCTOBER 1966

by Mark L. Oliphant

BOTH ERNEST RUTHERFORD and Ernest Lawrence led great laboratories and inspired the physicists who worked in them. Rutherford was personally involved in almost all of the work at the Cavendish Laboratory, dominating the laboratory by his sheer greatness as a physicist and providing for his colleagues only the barest minimum of equipment. Lawrence, on the other hand, created at the Radiation Laboratory, the first of the very large laboratories in which massive and expensive equipment was designed, built and used for investigations into basic problems in physics in which he played little part, personally. After the discovery and successful development of the cyclotron at his laboratory, Lawrence enthusiastically offered his assistance in the construction of cyclotrons at laboratories elsewhere.

The two men did not meet until the Seventh Solvay Congress, October 1933. At the meeting, Lawrence defended his hypothesis that the "deuton" (deuteron) was unstable, breaking up in nuclear collisions into a proton and neutron. By May of the following year, however, Lawrence was convinced by experiments in the Cavendish Laboratory that what he had actually observed were reactions of deuterons with deuterons. From that time onward, the contributions of Lawrence's laboratory were above reproach and of rapidly increasing importance as the energy and intensity of the beams available from the cyclotron increased.

Accurate mass measurements

One of the early results of more accurate observations of the energies released in nuclear reactions involving the light elements was realization that the relative masses of the atoms, as given by the mass spectrograph, were not sufficiently reliable to give consistent agreement. In the Cavendish Laboratory, we naturally used the mass determinations made there by Francis W. Aston, whose improved mass spectrometer was then in operation. We came to the conclusion that there was an appreciable error in Aston's value for the mass ratio of hydrogen to helium, a basic determination upon which many of his other mass values depended. Aston was a touchy person and reacted with characteristic violence to the suggestion that there were systematic errors in his list of isotopic masses. On 4 May 1935, Rutherford wrote to Lawrence:

"You will no doubt have heard from Cockcroft and others about what is going on here. We have given a complete account of our beryllium results in the P.R.S. [*Proceedings of the Royal Society*] which appears this month, and you will see that we have put forward a scheme of masses to fit in—practically along the same lines that [Hans] Bethe has independently suggested in your country. At first, Aston took a high line about the accuracy of his results, and the impossibility of any serious error between helium and oxygen, but when I told

Sir Mark was assistant director of research at Cavendish until 1937, when he became director of the physics department at Univ. of Birmingham. In 1950 he became director of the Research School of Physical Sciences, Australian National University, Canberra. He served for three years as president of the Australian Academy of Sciences.

KEY FIGURES in development and early use of the 60-in. Crocker cyclotron stand beside the machine during construction. Only the magnet yoke and the coils have been completed. Left to right: Luis Alvarez, William Coolidge (who was visiting), William Brobeck, Donald Cooksey, Edwin McMillan and Ernest Lawrence.

struction of an improved design. Perhaps the most interesting result is that the focusing action of the electric and magnetic fields is so nearly perfect that we can get just as large current of deuterons at 4.5 MV as at 2.5 MV. At the present time the apparatus delivers several microamperes of deuterons having a range of 16.7 centimeters (about 4.5 MV). We have bombarded several substances, using these energetic deuterons, and it appears that almost the whole periodic table can be activated, the type of nuclear reaction involved being that in which the neutron of the deuteron is captured by the bombarded nucleus. We have found that gold can be activated in this way, a result which is very surprising. We shall do a good deal more work yet on these things before we can have confidence in the experimental results and theoretical interpretations.

"We were all very much surprised to hear that Chadwick is leaving you to be professor at Liverpool. I suppose it is a promotion for him, but I am sure that if I were he I would be very loathe to leave you and the Cavendish Laboratory."

Cyclotrons for medical research

This letter mentions again Lawrence's readiness to develop the medical applications of the cyclotron and its products in order to obtain the funds required for the work of his laboratory. However, his interest in possible medical applications was not only financial. His early ambition to become a doctor and the fact that his younger brother, John, had qualified in medicine and had become an instructor at Yale Medical School had kept his genuine interest in the healing art. In the summer of 1935, John, who had broken his leg, went to California to stay with Ernest while he recuperated. He did some experiments while there, with the aid of Paul Aebersold, a young colleague of Ernest. They exposed rats to neutrons and gamma rays from the cyclotron. On 13 Aug. 1935 Lawrence wrote a letter to Rutherford that I quote in full:

"Dear Professor Rutherford:

"I am very, very grateful to you for the photograph of yourself which

him that if he did not get to work, I was going to put forward the correct mass scheme, he rapidly started in, and found that he had dropped one or two bricks of reasonable magnitude! I am not quite sure he is right yet, but no doubt he may amend his results later. As a matter of fact, it is obviously very difficult for mass-spectrographic methods to give the same accuracy as from transformations when we are sure of the reaction."

In his reply, Lawrence wrote:

"Your very much appreciated letter was forwarded to me in New Haven, Connecticut, late in May: I was in the East about two months, engaged in my annual task of raising money for the support of our work in the radiation laboratory. I rather expected considerable difficulty in raising needed funds this year, and indeed was rather worried that we might have to restrict our work a great deal, but fortunately matters turned out otherwise. In this country

medical research receives generous support, and it was the possible medical applications of the artificial radioactive substances and neutron radiation that made it possible for me to obtain adequate financial support. We are now able to produce several millicuries activity of radiosodium. We are devoting a good deal of attention to the further development of the magnetic resonance accelerator for considerably larger currents and also higher voltages. It is reasonable to expect that it will not be very long before we will be producing ten times as much radioactive substance as at present. However, according to the medical people, at the present time we can provide enough radiosodium for beginning clinical investigations, and we have agreed to begin supplying the University Hospital here early this fall.

"We have lately been making various tests of the performance of our apparatus with a view to the con-

I shall always treasure very highly. In asking Cockcroft to get a photograph of yourself for me and ask you to autograph it, I had in mind that he could purchase one in a bookstore and perhaps persuade you to write your signature on it. I appreciate very much your kindness in sending me the portrait.

"Work is going along quite satisfactorily in our laboratory, although at the moment we are bothered with cathode ray punctures of the insulators of the magnetic resonance accelerator, the result of increasing the voltage and current output. My brother, who is on the faculty of the Yale Medical School, is vacationing here, and I persuaded him to undertake a preliminary investigation of the biological effect of neutrons. He has been exposing rats to neutrons for periods of time from ten minutes to three hours, and has been observing the changes produced in the blood of the rats. The first rat was exposed for a period of three hours, and as a result died, and subsequent experiments indicate that neutron rays are considerably more lethal biologically than x rays. The immediate result is that we are taking rather greater precautions in the matter of exposing ourselves in the course of our work in the laboratory.

"I am very glad to hear that you are well, and again I want to thank you ever so much for your picture.

"With best wishes and highest personal esteem, I am

Respectfully yours,"

John tells me that in fact the rat died of suffocation, being too completely confined! However, an important result was that much more stringent precautions against neutron and gamma radiation were then instituted in the Radiation Laboratory. From then till 1937, John Lawrence visited Berkeley regularly, at intervals of about three months, taking with him biological experiments to be carried out with the aid of the cyclotron. In 1937 he moved to Berkeley permanently to take charge of the medical work with a 60-in. cyclotron provided through the generosity of Crocker. Direct treatment of patients with the neutron beam from the cyclotron began in 1938, in collaboration with Robert Stone of the University of California Medical School in San Francisco. Lawrence had encouraged Sloan to design, and get into operation, an x-ray equipment for about 1 MV, using a resonant transformer in a vacuum, and Stone was using this in the hospital. The mother of Ernest and John was treated for a malignant growth with this equipment by Stone in 1937, and the treatment was so successful that it reinforced the faith of the brothers in the possibility of developing still more effective uses of radiation in the treatment of cancer.

New equipment at Cavendish

A letter from Rutherford to Lawrence, of 22 Feb. 1936, contains the following passages:

"I was delighted to get your letter and to hear how your work is going on. I congratulate you on your success with your apparatus in getting high voltages and intense beams. The neutron photographs you sent me were certainly very impressive, and I can roughly estimate the strength of your artificial source of neutrons in terms of radium emanation.

"I was exceedingly interested to hear also that you [this work was done by John Livingood, under Lawrence's general direction] have been successful in producing radium E from bismuth—a great triumph for the new apparatus. I have a personal interest in this artificial product; for I do not know whether you know that I worked out the changes radium D-E-F long ago in Montreal, and showed that as the β rays decayed an α-ray product grew. The apparatus I used is now preserved in the Physical Laboratory in McGill. I shall be interested to hear the details of your experiments and how much radium E you manage to produce.

"I note what you say about the present stage of your apparatus. At present we are very busy transferring the apparatus from the Royal Society Mond Laboratory, and getting duplicates, and keeping the cryogenic work going as usual. We do not intend to get a duplicate of the big generator for producing strong magnetic fields, but have in view instead the installation of a large magnet for general purposes, and also probably for use as a cyclotron. We have not had time as yet to go into the matter, but I think probably Cockcroft will be writing to you soon to see whether you can give him any information of the best design of magnet to be used for the latter purpose.

"At present we are just beginning the new building for our high tension D.C. plant, and we hope with luck to reach 2 million volts positive and negative, and possibly higher, but no doubt we will find plenty of trouble before it is in working operation. We shall, of course, build up the component parts of the apparatus ourselves so as to keep down the expense.

"Aston will shortly be publishing the new values of the masses of the light elements obtained with his improved spectrograph, and these new values fit in very satisfactorily with transformation data, so that difficulty is removed. I have also heard from several sources that Bainbridge has also done very much the same thing with his new spectrograph, and it will be interesting to see how far these two independent sets of measurements agree. It will be an ultimate test of the accuracy of these two systems."

The reference to the Royal Society Mond Laboratory concerns equipment that had been provided for the work of Peter Kapitza, the Russian engineer-physicist who had joined the Cavendish Laboratory in 1921. He was in the habit of visiting Russia during the summer to see his old mother. In 1935 the Soviet government refused to allow him to return to Cambridge, but offered to buy his equipment from the university in order that he might continue his researches in Russia. With the able help of Cockcroft and others, Rutherford proved himself a better man of business than expected, and negotiated a good price for the equipment. Meanwhile, Rutherford's resistance to the idea of as complex a piece of apparatus as a cyclotron in the Cavendish Laboratory had been worn down, and he was willing to devote part of the sum received from Russia to the acquisition of a large magnet which could be used, inter alia, for a cyclotron.

The reply by Lawrence was characteristic of his generosity towards all who wished to build a cyclotron:

"Thank you ever so much for your good letter. I should have known that you were responsible for the radium D-E-F, but I must confess that I didn't. As regards the yields of radium E by bombarding bismuth with five-million-volt deuterons, I must say that they are quite small. If I remember correctly, several hours bombardment with several microamperes gives, after a few weeks, something like thirty alpha-particles count per minute when the bismuth target is placed near the ionization chamber of the linear amplifier. Measurements on the range distribution of the alpha particles from the bismuth indicate that the transmutation function is exceedingly steep (for nearly all of the alpha particles have very near the full polonium alpha-particle range). It is probable, therefore, that at six million volts, which is the voltage we are now using, the radium E and polonium yield should be very much greater; and doubtless in the near future Dr. Livingood will continue experiments at this higher voltage.

"We have recently made some alterations of the cyclotron which have made it possible to withdraw the beam completely from the vacuum chamber through a thin platinum window out into the air, and I assure you that we have got quite a thrill out of seeing the beam of six-million-volt deuterons making a blue streak through the air for a distance of more than twenty-eight centimeters. Our purpose in bringing the beam out and away from the cyclotron chamber is twofold: partly to make it convenient to carry on scattering experiments, and partly to bring the beam to a target at a considerable distance from the vacuum chamber in order to get rid of the annoying neutron background produced by the circulating ions in the chamber striking various parts of the ac-

celerating system. With this latest improvement in the design of the cyclotron, I think now we have an apparatus which closely approximates one's desires.

"I believe in my last letter I mentioned that we have been carrying on experiments on the biological action of neutron rays. During the past two months such biological matters have taken a good share of my attention, because I feel that such matters, as well as nuclear physics, are of great importance. My brother, Dr. John H. Lawrence of the medical faculty of Yale University, has been out here studying the effects of neutrons on a certain malignant tumor called 'mouse sarcoma 180.' He has compared the lethal effect of neutrons and x rays on the tumor and on healthy mice and has very impressive evidence that this malignant tumor is relatively much more sensitive to neutron radiation than to x-radiation. If this is generally true for malignant tumors, we have here a very important possibility for cancer therapy. I am sure that it will not be long before neutrons will be used in the treatment of human cancer. . . .

"I was interested to hear that you are beginning the new building for your two-million volt D.C. plant and that you are undertaking the construction of a large magnet.

"I received the letter from Cockcroft and in the next few days will be sending him detailed information.

"Several days ago I received an invitation to attend the meeting in September of the British Association for the Advancement of Science and I have written a tentative acceptance and I can arrange to be away from the laboratory at that time. I should like very much to come to England to spend two weeks. In the event that you should decide to build a cyclotron, it is possible that I could be helpful by going over in detail with you matters of design."

Unfortunately, the design of the cyclotron for the Cavendish Laboratory, and its brother for Chadwick, in Liverpool, did not follow the lines

JOHN LAWRENCE who used cyclotron for medical research, with Ernest, 1927.

ERNEST RUTHERFORD, by Birley.

developed in Berkeley. It was entrusted to a large electrical engineering firm, with no previous experience, while funds were too restricted to enable the magnets to be as large as was desirable. Much trouble was experienced with them, and they never performed as efficiently as the virtual copy of the 60-in. Crocker cyclotron built by us in Birmingham. However, they did useful work, and established the technique in Britain.

Biology and beam extraction

Lawrence wrote to Rutherford on 24 Nov. 1936:

"I had intended writing you some time ago regarding Dr. R. [Ryokichi] Sagane, who has been with us the past year and desired to spend this year in the Cavendish Laboratory. I am afraid that he has arrived, and therefore words in his behalf now are a bit late. However, I should like to say that we liked Sagane very much; he proved to be a self-reliant and competent experimenter and a congenial personality. I do hope that you will find him an agreeable person to have as a visitor in the Laboratory, for I know that he is very anxious to be with you and will profit a great deal by such a sojourn.

"All of us here are very busy with a number of things. In addition to the nuclear work, we are devoting a lot of attention to biological problems, as I feel that there is important work to be done in this direction as well as in nuclear physics. We are supplying various artificial radioactive substances to the chemists for investigations of chemical problems and to biologists, particularly physiologists, for use as tracers in biological processes. I do hope that in this way we shall be able to contribute to the elucidation of some biological questions. We are also investigating quite extensively the biological effects produced by neutrons. I think we can say pretty definitely now that neutrons do not parallel x rays in their biological action. Studies of the comparative effects of x rays and neutrons will doubtless shed light on the mechanisms whereby ionization produces effects in bi-

NEWS OF HIS NOBEL PRIZE brings joy to Ernest Lawrence, 9 November 1939.

ological systems, and of course also there are the possibilities of effective medical therapy with neutrons.

"In some preliminary experiments on a mouse sarcoma, we got indications that neutrons had a greater selective action in killing this tumor than x rays. Under separate cover I am sending you a reprint of this work. This fall, similar experiments have been carried out upon a mouse mammary carcinoma with similar indications. In these more recent experiments, many more tumors and mice were irradiated with neutrons and x rays than in the first experiments on the sarcoma, and the new data also indicate a greater selective action of the neutrons on tumor tissue. It seems to me quite probable that neutrons will prove to be valuable in the treatment of cancer.

"We are this year undertaking the establishment of a new laboratory, which might be called a laboratory of medical physics. The organization and planning of the new laboratory is taking a good share of my time this year, but of course I am glad to do it, although I regret I cannot spend full days in the laboratory. Friends of the University have given funds for a new building and equipment, and I hope that by late next fall, experimental work in the new building will get under way. The architects have practically finished the building plans and we are engaged in designing the new cyclotron. Many of us are

having pleasure in planning the new apparatus; although doubtless we are deluding ourselves into thinking that the new outfit will be all that a good cyclotron should be.

"For certain experiments in progress we recently further modified our present cyclotron to bring the beam entirely out of the magnetic field, and we are finding the new arrangement one of great convenience for many experiments. I am enclosing a photograph of six microamperes of six million volt deuterons emerging into the air through a platinum window at the end of a tube six feet long. The beam is quite parallel and can be brought out considerably farther if so desired without undue loss of intensity.

"I have heard from several sources that you are very well and very busy—and in view of the latter, I can hardly expect a letter from you, although, needless to say, I should be greatly delighted if you should find time to write a few lines.

"Professor and Mrs. Bohr are coming to Berkeley in March and we all are looking forward to their visit. I wish it were possible to persuade you to visit America also."

Rutherford replied with characteristic enthusiasm for Lawrence's success:

"I got your letter a few days ago, and was very interested to hear of your latest developments in getting a beam of fast particles well outside the chamber. I congratulate you on your success in this difficult task, and I gather you are hopeful to get even stronger beams in this way. The photograph you have sent me is a beautiful one, and I would be very grateful if you would allow me to reproduce it in a lecture I am just publishing called 'Modern Alchemy,' which is an expansion of the Sidgwick Memorial Lecture I gave in Cambridge a few weeks ago. Unless I hear from you to the contrary, I will assume that you agree to this.

"Dr. Sagane visited us this term and he then decided to go for a short tour to Germany and Copenhagen, and is returning here in the New Year to begin some work. He seems a pleasant fellow, but he writes to me that he is finding a difficulty in seeing some of the German laboratories, as it is necessary to get a special permit from the Government to do so. This state of affairs in Nazi-land is rather amusing, and when some of our men from the Cavendish wished to visit Berlin to see Debye's laboratory, he wrote to Cockcroft that official permission would have to be granted by the Government before he could admit them!

"As to our own work, we are going ahead as usual. The new High Tension Laboratory is nearly completed and we hope to get a D.C. potential of 2 million volts going. We are also making arrangements to run one of your cyclotrons in due course.

"We celebrated J. J. Thomson's 80th birthday on December 18th by giving him a dinner and presentation in Trinity and also an address with signatures from many of the Cavendsh people. He is still very alert intellectually, and he was much moved by our little homely address.

"I wish you good luck in the development of your new laboratory and success in your experiments."

Cyclotron radiations

It was on 11 Feb. 1937 that Lawrence wrote again to Rutherford:

"I greatly appreciate your very interesting letter received some time ago. I know that you are extremely busy and it is very kind of you to write at such length.

"Your account of the state of affairs in Germany is almost unbelievable. One would think with such a scientific tradition the German people could not adopt such an absurd course of action in scientific affairs.

"The dinner to J.J. Thomson must have been a very nice occasion. It is certainly fine that he has such vigor at his ripe old age.

"I am glad to hear that your new high tension laboratory is coming along nicely and that you are also constructing a cyclotron. As I have written Cockcroft, if we can be of assistance in any way we should be only too glad. I have just heard that he is coming over for some lectures at Harvard and I have written him a letter inviting him to come out to see us. I do hope it will be possible for him to do so. I think it is possible that he might be saved some unnecessary beginning troubles by spending a few days in our laboratory operating our cyclotron. Also, in a month or so we shall have our new cyclotron chamber for the present magnet practically completed in the shop. This new outfit has quite a few improvements which Cockcroft would probably want to consider in his design.

"During the past few weeks we have been bombarding with 11 million volt alpha particles, studying the radioactivities produced. In addition to those already reported we have been findng many new activities, especially on up the periodic table. Also we have been making some absorption measurements of the radiation from the cyclotron and find that there is a very penetrating component. We do not know what it is yet, but the indications are that the penetrating radiation consists simply of very energetic neutrons. A 7 inch thickness of lead does not cut it to half. According to Oppenheimer theoretical considerations indicate that the mean free paths of neutrons vary as their energy. Hence it may be that the 14 MV neutrons from Be + 5 MV D^2 have mean free paths of more than 50 cms—something like the penetration of the radiation observed. We are continuing with the experiments with the endeavor to get the experimental facts as clear-cut and definite as possible, and I am sure when this is done we shall understand what is going on. Under separate cover I am sending you several reprints."

He followed this with a further letter of 24 Feb., having received some reprints of lectures given by Rutherford:

"Thank you very much for the reprints of the lectures, which I have already read with much pleasure and profit. The history that you tell about is certainly absorbing. Your discussion of the essential role played by the development of new methods and techniques in the advance of science appealed to me very much, as I have always held similar views, and of course your mention of the cyclotron in this connection was to me the highest compliment. Your lectures, which I regard as models for us younger men, have a quality in common with your great experimental works, that is to say, they go to the heart of the matter and bring out the essential points with beautiful simplicity. . . .

"We have been pursuing the investigation of the radiations from the cyclotron, and have pretty well satisfied ourselves that there is nothing extraordinary about the radiations excepting that it is an extremely difficult matter to screen out all the neutrons and the gamma rays from any particular region. We have now quite a lot of water around most of the cyclotron, but in spite of that Professor Lewis in the Chemistry building next door is not able to carry on his experiments with his sources of neutrons consisting of a mixture of beryllium with 200 milligrams of radium, and we find that at a distance of 300 feet from the cyclotron the mixture of neutrons and gamma rays from the cyclotron produce an easily detectable ionization. We are now planning to have the cyclotron in the new laboratory in a

basement room rather than at
ground level in order to cut down
the amount of radiation getting out
into surrounding laboratories. I am
afraid that you will find your new
cyclotron something of a nuisance
in this regard also."

It is clear that these two enthusiastic men were developing a considerable understanding and respect for
one another. Lawrence absorbed more
than he realized of the spirit of the
father of nuclear physics, and he
was able to pass this on to others.
The center of gravity of the study
of the nucleus was already moving
across the Atlantic to the United
States, a move which was to become
almost complete by the end of the
second world war. Rutherford was to
write only once more, in reply to the
following invitation from Lawrence:

Invitation to Charter Day

"I have just been talking with
the President of the University, who
has asked me to write you informally as to whether there would
be any possibility that you might
be willing to come over here to
give a Charter Day address next
March or a year later.

"Charter Day here is regarded
as a very important occasion and
the speaker at the exercise is always someone of great distinction.
President [Robert] Sproul is aware
that you may be very reluctant to
come, but is most anxious to persuade you to do so, since he ap-
preciates your eminence, not only
with respect to your scientific contributions but also with respect to
your general scientific statesman-
ship and world-wide good influence.
I do hope you will entertain
thoughts of coming over, as quite
aside from the Charter Day exercises, all of us in the laboratory
would gain so much from your
visit, even though it were very
brief. Needless to say we would do
everything we could to make your
stay with us pleasant.

"The President is anxious to know
whether there is a possibility that
you will come, and so if it is not
too much trouble, I should appreciate a note from you at your early
convenience. In case you should

MARK OLIPHANT AND ERNEST LAWRENCE stand before 184-in. cyclotron, 1941.

consider coming, it would be helpful if you would give me some
informal indication of a suitable
financial arrangement which I could
transmit to President Sproul, as I
know it is customary to provide
a proper honorarium. . . .

"We enjoyed very much Cockcroft's
visit, brief though it was. I need not
describe here what we did when he
was with us, as doubtless he has
given you a complete report. . . .

"Hoping to hear from you soon
and again hoping that you will actually entertain thoughts of coming
over next March, and with highest
personal esteem, I am

Respectfully yours,"
Rutherford answered:

"I have just received your letter,
asking me whether I could visit
California next March, in order to
be present at your Charter Day
Exercises.

"Please convey my thanks to your
President for his very kind suggestion and invitation. I write, however, to let you know at once that
there is no possibility of a visit next
year, as I have already arranged

SKETCH of Ernest Rutherford in 1928.

to go to India in November and preside over a joint meeting of the British Assocation and the Indian Association of Science, in January, 1938. I shall not return until February, and I shall find great arrears of work to attend to. At this stage, I cannot make any promises about the following year. I have so many calls on my time, that it is difficult for me to make arrangements too far ahead. At the same time, I greatly appreciate the very kind invitation of the University and yourself. I should personally like to have the opportunity of visiting California again, and in particular of seeing something of the work of your laboratory. Cockcroft told me about his visit, and how kind you had been in helping him.

"We are now preparing the foundations for the cyclotron, which we hope will be ready for transmission to Cambridge in July.

"I am glad you were interested in the little book and the lectures I sent you.

         With best wishes,
                Yours sincerely,"

Lawrence was naturally disappointed that Rutherford could not accept the invitation to Berkeley, but wrote saying that he was glad that the possibility of a visit in the followng year was not ruled out.

Rutherford had looked forward with keen anticipation to the meeting in India. He believed implicitly in the British Commonwealth, and his political liberalism led to his welcoming the development of responsible self-government in India. He had had many Indian students and had known well that remarkable mathematical genius, Srinivasa Ramanujan, also a Fellow of Trinity College, who had died so young, leaving behind a series of intuitive mathematical theorems that intrigued the world of mathematics for the succeeding generation. He spent much time in preparing his presidential address for the occasion. This address contains two passages that are significant in the present context:

"It is imperative that the universities of India should be in a position not only to give sound theoretical and practical instruction in the various branches of science but,

what is more difficult, to select from the main body of scientific students those who are to be trained in the methods of research. It is from this relatively small group that we may expect to obtain the future leaders of research both for the universities and for the general research organisations. . . . This is a case where quality is more important than quantity, for experience has shown that the progress of science depends in no small degree on the emergence of men of outstanding capacity for scientific investigation and for stimulating and directing the work of others along fruitful lines. Leaders of this type are rare, but are essential for the success of research organisation. With inefficient leadership, it is as easy to waste money in research as in other branches of human activity. . . ."

Speaking of artificial radioactivity:

"As Fermi and his colleagues have shown, neutrons and particularly slow neutrons are extraordinarily effective in the formation of such radioactive bodies. On account of the absence of charge, the neutron enters freely into the nuclear structure of even the heaviest element and in many cases causes its transmutation. For example, a number of these radio-elements are produced when the heaviest two elements, uranium and thorium, are bombarded by slow neutrons. In the case of uranium, as Hahn and Meitner have shown, the radioactive bodies so formed break up in a succession of stages like the natural radioactive bodies, and give rise to a number of transuranic elements of higher atomic number than uranium (92). These radioactive elements have the chemical properties to be expected from the higher homologues of rhenium, osmium and iridium of atomic numbers 93, 94 and 95."

Rutherford's death

Rutherford was not destined to go to India. He had suffered for years from an umbilical hernia, to relieve which he wore a truss. On 14 Oct. 1937 he became unwell, and was sick enough in the night to be removed from his

home to a hospital next afternoon. An operation for Richter's hernia was performed at once, and the outlook appeared good. However, normal bowel movement was never reëstablished, and despite the efforts of his physicians, he died of intestinal paralysis and intoxication on 19 Oct. His great wish at the onset of his illness was to be well in time to fulfil his presidential task in India.

Cockcroft and I were in Italy, at the Galvani Celebrations when news of Rutherford's death reached us. We were very upset and sad. At the morning meeting on 20 Oct., before we left to return to England, Bohr, Rutherford's older student and colleague, who loved Rutherford as we did, spoke movingly of the great man. Afterwards, on 20 Dec., he wrote to Lawrence thanking him for the many kindnesses shown him, Mrs Bohr, and their son, on his recent visit to the Radiation Laboratory, and for his great help in the construction of the cyclotron in Copenhagen. His letter ended:

"When in spite of all this I have not written long before, it has, however, not least been due to the very sudden death of Rutherford which has caused, as you understand, so great upset among his friends. Only a few weeks before I attended his unforgetful dignified funeral in Westminster Abbey, I had visited him in Cambridge where he was as cheerful and enthusiastic over his work as ever. In some way it was the most beautiful end of his marvellous life, but at the same time it makes the feeling of loss ever so acute. Still, I know that the thought of Rutherford will be to you as to myself a lasting source of encouragement and inspiration and will be a close bond between all of us who admired and loved him."

To this Lawrence replied:

"Lord Rutherford's sudden passing . . . was a great shock and your remarks in your letter, which I appreciated so much, are very true. It is sad that Lord Rutherford could not have lived longer, but on the other hand we may rejoice in the memory of his great life. . . .

"These tragic events remind one that life is short and uncertain and that time is not to be wasted. I often think that, (perhaps more so now because of my mother's serious illness) that we know really so little about the biological processes, and we physicists should not pass by any opportunities to be of help in biological research, although perhaps our first inclination would be to devote ourselves to fundamental physical problems."

What happened to the neutron

Rutherford had predicted the existence of the neutron in his Bakerian Lecture to the Royal Society in June 1920. During the following years, sometimes with the aid of research students, he and Chadwick searched diligently for the particle which both were convinced was essential in the structure of the nucleus. Many experiments were made, and James Chadwick has given a charming personal account of these.[2] The elusive neutral particle was discovered by Chadwick in 1932, and its effectiveness as an agent producing nuclear transformations was established soon afterwards by Fermi and others. Rutherford was intrigued by the properties of the neutron, and in his last lecture, read posthumously by James Jeans at the joint congress in India, the passage that I have quoted shows how interested he was in the production by neutrons, in collision with uranium, of the transuranic elements of higher atomic number than any existing naturally on earth. He did not live to experience the excitement created by the discovery by Hahn and Strassmann in 1938, of the fission process, or the beautiful work of Otto Frisch and Lise Meitner, which established clearly that the uranium nucleus could indeed split into two parts when it absorbed a neutron. On 9 Feb. 1939 Lawrence wrote to Cockcroft: "We are having right now a considerable flurry of excitement following Hahn's announcement of the splitting of uranium."

He went on to say that within a day of reading about it in the newspapers, they had observed the heavy ionizing fragments produced in the fission of uranium, and had identified several radioactive species among them by chemical methods.

"We are trying to find out whether neutrons are generally given off in the splitting of uranium, and if so, prospects for useful nuclear energy become very real."

Lawrence was one of the few in the United States who rapidly appreciated the profound significance of the discovery of the fission process. In England the possibility that it had military significance was more quickly realized in particular by Frisch and Rudolf Peierls, and by Chadwick, who showed independently that a fast-neutron fission chain process in the uranium isotope of mass 235, leading to a super-explosion, was possible. In 1941 when I visited Lawrence again, the magnet for his giant cyclotron was being erected on the new site on a hillside above the campus of the university. We discussed the general problem, and in particular the methods that we had been considering in Britain for the separation of the isotopes of uranium. He was deeply impressed by the serious view of scientists in England that nuclear weapons were not only almost certainly possible, but that Germany might be working on the problem. Soon afterwards, he began his experiments upon the separation of the uranium isotope by means of the CALUTRON, a technique which we began to develop independently in my laboratory in Birmingham, using the magnet of the 60-in cyclotron, which was being built with the aid of information generously supplied by Lawrence during and after my visit to Berkeley in 1938. In 1943 this minor effort by us was abandoned in favor of coöperation with Lawrence, and under the arrangements for a joint attack on the problem of nuclear energy, made between the governments of our countries, we moved to Berkeley.

This is not the place to discuss subsequent events, in which Rutherford and his Cavendish Laboratory played no part. If he had lived, he would have rejoiced in the subsequent triumphs of Lawrence and his colleagues in the Radiation Laboratory. But he would have regretted that his nuclear atom had become of such practical importance that the main motives for the financial support of such work, in all countries, became other than the advance of knowledge of nature.

It was a great privilege to be the pupil and colleague of Rutherford, and to have known, and worked with that other Ernest who so ably took over the torch of nuclear physics from him, and carried it to further heights of achievement. Rutherford, the greater scientist, laid the foundations of modern physics. Lawrence, with his greater flair for technology and organization, showed how to build, on those foundations, the massive edifice of physics today. All who knew and worked with these great men shared deep respect for their genius. But they inspired more than that. The warmth of their natures, their generosity, and their simple, unassuming personalities, generated an abiding love that made our lives fuller and happier.

Acknowledgements

The author is grateful for the ready access given him to correspondence and papers in the Cambridge University Library and in the Lawrence Radiation Laboratory. He acknowledges the help given personally by Sir James Chadwick, Sir John Cockcroft, Mrs Molly Lawrence, John Lawrence, Robert Brode, Leonard Loeb, Raymond Birge, Edwin McMillan, Robert Thornton, Harold Fidler, Mrs Eleanor Davisson, Daniel Wilkes, and many others. Luis Alvarez suggested that the article be submitted for publication to PHYSICS TODAY.

The sketch of Rutherford by R. Schwabe is from a copy presented to the author by Lord Rutherford.

The portion of Birge's history of the Berkeley physics department covering the period 1868 to 1932 will be available in mimeograph form (in limited number) within the next few months.

(This is the second of two articles on Ernest Rutherford and Ernest Lawrence. The first appeared in the last issue.)

□

Reference

2. J. Chadwick, Ithaca 26 VIII, 2 IX (1962).

VAN VLECK AND MAGNETISM

Through his work on crystal field theory,
paramagnetism, resonance spectroscopy and quantum theory, one man
turned magnetism into a field too large for any one man.

PHILIP W. ANDERSON   PHYSICS TODAY / OCTOBER 1968

A FEW YEARS AGO there would have been little need to distinguish between the two halves of my title: John H. Van Vleck and magnetism, as a field of theoretical physics, were practically synonymous. Now the field has expanded so much that no one man can overwhelm all of its branches in the way Van did when my generation was being introduced to it. While I can safely assume that all physicists know some parts of Van's career in magnetism, I suspect few appreciate the whole of it.

John Hasbrouck Van Vleck is the son of the eminent mathematician Edward Burr Van Vleck. The only story about our Van that is not true is that the mathematics building at the University of Wisconsin is named after him: It is named for his father. Van's grandfather, John Monroe Van Vleck, was also an eminent mathematician: The observatory at Wesleyan University bears his name.

Our Van Vleck received his AB at Wisconsin in 1920 and his PhD at Harvard only two years later at the age of 23. He went to Minnesota in 1923, becoming a full professor in 1927. That same year he married Abigail Pearson. A year later he moved to his father's university, Wisconsin, as professor of theoretical physics; then in 1934 he returned to Harvard where he became a full professor the following year.

At Harvard during World War II he headed the theoretical group at the Radio Research Laboratory. After the war he became chairman of the physics department, serving until 1949. Then he became the first dean of engineering sciences and applied physics, a position he held until 1957. In 1951 he also became Hollis Professor of Mathematics and Natural Philosophy. He plans to retire next June.

Visiting professor

In the midst of this busy schedule, he found time to be a visiting lecturer on eight separate occasions, including the Eastman Chair at Oxford (one of the most universally envied positions in England, since it comes with a centrally heated house) and the Lorentz professorship at Leiden. He also has been a councillor and president of the American Physical Society and a vice-president of the American Academy of Sciences, the American Association for the Advancement of Science and the International Union of Pure and Applied Physics.

I shall not recite the full list of his remaining honors, merely noting that they are uniquely multinational: He is a foreign member of no less than five national academies. The Universities of Grenoble, Paris, Oxford and Nancy are among those that have awarded him honorary degrees; so is

Harvard, where he earned his real one. He was the first recipient of Case Institute's Michelson Award and the APS Langmuir Prize and last year received the National Medal of Science.

The reader may ask: "What has he done for us lately?" A great deal, it happens. For the past several years he has been working on the clathrate compounds in which the gas molecule is caught in a cavity or "cage" rather than chemically bonded, so that it can exhibit its free magnetic or other rotational behavior while conveniently trapped at the disposal of the experi-

Philip W. Anderson divides his time between a chair of theoretical physics at Cambridge and Bell Telephone Laboratories. A Harvard graduate and a member of the National Academy of Sciences, he works in solid-state physics and magnetism. This article is adapted from a talk he gave at a session honoring John H. Van Vleck during a Boston conference on magnetism last year.

menter. He also has been working on magnetism in the rare earths, among other things.

Enormous influence

My emphasis, however, is on the past: 'the enormous influence Van has had on the study of magnetism, viewed as an enterprise in the quantitative understanding of the real properties of magnetic materials in microscopic terms.

Van's first work was on optical spectra and dispersion relations in the old quantum theory, and his first book[1] is the most complete and elegant exposition of the old quantum theory ever produced. Unfortunately it was published in 1926, just as the new quantum theory appeared. This monumental piece of bad luck did not faze him; Van was already learning and using the new quantum theory as it came out, and his courses at Minnesota in that period are remembered by his students—

including among other notables Walker Bleakney and Walter Brattain—as the most scientifically exciting courses of their lives.

Apparently Van chose almost immediately to study electric and magnetic susceptibilities, an area in which the old quantum theory gave tantalizingly good results in many cases, and to see whether the new theory was definitely in better agreement. This point of view, though now unfamiliar to us, nonetheless gives his book[2] a sense of direction and cohesion that is very welcome. He did conclude that the new quantum theory is much better, incidentally. His ability to carry along an idea in each of three languages, classical and old and new quantum theory, is one of his greatest and most baffling strengths.

Bare bones fleshed out

It is a pleasant and fascinating task to go back and reread that book. One sees how even those basic ideas

originated by others are illuminated and their bare bones fleshed out by Van's special point of view. One of these is the "Heisenberg exchange Hamiltonian," so-called. It is true that Werner Heisenberg first pointed out the connection of statistics, electron exchange and ferromagnetism, and that P. A. M. Dirac introduced formally the connection between exchange and a dot product of spin operators, but really it is Van Vleck who is responsible for the Hamiltonian of the form $J \Sigma (S_i \cdot S_j)$ that we now use to describe magnetic insulators and who expanded this method into the Dirac–Van Vleck vector model, a method capable of treating the complicated coupling of the various angular-momentum vectors within atoms and molecules as well. Again, Felix Bloch's spin waves take on a much clearer form when discussed by Van.

Crystal field theory as introduced by Hans A. Bethe was mainly an abstruse exercise in group theory; as done by Van Vleck it took on the form we still use—an effective field acting again on those angular-momentum vectors, in this case the orbital angular momentum of d electrons. In that form it becomes delightfully clear why, in some cases, the orbital angular momentum cannot respond at all and the susceptibility is given by "spin only"—the ubiquitous idea of "quenching" orbital angular momentum.

Two things strike one in looking back at the book from today's point of view. Again and again the great developments of the next decade or so, in which Van himself often participated, were very specifically hinted at: Antiferromagnetism and the covalent explanation of crystal fields are two examples.

Van says he overlooked the possibility of an ordered state when the sign of the exchange-integral J is negative and so did not anticipate Louis Néel and Lev D. Landau by years in the theory of antiferromagnetism. As it is, the proper formal theory of this effect had to wait until he wrote it down in 1940, six or seven years after their rather obscure remarks.

Major contribution

The theory of crystal fields as originating from more-or-less weak covalent bonds of the magnetic ion to its ligands, now accepted as one of Van's most important contributions to magnetism (after some rather regrettable reversals), was foreshadowed in the

JOHN H. VAN VLECK, Hollis Professor of Mathematics and Natural Philosophy at Harvard, in a photo taken last year on his 68th birthday. He will retire next June.

THE ONLY AMERICAN at the Sixth Solvay Conference at Brussels in 1930 was Van Vleck. Here he is third from the right in the back row. Seated in front are, left to right, Théophile de Donder, Pieter Zeeman, Pierre Weiss, Arnold Sommerfeld, Marie Curie, Paul Langevin, Albert Einstein, Owen Richardson, Blas Cabrera, Niels Bohr and Wander de Haas. In back are E. Herzen, E. Henriot, Jules Verschaffelt, C. Manneback, Aimé Cotton, Jacques Errera, Otto Stern, Auguste Piccard, Walther Gerlach, Charles Darwin, P.A.M. Dirac, Hans Bauer, Peter Kapitza, Léon Brillouin, Hendrik Kramers, Peter Debye, Wolfgang Pauli, Jakov Dorfman, Van Vleck, Enrico Fermi and Werner Heisenberg.

book and very soon formally written down by Van Vleck, though it had to wait 15 years to be picked up again.

Of course, I have not mentioned many useful and important things that *are* in the book. Van Vleck paramagnetism is one. Another is that the book is not "dated;" it does not attempt or accept a wrong explanation for anything, but leaves subjects open for further ideas. A contribution deserving special mention is the only really clear exposition in existence of the meaning of Maxwell's equations in a medium, in terms of the actual atoms and molecules and the real microscopic electromagnetic fields.

A second look back at Van's career also came from my bookshelf: a Japanese reprint collection on the origins of the field of microwave and resonance spectroscopy. Virtually all of the basic papers that were not written by Van acknowledge his advice and contribution of ideas, and of course these we are using still. Van Vleck in his contacts with the Dutch low-temperature group was clearly the most important figure in understanding the nature of the relaxation of magnetic vectors, spin–spin and spin–lattice. He was the central figure in carrying these concepts into radiofrequency spectroscopy after the war.

His great papers applying Ivar Waller's moment method to spin–spin relaxation and pointing out the vital phenomenon of exchange narrowing were of great importance; so was his recognition of the importance to magnetism of the abstruse ideas of Kramers (time-reversal) degeneracy—that an odd number of electrons always has a free spin—and of the Jahn–Teller effect of distortion of the system in the presence of orbital degeneracy.

Information, stimulation

Most impressive during and after the war was his role as an information post and stimulant for this immensely important new field. To chat with him at a meeting in those days was to be interrupted by an endless parade of experimentalists asking for an idea or two on their latest results—and getting them.

Three separate efforts stand out in a very long career. All of them are relevant to today's events, and two of them are cases in which Van played a role he particularly likes: that of mediator between two valid points of view, emphasizing that common results of the two approaches indicated that in the end they would turn out to be compatible.

The first of these mediation efforts

occurred in the early days of what is now called "quantum chemistry." Seemingly, two ideas that earned Nobel Prizes many years apart—the Slater–Pauling valence-bond idea, now often called Heitler–London, and the Hund–Mulliken molecular-orbital concepts—conflicted in their explanations of the chemical bond. Van Vleck played a considerable role not only in emphasizing that this incompatibility need not be absolute, but in demonstrating that both schemes could be used to explain certain results. One of the most important of these is that carbon exhibits tetrahedral bonding: The only valid discussion of this vital fact to this day, in my opinion, was in his papers in the *Journal of Chemical Physics* during this period.

Results not incompatible

A second instance involved the series of papers and reviews he produced, from the late 1930's onward, emphasizing that the local spin and itinerant models of ferromagnetism need not be incompatible, and gave many similar results. In fact, with our present understanding of spin waves as collective excitations of the itinerant model and of the nature of spin phenomena in metals and insulators, this must be described as the only possible point of view. Either perfect itinerancy or perfect localization is very rare, and no magnetic phenomenon should be looked at exclusively from one point of view or the other. As Conyers Herring put it in a famous review, it is all a matter of how you mix your cocktails, but neither pure gin nor pure vermouth is very satisfactory.

A final contribution that crops up these days in such diverse areas as intergalactic hydroxyl-maser effects and satellite communications is Van Vleck's yeoman work over the years on molecular spectra in all their fascinating complication. It is his work on lambda doubling on which our knowledge of the hydroxyl spectrum is based, and it was his calculations on O_2 that explained the opacity of the atmosphere in certain millimeter-wavelength regions of the spectrum that otherwise would be ideal for satellite communications.

The teacher and the person

Then there is Van the teacher and Van the person. Almost all the stories about him are true: He does own a great collection of Japanese woodblock prints, inherited from his father, many acquired from Frank Lloyd Wright. He was for years the world's greatest expert on obscure railway timetables.

It is true that he once rode to Bell Laboratories from New York in the cab of the Phoebe Snow. I am particularly grateful for one of his rides: When I graduated from Harvard, Van took the Phoebe Snow to Bell Labs and talked them into taking a chance on me. It is also true that he has had two papers published in the *Annals of the National University of Tucuman* in Mexico.

Students challenged

Many of his students remember his habit of asking the class for a response —typical was the day when he came into class and started his lecture with: "A clever trick is what?" We also remember his group-theory course in which we learned most of the subject through a diabolical series of problems. He once wrote a problem on the blackboard as DOOCS and it took us a while to realize that he meant: "Do the molecule OCS."

The problems usually were short but had terribly long hints. We learned early that you should read the hint only after doing the problem, when it very likely would give you an entirely new insight in what you had done. But we ordinary mortals seldom were able to do the problems that way.

No one I have ever heard of came out of a Van Vleck course without having learned, usually having learned a great deal. The list of his students is so large and so eminent that it is hardly fair to pick out any suitable subset of names: Let me drop the names of, say, Robert Serber, John Bardeen and Harvey Brooks at random.

John Van Vleck laid the foundations in a field that has kept a generation of physicists busy. Many in the field owe the beginnings of their careers to him. This article is an attempt to show our appreciation.

AT MINNESOTA in 1925, Van Vleck (second from left in second row) is flanked on the left by Joseph Valasek and on the right by John Tate. Others in the picture include J. William Buchta (eighth from left in back row), Elmer Hutchinson (ninth from left, back row) and Walker Bleakney (tenth from left, back row).

AT WISCONSIN about 1929, Werner Heisenberg (first row, center) posed with the physics faculty. Van Vleck sits next to him at left. Also in picture are Leland Howarth (seventh from left, back row) and Albert Whitford (second from right, back row).

References

1. Van Vleck, *Quantum Principles and Line Spectra*, National Research Council, Washington (1926).
2. Van Vleck, *Theory of Electric and Magnetic Susceptibilities*, Clarendon Press, Oxford (1932). □

Alfred Lee Loomis— last great amateur of science

This multimillionaire banker, who for years led a double life, spending days on Wall Street and evenings and weekends in his private physics laboratory, became one of the most influential physicists of the century.

Luis W. Alvarez

PHYSICS TODAY / JANUARY 1983

The beginning of this century marked a profound change in the manner in which science was pursued. Before that time, most scientists were independently wealthy gentlemen who could afford to devote their lives to the search for scientific truth—Lord Cavendish, Charles Darwin, Count Rumford, and Lord Rayleigh come to mind. But after the turn of the century, university scientists found it possible to earn a living teaching students, while doing research "on the side." So the true amateur has almost disappeared— Alfred Loomis may well be remembered as the last of the great amateurs of science. He had distinguished careers as a lawyer, as an Army officer and as an investment banker before he turned his full energies to the pursuit of scientific knowledge, first in the field of physics and later as a biologist. By any measure that can be employed, he was one of the most influential physical scientists of this century:

▶ He was elected to the National Academy when he was 53 years old
▶ He received many honorary degrees from prestigious universities
▶ He played a crucial role as director of all NDRC-OSRD radar research in World War II.

Family background

Loomis was born in New York City on 4 November 1887. His father was Dr. Henry Patterson Loomis, a well-known physician and professor of clinical medicine at New York and Cornell medical colleges. His grandfather, for whom he was named, was a great nineteenth-century tuberculosis specialist. His maternal uncle was also a physician, as well as the father of Alfred Loomis' favorite cousin, Henry

Alfred Lee Loomis in his early seventies at the Rand Corporation circa 1963.

L. Stimson, who was Secretary of State under Herbert Hoover, and Secretary of War throughout World War II.

From Alfred Loomis' educational background, one would correctly judge that he came from a prosperous, but not exceedingly wealthy family. He attended St. Matthew's Military Academy in Tarrytown, New York, from the age of nine until he entered Andover at thirteen. His early interests were chess and magic; in both fields, he attained near-professional status. He was a child prodigy in chess, and could play two simultaneous blindfold games. He was an expert card and coin manipulator, and he also possessed a collection of magic apparatus of the kind used by stage magicians. On one of the family summer trips to Europe, young Alfred spent most of his money on a large box filled to the brim with folded paper flowers, each of which would spring into shape when released from a confined hiding place. His unhappiest moment came when a customs inspector, noting the protective manner in which the box was being held, insisted that it be opened—over the strong protests of its owner. It took a whole afternoon to retrieve all the flowers.

The story of the paper flowers is the only story of Loomis's childhood I can remember hearing from him. In the thirty-five years during which I knew him rather intimately, I never heard him mention the game of chess, and his homes contained not a single visible chessboard or set. (When I checked this point recently with Mrs. Loomis, she wrote, "Alfred kept a small chess set in a drawer by his chair and would use it, on and off, to relax from other intellectual pursuits. He preferred solving chess problems or inventing new ones to playing games with other people.")

He loved all intellectual challenges and most particularly, mathematical puzzles. He made a serious attempt to learn the Japanese game of Go, so that he could share more fully in the life of his son Farney, who was one of the best Go players in the US. But his chess background wasn't transferable to the quite different intricacies of Go, and he had to be content to collaborate with his son in their researches on the physiology of hydra. As he grew older his manual dexterity lessened, but he still enjoyed showing his sleight-of-hand tricks to the children of his friends and to his grandchildren—but never to adults.

It was characteristic of Loomis that he lived in the present, and not in the past the way so many members of his generation do. He apparently felt it

would sound as though he were bragging if he alluded to the great power he once wielded in the financial world when in the company of a university professor. In 1940, I casually asked him what he thought of Wendell Willkie, the Republican presidential candidate, and he said, "I guess I'll have to say I approve of him because I appointed him head of Commonwealth and Southern." Loomis was the major stockholder of that utility, so there was certainly an element of truth in his flip and very uncharacteristic remark. He was immediately and obviously embarrassed by what he had said, and it would be another twenty years before he made another reference to his financial career in my presence.

Loomis entered Yale in 1905, where he excelled in mathematics, but he was not interested enough in the formalities of science to enter Sheffield Scientific School. He took the standard gentlemen's courses in liberal arts, and without giving much thought to his career, felt he would probably engage in some kind of scientific work after he graduated. But one afternoon, a close friend came to him for advice on choosing a career. Loomis strongly urged him to go to law school, pointing out that a broad knowledge of the law was a wonderful springboard to a

variety of careers: In addition to formal legal work, a lawyer was well prepared for careers in business, politics, or government administration. Loomis was so impressed by the arguments he marshaled for his friend that he enrolled in Harvard Law School. He never regretted that decision, because it gave him a breadth of vision that he applied to many fields.

In his senior year at Yale, he was secretary of his class, but he had the time and the financial resources to pursue his life-long hobby of "gadgeteering." His extracurricular activities involved technical matters such as the building of gliders, model airplanes, and radio-controlled automobiles. He was fascinated by artillery weapons, and we shall learn that the great store of information he accumulated in that field played a crucial role in changing the major focus in his life from business to the world of science. A glider he built and tested from the dunes near his summer home at East Hampton stayed in the air several minutes. It was obvious to his friends that he was distinguished by a wide-ranging mind

R. W. Wood became a close friend of Loomis in the 1920s and served in effect as his private tutor in physics. (Photo courtesy of the AIP Niels Bohr Library.)

Luis W. Alvarez is professor emeritus in the department of physics at the University of California, Berkeley, California.

and the ability to learn all about a completely new field in a remarkably short time through independent reading. That facet of his personality and intellect was the most immutable throughout his life—a life that would be characterized by periodic and major changes of interest.

Loomis's decision to become a lawyer was certainly influenced by his cousin, Henry Stimson, in whose firm of Winthrop and Stimson he was assured a clerkship. But after his distinguished performance at Harvard Law School—where he was in the "top ten," helped edit the *Harvard Law Review*, and graduated *cum laude* in 1912—he would have been welcomed in any New York law firm. As one would guess from his later interests, he specialized in corporate law and finances.

Early career

Loomis's career as a young corporation lawyer was interrupted by World War I. When he joined the Army, his fellow officers were surprised to learn that he knew much more about modern field artillery than anyone they had ever met. He had made good use of the special communication channels available to Wall Street lawyers, and had accumulated a vast store of up-to-the-minute data on the latest ordnance equipment available to the warring European powers. His expertise in such matters led to his assignment to the Aberdeen Proving Grounds, where he was soon put in charge of experimental research on exterior ballistics, with the rank of major. At Aberdeen, he was thrown into daily contact with some of the best physicists and astronomers of this country, and he and they benefited from each other's talents.

In those days, before photoelectric cells and radar sets came to the aid of exterior ballisticians, there was no convenient way to measure the velocity of shells fired from large guns. Loomis invented the Aberdeen Chronograph, which satisfied that need for many years after its invention. It is hard for someone like me, who came into a scene long after an ingenious device had been invented, and later supplanted, to appreciate what made that device so special. But the fact that Loomis singled out the Aberdeen Chronograph for mention in his entries in *Who's Who* and *American Men and Women of Science*, and mentioned it on a number of occasions in conversations with me, makes me believe that it must have been a remarkably successful and important invention. Loomis set such high standards for his own performance that no other interpretation of the value of the Aberdeen Chronograph would be consistent with his pride in it.

One of the friends Loomis made at Aberdeen was Robert W. Wood, who was considered by many to be the most brilliant American experimental physicist then alive. They had known each other casually from the circumstance that each of their families had summer homes at East Hampton, on Long Island. But at Aberdeen, they initiated a symbiotic relationship that lasted many years. Wood became, in effect, Loomis's private tutor, and he responded by becoming Wood's scientific patron. The following paragraphs from Wood's biography, tell of this relationship better than anyone of the present era could:[1]

It was a consequence of Wood's scientific zest and social strenuousness that fate brought him, about this time, the facilities of a great private laboratory backed by a great private fortune. He had met Alfred Loomis during the war at the Aberdeen Proving Grounds, and later they became neighbors on Long Island. Loomis was a multimillionaire New York banker whose lifelong hobby had been physics and chemistry. Loomis was an *amateur* in the original French sense of the word, for which there is no English equivalent. During the war, he had invented the "Loomis Chronograph" for measuring the velocity of shells. Their relationship, resulting in the equipment of a princely private laboratory at Tuxedo Park, was a grand thing for them both.

A happy collaboration began, which came to its full flower in 1924. Here is Wood's story of what happened.

"Loomis was visiting his aunts at East Hampton and called on me one afternoon, while I was at work with something or other in my barn laboratory. We had a long talk and swapped stories of what we had seen or heard of science in warfare. Then we got onto the subject of postwar research, and after that he was in the habit of dropping in for a talk almost every afternoon, evidently finding the atmosphere of the old barn more interesting if less refreshing than that of the beach and the country club.

"One day he suggested that if I contemplated any research we might do together which required more money than the budget of the physics department could supply, he would like to underwrite it. I told him about Langevin's experiments with supersonics [what is now called "ultrasonics"] during the war and the killing of fish at the Toulon Arsenal. It offered a wide field for research in physics, chemistry, and biology, as Langevin had studied only the high-frequency waves as a means of submarine detection. Loomis was enthusiastic, and we made a trip to the research laboratory of General Electric to discuss it with Whitney and Hull.

"The resulting apparatus was built at Schenectady and installed at first in a large room in Loomis' garage at Tuxedo Park, New York, where we worked together, killing fish and mice, and trying to find out whether the waves destroyed tissue or acted on the nerves or what.

"As the scope of the work expanded we were pressed for room in the garage and Mr. Loomis purchased the Spencer Trask house, a huge stone mansion with a tower, like an English country house, perched on the summit of one of the foothills of the Ramapo Mountains in Tuxedo Park. This he transformed into a private laboratory deluxe, with rooms for guests or collaborators, a complete machine shop with mechanic and a dozen or more research rooms large and small. I moved my forty-foot spectrograph from East Hampton and installed it in the basement of the laboratory so that I could continue my spectroscopic work in a better environment..."

Loomis, who was anxious to meet some of the celebrated European physicists and visit their laboratories, asked Wood to go abroad with him. They made two trips together, one in the summer of 1926, the other in 1928.

Business career

After World War I, Loomis formed a lifelong business partnership with Landon K. Thorne, his sister Julia's husband: In the thirty-five years I was so personally close to Loomis, I met Thorne on only two occasions. Loomis kept his business friends and his scientific friends quite separate. For a long time, he apparently reasoned that while his broad range of interests made both groups exceedingly interesting to him, the two disparate groups might not feel about each other as he did about them. As he grew older, Loomis's personal ties to the scientific world became the dominant ones, and I find that his last entry in *Who's Who in America* lists his occupation simply as "Physicist."

Loomis was proud of the fact that he and Thorne were in many kinds of business deals, and in every one of them, they were equal partners. First of all, they had equal shares in the very profitable Bonbright and Co., the investment banking firm of which Landon was the president, and Loomis the vice-president. This firm was instrumental in putting together and financing many of the largest public utilities in the country.

The two partners also built a very innovative racing sloop of the J-class, which they hoped would win the right to race against Sir Thomas Lipton in one of his periodic attempts to capture the America's Cup from the New York Yacht Club. To cut down on wind resistance, the partners arranged to have most of the crew below decks at all times, working levers in the fashion of galley slaves, rather than hauling on wet lines on the deck. With the help of the MIT naval architecture department, they did a thorough study of hull shapes, and there were several changes in the location of the mast—made of strongest and lightest aluminum alloy—during the test program. But in spite of all these efforts, *Whirlwind* wasn't a success. Perhaps the best indicator of Loomis's financial state at that time is that J-boats were then almost always built by "syndicates" of wealthy men such as the Vanderbilts. But to have complete control of their J-boat, Loomis and Landon paid for the whole project, 50–50 as always. After World War II, J-boats became too expensive even for syndicates of rich men, so the America's Cup races are now sailed in the smaller "12-meter" boats.

Another of Loomis and Thorne's partnership was the ownership of Hilton Head, an island off the coast of South Carolina. Hilton Head is now a famous resort area, with luxurious hotels and golf courses. But when Loomis and Landon owned it, it was completely rustic. They used it only for riding and hunting, and invited their friends to share the beauties of the place with them. They also owned a large oceangoing steam yacht, which they donated to the Navy at the start of World War II. I can count on the fingers of one hand the number of times I've seen Loomis's name in the public press—he believed that the ideal life was one of "prosperous anonymity." The first time I saw his name in print was when *Time* identified him as a "dollar-a-yacht man," one of several who had given their yachts to the Navy in return for a dollar. Recently, I've found in the library two old articles about Loomis. The first was a popular article on the unusual J-boat and its owners. The second was an article in the very first issue of *Fortune* concerning Wall Street firms, and telling of the great success of Bonbright and Co., its well-known president, Landon Thorne, and its shadowy and brilliant vice-president, Alfred Loomis, who kept in the background and planned their financial coups. According to the article, "Bonbright . . . rose in the twenties

from near bankruptcy to a status as the leading US investment-banking house specializing in public-utility securities."

Tuxedo Park laboratory

When the *Fortune* article appeared, Loomis was leading a double life; his days were spent on Wall Street, but his evenings and weekends were devoted to his hilltop laboratory in the huge stone castle in Tuxedo Park. The laboratory was abandoned in November 1940, so those who worked in it could join the newly established MIT Radiation Laboratory that Loomis was instrumental in founding, and which reported directly to him, in his wartime role as head of the NDRC's Radar Division. I arrived at MIT at the same time, so I learned much about the Tuxedo Park laboratory from the young scientists and engineers who had worked there throughout the year, and from the former laboratory manager, P. H. Miller. The following account of a laboratory I never visited is based on those recollections, and on stories I heard from older physicists who had been Loomis's guests during summers at Tuxedo, and finally on the countless reminiscences of Loomis and other members of his family.

Because of Wood's strong influence, the laboratory concentrated at first on problems that interested him. As the quotations from his biography tell, the first major work was in ultrasonics. Loomis and Wood are still mentioned in the introductory chapters of textbooks on ultrasonics and sometimes referred to as the "fathers of ultrasonics." The field has grown enormously since they did their pioneering work, and it now has practical applications in industrial cleaning, emulsifying, and most recently in medical imaging, in place of x rays when the required moving pictures would involve excessive radiation doses. Imaging ultrasonic scanners are now in common use to watch the motion of heart valves, to observe fetuses, and at the highest frequencies, they serve as high resolution microscopes.

A bound volume of the "Loomis Laboratory Publications" (1927–1937) includes reprints of sixty-six scientific papers, of which twenty-one were on ultrasonics; Loomis was a co-author of the first four, and of four later ones. The first is the classic 1927 paper by Wood and Loomis, some of whose results are described by Wood in the quotation above.

The laboratory was well equipped for work in Wood's specialty of optical spectroscopy. Ten papers in this field came from the laboratory, including one by Loomis and George B. Kistiakowsky entitled "A Large Grating Spectrograph," which illustrates Loo-

mis' talents as an innovative designer of precision mechanical devices. None of the spectroscopic papers bear his name; it wasn't in his nature to publish in a mature field. Although Loomis admired those who could do the involved spectroscopic analyses that came from his laboratory, he preferred to do the pioneering work in some new field. His admiration for the real professionals of this era is shown by the fact that he arranged a series of conferences in honor of visiting European physicists. Guests at the conferences were transported to Tuxedo Park in a private train, and entertained in lavish style at the laboratory. The *Journal of the Franklin Institute*, in the issue of April 1928, has a sixty-five page section entitled "Papers Read at a Conference in Honor of Professor [James] Franck, at the Loomis Laboratories, Tuxedo, New York, January 6, 1928." Included are papers by Franck, Wood, Karl Taylor Compton, and several others.

I have no records of the other conferences, but Loomis once showed me the guest book from the laboratory. (It had just been returned to him by his son, Farney, when the latter had closed his "Loomis Laboratory" to join the Brandeis University faculty.) The book showed the names of most of the well-known American and European physicists of the period. On some occasions, a page with many famous names would be headed by the name and the man in whose honor the group had assembled. Most often such an honored guest was a visiting European physicist, for example, Einstein, Bohr, Heisenberg, or Franck.

Loomis's main interest at that time was in accurate time-keeping. The following quotation from Wood's biography will serve to introduce that subject:[2]

Wood's second trip abroad with Alfred Loomis was made in 1928. They called first on Sir Oliver Lodge, who presented each of them with an autographed copy of his latest book, *Evidence of Immortality* . . .

One of the things Loomis hoped to obtain in England was an astronomical "Shortt clock," a new instrument for improving accuracy in measurement of time. It had a "free pendulum" swinging in a vacuum in an enormous glass cylinder—and was so expensive that only five of the big, endowed observatories yet possessed one. Says Wood:

"I took Loomis to Mr. Hoke-Jones, who made the clocks. His workshop was reached by climbing a dusty staircase, and there was little or no machinery in sight, but one of the wonderful clocks was standing in the corner, almost

Ernest O. Lawrence and Loomis developed quick friendship when they first met in 1939 at Berkeley. Loomis helped obtain backing of scientific establishment for Lawrence's 184-inch cyclotron and $2.5 million funding from the Rockefeller Foundation. (Courtesy of Watson Davis, Science Service.)

placed at the corners of an equilateral triangle, facing inward, and the coupling was broken.

The Bell Telephone Laboratories had at the time been developing quartz crystal oscillators with low temperature coefficients, and they came to surpass the Shortt clocks for short-term accuracy, but not for periods greater than a day. Loomis had a private line installed to carry the Bell oscillator signals to his horological laboratory, and he designed an ingenious chronograph to compare the timekeeping abilities of the Shortt pendulum clocks with the quartz oscillators. Because the first of these types was sensitive to gravity but the second was not, Loomis used his chronograph to demonstrate the expected but previously undetected effect of the moon on pendulum clocks. Loomis accumulated the observational data himself, but the data analysis required the services of a battery of "computers"— women who operated desk-top computing machines, and whose salaries were paid by Loomis. The results of the analysis were published by Ernest W. Brown and Dirk Brouwer in a paper immediately following Loomis's "The Precise Measurement of Time," in the *Monthly Notices of the Royal Astronomical Society*, March 1931.

Loomis published several papers on biology and physiology with E. Newton Harvey and Ronald V. Christie. I never heard him speak of the physiological work, but he was obviously proud of the microscope–centrifuge he developed with Harvey. This was typical Loomis "gadget" of the kind he enjoyed building all his life. The device made is possible for a biologist to watch for the first time the deformation of cells under high "g-forces." As Harvey and Loomis said in the introduction to their first paper on the subject.[3]

The previous procedure has been to centrifuge the cell in a capillary tube, remove it from the tube and observe it under a microscope to determine what happens. It would obviously be far better to observe the effect of centrifugal force while the force was acting.... Our communication describes a practical means of attaining this end.

In typical Loomis fashion, his name appears on only the first of thirteen papers on the microscope–centrifuge that are to be found in the collected reprints of the laboratory.

In the mid-thirties, Loomis turned his attention to the newly discovered brain waves. Hans Berger had published his observations in the German literature, but American physiologists were unable to duplicate his results, and most of them apparently doubted the existence of the very low voltage signals that Berger described. From

completed, which made the total production to date six. Mr. Loomis asked casually what the price of the clock was, and on being told that it was two hundred and forty pounds (about $1200), said casually. 'That's very nice. I'll take three,' Mr Jones leaned forward, as if he had not heard, and said, 'I beg your pardon?' 'I am ordering three,' replied Mr. Loomis. 'When can you have them finished? I'll write you a check in payment for the first clock now.'

Mr. Jones, who up to then had the expression of one who thinks he is conversing with a maniac, became apologetic. 'Oh no,' he said, 'I couldn't think of having you do that, sir. Later on, when we make the delivery, will be quite time enough.' But Loomis handed him the check nevertheless."

Back in America, they learned that Professor James Franck, Nobel prize winner, was coming over in January to give lectures at various universities. Wood suggested to Loomis that he hold a congress of physicists in his Tuxedo Park labora-

tory in Franck's honor. Franck accepted and the meeting was held in the library, a room of cathedral-like proportions, with stained-glass windows. Franck gave his first lecture in America there; Wood, Loomis, and others made subsequent addresses. The visiting American physicists were conducted through the laboratory and shown the supersonic and other experiments. The congress in this palace of science proved such a success that it was repeated the following year.

Loomis's interest in accurate time-keeping probably resulted from his seagoing background, and his fascination with the art and science of navigation. He installed the three Shortt clocks on separate brick piers that were isolated from the laboratory structure, and extended down to bedrock. He was surprised to find that the clocks beat for long times in exact synchronism, and thought at first that they were locked together by gravitational interactions between the pendula. But he found that the coupling was through the bedrock, so the clocks were then

his contacts with industry, Loomis had available the best amplifiers, and he did his work inside "a screen cage," to eliminate interfering electrical noise. He had by this time retired from his Wall Street firm, and was devoting his full attention to his scientific work. For this reason, his name appears on all of the laboratory papers on brain waves, many of which were of great importance. His work erased any lingering doubts concerning the value of Berger's discovery; electroencephalograms are now used routinely in the diagnosis of epilepsy and many other diseases. In fact, one finds advertisements in magazines for "bio-feedback devices" that let the user observe his Berger "alpha waves," and learn to control them, "leading to greater creativity." (In kit form, $34.95.)

Loomis and his coworkers investigated many aspects of brain waves and did particularly important work with sleeping subjects that involved the abrupt changes in the character of the waves as the subject underwent "quantum jumps" in his "depth of sleep." It was then possible to tell precisely when a subject dropped from one of five states of sleep from which he could instantly be awakened by a small disturbing noise, into one in which he would fail to respond to the loudest noises that Loomis's high-fidelity amplifiers could produce. (Loomis was one of the first "hi-fi buffs"; his homes were always filled to overflowing with a changing parade of the latest and most advanced high-fidelity sound-reproducing equipment. Avery Fisher and Loomis were personally close, and on at least one occasion, Fisher improved his superb product line with an idea that

Loomis had devised to make the fidelity even higher.)

The only formal scientific talk I ever heard Loomis give was the the weekly physics-department colloquium in Berkeley, in 1939. He described his important brain-wave experiments on sleeping, hypnotized, and blind subjects.

In 1939, Loomis's scientific interests changed drastically. He became deeply involved in Ernest Lawrence's projects and he shifted the emphasis of his own laboratory from pure science to war-related technology, by starting the construction of a microwave radar system to detect airplanes. The Sperry Gyroscope Company had brought an interest in the klystron patents that were owned by the Varian brothers, who invented the klystron, and Stanford University, which had supported the development work. Sperry built a small klystron plant in San Carlos, near Stanford, and their first customer was Alfred Loomis, who appeared, checkbook in hand, as he had years before at the small plant making Shortt clocks.

Work with Lawrence

I was not surprised to meet Loomis in Berkeley, on his first visit to the Radiation Laboratory, in 1939. Francis Jenkins of the Berkeley physics department had spent a summer at Tuxedo as Loomis's guest, and he had told me in wide-eyed amazement about the fantastic laboratory at Tuxedo Park, and about the mysterious millionaire-physicist who owned it. Everyone who had submitted an article to the *Physical Review* in the depression years had received a bill for page charges togeth-

er with a note saying that in the event the author or his institution was unable to pay the charges, they would be paid by an "anonymous friend" of the American Physical Society. There was of course no way to break the veil of secrecy surrounding the "anonymous friend," but Jenkins told me in confidence that he felt sure that Loomis was the Society's benefactor. (That was a correct surmise.) Jenkins told me that Loomis was a wonderful person, but he didn't like the other residents of Tuxedo Park. He thought they were too "snooty," and looked down on the scientists as barbarians who "didn't even dress for dinner."

The relationship that quickly developed between Loomis and Ernest Lawrence had all the earmarks of a "perfect marriage": They were completely compatible in every sense of the word, and their backgrounds and talents complemented each other almost exactly. Lawrence was a country boy from South Dakota and the first faculty member of a state university to win a Nobel prize. He had developed an entirely new way of doing what came to be called "big science," and that development stemmed from his ebullient nature plus his scientific insight and his charisma; he was more the natural leader than any man I've met. These characteristics attracted Loomis to him, and Loomis in turn introduced Lawrence to worlds he had never known before, and found equally fascinating. Anyone who was in their company from 1940 until Lawrence died in 1958 would have thought that they were lifelong intimate friends with all manner of shared experiences going back to childhood.

Meeting of top physicists in 1940 to discuss plans for 184-inch cyclotron included (left to right) Ernest O. Lawrence, Arthur H. Compton, Vannevar Bush, James B. Conant, Karl T. Compton and Loomis. (Courtesy of Lawrence Radiation Laboratory.)

I was impressed by the way Loomis would seek out the younger members of the laboratory to learn everything he could about us and what we were doing and planning to do in our next round of experiments. I had never before had any serious discussions of physics with anyone as old as Loomis, and I was pleased that he liked to visit with me after I had taught a freshman class and was sitting out my required "office hour"—waiting to talk with the students who seldom came by. We talked a lot about physics, and found we were *simpatico*. He taught me an important lesson that I have put to good use in my life: The only way a man can stay active as a scientist as he grows older is to keep his communication channels open to the youngest generation—the front-line soldiers.

Although Loomis's real mission in coming to Berkeley was to help Lawrence raise the funds to build the 184-inch cyclotron, he also used the time to learn everything he could about cyclotron engineering and nuclear physics. I remember one occasion when I mentioned in passing that because of the war in Europe, the price of copper had risen to almost twice that of aluminum, for a given volume. Since aluminum had only 60 percent more specific resistivity than copper. I suggested to Loomis that aluminum might now be the preferred metal for the magnet windings of the 184-inch cyclotron. It seemed obvious to me, from elementary scaling laws, that an aluminum coil would be larger but would cost less. I had completely forgotten the suggestion, when a few days later, Loomis showed me a long set of calculations based on several altered designs of the 184-inch cyclotron that proved my snap judgment wrong. I came to appreciate for the first time the difference between the world of business, where a 20 percent decrease in cost was a major triumph, and the world of science, where nothing seemed worth doing unless it promised an improvement of a factor of ten. I hadn't done the calculations concerning the cyclotron cost because they obviously didn't permit a "large" savings in cost. But Loomis considered it worth a day or two of his time to see if he could cut the cost of the magnet windings by $50 000.

Lawrence once told me of spending some time with Loomis in New York, after the Rockefeller Foundation had allocated $2.5 million to build the 184-inch cyclotron. Earlier, Loomis had been instrumental in securing the virtually unanimous backing of the "scientific establishment" for the proposal, thus relieving the Rockefeller Foundation of any necessity for acting as a judge between factions competing for the largest funds ever given to any physics project. So after acting as a

senior statesman in the worlds of science and philanthropy, Loomis was ready to help Lawrence obtain the best possible bargains in the purchase of iron and copper for the giant cyclotron. Lawrence recalled that after spending some time with the Guggenheims, during which a favorable price for copper was negotiated, Loomis said, "Well, now we have to go after the iron. I think Ed Stettinius is the right man." (Stettinius was then Chairman of US Steel, and later Secretary of State.) Lawrence was impressed when a call was put through and Loomis said, "Hello Ed, this is Alfred, I have someone with me I think you'd like to meet. When can we come over?" They were soon in Stettinius' office, and shortly after Lawrence had given him a pitch on the great cyclotron, he and Loomis were in the latter's apartment celebrating their success with a drink.

Radar development

In early 1940, Loomis was back in Berkeley, and he told me that his next big project was to arrange for the funding of Enrico Fermi's embryonic plans to build a nuclear chain reactor. I hadn't given any thought to the problems involved in designing or building such a device, so everything Loomis told me was most interesting. But his involvement in reactors was cut short in the summer of 1940 by the dramatic appearance in Washington of the "Tizard Mission." The purpose of this group of visiting British scientists was to enlist the help of the United States in developing and building the new devices needed to meet the military requirements of a war that had become technologically oriented to a degree quite unappreciated by our military–industrial–scientific establishment. As an example, radar had been invented independently in the United States by the Navy and the Army, and in England by Robert Watson-Watt. The US military departments treated the subject with such excessive secrecy that no "outsiders" learned of it. Since the outsiders were the real professionals in radio engineering, they were the ones who could have developed American radar into the useful military tool that the insiders didn't manage to achieve. (The dismal state of US radar was demonstrated at Pearl Harbor, a year and a half after the Tizard Mission had revealed all the British successes to the US armed forces.)

The world now knows that the operational success of the long-wave British radar was the foundation on which the RAF triumphs of the Spitfire and Hurricane pilots were based. A second generation of VHF radar, in the 200-MHz (1.5-m) band, could be fitted into planes to turn them into night fighters and anti-submarine patrols. Everyone

agreed that microwave radar in the 3000-MHz (10-cm) band would be vastly superior to the 1.5-m equipment then available. But there appeared to be little chance that a powerful generator of such pulsed microwaves could be developed.

When John T. Randall and Henry Boot made their breakthrough with the cavity magnetron in Mark Oliphant's laboratory in Birmingham, it was suddenly clear that microwave radar was there for the asking, but Britain had no spare "bodies" who could be asked to do the development—everyone with applicable skills was working at breakneck speed on the immediate problems of a desperate war that could be lost any day by the starvation of the submarine-blockaded British people. So, in a great and successful gamble, Winston Churchill made the decision to share all of his country's technical secrets with the United States, in the hope that the potential gain would offset the loss in compromised security. Sir Henry Tizard was sent to Washington with a committee of experts, including such luminaries as Sir John Cockcroft, to brief their American counterparts on all aspects of the scientific war.

Loomis was included in the briefings not only because of his unique position in the scientific establishment, but because his laboratory had built one of the two microwave radar sets then existing in the United States. Both were based on the klystron tube recently invented by Russell and Sigurd Varian at Stanford University, and both were "continuous-wave Doppler radars" of the type now used by police departments to apprehend speeders. William Hansen, who designed the first of these microwave radar sets, attempted for the next few years to find a wartime niche for such a device, but without much success. Loomis immediately sensed the great superiority of the pulsed microwave radar devices that could be based on the new magnetron, so he dropped his work on the klystron-powered radar set, and devoted all his energies to pulsed microwaves for the next five years. But his klystron radar could detect planes, as he demonstrated to the "founding fathers" of the MIT Radiation Laboratory in the winter of 1940—in fact, it was the first working radar set that any of us had ever seen. But immediately after that demonstration, it was junked.

The Tizard Committee spent some time in Tuxedo Park as guests of Loomis, and on that occasion, he brought a number of friends, including Lawrence, into the newly formed Microwave Committee of the National Defense Research Committee, which had just been established by President Roosevelt on the advice of Vannevar Bush. Loomis was chairman of the

Committee, which took the responsibility for establishing the MIT Radiation Laboratory, one of the world's most successful scientific and engineering undertakings. Loomis made the arrangements with industry for equipping the laboratory with the necessary hardware to make several flyable night-fighter intercept radar sets, and Lawrence took the responsibility of staffing the laboratory, mostly with young nuclear physicists. (The Tizard Mission suggested this because the British had found nuclear physicists to be more quickly adaptable to a radically new set of "ground rules" than were professional radio engineers.) Lawrence persuaded Lee DuBridge to become the director of the new laboratory, and that was a most fortunate choice. He also traveled all over the country, recruiting his former students and their colleagues from the cyclotron laboratories they had modeled after his own, and he didn't spare his own laboratory; Edwin McMillan, Winfield Salisbury and I all rushed off to Cambridge in November of 1940, and didn't return to Berkeley for five years.

But this is the story of Alfred Loomis, and not that of his friends, nor of the great laboratory he founded and guided so successfully with a loose rein. So I will single out from the many successes of the laboratory only two projects, one invented by Loomis and the other invented by Lawrence Johnston and me, but in which he played a key role. The first was Loran (for Long Range Navigation), which was of great importance during the war, and is still a major navigational aid in use all over the world. Loran is a pulsed, "hyperbolic system," and in its original form made use of Loomis's great store of knowledge about accurate timekeeping. In fact, the Loran concept of a master station and two slave stations can be traced to the Shortt clocks, which had a master pendulum swinging in a vacuum chamber, and a heavy-duty pendulum "slaved" to it, oscillating in the air.

To obtain a navigational "fix" with Loran requires the measurement of the time difference in arrival of pulses from two pairs of transmitting stations. Each such time difference places the observer on a particular hyperbola. The observer's position is fixed by the intersection of two such hyperbolas, each derived from signals originating from a pair of long-wave transmitting stations. It is common for a Loran fix to derive from only three transmitters, with the middle one serving as a member of two different transmitter pairs. All the wartime Loran stations operated at the same radio frequency, and different pairs of transmissions were distinguished by characteristic repetition rates for their pulses. The

techniques for separating the signals and for measuring their differences in arrival time were "state of the art" at that time, but the problem of synchronizing the transmissions to within a microsecond, at points hundreds of miles apart, was a new one in radio engineering. Loomis proposed the following solution: The central station was to be the master station, and its transmissions were timed from a quartz crystal. The other stations also used quartz crystals, but in addition, monitored the arrival times of the pulses from the master station. When the operators noted that the arrival time of the master pulses was drifting from its correct value, relative to the transmitting time at that particular "slave station," the phase of the slave's quartz crystal oscillator was changed to bring the two stations back into proper synchronization. This procedure was able to bridge over periods when the signals at one station "faded out," and it was also what made Loran a practical system during World War II, rather than an interesting idea that would have to await the invention of cesium beam clocks, which were introduced in the 1950s.

The second project of interest in this biographical sketch is Ground Controlled Approach, the "radar talkdown" system for landing planes in bad weather. The basic idea behind GCA came to me one day in the summer of 1941 as I watched the first microwave fire-control radar track an airplane, automatically, from the roof of MIT. It occurred to me that if a radar set could track a plane accurately enough in range, azimuth and elevation to shoot it down, it could use that same information to give landing instructions to a friendly plane caught up in bad weather.

Starting from the simple concept, my associates and I, with strong backing from Loomis, showed that the technique would work if the radar set gave angular information that was as reliable as the optical information we used in our tests. We had to wait several months for the radar set to become available for landing tests, but in one early demonstration, the radar did track several planes successfully as they executed their approach and landing. But in the scheduled radar tests, the equipment was found to be quite unable to track planes near the ground; it would suddenly break away from the line of sight to the plane, and point instead down at the image of the plane that was reflected in the surface of the ground.

At the conclusion of this disastrous set of tests, Loomis invited me to have dinner with him in his suite at the Ritz-Carlton in Boston and he did an amazing job in restoring my morale, which

was at is lowest ever. He said, "We both know that GCA is the only way planes will be blind-landed in this war, so we have to find some way to make it work. I don't want you to go home tonight until we're satisfied that you've come up with a design that will do the job." We both contributed ideas to the system that eventually worked, and that involved a complete departure from all previous antenna configurations. I'm sure that had it not been for Loomis's actions that night, there would have been no effective blind landing system in World War II, and many lives would have been lost unnecessarily. I would have immersed myself in the other interesting projects that concerned me, and would soon have forgotten my disappointment and my embarrassment.

Loomis played another interesting role in GCA by ordering ten preproduction models of the embryonic device we had invented at the Ritz-Carlton from a small radio company on the West Coast. He did this for two reasons: in the first place, the laboratory had failed badly in transferring its first airborne radar set to industry for production. The industrial engineers predictably developed a bad case of NIH (Not Invented Here), and promptly decided that everything had to be re-engineered. The final product came out so late and was so heavy that it never saw any action. Because of that experience, Loomis and Rowan Gaither (later the first president of the Ford Foundation) set up the "Transition Office," whose job was to avoid just such problems. Rowan became head of the Transition Office, and GCA was selected as the first test case of the new technique. Its basic idea was that a company would be selected to produce a new radar set before the original ideas had been worked out in any detail. The chief engineer of the designated company, plus a few of his assistants, would come to the laboratory and participate in the design and testing of the new device, as members of an MIT–company team. In this way, when they returned to their factory to produce the device, everything in it would be "our ideas" and "our design." The Transition Office was a spectacular success, and in the process, Rowan Gaither became extraordinarily close, personally, both to Loomis and me.

The second reason that Loomis ordered the ten preproduction sets, using NDRC-OSRD funds, was that the Army and Navy as well as the RAF had all said, independently, that their pilots would "never obey landing instructions from someone sitting in comfort on the ground," and that they would continue pressing for something like the ILS (Instrument Landing System) that is now in general use throughout the

Loomis at Schenectady about 1960 while visiting Guy Suits at G. E. Research Labs.

world. Loomis was confident that as soon as the three services saw GCA work, they would immediately accept it, and want working models to test, "yesterday."

After some very successful tests at Washington National Airport, in which high service officials watched pilots land "under the hood," when those pilots had never even heard of the system until after they were in the air, there was a rush to order several hundred GCA sets. When the three services learned that NDRC had ten sets almost built, they called a meeting at the Pentagon to allocate them for tests in this country and in England. Loomis was invited, and he asked me to sit in. Neither of us said a word as the admirals, generals, and air marshalls engaged in a horse-trading session that ended up with all ten sets allocated to the services, and none to MIT or to the NDRC. The meeting was about to break up when Loomis said quietly, "Gentlemen, there seems to be some misapprehension concerning the ownership of these radar sets; it is my understanding that they belong to NDRC, and I am here to represent that organization." His training as a lawyer was immediately apparent, and after he had shown in his gentle manner that he held all the cards, an allocation that was satisfactory to all concerned was quickly worked out. And NDRC even ended up with one of its own GCA sets!

At the end of the war, Lawrence gave this evaluation of Loomis's contribution to radar:[4]

> He had the vision and courage to lead his committee as no other man could have led it. He used his wealth very effectively in the way of entertaining the right people and making things easy to accomplish. His prestige and persuasiveness helped break the patent jams that held up radar development. He exercised his tact and diplomacy to overcome all obstacles. He's that kind of man, I've never seen him lose his temper or heard him raise his voice. He steers a mathematically straight course and succeeds in having his own way by force, logic and by being right. I am perfectly sure that if Alfred Loomis had not existed, radar development would have been retarded greatly, at an enormous cost in American lives.

Loomis's other important role during the war is so little known that its only mention in print is in a brief obituary notice I wrote for PHYSICS TODAY.[5] Many authors have commented on the remarkable lack of administrative roadblocks experienced by the Army's Manhattan District, the builders of the atomic bombs. In my opinion, this smooth sailing was due in large part to the mutual trust and respect that Secretary of War Stimson and Loomis had. Loomis was in effect Stimson's minister without portfolio to the scientific leadership of the Manhattan District—his old friends Lawrence, Compton, Fermi, and Robert Oppenheimer. Loomis maintained a hotel room in Washington throughout the war, which his friends used when they couldn't find other accommodations, and one of the reasons for this was so that he could be available to talk with the Secretary on short notice. Loomis was also a member of a small committee set up by the Secretary to advise him concerning the V-1 and V-2 weapons being developed by the Germans, and just coming to the attention of military intelligence. At the committee's sugestion, the V-1 menace was largely blunted by a combination of the SCR-584 developed in Loomis's laboratory, an advanced computer developed by the Bell Telephone Laboratory, the proximity fuses developed by Merle Tuve and his associates working under NDRC sponsorship, and the Army's anti-aircraft guns. The V-2 rockets could not be defended against, and the committee recommended the only course of action possible, and the one that was followed—capture of the firing sites.

Later years

Toward the end of the war, Loomis was able to relax for the first time in five years, and he concurrently made an important change in his personal life. He and Ellen were divorced, and he married Manette Seeldrayers Hobart. They had an extraordinarily happy time together during the final 32 years of Alfred's life. His lifestyle underwent a dramatic change from one of multiple homes staffed by many servants to a very simple one, in which he and Manette cooked dinner every evening in East Hampton, side by side in the kitchen. Alfred designed a special rolling cart that brought the food to one end of the table, where he and Manette sat opposite each other, and served themselves from the cart. If there were guests, the plates were passed down each side of the table to them, from the cart. This new style of servantless elegance was written up in a magazine devoted to "good living."

Loomis's principal scientific interests changed at this time from the physical to the biological. As an example, I've mentioned his contributions to research on hydra. In that period, one of the bathrooms in his Park Avenue apartment was filled with petri dishes containing hydra. Loomis spent hours each day examining the hydra under a microscope, and comparing his observations with those of his son, Farney. He and Farney organized small meetings to which they invited specialists in subjects about which they wished to learn more. As in the old Loomis Laboratory days, the invitations included first-class round-trip transportation, plus luxurious living at the resorts where the meetings were held.

Loomis enjoyed introducing his scientific friends to the pleasures that are normally known only to the very wealthy. For many years, he and Manette visited California each spring, and invited several couples from Lawrence's laboratory to be their guests at the Del Monte Lodge at Pebble Beach, and to play golf at the Cypress Point Golf Club. In later years, the Loomises spent their winters in Jamaica, where their friends were invited, a week at a time, to share with their hosts the sun, the beach, and good food and good conversation. As often happens with men as they grow older, Loomis's circle of closest friends shrank to those he called "my other sons." I was fortunate to be included, along with John S. Foster Jr, Walter O. Roberts, Ronald Christie and Julius A. Stratton. Had Lawrence and Rowan Gaither outlived Alfred, they would have continued to visit the Loomises each winter in Jamaica, as members of the "other sons."

I can think of no better way to end this biographical memoir than by quoting myself[5]

For those of us who were fortunate to know him well, he will be remembered as a warm and wise friend, always interested in learning new things. I was his guest for three days in May of this year, and what he most wanted to learn from me concerned programming tricks for the Hewlett-Packard model 65 hand-held computer that was his constant companion. I think it most fitting that my last visual memories of this renaissance man, whose life encompassed and contributed much to the electronic age, should have him operating a hand-held electronic computer containing tens of thousands of transistors.

* * *

This article was adapted from Biographical Memoirs 51, *The National Academy of Sciences (1980).*

References

1. W. B. Seabrook, *Dr. Wood, Modern Wizard of the Laboratory*, New York, Harcourt, Brace, New York (1941), page 213.
2. *Ibid*, page 221.
3. E. N. Harvey, A. L. Loomis, Science **72**, 42 (1930).
4. "Amateur of the Sciences," Fortune, March 1946, page 132.
5. L. W. Alvarez, PHYSICS TODAY, November 1975, page 84. ☐

Harold Urey and the discovery of deuterium

Chemistry, nuclear physics, spectroscopy and thermodynamics came together to predict and detect heavy hydrogen before the neutron was known.

Ferdinand G. Brickwedde

PHYSICS TODAY / SEPTEMBER 1982

It was on Thanksgiving day in 1931 that Harold Clayton Urey found definitive evidence of a heavy isotope of hydrogen. Urey's discovery of deuterium is a story of the fruitful use of primitive nuclear and thermodynamic models. But it is also a story of missed opportunity and errors—errors that are particularly interesting because of the crucial *positive* role that some of them played in the discovery. A look at the nature of the theoretical and experimental work that led to the detection of hydrogen of mass 2 reveals much about the way physics and chemistry were done half a century ago.

Although George M. Murphy and I coauthored with Urey the papers[1-3] reporting the discovery, it was Urey who proposed, planned and directed the investigation. Appropriately, the Nobel Prize for finding a heavy isotope of hydrogen went to Urey.

In this article we will look first at the research that led to the discovery, as that work was understood at the time. Then we will look at some of the same activity with the understanding that only hindsight can give. Throughout the discussion I will include fragments from my memory—illustrative episodes connected with the discovery.

Urey's career

Urey died last year at 87 years of age, after a remarkably productive and interesting life. He was a chemist with very broad interests in science, reminiscent of the natural philosophers of the eighteenth and nineteenth centuries. Murphy[4], who went on to become professor and head of the department of chemistry at New York University, died in 1968.

Urey was born on a farm in Indiana in 1893, and in childhood moved with his family to a homestead in Montana. After graduating from high school, he taught for three years in public schools, and then entered Montana State University as a zoology major and chemistry minor. Money was tight for him as a college student. During the academic year he slept and studied in a tent. During his summers he worked on a road gang laying railroad track in the Northwest.

Urey graduated with a BS degree in 1917, when there was a need for chemists in the war effort. He worked for the Barrett Chemical Company in Philadelphia on war materials. After the war, Urey taught chemistry for two years at Montana State University, and in 1921 entered the University of California at Berkeley as a graduate student in chemistry, working under the guidance of the renowned chemical thermodynamicist Gilbert N. Lewis. As a graduate student, Urey was a pioneer in the calculation of thermodynamic properties from spectroscopic data. He received a PhD in 1923 and spent the next academic year as an American–Scandinavian Foundation Fellow in the Physical Institute of Niels Bohr in Copenhagen.

After Copenhagen, Urey joined the faculty at Johns Hopkins University. Although in the chemistry department, he attended the physics department's regular weekly "journal" meetings for faculty and graduate students, and he participated in the discussions. It was at these meetings that I, as a graduate student in physics, became acquainted with Urey. While Urey was at Hopkins, he and Arthur E. Ruark coauthored the classic textbook, *Atoms, Molecules, and Quanta*, which was the first comprehensive text on atomic structure written in English. I proofread the entire book in galley for the authors.

Urey's work bridged chemistry and physics. In 1929 he was appointed associate professor of chemistry at Columbia University, and from 1933 to 1940 he was the founding editor of the American Institute of Physics publication, *Journal of Chemical Physics*. When the biographical publication "American Men of Science" took note of scientists selected for recognition by their peers, Urey was elected in physics. In 1934—only three years after the discovery of deuterium—Urey was awarded the Nobel Prize in chemistry.

Before the search

In 1913, Arthur B. Lamb and Richard Edwin Lee, working at New York University, reported[5] a very precise measurement of the density of pure water. Their measurements were sensitive to 2×10^{-7} g/cm^3. Various samples of water, which were carefully prepared using the best purification techniques and temperature controls, varied in density by as much as 8×10^{-7} g/cm^3. They concluded that pure water does not possess a unique density.

Today we know that water varies in isotopic composition, and that samples of water with different isotopic compositions have different vapor pressures, making distillation a fractionating process. The Lamb–Lee investigation is interesting because it was the first reported experiment in which an isotopic difference in properties was clearly in evidence. It is the earliest recognizable experimental evidence for isotopes. (The existence of isotopes was proposed independently by Frederick Soddy, in England, and by Kasimir Fajans, in Germany, in 1913.) Think what the result might have been had Lamb and Lee pursued a progressive fractionation of water by distillation and separated natural water into fractions with different molecular weights.

Less than two decades later, by the time of the discovery of deuterium, isotopes were an active field of research. The rapid development of nuclear physics after 1930 was initiated by isotope research. It was a time of search for as-yet undiscovered isotopes,

Ferdinand G. Brickwedde is Evan Pugh Research Professor of Physics emeritus at Pennsylvania State University, in University Park, Pennsylvania.

especially of the light elements, hydrogen included, and Urey was very much a participant.

I remember a conversation in 1929 with Urey and Joel Hildebrand, a famous professor of chemistry at Berkeley. It took place during a taxi ride between their hotel and the conference center for a scientific meeting we were attending in Washington. When Urey asked Hildebrand what was new in research at Berkeley, Hildebrand replied that William Giauque and Herrick Johnston had just discovered that oxygen has isotopes with atomic weights 17 and 18, the isotope of weight 18 being the more abundant. Their paper[6] would appear shortly in the *Journal of the American Chemical Society*. Then Hildebrand added, "They could not have found isotopes in a more important element." Urey responded: "No, not unless it was hydrogen." This was two years before the discovery of deuterium. Urey did not remember this remark, but I did.

At the time, answers were being sought to questions such as: Why do isotopes exist, and what determines their number, relative abundances and masses (packing fractions)?

Urey, along with others, constructed charts of the known isotopes to show relationships bearing on their existence. The figure on page 36 is one of Urey's charts. At the time, the neutron had not been discovered—it was discovered in 1932, the year after deuterium. The chart was based on the theory that atomic nuclei were composed of protons, plotted here as ordinates, and nuclear electrons, plotted as abscissae—the number of protons was the nuclear mass number, and the number of nuclear electrons was the number of protons minus the atomic number of the element. In Urey's chart, the filled circles represent the nuclei from H^1 to Si^{30} that were known to exist before 1931. The open circles represent nuclei unknown before 1931. The chart's pattern of staggered lines, when extended down to H^1, suggested to Urey that the atoms H^2, H^3 and He^5 might exist because they are needed to complete the pattern.

Urey had a copy of this chart hanging on a wall of his laboratory. The isotope helium-5 does not exist, and the staggered line does not provide a place for the isotope helium-3, which was discovered later. The diagram is only of historical interest now, but it was an incentive to Urey to look for a heavy isotope of hydrogen.

Prediction and evidence

In 1931—the year of the discovery of deuterium—Raymond T. Birge, a professor of physics at the University of California, Berkeley, and Donald H. Menzel, professor of astrophysics at Lick Observatory, published[7] a letter to the editor in *Physical Review* on the relative abundances of the oxygen isotopes in relation to the two systems of atomic weights that were then in use—the physical system and the chemical system. Atomic weights in the physical system were determined with the mass spectrograph and were based on setting the atomic weight of the isotope O^{16} at exactly 16. In the chemical system, atomic weights were determined by

Harold Clayton Urey and a country schoolhouse in Indiana where he taught after graduating from high school. Urey taught for three years in public schools in Indiana and Montana before he entered Montana State University. (Schoolhouse photo from the Urey collection, AIP Niels Bohr Library.)

DR. HAROLD C. UREY
NOBEL PEACE PRIZE
IN SCIENCE 1934
TAUGHT HERE 1911

bulk techniques, and the values were based on setting at 16 the atomic weight of the naturally occurring *mixture* of oxygen isotopes, O^{16}, O^{17} and O^{18}. Thus the atomic weights of a single isotope or element on the two scales should differ. The weight numbers should be greater on the physical scale.

However, in 1931 the atomic weights of hydrogen on the two scales were the same within the claimed experimental errors. The chemical value was 1.00777 ± 0.00002. The mass-spectrographic value, determined by Francis W. Aston of the Cavendish Laboratory, was 1.00778 ± 0.00015. Birge and Menzel pointed out that the near coincidence of these two atomic weights leads to the conclusion that normal hydrogen is a mixture of isotopes—H^1 in high concentration and a heavy isotope in low concentration. The atomic weight was not higher on the physical scale because the mass-spectroscopic techniques saw only the light isotope.

To the heavy isotope they gave the symbol H^2, perhaps the first time this symbol occurred in the literature. Assuming the atomic weight of heavy hydrogen to be two, Birge and Menzel calculated its relative abundance from the supposed equivalence of the atomic weights of hydrogen-1 on the physical scale and the normal mixture of hydrogen isotopes on the chemical scale. They obtained 1/4500 for the abundance of H^2 relative to H^1.

Within a day or two at most after receiving the 1 July 1931 issue of the *Physical Review*, Urey proposed and planned an investigation to determine if a heavy isotope of hydrogen did really exist.

Urey and Murphy, working at Columbia, identified hydrogen and its isotope spectroscopically, using the Balmer series lines. The atomic spectrum was produced with a Wood's electric discharge tube operated in the socalled black stage—the configuration of current and pressure that most strongly excites hydrogen's atomic spectrum relative to its molecular spectrum. They observed the spectra with a 21-foot grating, in the second order. The dispersion was 1.3 Å per mm. The expected shifts, then, were of the order of 1 mm, as the numbers in the table indicate. The vacuum wavelengths of deuterium's lines were calculated using the Balmer series formula

$$1/\lambda_{\mathrm{H}} = R_{\mathrm{H}}(1/2^2 - 1/n^2) \qquad (1)$$
$$n = 3, 4, 5, \ldots$$
$$R_{\mathrm{H}} = (2\pi^2 e^4/h^3 c)m_{\mathrm{e}} m_{\mathrm{H}}/(m_{\mathrm{e}} + m_{\mathrm{H}})$$

and the "best" values for the atomic constants. The Balmer α-lines of hydrogen and deuterium are separated by 1.8 Å, the β-lines by 1.3 Å, and the γ-lines by 1.2 Å. The concentrations of deuterium relative to hydrogen are de-

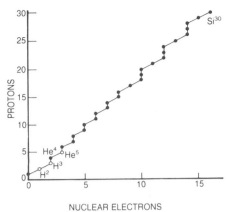

Protons versus "nuclear electrons" for atomic nuclei from H^1 to Si^{30}. The plot shows a pattern that led Urey to look for a heavy isotope of hydrogen. Open circles represent nuclei that were unknown in 1931, when the chart was produced.

termined by comparing the measured times required to produce plate lines of H and D of equal photographic densities. The exposure times for $H\beta$ and $H\gamma$ were about 1 second.

Using cylinder hydrogen, Urey and Murphy found very faint lines at the calculated positions for $D\beta$, $D\gamma$ and $D\delta$. The lines were faint because of the low concentration of deuterium in normal hydrogen. There was a possibility that the new lines arose from impurities, or were grating ghost lines arising from the relatively intense hydrogen Balmer spectrum.

Clinching evidence

Urey decided *not* to rush into print to stake a claim to priority in this important discovery; he decided to postpone publication until he had conclusive evidence that the "new" spectral lines attributed to the heavy isotope were authentic and not impurity or ghost lines. This evidence could be obtained by increasing the deuterium concentration in the hydrogen filling the Wood's tube and looking for an increase in intensity of the deuterium Balmer lines relative to the hydrogen Balmer lines.

After careful consideration of different methods for increasing the deuterium concentration, Urey decided on a distillation that would make use of an anticipated difference in the vapor pressures of liquid H_2 and liquid HD. He made a statistical, thermodynamic calculation of the vapor pressures of solid H_2 and solid HD at the triple point of H_2, 14 kelvins, where the liquid and crystal phases of H_2 are in equilibrium and have the same vapor pressure. The calculation was based on the Debye theory of solids and the zero-point vibrational energy of the solid, $9R\theta/8$ in the Debye notation. At 14 K, the calculated ratio of vapor pressures, $P(HD)/P(H_2)$, is 0.4, indicating a large differ-

ence in the vapor pressures of solid H_2 and HD. On this basis Urey expected a sizeable difference in the vapor pressures of *liquid* H_2 and HD at 20.4 K, the boiling point of H_2.

Urey approached me at the National Bureau of Standards in Washington, inviting me to join the search for a heavy isotope of hydrogen by evaporating 5- to 6-liter quantities of liquid hydrogen to a residue of 2 cm^3 of liquid, which would be evaporated into glass flasks and sent by Railway Express to Columbia University for spectroscopic examination. At the time, 1931, there were only two laboratories in the United States where liquid hydrogen was available in 5- or 6-liter quantities. One was the National Bureau of Standards in Washington and the other was Giauque's laboratory at the University of California, Berkeley. I was happy to cooperate, and I prepared—by distilling liquid hydrogen at the Bureau of Standards—the samples of gas in which the heavy isotope was identified.

The first sample I sent to Urey was evaporated at 20 K and a pressure of one atmosphere. It showed no appreciable increase in intensity of the spectral lines attributed to heavy hydrogen. This was unexpected.

The next samples were evaporated at a lower temperature—14 K at 53 mm of mercury pressure, the triple point of H_2—where the relative difference in the vapor pressures of H_2 and HD was expected to be larger than at 20 K, and the rate at which heavy hydrogen is concentrated was expected to be more rapid.

These samples showed 6- or 7-fold increases in the intensities of the Balmer lines of deuterium. On this basis, it could be concluded that the lines in the normal hydrogen spectrum attributed to deuterium were really deuterium lines, but the clinching evidence was finding that the photographic image of the $D\alpha$ line—the most intense D-Balmer line—was a partially split doublet as predicted by theory for the Balmer series spectrum.

From measurements of the relative intensities of the H and D Balmer series lines, Urey estimated that there was one heavy atom per 4500 light atoms in normal hydrogen. Later measurements showed it to be nearer one in 6500.

Unraveling a comedy of errors

It is now clear why the first distilled hydrogen sent to Urey did not show the expected increase in the deuterium concentration, and maybe even showed

News story on the awarding of the 1934 Nobel Prize in chemistry. Article appeared 16 November 1934. (Copyright *The New York Times.* Reprinted by permission.)

a small decrease. The explanation came with the discovery of the electrolytic method for separating H and D, suggested by Edward W. Washburn, chief chemist at the National Bureau of Standards, and verified[8] experimentally by Washburn and Urey just after the publication of our April 1932 paper.[3]

When Urey considered different methods for concentrating deuterium, he included the electrolytic method, and discussed it with Victor LaMer, a colleague at Columbia, and a world authority on electrochemistry. LaMer was so discouraging about success in separating hydrogen isotopes by electrolysis that Urey abandoned the electrolytic method and adopted the distillation method. LaMer reasoned that the differences in equilibrium concentrations of isotopes at the electrodes of a cell at room temperature would be very small and hence a fractionation of the isotopes would be negligible.

Washburn viewed the situation differently. He pointed to the large relative difference in atomic weights of the hydrogen isotopes—a relative difference that is much larger for the hydrogen isotopes than for the isotopes of any other element. Hence, thought Wash-

Calculated Balmer series wavelengths

Line	$\lambda(H^1)$ (Å)	$\lambda(D)$ (Å)	$\Delta\lambda(H^1 - D)$ calculated (Å)	observed (Å)
α	6564.686	6562.899	1.787	1.79
β	4862.730	4861.407	1.323	1.33
γ	4341.723	4340.541	1.182	1.19
δ	4102.929	4101.812	1.117	1.12

These values were calculated using equation 1 with $M_H = 1.007775$ g, $M_D = 2.01363$ g, $m_e = 5.491 \times 10^{-4}$ g and $R_H = 109677.759$ cm^{-1}.

burn, the hydrogen isotopes might behave differently from the isotopes of other elements.

Washburn, the empiricist, was right; the isotopes of hydrogen are separated relatively easily by electrolysis, but this was not realized until after the discovery of deuterium.

The hydrogen we liquefied and distilled for Urey was generated electrolytically. Before preparing the first sample for Urey, the electrolytic generator was completely dismantled, cleaned and filled with a freshly prepared solution of sodium hydroxide. Because deuterium becomes concentrated in the electrolyte in the generator, the first gaseous hydrogen to be discharged was deficient in deuterium. The concentration of deuterium in the hydrogen evolved was about one sixth the concentration of deuterium in the electrolyte, and hence about one sixth the concentration of deuterium in normal hydrogen. Distillation of the deuterium-deficient liquid hydrogen increased the concentration of D relative to H and restored in the first sample approximately the original concentration of deuterium in normal hydrogen.

As electrolysis progressed, water was added to replace that which was consumed. The concentration of deuterium in the electrolyte increased to the point where the rate at which deuterium left the generator balanced the rate at which it arrived in the added water. Hence, after the electrolytic generator had been in use for some time, there was a dynamic equilibrium; so the hydrogen evolved from the generator for our second and third samples for Urey had approximately the normal concentration of deuterium. When we liquefied this hydrogen and evaporated 5 or 6 liters down to 2 cm^3, the concentration of deuterium in the residue was increased by a factor of about six.

Here we lower the curtain on a "comedy of errors"—LaMer's error of not understanding better the principles that govern isotopic fractionation during electrolysis, and my error of attributing to sloppy technique our failure to effect an increase of deuterium concentration in the first sample we sent to Urey. Had I analyzed our part of the process, I think we might have discovered the electrolytic concentration of deuterium. Had LaMer been more knowledgeable, Urey would have made his own concentration of deuterium electrolytically and I should have had no part in the discovery of deuterium.

Reporting the result

After the discovery of deuterium, Urey faced a very practical problem in reporting it—a problem characteristic of the status of research before World War II. Urey's research at Columbia, and ours at the National Bureau of Standards, where I was chief of the low temperature laboratory, was carried out without the support of any government research grant. It was said that

research in that period was done with string and sealing wax; it was in fact done mostly with homemade apparatus. The US government policy of grants in support of research dates from a later time—from World War II.

Before the War it was a problem to find funds for travel to scientific meetings. I received a telephone call from Urey, telling me that it appeared he was not going to get funds to travel to the December 1931 American Physical Society meeting at Tulane University, where he planned to present a paper reporting the discovery of deuterium. He asked me if I could get travel funds and present the paper. For this I had to see Lyman J. Briggs, assistant director of research and testing at the Bureau of Standards. Briggs, soon to be named NBS director, was an understanding and considerate physicist who, on learning of the work to be reported, made funds available for my travel. In the meantime, Bergen Davis, a prominent physicist at Columbia, heard of Urey's problem and went to see Columbia president Nicholas Murray Butler, who made funds available for Urey's travel. So we both went to Tulane for the APS meeting, and Urey presented the ten-minute paper.[1] Over the next few months we published more detail in a letter[2] to the editor and a full-length paper[3] in *Physical Review*.

I remember asking Birge at a later APS meeting why he and Giauque had not followed up on his prediction[7] of the existence of heavy hydrogen. They might have demonstrated the existence of deuterium by concentrating the heavy isotope through distillation of a large quantity of liquid hydrogen as Urey and I had done. Giauque had a very fine, large-capacity hydrogen liquefier suitable for this. Birge's reply was that he was busily engaged on other important work that demanded his attention. When I told Urey of this discussion, his comment was: "What in the world could Birge have been working on that was so important?"

Apropos of the above, I quote here from a letter of 6 May 1981 from Robert W. Birge, son of Raymond T. Birge, and also a physicist:

After reading some more about my father's life, I think I know why he didn't try to concentrate deuterium. I believe he was an analyst more than a hardware builder and it probably never occurred to him to do it that way. He said that at the time several people were trying to see the deuterium lines in spectra, but they [Urey, Brickwedde and Murphy] did it first. But as you know, the important point was that Urey realized that [the concentration of] deuterium could be enhanced.

The two men remained friends

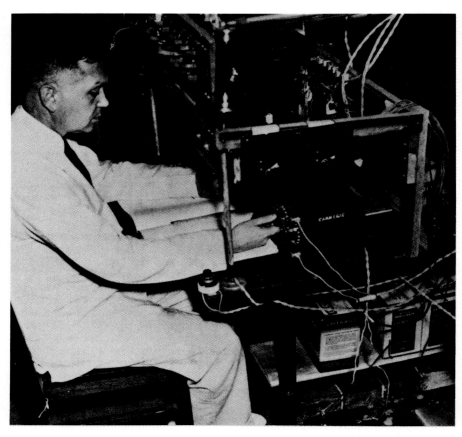

Mass spectrometer with Urey at the controls, after the discovery of deuterium. (Photograph courtesy of King Features Syndicate.)

throughout their lifetime.

Frederick Soddy, the English chemist who received the 1921 Nobel Prize in Chemistry for discovering the phenomenon of isotopy, did not accept the notion that deuterium was an isotope of hydrogen. Soddy worked with isotopes of the naturally radioactive elements, whose atomic weights are large and whose isotopic relative mass differences are small. These isotopes showed no observable differences in chemical properties and were inseparable chemically. When Soddy coined the word isotope he gave it a definition that included chemical inseparability of isotopic species of the same element. This was generally accepted before the discovery of the neutron in 1932.

After the discovery of the neutron, isotopes were defined as atomic species having the same number of protons in their nuclei but different numbers of neutrons. But Soddy stuck to chemical inseparability as a criterion for isotopes and therefore refused to recognize deuterium as an isotope of hydrogen. For Soddy, deuterium was a species of hydrogen, with different atomic weight, but not an isotope of hydrogen.

A fortunate mistake

Four years after the discovery of deuterium, Aston reported[9] an error in

his earlier mass-spectrographic value of 1.00778 for the atomic weight of hydrogen-1 on the physical scale—the value used by Birge and Menzel in their 1931 letter.[7] The revised value on the physical scale was 1.00813, which corresponds to 1.0078 on the chemical scale, in agreement with the then current value for the atomic weight of hydrogen (1.00777) on the chemical scale. There was then no need or place for a heavy isotope of hydrogen. The conclusion of Birge and Menzel was thus rendered invalid. Indeed, on the basis of Aston's revised value, Birge and Menzel would have been obliged to conclude that, if anything, there was a lighter—not a heavier—isotope of hydrogen.

The prediction of Birge and Menzel of a heavy isotope of hydrogen was based on *two* incorrect values for the atomic weight of hydrogen, namely Aston's mass-spectrographic value and the chemical value, which also should have been greater. We are obliged to conclude that the experimental error in the determination of the atomic weights exceeded the difference in the atomic weights on the two scales.

Urey was not aware of this when he planned his experiment. It was not until 1935 when Urey's Nobel lecture was in proof that Aston published his revised value. Urey added the following to the printed Nobel lecture:

Addendum

Since this [Nobel lecture[10]] was written, Aston has revised his mass-spectrographic atomic weight of hydrogen (H) to 1.0081 instead of 1.0078. With this mass for hydrogen, the argument by Birge and Menzel is invalid. However, I prefer to allow the argument of this paragraph [the third paragraph of Urey's Nobel lecture] to stand, even though it now appears incorrect, because this prediction was of importance in the discovery of deuterium. Without it, it is probable we would not have made a search for it and the discovery of deuterium might have been delayed for some time.

Needless to say, Urey and his colleagues were very glad that an error of this kind had been made. Aston said that he did not know what the moral of it all was. He would hardly advise people to make mistakes intentionally, and he thought perhaps the only thing to do was to keep on working.

Impact of the discovery

It has been said that Nobel prizes in physics and chemistry are awarded for work, experimental or theoretical, that has made a significant change in ongoing work and thinking in science. The announcement that Urey was chosen as the 1934 laureate in chemistry came less than three years after that ten-minute paper in New Orleans announcing the discovery of deuterium. This uncommonly early award followed a spectacular display in deuterium-related research. In the first two-year period following the discovery, more than 100 research papers were published on or related to deuterium and its chemical compounds, including heavy water. And there were more than a hundred more[11] in the next year, 1934.

The use of deuterium as a tracer made it possible to follow the course of chemical reactions involving hydrogen. This was especially fruitful in investigations of complex physiological processes and in medical chemistry, as in the breakdown of fatty tissue and in cholesterol metabolism.

Also, the discovery of heavy hydrogen provided a new projectile, the deuteron, for nuclear bombardment experiments. The deuteron proved markedly efficient in disintegrating a number of light nuclei in novel ways. As the deuteron, with one proton and one neutron, is the simplest compound nucleus, studies of its structure and of its proton–neutron interaction took on fundamental importance for nuclear physics.

Many of the early research papers dealt with isotopic differences in physical and chemical properties. Theories developed for the atomic mass dependence of physical and chemical properties were tested experimentally. These investigations were especially interesting because, before the discovery of deuterium, chemical properties were generally supposed to be determined by the number and configuration of the extranuclear electrons, quantities that are identical for isotopes of the same element. It had not been realized that chemical properties are also affected—but to a lesser degree—by the mass of the nucleus.

In thinking about Urey's search for deuterium, beginning with his early diagram of the isotopes, I am reminded of the Greek inscription on the facade of the National Academy of Sciences building in Washington, taken from Aristotle:

The search for truth is in one way hard and in another easy, for it is evident that no one can master it fully or miss it wholly. But each adds a little to our knowledge of nature, and from all the facts assembled there arises a certain grandeur.

* * *

I wish to acknowledge the valuable assistance of my wife, Langhorne Howard Brickwedde, especially for her help in recalling incidents of the early thirties connected with the discovery of deuterium. This article is based on a paper I presented 22 April 1981 in Baltimore, Maryland, at the inaugural session of the American Physical Society's Division of History of Physics.

References

1. The thirty-third annual meeting of the American Physical Society at Tulane University, 29–30 December 1931. Abstracts of papers presented: Phys. Rev. **39**, 854. Urey, Brickwedde and Murphy abstract #34.
2. H. C. Urey, F. G. Brickwedde, G. M. Murphy, Phys. Rev. **39**, 164 (1932).
3. H. C. Urey, F. G. Brickwedde, G. M. Murphy, Phys. Rev. **40**, 1 (April 1932).
4. For an interesting account of the discovery of deuterium, see G. M. Murphy, "The discovery of deuterium," in *Isotopic and Cosmic Chemistry*, H. Craig, S. L. Miller, G. J. Wasserburg, eds., North-Holland, Amsterdam (1964). (Dedicated to Urey on his seventieth birthday.)
5. A. B. Lamb, R. E. Lee, J. Am. Chem. Soc. **35**, part 2, 1666 (1913).
6. W. F. Giauque, H. L. Johnston, J. Am. Chem. Soc. **51**, 1436, 3528 (1929).
7. R. T. Birge, D. H. Menzel, Phys. Rev. **37**, 1669 (1931).
8. E. W. Washburn, H. C. Urey, Proc. Nat. Acad. Sci. US **18**, 496 (1932).
9. F. W. Aston, Nature **135**, 541 (1935); Science **82**, 235 (1935).
10. H. Urey in *Nobel Lectures in Chemistry, 1922–1941*, published for the Nobel Foundation by Elsevier, Amsterdam (1966).
11. Industrial and Engineering Chemistry, News Edition **12**, 11 (1934). □

Pyotr Kapitza, octogenarian dissident

Despite years of working under
house arrest in his native land, Kapitza has remained
the outspoken dean of Soviet science.

Grace Marmor Spruch

PHYSICS TODAY / SEPTEMBER 1979

Four of the five Soviet academicians attending a Pugwash Conference in 1973 had, earlier that year, signed a condemnation of Andrei Sakharov for his political utterances. The fifth, Pyotr L. Kapitza, had not signed. Kapitza—the dean of Soviet science, winner of two Stalin prizes, four times awarded the Order of Lenin and last year awarded the Nobel Prize—had previously played Sakharov's role as the most outspoken Soviet scientist. In fact, he was also thought to have played Sakharov's role as father of the Soviet H bomb. Nobel Laureates are usually known solely for their work; a small number, however, are known as much for their personalities

or the circumstances under which they worked. Kapitza is one of the latter. Most scientists have seen abstracts of his life, but few are familiar with the entire article.

The early years

Kapitza was born in 1894 in Kronstadt, famous as the site of a sailors' uprising in 1921. His father was a general in the tsarist army engineering corps (said to have worked on the defenses of Kronstadt), and his mother was the daughter of a general. Kapitza received his secondary school education in Kronstadt and electrical-engineering training at the

Petrograd Polytechnic Institute. After his graduation in 1918, he stayed on as a lecturer at Petrograd and by the time he left for England in 1921, had six scientific papers to his credit.

Kapitza arrived in England at age twenty-seven—thin, unhappy, unknown, looking like "a tragic Russian prince," according to Cambridge don G. Kitson Clark. Kapitza's native country lay rent by civil war, disease and famine, his wife and two small children victims. In extreme depression, he had left on the recommendation of the eminent scientist Abram F. Ioffe and, it is said, the intercession of writer Maxim Gorky, as part of

DAVID SCHOENBERG

DAVID SCHOENBERG

an official Soviet delegation to the UK.

In England Kapitza made for Cambridge and the Cavendish Laboratory, headed at that time by Ernest Rutherford. The story is told that Kapitza asked to work in the Cavendish as a research student but was told by Rutherford that there were no openings. Kapitza inquired about the accuracy of Rutherford's measurements and was told it was roughly ten per cent. Kapitza then pointed out that one additional student was less than ten per cent of the total of thirty and that Rutherford would be within the limits of his experimental error if he accepted him—which Rutherford did. (There is a "second-order" error here in that some versions of the incident have Rutherford stating three per cent as his experimental error—in which case Kapitza's entry into the Cavendish was a tighter squeeze.)

The ancedote illustrates the character of both men and the relationship that was to develop between them. Kapitza had a *chutspah* and vitality seldom found in Englishmen, and Rutherford, no Englishman himself but a New Zealander—a loud, brusque, "Colonial"—responded.

Kapitza's letters

Not everyone responded positively to Kapitza. One Englishman, offering objective criteria by which to assess Kapitza's character, recommended I read letters Kapitza had written to his mother during his early days in Cambridge. These, in a

Grace Marmor Spruch is professor of physics at the Newark Campus of Rutgers University.

Russian biography of Rutherford, would show Kapitza's "Napoleonic ambition." W. H. Auden once wrote:

. . . to me, at least, who was born and bred a British Pharisee, Russians are not quite like other folk. If their respective literatures in the nineteenth century are a guide, no two sensibilities could be more poles apart than the Russian and the British . . . Time and time again, when reading the greatest Russian writers, like Tolstoy and Dostoievsky, I find myself exclaiming, "My God, this man is bonkers!"

I read the letters, and to me, at least, an American of Russian descent, though they seemed somewhat Oedipal, they were within the bounds of normal boasting to one's mother.

Dated every few days at first, the letters overflow with terms of endearment and concern over how his mother will get along without him. (According to a British woman who met her years later, Kapitza's mother, a collector of folklore and writer of children's stories, was an intelligent, capable woman who gave the impression she could more than cope.) The letters tremble with Kapitza's own insecurity. He fears that inadequate knowledge of English hampers him in the expression of ideas and writes that even in Russian he expresses himself poorly. He refers to his manners as crude.

The interval between letters becomes longer and the mood more confident as he gets more involved in his work. After three months in the laboratory he writes:

The famous crocodile marks the entrance to the Royal Society Mond Laboratory in Cambridge. Sculptor Eric Gill carved the design, on Kapitza's commission, for the opening of the laboratory in 1933. During the following year Kapitza was detained while on a visit to the Soviet Union, and he was not allowed to leave the country until 1966. The photographs to the right and left show him in 1937 and during the 1960's.

. . .Rutherford is increasingly aimiable to me . . . But I am somewhat afraid of him. I work right next door to his office. This is bad, as I must be careful about smoking. If he should see me with a pipe in my mouth, there would be trouble. But thank God he has a heavy step, and I can distinguish it from others . . .

In the next letter Kapitza calls Rutherford "Crocodile." (One is tempted to associate Rutherford's heavy tread with the ticking clock in Captain Hook's crocodile, but Kapitza's friends doubt that he had read *Peter Pan.*)

The letters trace Kapitza's progress and tell of the increasing amount of space he is occupying in the Cavendish, with sociological comments interspersed:

. . . Englishmen get drunk easily. And it is noticeable immediately. Their features become lively and animated; they lose their stoniness . . . Apparently my Russian belly is better adapted to alcohol than an English one.

Here—it's funny—if the professor is nice to you, it immediately affects everyone else in the laboratory; they also show you consideration.

The letters also relate Kapitza's attachment to his motorcycle and one instance of his lack of attainment when he and James Chadwick were sent flying, Kapitza pointing out that it was Chadwick in the driver's seat. Kapitza's face bore the brunt of the experience; it was so swollen and discolored he was ashamed to show it in the laboratory until informed by a friend that at Cambridge such a face, when associated with sport rather than with alcohol, was considered chic.

About Rutherford he wrote:

You can't imagine what a great and wonderful man he is,

and about Rutherford's solicitude for him:

[it] must surely equal that from one's own father . . . his kindness to me is boundless.

Although Auden might attribute some of Kapitza's statements to Russsian effusiveness, most scientists would agree that Cambridge was "the world's foremost school," and, if qualified by "experimental," that Rutherford was "the world's foremost physicist and organizer." In addition, the letters cite facts that support some effusions. For example, after receiving his PhD, Kapitza, effectively "broke," was lent money by Rutherford with which to go away for a rest. Rutherford also offered Kapitza the Clerk Maxwell Prize, a three-year stipend normally awarded to the best young student to help him through his degree, despite Kapitza having already completed his doctorate.

In the final letter of the group Kapitza tells his mother that Rutherford has asked him to stay on for about five years, after which he could dictate his own terms in seeking employment.

At the Cavendish

Kapitza started out in nuclear physics at the Cavendish, but soon his natural inclination toward engineering began to assert itself. Some say engineering is his real métier. (When he had enough money to buy a new car, he would consider only one for which the manufacturer would supply a set of blueprints; Vauxhall was the only company to comply.) Kapitza set to work on the problem of obtaining very strong magnetic fields for investigations of atomic properties. He sent much greater currents through the coil of his electromagnet than the coil could sustain—but for less time than it would take the coil to burn out. Here Kapitza utilized his engineering background in designing apparatus that would permit currents of about 10 000 amperes to be switched off after 0.01 second. He wrote to Rutherford, then on vacation:

We managed to obtain fields over 270 000 [gauss] . . . we could not go further as the coil bursted with a great bang . . . The power in the circuit was about 13½ kilowatts . . . approximately three Cambridge supply stations con-

DAVID SCHOENBERG

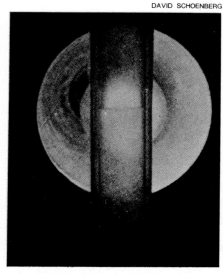

The first flask of liquid helium made with Kapitza's liquefier at the Mond Laboratory, 1934. Kapitza's inexpensive method for helium liquefaction led to the Collins Liquefier.

nected together, but the result of the explosion was only the noise, as no apparatus has been damaged, except the coil . . . The accident was the most interesting of all the experiments . . . as we know exactly what has happened when the coil bursted. We know just what an arc of 13 000 amperes is like. Apparently it is not at all harmful for the apparatus and the machine, and even for the experimenter if he is sufficiently far away.

Kapitza began to put down roots in Cambridge, within the limits any foreigner can put down roots in England and within his own limits: He was a Soviet citizen, and a loyal one. He became Assistant Director of Magnetic Research at the Cavendish, a Fellow of Trinity College and of the Royal Society, took to smoking a special tobacco carried by a local Cambridge tobacconist, and made some deep, enduring friendships.

One such friendship was with John Cockcroft. Lady Cockcroft related to me her first impression of Kapitza:

A wild kind of character, untidy, his overcoat fastened with safety pins, bursting with energy, his words tumbling out . . . He drove a high-powered Lagonda, the sporting car of the day.

Another friendship was with James Chadwick. Kapitza was best man at Chadwick's wedding. He wore his everyday clothes for the occasion, upgraded with a borrowed top hat.

Kapitza married Anna Krylova, daughter of the prominent Soviet mathematician, Aleksei N. Krylov. Krylov and his wife were separated, and Anna had been living with her mother in Paris where Kapitza met her while on a holiday. A warm, sympathetic person, Anna soon won the heart of Cambridge. J. J. Thomson, Master of Trinity College at

the time, assigned his daughter Joan the task of getting to know the wives of the Fellows of the college, particularly the younger ones. Joan and Anna became fast friends. When Joan later married a Russian named Charnock, the friendship became a foursome. Joan Charnock commented on the Kapitzas' very happy marriage:

Some women are essentially wives, and some women are essentially mothers. Anna was much more a wife than a mother.

The Kapitzas' first son was christened in the Russian Orthodox church for the sake—or under the influence—of Anna's mother. Kapitza entreated the priest not to make the boy an "Anglosax."

Kapitza, too, made contact easily. He had long conversations with the poet A. E. Housman, with whom others had great difficulty conversing because for extended periods Housman would say nothing. People wondered what it was they talked about. When asked, Kapitza finally replied, "The Church of England."

Kapitza was also very close to P. A. M. Dirac. The two spent a great deal of time in Kapitza's laboratory, Kapitza teaching Dirac such arts as how to grind rough edges off a piece of glass (startling information for those who consider Dirac the ultimate theorist). Together, they wrote a paper on the reflection of electrons by standing light waves, an experiment that could not be performed at the time, having to await the laser. Kapitza started a discussion club, which met Tuesday evenings after dinner, and, while in Cambridge more than thirty years later, I listened to talks at "The Kapitza Club."

Kapitza's experiments

In the laboratory Kapitza worked very hard and expected others to do likewise. These "others" included three technicians: an Estonian named Laurmann who had come with Kapitza from the Soviet Union; Pearson, the senior technician, and Frank Sadler, a former apprentice who, at age twenty, was completely won over at his hiring interview by Kapitza's statement, "I'm looking for a craftsman." Kapitza warned that the job was to come before anything else. Sadler told me:

It was a nice life. I had no ties and could work all hours. But poor Mrs Pearson, she never saw her husband. Kapitza would be in the workshop saying, "Pearson you cut one, Sadler another, me the third." We all mucked in.

Sadler also recalled evenings when Anna would come to the laboratory to drag her husband off to a dinner party he had forgotten. As Kapitza was going out the door, dusting himself off, he would call back, "I'll look in afterwards, to see how you're doing."

Sadler confessed he never understood Kapitza's many humorous stories, told in

The Institute of Physical Problems built for Kapitza in the Lenin Hills near Moscow. Part of Kapitza's twelve-roomed "cottage" is visible at the right. Housing for staff is also on the site.

"Kapitzarene"—a language said to be equidistant from Russian, English and French. Sadler always knew when to laugh, however, because Kapitza would burst into gales of laughter. Sadler was sure no one else understood the stories either, except Cockcroft who knew Russian.

Kapitza carried out pioneer experiments on properties of matter, such as the electrical resistance of metals, in strong magnetic fields. As some of the effects are more pronounced at low temperatures, cryogenic research was added to the magnetic investigations. Kapitza invented a new and simpler apparatus with which to liquify helium in quantity at relatively low cost. His helium liquifier led to the commercial Collins liquifier. More laboratory space was needed for his experiments. The Royal Society provided funds from a bequest of multimillionaire chemist Ludwig Mond to build a new laboratory, named after Mond, and to be directed by Kapitza. At the same time, Kapitza was appointed Royal Society Messel Professor.

The new laboratory was opened in 1933 by Stanley Baldwin, former (and later) Prime Minister and chancellor of the university at the time. On the facade, to the right of the door, was a crocodile. It had been carved by British sculptor Eric Gill on commission from Kapitza. There is considerable commentary on the crocodile. In *Brighter than a Thousand Suns* by Robert Jungk, Kapitza says, "Mine is the crocodile of science. The crocodile cannot turn its head. Like science it must always go forward with all-devouring jaws," although everyone knew the crocodile represented Rutherford, except Rutherford himself. The latter's biographer, A. S. Eve, wrote that it was said the crocodile never turns back and was accordingly regarded as a symbol of Rutherford's scientific acumen and career. The crocodile, Eve wrote, is regarded in Russia with mingled awe and admiration. I queried several Russians on the subject, and one pointed out that the Russian humor magazine is named *Krokodil*. Another, who worked with Kapitza, said "crocodile is slang for 'boss' in Russian." The London *Times*, reporting the opening of the laboratory said, "The entrance is guarded by a dragon."

Kapitza showed Baldwin around, explained how things worked and pointed out the special design that ensured the roof would not blow off in an explosion.

Kapitza addressing a 1956 meeting in Moscow to honor the 250th anniversary of the birth of Ben Franklin. On such occasions he made indirect appeals for more contact with foreign scientists.

At one point Baldwin asked, "Is that so?" to which Kapitza replied, "Oh, yes you can believe me. I'm not a politician."

Return to the Soviet Union

In the summer of 1934 the Kapitzas went to the Soviet Union, as they had a number of previous summers. The first time had been at the invitation of the Soviet government. George Gamow states in his autobiography that, as a precaution, Rutherford had written the Soviet Ambassador to Britain for assurance that Kapitza would return to Cambridge in September. The assurance was granted and Kapitza returned at the specified time. The same routine was followed for subsequent visits, except for 1934. Gamow says that Kapitza told Rutherford the letter of guarantee was not needed. Cambridge friends say the letter was slow in coming and Kapitza did not want to delay his departure. In any case, Kapitza was certain he would not be detained in the Soviet Union.

When the time came for Kapitza to return to Cambridge, a telegram arrived instead. Time passed. When it became abundantly clear that Kapitza would not be allowed to return, Anna returned by herself to seek aid from influential scientists. She devoted months to the effort.

In discussing Kapitza's detention, one Englishman told me Rutherford had said all along, "They'll getcha." Dirac said that incidents and remarks such as the one Kapitza made to Baldwin may have kept the English from trying harder to get him back. "He had trodden on too many toes," Dirac said. "He was always impatient with important people who wanted to see his lab. He wasn't too polite."

Once Kapitza had been too busy to show two visiting Russians around. It is believed they reported that Kapitza was doing secret war work, seeing no other reason for their not having been shown the laboratory.

Rutherford wrote to Baldwin that Soviet authorities had commandeered Kapitza in the belief he would give important aid to their electrical industry, and "they have not found out they were misinformed." Rutherford also appealed to the Soviet government for Kapitza's return to complete his work "in the interests of science." The Soviet reply was eminently reasonable: It was understandable that England should want Kapitza, and the Soviet Union, for its part, would equally like to have Rutherford. The Soviet Embassy statement said:

> As a result of the extraordinary development of the national economy of the U.S.S.R., the number of scientific workers available does not suffice and in these circumstances the Soviet Government has found it necessary to utilize for scientific activity within the country the services of Soviet scientists who have hitherto been working

The Lomonosov Medal was presented to Kapitza in Moscow, February 1960, to recognize his achievements in low-temperature physics. Eighteen years later he received the Nobel Prize.

abroad. Kapitza belongs to this category.

Kapitza was in a deep depression. Despite attractive offers, he did not start work. The story among Russians is that Kapitza told Premier Molotov, "Don't you know a bird in a cage doesn't sing?" to which Molotov replied, "This bird will sing."

The Kapitza Institute

It was more than a year before the bird began to sing, and then with not much voice. When efforts to persuade the Soviet government to release Kapitza failed, E. D. Adrian and Dirac were dispatched to Russia. Adrian's visit was an official one; Dirac combined a lecture tour with a visit to "cheer up" Kapitza, though he later reported it took Kapitza several years to come out of his depression. Adrian and Dirac's mission was followed by the Soviet purchase of Kapitza's apparatus from the Mond Laboratory for £ 30 000—considered a fair price—and

Cockcroft had it packed and shipped.

Meanwhile, Kapitza's "cage" was being gilded to his specifications: a replacement for the Mond Laboratory was being built. Kapitza was given his choice of site on which to build and the spot he selected, the best in Moscow according to Nikita Khruschev's memoirs, had been designated for a new American Embassy. Stalin, however, became disillusioned with William C. Bullitt, Ambassador to Moscow from 1933 to 1936, and, infuriated by Bullitt's "hardline" politics, decreed that Kapitza's institute and not the US Embassy would be built on that choice location.

In addition to the Institute, a "cottage" consisting of twelve rooms and a terrace was built for Kapitza. A row of townhouses for scientific co-workers and smaller homes for the technical staff were also constructed. Tennis courts and a chauffered limousine completed the package.

The Institute for Physical Problems, as

it was called, took two years to build. Kapitza described the quality of construction as only satisfactory, but in equipment it was one of the foremost—not only in the Soviet Union, but in Europe as well. The bird was chirping, if not singing. Kapitza said of the precision lathes, "we can say with pride that the majority are of Soviet origin." Every effort was made to reduce the administrative staff: A simplified bookkeeping system allowed one accountant to do the work of five, and nine firemen were reduced to a volunteer brigade and an electric signalling system.

One townhouse occupant was Lev Landau. Kapitza's reputation as a master string-puller was evidenced in connection with Landau, who, despite his being Jewish, had been imprisoned in the 1930's as a German spy. Kapitza threatened to leave his institute if Landau were not freed; Landau was freed.

Kapitza went on to criticize relations between Soviet science and industry, criticism he kept up through the years. He argued that Soviet industry, although sufficiently advanced to make anything that could be made elsewhere, was geared to manufacture on a large scale and was ill-adapted to serve scientific needs of a smaller scale. The conditions Kapitza criticized still exist, according to Americans who deal with Soviet industry.

In 1941 the Institute was evacuated to Kazan, capital of the Tatar Republic. No scientific papers came out of the Institute between 1941 and 1944, presumably because it was engaged in war work. The buildings in Moscow suffered no damage, and by August 1943 most of the staff was able to return.

Papers on the superfluidity of liquid helium published in 1944 represent work begun before the war. Then, in 1949, Kapitza published a paper with a somewhat strange title: "On the Problem of the Formation of Sea Waves by the Wind." Similar papers followed: "Dynamic Stability of a Pendulum when its Point of Suspension Vibrates" and "On the Nature of Ball Lightning" (1955). It was not until 1959 that Kapitza was again publishing on subjects that seem appropriate, such as the liquefaction of helium.

Kapitza's friends in the West knew something was wrong during those postwar years. In fact, Kapitza was no longer at the Institute, but working in the garage of his dacha about twenty miles from Moscow—under house arrest. He had been abruptly fired from his position at the Institute in 1948 and reinstated only after Stalin's death.

Khrushchev's memoirs, published in 1974, fill in some details. In 1939 Kapitza had built apparatus to produce liquid oxygen in quantity. Krushchev states:

> ... as time went by, Stalin began expressing his displeasure—I'd even say his indignation—about Kapitza. He

said Kapitza wasn't doing what he was supposed to do; ... the bourgeois press started howling like a pack of mad dogs about how the Russians must have gotten their A-bomb from Kapitza because he was the only physicist capable of developing the bomb. Stalin was outraged. He said Kapitza had absolutely nothing to do with the bomb ...

Kapitza refused to cooperate on the Soviet bomb project and was accused of "premeditated sabotage of national defense." He lost his house, his car and the other perquisites of the directorship, but was still a member of the Academy of Sciences. A small salary from the Academy enabled him to live in his country house, but in a more proletarian manner.

Khrushchev describes his dealings with Kapitza after Stalin's death, when Kapitza again headed the Institute. Kapitza tried to impress upon Khrushchev the importance to the Soviet economy of his method for producing oxygen. Khrushchev had other ideas, stating:

> We wanted Kapitza actually to do what the bourgeois press said he had done: we wanted him to work on our nuclear bomb project ... The point is, he refused to touch any military research. He even tried to pursuade me that he couldn't undertake military work out of some sort of moral principle.

Khrushchev explains why Kapitza was not permitted to travel abroad. Khrushchev had asked M. A. Lavrentev, an influential mathematician, about Kapitza's loyalty and was assured that while Kapitza's thinking on military subjects might be "pretty original," he was still a loyal citizen. Khrushchev then asked if Kapitza might know anything about the military work of others. Lavrentev replied that scientists always talk to one another about their work, and Khrushchev, fearing not Kapitza's disloyalty but his talking too much, refused permission for Kapitza to travel abroad. Behind Khrushchev's decision was the desire to conceal Russia's lack of atomic weapons from the rest of the world.

Apart from Kapitza's direct appeals to travel abroad, he indirectly expressed his desire in addresses commemorating anniversaries of scientists such as Rutherford, Benjamin Franklin, and Lomonosov, regarded by Russians as the founder of their science. After stressing the international character of science, Kapitza would point out that Franklin, having come to science in his forties, was able to make contributions partly because of his contact with major scientific figures. Lomonosov, on the other hand, is virtually unknown outside the Soviet Union, his work not properly credited, owing to his isolation.

Kapitza's isolation, however, was not as extreme as that of Lomonosov. During the late 1950's and early 60's, foreign vis-

itors were allowed and copies of *Time, U.S. News and World Report* and *Playboy* could be found in the Institute. Eventually, Kapitza was even allowed to travel, but not until in his seventies when the Soviet Union had become a major nuclear power.

Visits to the West

Kapitza returned to England in 1966. It was a sentimental visit. Old friendships were renewed as if there had been no interruption. There was some debate, however, as to whether Kapitza's English had improved or deteriorated.

Kapitza offered reporters a wry peace plan involving an exchange of military scientists. "Then there would be no more secrets," he said. He went on to comment on the "brain drain" of British scientists to the United States, saying that Russia was in a more difficult situation, having no one to drain.

In Cambridge Kapitza stood before the Mond Laboratory and, gazing at the crocodile on its facade, put an end to years of speculation. He admitted that the crocodile represented Rutherford, saying "in Russia the crocodile represents the father of the family."

Kapitza came to the US for the first time in 1969 and received an honorary degree from Columbia University. Polykarp Kusch, then Vice-President and Dean of Faculties, called Kapitza's 1922 paper "On the possibility of an Experimental Determination of the Magnetic Moment of an Atom" (the last of the six papers he wrote before going to England) "clairvoyant." Kusch himself had been awarded a Nobel Prize for measuring a related property.

After the ceremonies came a reception. I prefaced my questions by telling Kapitza I had already spoken to his wife at length. "You did well," he replied. "She is authorized person." I questioned him on his role in the development of the Soviet H bomb. At that time Sakharov's name was unfamiliar and most people thought Kapitza had directed the project. "I never did! I never did!" he insisted. "I even suffered for it!"

Back in the Soviet Union, Kapitza no longer works on low-temperature physics, having gone to the other extreme—controlled thermonuclear fusion. The work comes out of his earlier work on ball lightning, done while under house arrest. Old physics apparently neither dies nor fades away.

Old physics won Kapitza his Nobel Prize last year. The citation was for his basic inventions and discoveries in the area of low-temperature physics. (A detailed account of this work was given by Gloria B. Lubkin in the December 1978 issue of PHYSICS TODAY.) In Stockholm, after an introduction by the Stockholm Philharmonic playing Glinka's overture to "Ruslan and Ludmilla," Kapitza accepted his prize in person. (We were told that a large fraction of the award money went to the purchase of a high-powered Mercedes. It is not known whether the manufacturer supplied blueprints.)

The present

Working for the older Kapitza is evidently pretty much the same as working for the younger Kapitza, as the longevity of his nickname—"Centaur"—attests. The name is said to have originated when someone asked a member of the Institute, "What sort of man is your boss?" and the reply was, "It's difficult to say—half man, half beast. Maybe we should call him a centaur." (The Russian word translated here as "beast" is used for a domestic laboring animal.)

Although he often polarizes people, everyone agrees Kapitza is cultivated, vital, witty, and above all, "outspoken." Over the years his criticisms have antagonized people in various high places; he once criticized Marxist philosophers for their rejection of cybernetics which, if followed by Soviet scientists, would have eliminated the Soviet Union from the space race. He later joined intellectuals in appealing to the Communist Party's Central Committee to allow Solzhenitsyn to live and work without interference. When Zhores Medvedev suffered detention in a mental hospital, protests came not only from Solzhenitsyn and Sakharov as expected, but from Kapitza as well.

In his eighties, the outspoken Kapitza now employs the weapon of silence. When academicians sign a condemnation of Sakharov for his political utterances, Kapitza's silence speaks out eloquently.

Bibliography

This article is based primarily on interviews with friends and colleagues of Kapitza from his Cambridge days. Other sources were:

- D. S. Danin, *Rutherford,* Young Guard Publishing House of the Central Committee of the Communist Youth League, Moscow (1966).
- A. S. Eve, *Rutherford,* Cambridge University Press, Cambridge (1939).
- G. Gamow, *My World Line: an Informal Autobiography,* Viking, New York (1970).
- R. Jungk, *Brighter Than a Thousand Suns: A Personal History of the Atomic Scientists,* (J. Cleugh trans.) Harcourt, Brace, Jovanovich, New York, (1970).
- N. Khrushchev, *Khrushchev Remembers, the Last Testament,* (S. Talbott, ed. and trans.) Little, Brown (1974).
- P. Kapitza, *Peter Kapitsa on Life and Science,* (A. Parry, ed. and trans.) Macmillan, New York (1968).
- D. Shoenberg, "Royal Society Mond Laboratory, Cambridge," Nature **171,** 458 (1953).
- P. Kapitza, *Collected Papers of P. L. Kapitza,* (D. Ter Haar, ed. and trans.) Pergamon Press, London, 1967.
- A. Wood, *The Cavendish Laboratory,* Cambridge University Press, Cambridge (1946).   □

The young Oppenheimer: letters and recollections

Correspondence with friends and colleagues and reminiscences
—his own and others'—give insights into the development and character
of an important physicist and public figure.

Alice Kimball Smith and Charles Weiner

PHYSICS TODAY / APRIL 1980

A prominent physicist before World War II, J. Robert Oppenheimer became the wartime director of the Los Alamos nuclear weapons laboratory. After the war he became an influential adviser to the government on atomic energy, but fell from favor during the McCarthy era. This story has become the stuff of myth and drama. Here we would like to present glimpses of the less familiar Oppenheimer—learning, playing, making friends, doing physics, winning recognition—as yet unburdened by the actuality of the bomb, by fame and by public responsibilities.

To many of his contemporaries Oppenheimer was a brilliant scientist, a dedicated public servant, and a fine human being in whom virtue far transcended defect and fully compensated for it. Others saw a man of flawed judgment, sometimes devious or affected in personal relations or in public posture, whose actual contributions did not match his reputation as a physicist. Oppenheimer is often described as complex, but complexity is not (in itself) a trait of personality; it indicates rather that the observer is puzzled. What can confidently be said on the basis of Oppenheimer's early letters is that, even when he was a young man, the world around him, the choices it offered, and the human beings with whom he associated were not simple or easily defined. As Oppenheimer's personality became an object of wider interest, he still maintained an air of privacy, suggesting an inner self withheld from public view. People reacted to this quality with either fascination

Alice Kimball Smith is Dean Emerita of the Bunting Institute at Radcliffe College and Charles Weiner is professor of history of science and technology at the Massachusetts Institute of Technology. This paper is adapted from *Robert Oppenheimer: Letters and Recollections* (Cambridge, MA: Harvard University Press, 1980).

or displeasure. Throughout his life he sometimes showed an uncanny ability to cut through confusion with clarity and precision. At other times he groped his way toward answers and spoke and acted with an ambiguity that puzzled or antagonized those of a different cast of mind.

Yet this man—so difficult to classify, so selective in his preoccupations and his friendships, most at home in abstruse reaches of mathematical physics, who never courted approval outside a small social and intellectual circle, who was considered unpredictable and temperamental even by admirers—became the disciplined leader of the project which built the atomic bombs dropped on Japan in August 1945, thereby revolutionizing warfare and international relations. Oppenheimer's subsequent role as weapons adviser and as a leading architect of American nuclear policy was likewise an unexpected one.

The correspondence

We have collected many of Oppenheimer's letters from 1922, when he entered Harvard, to 1945, when he resigned as director of the Los Alamos nuclear weapons laboratory. The letters help to explain this man who played no small part in shaping the events and character of an era. To supplement the letters we have drawn upon interviews with Oppenheimer and with many of his contemporaries. A particularly valuable source is the interview with Oppenheimer by Thomas S. Kuhn in November 1963, for the Archive for History of Quantum Physics.[1]

One hundred and sixty seven letters together with excerpts from the interviews and other materials were published by the Harvard University Press in 1980.[2] Here we present selections from the years 1926 to 1939.

Part of Oppenheimer's attraction, at

first for his friends and later for the public, was that he did not project the popularly held image of the scientist as cold, objective, rational, and therefore above human frailty, an image that scientists themselves fostered by underplaying their personal histories and the disorder that precedes the neat scientific conclusion. Oppenheimer's foibles, his vulnerability, his capacity for enjoyment and affection are fully apparent in the early letters. We see a sensitive, sometimes awkward young man growing in self-assurance and finding satisfaction in a widening circle of friends, especially when personal compatibility strengthened a bond in physics.

Later letters shed light on Oppenheimer's role in physics in the 1930's when his own interpretation of what he liked to call style in science was influencing colleagues and students. They show that certain qualities of Oppenheimer the charismatic leader did not appear overnight—an engaging blend of hedonism and asceticism, a tough-minded skepticism tempered at times by a compassion born of his own struggle into adulthood, and a hardwon capacity for self-command. Yet the precocious Harvard student and the graduate and postdoctoral worker in Cambridge, Göttingen and Zürich, making a place for himself in the new world of quantum physics, was very much father to the distinguished theoretical physicist and the successful wartime leader. Many of the prewar letters that we have located deal with science. Some of these convey Oppenheimer's sense of excitement and a growing confidence in his ability to understand and extend the new physics unfolding all about him; others express his frustration in attempting to resolve the difficult problems inherent in the theory of quantum mechanics. Mathematics was the powerful tool that promised to illuminate the fundamental nature of physical reality, and it was the international language spoken and written by Oppenheimer and the other young theorists of his generation.

Göttingen

After graduating *summa cum laude* in chemistry from Harvard in 1925, Oppenheimer went to Cambridge to work at the Cavendish Laboratory. His year there was not happy and his work on experimental physics was frustrating; in 1926 he accepted an offer from Max Born to continue his work in Göttingen.

To Oppenheimer looking back, this year represented his "coming into physics." As he told Kuhn,[1] "When I got to Cambridge, I was faced with the problem of looking at a question to which no one knew the answer but I wasn't willing to face it. When I left Cambridge I didn't know how to face it very well but I understood that this was my job; this was the change that occurred that year. I owe a great deal just to the existence of the place and the people who were there; specifically I owe a great deal to [Ralph H.] Fowler's sense and kindness...[By the time I decided to go to Göttingen] I had very great misgivings about myself on all fronts, but I clearly was going to do theoretical physics if I could...It didn't seem to me like foreclosing anything; it just seemed to me like the next order of business. I felt completely relieved of the responsibility to go back into a laboratory. I hadn't been good, I hadn't done anybody any good, and I hadn't had any fun whatever; and here was something I felt just driven to try."

The fulfillment of that passionate urge to contribute to the new physics helped him to resolve his personal and professional dilemmas. From that miserable year in Cambridge Oppenheimer emerged a theoretical physicist as well as his own best therapist.

By 1963 when Oppenheimer reminisced, the tentative nature of the move to Göttingen had seemingly been forgotten. At any rate, he did not burn his bridges when he notified the Board of Research Studies that he was leaving Cambridge (letter numbers are from our book):

Letter 51 to R.E. Priestley

Cambridge, [England]
August 18, 1926

Dear Sir:

I should like to apply to the Board of Research Studies for permission to spend two or three terms next year in Goettingen. My supervisor, Prof. Sir Joseph Thomson is not at present in Cambridge. But Prof. Sir Ernest Rutherford has kindly told me that he would be willing to assure you that my work here had been satisfactory, and that the work which I intended to do at Goettingen was an extension of that which I have started here. He also advised me to tell you that I would, at Goettingen, be under the supervision of Prof. Dr. Max Born, and that Prof. Born was particularly interested in the problems at which I hoped to work. It is now my intention to return to Cambridge immediately on the conclusion of my work in Goettingen.

Yours very sincerely,

J.R. Oppenheimer

The year 1926–27 spent at the University of Göttingen was as important to Oppenheimer's personal and professional growth as any comparable period in his young manhood. He shed the depression of the previous winter and obtained the PhD (under Born) and a postdoctoral fellowship for the year to follow. More important, his standing in the world of physics was transformed by day-to-day discussion with major participants in the development of new theoretical concepts and by his own contributions to this work.

Long after the details had faded he remembered the stimulation of the Göttingen experience: "In the sense which had not been true in Cambridge and certainly not at Harvard, I was part of a little community of people who had some common interests and tastes and many common interests in physics. I remember this more than I do lectures or seminars. I think it quite probable that I attended some of Born's lectures, but I don't remember."

CERN PHOTO (1962) FROM AIP NIELS BOHR LIBRARY

I'm sure I gave a seminar or two, but I don't remember. I met [Richard] Courant . . . I met [Werner] Heisenberg who came there and I had not met him before; [I also met Gregor] Wentzel, and [Wolfgang] Pauli in Hamburg or in Göttingen so that something which for me more than most people is important began to take place; namely I began to have some conversations. Gradually, I guess, they gave me some sense and perhaps more gradually, some taste in physics, something that I probably would not have ever gotten to . . . if I'd been locked up in a room."[1]

Oppenheimer's own surviving correspondence with other physicists during this period provides only a sporadic view of his day-to-day efforts to make physics comprehensible. Individual letters describe fragments of his work and these are difficult to understand and place in perspective today, even by his students. Some of the ideas he soon abandoned because they were wrong or because they did not lead to greater understanding; others emerged as publications in the scientific journals. Even much of the work that survived to the publication stage is now obsolete. Like most of the scientific literature more than a decade old, it has been superseded by new experimental discoveries and new theoretical formulations. As Robert Serber, one of Oppenheimer's students and close collaborators, recently reflected, "Things that are obvious now were not for the people doing it then. It all falls out once you know the answer. The problems they struggled through do not appear today. But there are other problems now."[3]

A letter to Edwin Kemble, who had been one of his physics professors at Harvard, shows how thoroughly Oppenheimer had become involved in specific problems then occupying the attention of scientists in Göttingen. This and subsequent letters to other physicists demonstrate growing familiarity with the mathematical language of quantum mechanics and the range of Oppenheimer's interest in it. They also provide vivid glimpses of the informal communication patterns of scientists as they gossiped and as they proposed solutions to the problems that concerned them.

Letter 53 to Edwin C. Kemble

Göttingen
Physikalisches Institut
Nov 27. [1926]

Dear Dr Kemble,

Many thanks for your kind letter. As I shall not see Mr. Fowler for some time, I have taken the liberty of quoting a paragraph from your letter in a note I sent him.

This term I am spending at Göttingen. It is a very nice place, and I think that you will surely like it. Even now there are quite a few American physicists here, and some will be staying on until the Spring. I expect to be here until March, & then go back to Cambridge; and I hope that I shall have the opportunity of seeing you either here or there.

Almost all of the theorists seems to be working on q-mechanics. Professor Born is publishing a paper on the Adiabatic Theorem, & Heisenberg on "Schwankungen [fluctuations]." Perhaps the most important idea is one of Pauli's, who suggests that the usual Schroedinger ψ-functions are only special cases, & only in special cases—the spectroscopic ones—give the physical information we want. He considers the ψ-solutions when any set of canonical variables is chosen as independent. But of all this you probably know more than I do. People here are also very anxious to apply the q-mechanics to molecules; but so far the only attempt, Alexandrow's paper on the H_2^+-ion, seems to be completely wrong.

I have been working for some time on the quantum theory of aperiodic phenomena. It is possible to get the intensity distribution in continuous spectra on the new theory—and without any special assumption. And in fact the theory gives, when applied to a simple Coulomb model, a very good approximation to the X-ray absorption law. For K electrons, for instance, the absorption per electron is of the form $\lambda^{\alpha} z^3$, where α lies, except just near the limit, between 2.5 and 3.1.

Another problem on which Prof. Born and I are working is the law of deflection of, say, an α-particle by a nucleus. We have not made very much progress with this, but I think we shall soon have it. Certainly the theory will not be so simple, when it is done, as the old one based on corpuscular dynamics.

Please remember me to Professor Bridgman. And thank you again for your letter.

J R Oppenheimer

Although at the end of November Oppenheimer had not yet eliminated the possibility of returning to the Cavendish, the collaboration with Born was so satisfying and productive that he soon decided to complete his doctorate in Göttingen. As indicated in the letter to Kemble, Oppenheimer was continuing work started in England on the application of quantum theory to transitions in the continuous spectrum. This research was embodied in the dissertation for which he received the PhD degree from the University of Göttingen in the spring of 1927. Meanwhile, he also employed quantum mechanics to explain scattering. An important contribution to theoretical

Oppenheimer in 1926 or 1927. (Photo courtesy of Frank Oppenheimer.)

physics was a joint paper with Born on the quantum theory of molecules. The "Born–Oppenheimer approximation" remains in use today.

Born's favorable view of Oppenheimer is recorded in a letter of February 1927 to S.W. Stratton, president of the Masschusetts Institute of Technology. "We have here a number of Americans, five of them working with me. One man is quite excellent, Mr. Oppenheimer, who studied at Harvard and in Cambridge–England. The other men did not surpass the average, but I hope, that not only Oppenheimer, but also some of the other fellows will get their doctor's degree during the next term."[4]

Oppenheimer looked back with mixed feelings upon aspects of the Göttingen experience other than physics: "Although this society was extremely rich and warm and helpful to me, it was parked there in a very miserable German mood . . . bitter, sullen, and, I would say, discontent and angry and with all those ingredients which were later to produce a major disaster. And this I felt very much."[1]

When Edwin Kemble visited Göttingen in June he was able to report to his colleague Theodore Lyman that Harvard's odd duckling was looking more and more like a swan. "Oppenheimer is turning out to be even more brilliant than we thought when we had him at Harvard. He is turning out new work very rapidly and is able to hold his own with any of the galaxy of young mathematical physicists here. Unfortunately Born tells me that he has the same difficulty about expressing himself clearly in writing which we observed at Harvard."[5]

Berkeley and Cal Tech

After postdoctoral work at Harvard, the California Institute of Technology, and several European universities, Op-

Berkeley, mid-1930's. Oppenheimer with Enrico Fermi and Ernest O. Lawrence. (Courtesy of AIP Niels Bohr Library: Fermi Film.)

penheimer accepted, in 1929, a joint appointment at the University of California in Berkeley and Cal Tech. At each school he was regarded as the authority on the new developments in quantum theory, and soon became an influential teacher and leader of a major school of theoretical physics. As he recalled later:

"I think that the whole thing has a certain simplicity. I found myself entirely in Berkeley and almost entirely at Caltech as the only one who understood what this was all about, and the gift which my high school teacher of English had noted for explaining technical things came into action. I didn't start to make a school; I didn't start to look for students. I started really as a propagator of the theory which I loved, about which I continued to learn more, and which was not well understood but which was very rich. The pattern was not that of someone who takes on a course and who teaches students preparing for a variety of careers but of explaining first to faculty, staff, and colleagues and then to anyone who would listen what this was about, what had been learned, what the unsolved problems were."[1]

Among the faculty at Berkeley at the time of Oppenheimer's appointment was Ernest O. Lawrence, who was to play a dynamic role in making the department a center for nuclear-physics research. He and Oppenheimer became close friends. A letter to Lawrence captures some of the spirit of the time. It was written shortly after an APS meeting in New Orleans (held simultaneously with an AAAS meeting), as Oppenheimer and his father were travelling to Pasadena. Oppenheimer's brother Frank had also been in New Orleans for the Christmas holidays. (Frank, eight years younger, was then a sophomore at Johns Hopkins.)

Letter 79 to Ernest O. Lawrence

Texas
Sunday [3 January 1932]

Dear Ernest,

This is an entirely gratuitous little note, written only to compensate for the brevity and sketchiness of our time together in New Orleans, and to thank you for certain very generous things to which in that time I did not do full justice. My brother was very happy at last to meet you, sorry only that the times had been so short. He asked me to tell you this, to send you his greetings, to tell you too with what eagerness he was looking to your visit next summer. We had a fine holiday together; and I think that it settled definitely Frank's vocation for physics. Seeing so together a good number of physicists, it is impossible not to conceive for them a great liking and respect, and for their work a great attraction. We went Thursday with [George] Uhlenbeck and [L.H.] Thomas to a joint session of biochemistry and psychology, it was enormously rowdy and very funny; and it discouraged an excessive faith in either of these sciences...

I hope that in this week before term you will be able to get a good deal of work done. I suppose that it is too much to hope that by the beginning of term the big magnet [for the cyclotron that Lawrence was building] will be ready; but perhaps by then your contractors will be done. If there are any minor theoretical problems to which you need urgently the answer, tell them to [J. Franklin] Carlson or [Leo] Nedelsky; and if they are stumped let me have a try at them. When you see [David H.] Sloan please give him my wishes for a good recovery; and to Berkeley my greetings.

Thank you again for your fine Christmas present. Let things go well with you. à bientôt

Robert

In his interview with Kuhn, Oppenheimer discussed a change he underwent during these years:

"I would think that the transition was... from that of a person who had been learning and also explaining in European centers and in Harvard and Caltech to someone who couldn't much any longer learn from masters but could learn from the literature and from what he did himself, and who had a lot of explaining to do because there was no one else. I think it was not such a sudden transition. Living a life in which you lecture three hours a week and have a seminar [or] another lecture two hours a week leaves you a lot of time for physics and for lots of other things and I wasn't an altered character. I was still primarily a student in terms of what I spent time on...

[Lecturing] took energy, a great deal of energy, but I didn't have to look much up in the book and it was more a question of keeping the presentation fresh and making it sharper and richer. I would think that the big change was that I wasn't an apprentice any longer and I had decided where to make my bed... In a certain sense I had not grown up but had grown up a little, and I think if circumstances had been such that I had had to teach to make a living earlier it probably would have been better for me. I don't think it would have derailed my interest in physics but I think it would perhaps first of all have made it necessary for me to learn what I wanted to know."[1]

Brotherly advice

Oppenheimer wrote many long letters to his brother, reporting on his own life, offering advice and sharing ideas with him. They spent happy summers together at the Oppenheimer family's ranch, which the brothers named Perro Caliente, in the upper Pecos Valley, New Mexico, near Santa Fe. Frank studied at Johns Hopkins, graduating (after three years) in 1933. Following Robert's advice, Frank started with biology, but found himself seduced by physics.

Letter number 83 to Frank Oppenheimer

Berkeley
Sunday [ca. fall 1932]

Dear Frank,

There has never been so long a time which I have let pass without a letter; and never a time when so constantly I have enjoyed your company. Our common life last summer left in me a fine deposit which I have been tapping all these months, a great repository of your words and gestures and of the good hours which we shared. Even now, perhaps, with an answer to your marvellous letters so long overdue, I should not be writing to you if I had not the hope and project of another common holiday in mind. And that is Christmas . . . My suggestion for a half way meeting is perhaps foolish: but how would New Mexico do, that we have neither of us seen in winter, that would be friendly and not quite so far for you as this coast? The only certain point for me is that we should be together, and that we should make the time as pleasant and as right for father as we can.

Your courses sound swell; only I am distressed by this, that they are covering an area very much like the one I cover in my introduction to theoretical physics; for I have an arrogant and stubborn wish that you might be learning these beautiful things from me. I feel sure that your three courses would profit by union: that the reciprocal illumination which function theory, vector analysis, and potential theory give each other is indispensable to a profound understanding of any one of them. Maybe you can try to fill in the bridges for yourself; and I shall try to get you a set of notes for next summer that will help to anchor them. You know of course that I have pretty mixed feelings about this program of yours in which theoretical physics plays such a large part: it is the most delightful and rewarding study in the world, and I can be only glad that you are enjoying it, glad too that we shall always be able to share this treasure. It is only the implications of the course that trouble me: the possibility that you are more and more deeply committing yourself to a vocation which you will regret; the possibility that your motives in this choice—and I wish that I might dismiss this, but only you can—are not wholly in physics and your liking for it. I take it that the biology at Hopkins is abominable—this from many sources: and that the only other academic study of any consequence, that of hard languages, leaves you pretty cold; that does not leave much but vectors and Cauchy's theorem for you to try. But let me urge you with every earnestness to keep an open mind: to cultivate a disinterested and catholic interest in every intellectual discipline, and in the non academic excellences of the world, so that you may not lose that freshness of mind from which alone the life of the mind derives, and that your choice, whatever it be, of work to do, may be a real choice, and one reasonably free. Just yesterday I was over in Marin, the country on the northern seaward arm of San Francisco bay; it was a grey day, with heavy fog blowing in from the sea; and the little lighthouses at all the

perilous points cut off from...the world by the mountains behind and the fog banks out to sea. I suppose that only very gifted and industrious lighthouse keepers get to live in such places; but their mere existence makes me wonder how any man of sense can ever adopt any other vocation... .

The work is fine: not fine in the fruits but the doing. There are lots of eager students, and we are busy studying nuclei and neutrons and disintegrations; trying to make some peace between the inadequate theory and the absurd revolutionary experiments. Lawrence's things are going very well; he has been disintegrating all manner of nuclei, apparently with anything at all that has an energy of a million volts. We have been running a nuclear seminar, in addition to the usual ones, trying to make some order out of the great chaos, not getting very far with that. We are supplementing the paper I wrote last summer with a study of radiation in electron electron impacts, and worrying about the neutron and Anderson's positively charged electrons, and cleaning up a few residual problems in atomic physics. I take it that there will be a lull in the theory for a time; and that when the theory advances, it will be very wild and very wonderful indeed. —I am reading the Cakuntala with Ryder; and at our next meeting shall afflict you with clumsy translations of the superb poems . . .

Write to me pretty soon, if only to tell me what plans for the holidays you like best. God keep you; and let the days be rich and sweet.

Robert

Oppenheimer's remark about "trying

to make some peace between the inadequate theory and the absurd revolutionary experiments" should be viewed against a background of important achievements in 1932 that focused the attention of the physics community on nuclear and cosmic ray research. In January Harold C. Urey at Columbia University discovered a heavy isotope of hydrogen (deuterium). In February James Chadwick at the Cavendish Laboratory demonstrated the existence of the neutron, a new nuclear particle. In April John Cockcroft and E.T.S. Walton, also of the Cavendish, disintegrated the nuclei of light elements by bombarding them with artificially accelerated protons. In August, at Caltech, Carl D. Anderson's photographs of cosmic ray tracks showed the existence of the positron, the positively charged electron. Soon after, at Berkeley, Ernest Lawrence and his students Stanley Livingston and Milton White used their new particle accelerator, the cyclotron, to disintegrate nuclei. These discoveries and techniques provided theorists with exciting challenges and opportunities.[2]

After Johns Hopkins, Frank followed Robert's footsteps and continued his studies at the Cavendish, where he worked on problems related to nuclear physics, a subject of mutual interest. Robert wrote him about these and other problems.

Letter 93 to Frank Oppenheimer

Pasadena
June 4 [1934]

Dear Frank,

Only a very long letter can make up for my great silence, and for the many sweet things for which I have to thank you, letters and benevolences stretch-

Oppenheimer and Lawrence on a visit to the Oppenheimer family's ranch, Perro Caliente, near Cowles, New Mexico, before 1932. (Courtesy of Molly B. Lawrence.)

ing now over many months...

Have you seen what Gamow has done about angular momentum quantum numbers for the levels of the radioactive series? It will be wrong in detail but right in principle. My own labors have been largely devoted to disentangling the still existing miseries of positron theory; and Furry and I have just published another manifesto after which I hope to be able to forget the subject for a time. All of us have been working quite hard, and if you were here I should have a good many minor things of which to tell you; but only in conversation could I do sufficiently casual justice to them. As you undoubtedly know, theoretical physics—what with the haunting ghosts of neutrinos, the Copenhagen conviction, against all evidence, that cosmic rays are protons, Born's absolutely unquantizable field theory, the divergence difficulties with the positron, and the utter impossibility of making a rigorous calculation of

anything at all—is in a hell of a way.

In a fortnight I shall be driving to Ann Arbor, to have three weeks there, exposing positrons. Gamow will be there, and Uhlenbeck, and it should be pleasant. They asked me next year to go to Princeton, where Dirac will be, and permanently to Harvard. But I turned down these seductions, thinking more highly of my present jobs, where it is a little less difficult for me to believe in my usefulness, and where the good California wine consoles for the hardness of physics and the poor powers of the human mind.

[Robert]

Nuclear fission

Glenn T. Seaborg, then an instructor in chemistry, later described the response at Berkeley to the discovery of nuclear fission: "I remember...a seminar in January 1939 when new results...on the splitting of uranium with neutrons were excitedly discussed;

I do not recall ever seeing Oppenheimer so stimulated and so full of ideas."[6]

News of research on nuclear fission and interest in the enormous amounts of energy that might be released were the focus of a letter to his good friend George Uhlenbeck.

At the same time, Oppenheimer was working on the application of general relativity and nuclear physics to theoretical astrophysics. The significance of his work on neutron stars and gravitational contraction became evident in the 1960's and 1970's when the reality of neutron stars, pulsars, and black holes was established through new astronomical research techniques. This work and his continuing interest in the radioactivity of the "mesotron" (the muon) is also mentioned in the letter.

Letter 106 to George Uhlenbeck

Berkeley
Feb. 5 [1939]

Dear George,

I want to answer your fine long very welcome letter at once, partly to show how happy I was to have it...

Here too there is further evidence for the bursting U. They have recorded the heavy tracks in a differential chamber, and seen them, very prominent

Dinner at the International House at Berkeley, around 1939. With Oppenheimer are Chien-Shiung Wu and Emilio Segré. (Courtesy of Emilio Segré; AIP Niels Bohr Library.)
A sailing expedition on the Zürichsee, circa 1930: Oppenheimer, I.I. Rabi, H.M. Mott-Smith and Wolfgang Pauli. (Photo by Rudolph Peierls. Courtesy of AIP Niels Bohr Library.)

Los Alamos and after

"I think that the world in which we shall live these next 30 years will be a pretty restless and tormented place; I do not think that there will be much of a compromise possible between being of it, and being not of it."

Robert Oppenheimer
to Frank Oppenheimer, 10 August 1931

After the start of World War II, Robert Oppenheimer was very much of the world. In 1942 he became coordinator of all fast-neutron research in the US project to develop an atomic bomb. During 1942 and 1943 much of the work on nuclear weapons was centralized, becoming the Manhattan project, and a laboratory was established at Los Alamos, with Oppenheimer as its scientific director.

After the war, Oppenheimer accepted an offer to become director of the Institute for Advanced Study in Princeton. He continued his government service, giving advice on atomic energy, and became chairman of the General Advisory Committee to the Atomic Energy Commission. His ability to give succinct summaries of long discussions and clear statements of complex technical issues made him a valuable member of many advisory panels.

By the 1950's he was a highly respected figure, admired for his unique wartime contribution and for his unstinting service to his country in time of peace. In the meantime, however, the political climate in the US had changed, and in 1954 the AEC suspended Oppenheimer's security clearance and convened a special board to determine the validity of charges that his left-wing activities and associations in the late 1930's—which he had put aside as the war started—made it unwise to trust him with classified information. A further charge involved his postwar work: When the AEC rejected the General Advisory Committee's unanimous recommendation against a crash program to develop a hydrogen bomb, Oppenheimer's lack of enthusiasm for the project was said to have deterred some scientists from working on it.

After lengthy hearings, the panel and the AEC voted to revoke Oppenheimer's security clearance. Many of the politicians and scientists who had frequently sought Oppenheimer's views had been replaced by others, who sought advice more consistent with their own values and priorities, and the votes reflect that shift in power. Attempts to discredit Oppenheimer, however, were vigorously denounced and never fully succeeded.

The hearings caused Oppenheimer a great deal of anguish, but he survived the ordeal surprisingly well. Although one could have expected him to have felt defeated or disgraced, he maintained his dignity after the AEC verdict and showed no rancor or bitterness in his public statements. In 1963 he received the AEC's Enrico Fermi Award for outstanding contributions to atomic energy. In accepting the award from President Johnson he said "I think it is just possible, Mr. President, that it has taken some charity and some courage for you to make this award today. That would seem to be a good augury for all our futures."

After the war, he continued to work on theoretical physics, although his productivity, as measured by papers published, was much less than earlier. But he "was always there to stimulate, to discuss, to listen to ideas."[8] For physicists he became a catalyst and critic, organizing conferences, encouraging younger scientists and new ideas (although he was also often intolerant of views that differed from his).

For the general public he became an interpreter of the atomic age and a spokesman for the cultural values of science.

He retired from the Institute in 1966, when a malignant throat tumor required surgery, but he maintained, as much as possible, his connections with his friends and colleagues and his commitments to organizations. On 18 February 1967, Robert Oppenheimer, 62 years old, died at his home in Princeton. —T. von Foerster

---

over the haze of recoil protons and the faint alpha tracks, from a U foil bombarded by neutrons in a very low pressure cloud chamber. Also Abelson showed that the 72 hour period follows chemically Te, and emits an X ray which by differential critical absorption can be positively identified as the K alpha, and a little K beta, of Iodine. The next activity after Te separates out with I chemically. We too of course have been thinking of the 10^{18} ergs per gram. It seems to me that the pieces after parturition must be highly excited, if only because of their anomalous charge distribution. Some of that must go into radiation, but one would expect neutrons too. So I think it really not too improbable that a ten cm cube of uranium deuteride (one should have something to slow the neutrons without capturing them) might very well blow itself to hell.

There would be much physics to tell, in exchange for your good account, perhaps too much for a letter...We have been working here too on static and nonstatic solutions for very heavy masses that have exhausted their nuclear energy sources: old stars perhaps which collapse to neutron cores. The results have been very odd, will be in part out so soon that I won't bother to write them here—I have gradually talked myself into believing the mesotron decay, although the evidence is not much better than it was two years ago

when we first were thinking of it. The Pasadena people promise to do a really clean experiment with ionization chambers in lakes 4000 m apart next summer.

Two more points, and I shall write soon again. For the first, we have been hoping to get to Perro Caliente quite early in June this year. It is a very beautiful month there, without rain, but with snow in the peaks and very green. How is it: could you and Else come? Don't forget: not the time nor the place nor your welcome...

Say a warm greeting from me to Else, whose generosity reopened this long dormant correspondence; tell her to take good care of herself so she can ride a horse next June.

hasta luegito

Robert

(Philip H. Abelson, a doctoral candidate at Berkeley, was an assistant in the Radiation Lab. When the news of fission reached Berkeley, Abelson immediately saw that the research he was doing for his dissertation might have led to the discovery. As he later recalled,[7] "I almost went numb as I realized that I had come close but had missed a great discovery.")

Oppenheimer's involvement in research on nuclear fission was, at first, only incidental and theoretical. Only in 1941 did he get involved in the wartime effort, starting as director of fast-neutron research in Berkeley. In the course of 1943 the Los Alamos lab was established, with Oppenheimer as its director, in the hills overlooking Santa Fe, near the country where he had earlier spent such happy vacations. That appointment marks the end of the private Oppenheimer.

References

1. Interview with J. Robert Oppenheimer by Thomas S. Kuhn, 18 November 1963, Archive for History of Quantum Physics, AIP Niels Bohr Library, and other repositories.
2. *Robert Oppenheimer: Letters and Recollections*, Alice Kimball Smith, Charles Weiner, eds., Harvard U.P., Cambridge Mass. (1980), pap. ed. 1981.
3. Interview with Robert Serber by Charles Weiner, 25 May 1978, AIP Niels Bohr Library.
4. Born to Stratton, 13 February 1927, Institute Archives and Special Collections; MIT Libraries. Quoted in K. Sopka, "Quantum Physics in America," PhD thesis, Harvard, 1976.
5. Kemble to Lyman, 9 June 1929, Harvard University Archives. Quoted in K. Sopka, ref. 4.
6. G.T. Seaborg, in I.I. Rabi *et. al.*, *Oppenheimer*, Scribner's, New York (1969), page 48.
7. P.H. Abelson, in *All in Our Time: The Reminiscences of Twelve Nuclear Pioneers*, J. Wilson, ed., Bulletin of the Atomic Scientists, Chicago (1975), page 28.
8. H. Bethe, Science **155**, 1081 (1967). □

Robert G. Sachs

When in 1963 she received the Nobel Prize in Physics, Maria Goeppert Mayer was the second woman in history to win that prize—the first being Marie Curie, who had received it sixty years earlier—and she was the third woman in history to receive the Nobel Prize in a science category. This accomplishment had its beginnings in her early exposure to an intense atmosphere of science, both at home and in the surrounding university community, a community that provided her with the opportunity to follow her inclinations and to develop her remarkable talents under the guidance of the great teachers and scholars of mathematics and physics. Throughout her full and gracious life, her science continued to be the theme about which her activities were centered, and it culminated in her major contribution to the understanding of the structure of the atomic nucleus, the spin–orbit-coupling shell model of nuclei.

Göttingen

Maria Goeppert was born on 28 June 1906 in Kattowitz (now Katowice), Upper Silesia (then in Germany), the only child of Friedrich Goeppert and his wife, Maria, née Wolff. In 1910 the family moved to Göttingen, where Friedrich Goeppert became Professor of Pediatrics. Maria spent most of her life there until her marriage.

On 19 January 1930 she married Joseph E. Mayer, a chemist, and they had two children: Maria Ann, now Maria Mayer Wentzel, and Peter Conrad. Maria Goeppert Mayer became a citizen of the United States in 1933. She died on 20 February 1972.

Both her father's academic status and his location (Göttingen) had a profound influence on her life and career. She was especially proud of being the seventh straight generation of university professors on her father's side. Her father's personal influence on her was great. She is quoted as having said that her father was more interesting than her mother: "He was after all a scientist."[1] She was said to have been told by her father that she should not grow

Robert G. Sachs is professor in the Enrico Fermi Institute and the physics department of the University of Chicago. He was Maria Mayer's first graduate student and was director of the theoretical physics division of Argonne National Laboratory in 1946 when Mayer received an appointment to the lab.

COURTESY AIP NIELS BOHR LIBRARY

Maria Goeppert Mayer —two-fold pioneer

PHYSICS TODAY / FEBRUARY 1982

Although Maria Mayer made significant contributions (leading to the Nobel Prize) starting in 1930, it was 30 years before she received a full-time faculty appointment.

up to be a woman, meaning a house-wife, and therefore decided, "I wasn't going to be *just* a woman."[2]

The move to Göttingen came to dominate the whole structure of her education, as might be expected. Georgia Augusta University, better known simply as "Göttingen," was at the height of its prestige, especially in the fields of mathematics and physics, during the period when she was growing up. She was surrounded by the great names of mathematics and physics. David Hilbert was an immediate neighbor and friend of the family. Max Born came to Göttingen in 1921 and James Franck followed soon after; both were close friends of the Goeppert family. Richard Courant, Hermann Weyl, Gustav Herglotz, and Edmund Landau were professors of mathematics.

The presence of these giants of mathematics and physics naturally attracted the most promising young scholars to the institution. Through the years, Maria Goeppert came to meet and know Arthur Holly Compton, Max Delbrueck, Paul A. M. Dirac, Enrico Fermi, Werner Heisenberg, John von Neumann, J. Robert Oppenheimer, Wolfgang Pauli, Linus Pauling, Leo Szilard, Edward Teller and Victor Weisskopf. It was the opportunity to work with James Franck that led to Joseph Mayer's coming to Göttingen and gave him the chance to meet and marry her.

Maria Goeppert was attracted to mathematics very early and planned to prepare for the university, but there was no public institution in Göttingen serving to prepare girls for this purpose. Therefore, in 1921 she left the public elementary school to enter the Frauenstudium, a small private school run by sufragettes to prepare those few girls who wanted to seek admission to the university for the required examination. The school closed its doors before the full three-year program was completed, but she decided to take the university entrance examination promptly in spite of her truncated formal preparation. She passed the examination and was admitted to the university in the spring of 1924 as a student of mathematics. Except for one term spent at Cambridge University in England, her entire career as a university student was completed at Göttingen.

In 1924 she was invited by Max Born to join his physics seminar, with the result that her interests started to shift from mathematics to physics. It was just at this time that the great developments in quantum mechanics were taking place, with Göttingen as one of the principal centers; in fact, Göttingen might have been described as a "cauldron of quantum mechanics" at that time, and in that environment Maria Goeppert was molded as a physicist.

As a student of Max Born, a theoretical physicist with a strong foundation in mathematics, she was well trained in the mathematical concepts required to understand quantum mechanics. This and her mathematics education gave her early research a strong mathematical flavor. Yet the influence of James Franck's nomathematical approach to physics certainly became apparent later. In fact, a reading of her thesis reveals that Franck already had an influence at that stage of her work.

She completed her thesis and received her doctorate in 1930. The thesis was devoted to the theoretical treatment of double-photon processes. It was described many years later by Eugene Wigner as a "masterpiece of clarity and concreteness." Although at the time it was written the possiblity of comparing its theoretical results with those of an experiment seemed remote, if not impossible, double-photon phenomena became a matter of considerable experimental interest many years later, both in nuclear physics and in astrophysics. Now, as the result of development of lasers and nonlinear optics, these phenomena are of even greater experimental interest.

Johns Hopkins

After receiving her degree, she married and moved to Baltimore, where her husband, Joseph Mayer, took up an appointment in the chemistry department of Johns Hopkins University. Opportunities for her to obtain a normal professional appointment at that time, which was at the height of the Depression, were extremely limited. Nepotism rules were particularly stringent then and prevented her from being considered for a regular appointment at Johns Hopkins; nevertheless, members of the physics department were able to arrange for a very modest assistantship, which gave her access to the university facilities, provided her with a place to work in the physics building, and encouraged her to participate in the scientific activities of the university. In the later years of this appointment, she also had the opportunity to present some lecture courses for graduate students.

At the time, the attitude in the physics department toward theoretical physics gave it little weight as compared to experimental research; however, the department included one outstanding theorist, Karl Herzfeld, who carried the burden of teaching all of the theoretical graduate courses. Herzfeld was an expert in classical theory, especially kinetic theory and thermodynamics, and he had a particular interest in what has come to be known as chemical physics. This was also Joseph Mayer's primary field of interest, and under his and Herzfeld's guidance and influence Maria Mayer became actively involved in this field, thereby deepening and broadening her knowledge of physics.

However, she did not limit herself to this one field but took advantage of the various talents existing in the Johns Hopkins department, even going so far as to spend a brief period working with R. W. Wood, the dean of the Johns Hopkins experimentalists. Another member of the department with whom she had a substantial common interest was Gerhard Dieke. The Mathematics Department, which was quite active at that time, included Francis Murnaghan and Aurel Wintner, with whom she developed particularly close connections. However, the two members of the Johns Hopkins faculty who had the greatest influence were her husband and Herzfeld. Not only did she write a number of papers with Herzfeld in her early years there, but also they became close, lifelong friends.

The rapid development of quantum mechanics was having a profound effect in the field of chemical physics in which she had become involved, and the resulting richness and breadth of theoretical chemical physics was so great as to appear to have no bounds. She was in a particularly good position to take advantage of this situation, since no one at Johns Hopkins had a background in quantum mechanics comparable to hers. In particular, she became involved in pioneering work on the structure of organic compounds with a student of Herzfeld's, Alfred Sklar; and in that work she applied her special mathematical background, using the methods of group theory and matrix mechanics.

During the early years in Baltimore, she spent the summers of 1931, 1932 and 1933 back in Göttingen, where she worked with her former teacher, Max Born. In the first of those summers she completed with him their article in the

Handbuch der Physik, "Dynamische Gittertheorie der Kristalle." In 1935 she published her important paper on double beta-decay, representing a direct application of techniques she had used for her thesis, but in an entirely different context.

Later, James Franck joined the faculty at Johns Hopkins and renewed his close personal relationship with the Mayers. Also in that later period, Edward Teller became a member of the faculty of George Washington University, in nearby Washington, D. C., and she looked to him for guidance in the developing frontiers of theoretical physics. At about the same time, she became deeply involved in a collaboration with Joseph Mayer in writing the book *Statistical Mechanics,* published in 1940.

When, as her first *bona fide* student, I turned to her for guidance in choosing a research problem, nuclear physics was on the rise; she told me that it was the only field worth considering for a beginning theorist. She took me to Teller to ask his advice about possible research problems. Our resulting joint work was her first publication in the field of nuclear physics. My thesis problem on nuclear magnetic moments was also selected with Teller's help, and she gave her guidance throughout that work, suggesting application to this problem in nuclear physics of techniques of quantum mechanics in which she was so proficient. These two forays into the field were her only activities in the physics of nuclear structure until after World War II.

Her approach to quantum mechanics, having been so greatly influenced by Born, gave preference to matrix mechanics over Schrödinger's wave mechanics. She was very quick with matrix manipulations and in the use of symmetry arguments to obtain answers to a specific problem; this ability stood her in good stead in her later work on nuclear shell structure, which led to her Nobel Prize. She appeared to think of physical theories, in general, and quantum mechanics, in particular, as tools for solving physics problems and was not much concerned with the philosophical aspects or the structure of the theory.

When she had the opportunity to teach graduate courses, her lectures were well organized, very technical, and highly condensed. She spent little time on background matters of physical interpretation. Her facility with the methods of theoretical physics was overwhelming to most of the graduate sutdents, in whom she inspired a considerable amount of awe. At the same time, the students took a rather romantic view of this young scientific couple, known as "Joe and Maria," and felt that it was a great loss when they left Johns Hopkins to go to Columbia University in 1939.

Columbia

At Columbia University, where Joseph Mayer had been appointed to an associate professorship in chemistry, Maria Mayer's position at first was even more tenuous than at Johns Hopkins. The chairman of the physics department, George Pegram, arranged for an office for her, but she had no appointment.

This was the beginning of a close relationship between the Mayers and Harold Urey and his family, a relationship which was to continue throughout her life, as they always seemed to turn up in the same places in later years. Willard Libby became a good friend, and it was at Columbia that she first began to come under the influence of Enrico Fermi, although she had already met him in her first summer in the United States (1930) at the University of Michigan Special Summer Session in Physics. The Mayers also saw much of I. I. Rabi and Jerrold Zacharias during their years at Columbia.

She quickly put to work her talent for problem solving when Fermi suggested that she attempt to predict the valence-shell structure of the yet-to-be-discovered transuranium elements. By making use of the very simple Fermi–Thomas model of the electronic structure of the atom, she came to the conclusion that these elements would form a new chemical rare-earth series. In spite of the oversimplifications of the particular model, this subsequently turned out to be a remarkably accurate prediction of their qualitative chemical behavior.

In December 1941, she was offered her first real position: a half-time job teaching science at Sarah Lawrence College; she organized and presented a unified science course, which she developed as she went along during that first presentation. She continued, on an occasional basis, to teach part-time at Sarah Lawrence throughout the war.

She was offered a second job opportunity in the spring of 1942 by Harold Urey, who was building up a research group devoted to separating U-235 from natural uranium as part of the work toward the atomic bomb. This ultimately became known as Columbia University's Substitute Alloy Materials (SAM) Project. She accepted this second half-time job, which gave her an opportunity to use her knowledge of chemical physics. Her work included research on the thermodynamic properties of uranium hexafluoride and on the theory of separating isotopes by photochemical reactions, a process that, however, did not develop into a practical possibility at that time. (Much later, the invention of the laser reopened that possibility.)

Edward Teller arranged for her to participate in a program at Columbia referred to as the Opacity Project, which concerned the properties of matter and radiation at extremely high

Goeppert-Mayer and husband, Joseph Mayer, were married in 1930; Maria was 24 and Joseph was 26. Mayer is emeritus professor of chemistry at the University of California, San Diego.

temperatures and had a bearing on the development of the thermonuclear weapon. Later, in the spring of 1945, she was invited to spend some months at Los Alamos, where she had the opportunity to work closely with Teller, whom she considered to be one of the world's most stimulating collaborators.

Chicago

In February of 1946, the Mayers moved to Chicago where Joe had been appointed professor in both the chemistry department and the newly formed Institute for Nuclear Studies of the

Goeppert-Mayer and colleagues—with Max Born (left); with husband and Karl Herzfeld (at right above) and (right) Robert Atkinson (extreme left) and Enrico Fermi (center).

University of Chicago. At the time, the university's nepotism rules did not permit the hiring of both husband and wife in faculty positions, but Maria became a voluntary associate professor of physics in the institute, a position which gave her the opportunity to participate fully in activities at the university.

Teller had also accepted an appointment at the University of Chicago, and he moved the Opacity Project there, giving Maria Mayer the opportunity to continue with this work. It was accommodated in the postwar residuum of the Metallurgical Laboratory of the university where, in its heyday during the war, the initial work on the nuclear chain reaction had been carried out. She was hired as a consultant to the Metallurgical Laboratory so that she could continue her participation in this project, and several students from Columbia who had become graduate students at Chicago worked under her guidance.

The Metallurgical Laboratory went out of existence to make way for establishing Argonne National Laboratory on 1 July 1946, under the aegis of the newly formed Atomic Energy Commission. She was offered and was pleased to accept a regular appointment as senior physicist (half-time) in the theoretical physics division of the newly formed laboratory. The main interest at Argonne was nuclear physics, a field in which she had had little experience, and so she gladly accepted the opportunity to learn what she could about the subject. She continued to hold this part-time appointment throughout her years in Chicago, while maintaining her voluntary appointment at the university. The Argonne appointment was the source of financial support for her work during this very productive period of her life, a period in which she made her major contribution to the field of nuclear physics, the nuclear shell model, which earned her the Nobel Prize.

Since the mission of Argonne National Laboratory at the time was, in addition to research in basic science, the development of peaceful uses of nuclear power, she also became involved in applied work there. She was the first person to undertake the solution by electronic computer of the criticality problem for a liquid metal breeder reactor. She programmed this calculation (using the Monte Carlo method) for ENIAC, the first electronic computer, which was located at the Ballistic Research Laboratory, Aberdeen Proving Ground. A summary of this work was published in 1951 (US Department of Commerce, Applied Mathematics, Series 12:19–20).

While carrying on her work at Argonne, she continued her voluntary role at the University of Chicago by lecturing to classes, serving on committees, directing thesis students, and participating in the activities at the Institute for Nuclear Studies (now known as the Enrico Fermi Institute). The university had pulled together in this institute a stellar assembly of physicists and chemists, including Fermi, Urey, and Libby, as well as Teller and the Mayers, Gregor Wentzel joined the faculties of the physics department and institute later, and the families quickly became very close, one outcome being the joining of the families by marriage of Maria Ann to the Wentzels' son.

Subrahmanyan Chandrasekhar, who had been on the faculty of the astronomy department for many years, also joined the institute. A stream of young and very bright physical scientists poured into the institute, and the atmosphere was stimulating to the extreme. To add to this exciting atmosphere, which in some ways must have been reminiscent of Göttingen in the early days, her former teacher and friend, James Franck, was already a member in the university's chemistry department.

The activities in the institute reflected the interests of the leading lights, interests that were very broad indeed, ranging from nuclear physics and chemistry to astrophysics and from cosmology to geophysics. The interdisciplinary character of the institute was well suited to the breadth of her own activities in the past, so that her Chicago years were the culmination of her variety of scientific experience. In keeping with this, she turned her attention at first to completing and publishing some earlier work in chemical physics, including work with Jacob Bigeleisen on isotopic exchange reactions. Bigeleisen had collaborated with her in other work at Columbia University and at this time was fellow of the institute. At the same time, she began to give attention to nuclear physics.

The shell model

Among the many subjects being discussed at the institute was the question of the origin of the chemical elements. Teller was particularly interested in

this subject and induced Maria Mayer to work with him on a cosmological model of the origin of the elements. In pursuit of data required to test any such model, she became involved in analyzing the abundance of the elements and noticed that there were certain regularities associating the highly abundant elements with specific numbers of neutrons or protons in their nuclei. She soon learned that Walter M. Elsasser had made similar observations in 1933, but she had much more information available to her and found not only that the evidence was stronger but also that there were additional examples of the effect. These specific numbers ultimately came to be referred to as "magic numbers," a term apparently invented by Wigner.

When she looked into information other than the abundance of the elements, such as their binding energies, spins, and magnetic moments, she found more and more evidence that these magic numbers were in some way very special and came to the conclusion that they were of great significance for the understanding of nuclear structure. They suggested the notion of stable "shells" in nuclei similar to the stable electron shells associated with atomic structure, but the prevailing wisdom of the time was that a shell structure in nuclei was most unlikely

because of the short range of nuclear forces as compared to the long-range Coulomb forces holding electrons in atoms. There was the further difficulty that the magic numbers did not fit simple-minded ideas associated with the quantum mechanics of shell structure.

Maria Mayer persisted in checking further evidence for shell structure, such as nuclear beta-decay properties and quadrupole moments, and in trying to find an explanation in terms of the quantum mechanics of the nuclear particles. In this she was greatly encouraged by Fermi and had many discussions with him. She was also strongly supported by her husband, who acted as a continual sounding board for her thoughts on the subject and provided the kind of guidance that could be expected from a chemist who, in many ways, was better equipped to deal with phenomena of this kind than a physicist. The systematics of regularities in behavior with which she was faced had great similarity to the systematics in chemical behavior that had led to the classical development of valence theory in chemistry, and whose fundamental explanation had been found in the Pauli Exclusion Principle.

It was Fermi who asked her the key question, "Is there any indication of spin–orbit coupling?" whereupon she immediately realized that that was the answer she was looking for, and thus was born the spin–orbit-coupling shell model of nuclei.

Her ability to recognize immediately spin–orbit coupling as the source of the correct numerology was a direct consequence of her mathematical understanding of quantum mechanics and especially of her great facility with the numerics of the representations of the rotation group. This ability to identify instantly the key numerical relationships was most impressive, and even Fermi was surprised at how quickly she realized that his question was the key to the problem.

Joseph Mayer gives the following description of this episode:

> Fermi and Maria were talking in her office when Enrico was called out of the office to answer the telephone on a long distance call. At the door he turned and asked his question about spin-orbit coupling. He returned less than ten minutes later and Maria started to 'snow' him with the detailed explanation. You may remember that Maria, when excited, had a rapid-fire oral delivery, whereas Enrico always wanted a slow detailed and methodical explanation. Enrico smiled and left: 'Tomorrow, when you are less excited, you can explain it to me.'

While she was preparing the spin–

Other colleagues included Harold Urey (left), Edward Teller (bottom left) and Hans Jensen (below).

orbit-coupling model for publication she learned of a paper by other physicists presenting a different attempt at an explanation and, as a courtesy, she asked the editor of the *Physical Review* to hold her brief letter to the editor in order that it appear in the same issue as that paper. As a result of this delay, her work appeared one issue following publication of an almost identical interpretation of the magic numbers by Otto Haxel, J. Hans D. Jensen, and Hans E. Suess. Jensen, working completely independently in Heidelberg, had almost simultaneously realized the importance of spin–orbit coupling for explaining the shell structure, and the result had been this joint paper.

Maria Mayer and Jensen were not acquainted with one another at the time, and they did not meet until her visit to Germany in 1950. In 1951 on a second visit, she and Jensen had the opportunity to start a collaboration on

further interpretation of the spin–orbit-coupling shell model, and this was the beginning of a close friendship as well as a very productive scientific effort. It culminated in the publication of their book, *Elementary Theory of Nuclear Shell Structure* (1955). They shared the Nobel Prize in 1963 for their contributions to this subject.

After Fermi's death in 1954, other members of the Institute for Nuclear Studies who had provided so much stimulation for her left Chicago. Teller had gone earlier in 1952, Libby left in 1954, and Urey in 1958. In 1960 she accepted a regular appointment as professor of physics at the University of California at San Diego when both she and her husband had the opportunity to go there.

Her appointment as a full professor in her own right at a major university was very gratifying to her, and she looked forward to the stimulation of

this newest interdisciplinary group of scientists that was being drawn together there. However, shortly after arriving in San Diego, she had a stroke, and her years there were marked by continuing problems with her health. Nevertheless, she continued to teach and to participate actively in the development and exposition of the shell model. Her last publication, a review of the shell model written in collaboration with Jensen, appeared in 1966; and she continued to give as much attention to physics as she could until her death in early 1972.

* * *

This article was adapted from Biographical Memoirs **50**, The National Academy of Sciences (1979).

References

1. Joan Dash, *A Life of One's Own* (New York: Harper and Row, 1973), page 231.
2. *Ibid.*   □

Philip Morrison— a profile

PHYSICS TODAY / AUGUST 1982

Valued for his scientific contributions to the Manhattan Project, to theoretical physics and to astrophysics, he has also contributed to the public understanding of science and has been one of the most thoughtful advocates of arms control.

Anne Eisenberg

When Philip Morrison, Institute Professor at MIT, came to the Polytechnic Institute of New York recently to give the Sigma Xi lecture, a diverse group attended. The group included physicists, chemists, engineers; people who admired Morrison for his sustained fight against red-baiting in the 1950s (in 1953 a national newsletter called him "the man with one of the most incriminating pro-Communist records in the entire academic world"); and people in the humanities who had enjoyed his book reviews, films, articles and textbooks. The diversity of the audience reflected the diversity of Morrison's career.

Morrison is valued in the scientific community for his gift of language, for his wide-ranging intellect, and for his ability to pull together insights from different fields to shed light on a subject. Because he has spent considerable time writing about science—explaining and interpreting it for the public—he exists also in the imaginations of people outside science. He possesses what historian Alice Kimball Smith has called[1] a "rare sensitivity of spirit."

His career has included Los Alamos and Hiroshima in the 1940s, McCarthyism in the 1950s, the Peace Movement in the 1960s, and arms control from 1945 to the present. It began in Pittsburgh where he was reared and attended Carnegie Tech. After an initial interest in radio engineering, he majored in physics, and went on to do his doctoral work in theoretical nuclear physics with Oppenheimer at the University of California at Berkeley. They got along well; Morrison admired Oppenheimer and reminisces today about him: "There was only one difficulty most of us had with Robert. You had to be very careful with him, you couldn't give him too much of your problem, or he would solve it before you."

The Manhattan Project

Morrison had just gone to the University of Illinois at Urbana when the war broke out. Hired by the Manhattan Project, he went to Chicago to work with Fermi, and stayed there until 1944. Morrison became leader of the group that tested neutron multiplication in successive design studies for the Hanford reactors.

Then, in 1944, he was recruited for the Los Alamos effort by Robert Bacher. Morrison worked at Los Alamos in the group headed by Robert Frisch, who, with his aunt Lise Meitner, had pioneered in fission a few years earlier. His job at Los Alamos was to extend work done at Chicago at which he was expert. "We made small critical assemblies to test the neutron behavior of the new plutonium and uranium fission materials being produced at the main plants and shipped to Los Alamos, in preparation for use in the two bombs. Our job was to study chain reactions in that stuff."

It was here that Morrison and his group did the famous experiments later characterized by Feynman as "tickling the dragon's tail." "No one had ever made a chain reaction that had so many prompt neutrons in it," Morrison comments. "All the chain reactions of reactors are mediated in part by delayed neutrons; otherwise they aren't controllable at all. The bomb, on the other hand, is made by fast, prompt neutrons, which of course are uncontrollable."

Morrison was concerned with building up experience on the passage from the controlled state to the uncontrolled state. This meant keeping the reaction in a partially contained state under active control, instead of relying on the inherent stability of the system. "We moved the system so carefully, but so rapidly, that it had no chance to build up on us—we hoped. We came very close to making explosions, stopping just in time. Feynman said this was like tickling the tail of a dragon, and so it was."

In *Disturbing the Universe*, Freeman Dyson characterizes[2] the spirit of Los Alamos as the "shared ambition to do great things in science without any personal feeling of jealousy." Morrison says that for himself the motivation was not science, but victory over the Germans.

In my group, two people died. We had the feeling of front-line soldiers with an important campaign at hand.

To begin with, we felt we were well behind the Germans. Rightly or wrongly, we were seized by the notion of this terrible weapon in the hands of the Germans, whose scientists we respected, admired, and feared greatly because they had been the teachers of our teachers and colleagues.

We felt ourselves a little like the English in 1940—a small band standing in the way. Could we possibly beat them? At first there was this terrible responsibility, and then in the end we became

more and more flushed with the fact that we had overcome them. But it wasn't a question of science. It was one of victory. I remember very well.

Morrison conveyed this atmosphere to us with a story of John Wheeler in Chicago: "When noontime came and the 12:00 o'clock bell rang, most of us would go to lunch at the nearby cafeteria. We'd learned, though, not to bother Wheeler. He brought his lunch and when the bell rang he took it and his Princeton notebook out. Then he went ahead to do what he regarded as his 'real work'. He was so conscientious he would never do this during work hours, only during lunch. And that was the attitude at Los Alamos as well."

The absorption in the immediate task was complete. Only as work on the bomb drew to a climax did Morrison consider how it would be used against the Japanese. "We knew there would have to be a trial, but we thought suitable conditions could be made. For instance, I thought, as did many other people, that there was going to be a warning." But no explicit warning was given. The bomb was tested at Alamogordo 16 July and used on Hiroshima 6 August. Morrison says, "The military authorities rejected any demonstration as impractical. They felt Japan would not be deterred by the sight of a patch of scorched earth in the desert. The military had made up its mind. It would have taken a very powerful political presence—one that wasn't available—to sway them. The United States therefore gave no explicit warning. I think this was a moral failure."

Was Morrison surprised the scientists at Los Alamos were not more concerned with the implications of the bomb they were building? "Not at all. There was much discussion about this in the labs, quieter, of course, than those at the Met Labs in Chicago. But we were seized with a terrible responsibility, and our leaders were trying to make sure our attention was not diverted."

After the Trinity test, Morrison, who had been responsible for the design and final deployment of the plutonium core, again prepared and packed the equipment, this time to go to the Mariana Islands. When the bombs were dropped on Japan he was on the island of Tinian, from which the planes for both atomic attacks set off.

He was among the first Americans to visit Hiroshima after the war. "I had earlier decided that the most useful thing one could do would be to try to go through the entire process as a historical witness." At the invitation of General Thomas Farrell, assistant to General Leslie Groves, Morrison joined the 12-man group that went to Hiroshima just 31 days after the explosion to determine the effects of the atomic bomb released by the *Enola Gay*. They arrived in Yokohama the day after MacArthur, and followed him to Tokyo. "For me," Morrison said in an interview[3] with Daniel Lang, "The first and main impact of Hiroshima's destruction had come . . . when we were flying down there from Tokyo. First we flew over Nagoya, Osaka, and Kobe, which had been bombed in the conventional manner, and they looked checkered—patches of red rust where fire bombs had hit intermingled with the gray roofs and green vegetation of undamaged sections. Then we circled Hiroshima, and there was just one enormous, flat, rust-red scar, and no green or gray, because there were no roofs or vegetation left."

Morrison walked through the city with Geiger counters and Lauritzen electroscopes and aided by an interpreter, a guide, and a policeman. "It had burst precisely at the spot we wanted it to, high over Hiroshima. There had been a minimum of radioactivity."

Arms control

After the war, Morrison returned to the US to find himself at the heart of the movement for international arms control, whose advocates operated in diverse ways—in arenas ranging from guarded offices to hearing chambers and press conferences at the Senate Office Building; dispensing the message through coded teletypes and rushed press statements; disputing with colonels and reconnaisance experts; persuading congressmen and reporters. A large number of concerned scientists—many of them organized into groups such as the Manhattan Project Scientists, the Association of Los Alamos Scientists, the Association of Oak Ridge Scientists, Atomic Scientists of Chicago—met in Washington in the fall of 1945. Out of this meeting the Federation of American Scientists was eventually formed. The Federation began operating in January of 1946, with Morrison as a member of the administrative committee.

Morrison described their original goals to us:

We—the people the press soon characterized as atomic scientists—wanted to turn over technical details of bomb production to a world authority under adequate controls. We sought to prevent a

---

Anne Eisenberg teaches science writing at the Polytechnic Institute of New York in Brooklyn.

nuclear arms race by establishing this worldwide authority.

The Federation believed that a continuing monopoly of the atom bomb by the United States was impossible. Without staff or salary, Federation members worked in Washington preparing reports on how to establish a worldwide atomic authority.

It seems to me that one finds in the story two distinct ways of meeting the sense of responsibility—indeed, of grave duty—that the Manhattan-project scientists as a whole felt then and feel still.

One of these is the way of the "insider." Oppenheimer—lucid, persuasive, wonderfully analytical—worked in secret with generals and diplomats, trying in a thousand ways to demonstrate what the facts implied. Szilard lived by the phone, buttonholing lobbyists and becoming himself the lobbyist *par excellence*. Both men acted inside the government, personally bringing their schemes before the individuals who had power, who wrote and passed laws.

And then there were the rest of us: younger, less famous and less able. Ours was the way of the dissenter. In the way we acted there was a sense less of knowledge than of commitment. William Higinbotham, Joseph Rush, Louis Ridenour, John Simpson and scores of others in Washington spoke and wrote publicly for 3000 scientists back home at the project laboratories or crowding back into the universities, and also for the physicists and chemists who had not been in the project at all but felt about as we did. From shabby rented offices overcrowded and littered with mimeographed statements and pamphlets the 'atomic scientists' floated in the eddying stream of American public opinion.[4]

Morrison comments that "mutual deterrence was not the vision of 1946. The scientists sought true stability then, not metastability, not the top-heavy balancing rock on which we all breathlessly sit."

Morrison played many roles during the period, roles that called both upon his fertile mind and upon his considerable ability as a speaker. He worked for the Bulletin of the Atomic Scientists, composed FAS policy drafts, and appeared as a principal witness at hearings on atomic bomb policy. He worked on a report of ways to detect atomic bomb laboratories, testing sites and assembly plants. But no matter how carefully he and others stressed how an international authority could operate under adequate controls—indeed, no matter how many times they explained

what they meant by "under adequate controls"—they were accused of wanting to give away the bomb.

The arms race that Morrison had predicted grew as the scientists' movement for international controls waned after 1946. Morrison, who joined the faculty of Cornell University in 1946, remained in the fight for international arms control even as the public acclaim for scientists began to ebb.

McCarthyism

He was soon in need of defense himself. As an undergraduate at Carnegie Tech Morrison had joined the Communist Party, and he remained a member when he went to graduate school at the University of California at Berkeley, a school known at that time for its free-thinking, socialistic atmosphere. By 1941, Morrison was out of the party, but his political activities continued. At Cornell, he was deeply involved in the Peace Movement and in a variety of radical intellectual activities. It was not the involvement that was so noteworthy as much as the level of activity: a continuous string of speeches and appearances made Morrison one of the most politically active scientists throughout the fifties.

During this period there were many attempts to fire him. "What has Cornell University done about Morrison?" the right-wing newsletter *Counterattack* asked[5] in March 1953, answering "Nothing!" In part the attempts were foiled by his situation, because, as a private school, Cornell was not quite as vulnerable to pressure as public schools. Nonetheless, considerable forces were exerted on Cornell, where his promotion from associate to full professor was held up for so long that the Physics Department began to talk of refusing to submit any further proposals for promotions until Morrison's was acted on.

His promotion finally became an issue before the Cornell Board of Trustees, who had him summoned. Even in those times, with Morrison the center of a series of attacks for such charges as "urging clemency for the Rosenbergs," the trustees were charmed by Morrison's intelligence and grace; they granted his promotion.

Morrison was also called before Senator William Jenner's Internal Security Subcommittee, where he talked frankly about himself and his early involvement with the Communist Party without naming other names; unsatisfied, the subcommittee continued to pry. For instance, they summoned another physicist for a special security clearance. This physicist was surprised but somewhat flattered to be called for special clearance. When he got there, he was taken aback to discover the committee had no interest in him; they were only using the occasion as an opportunity to pump him about Morrison.

Morrison spent 19 years on the Cornell faculty before going to MIT. At Cornell, Morrison was famous not only for his social activism but also for his teaching. "Phil's a born teacher," Dyson, who was a colleague of Morrison's at Cornell, comments. "Whenever one didn't know what to do with a student, one sent the student along to Phil. He had an infinite supply of patience." Dyson says that it often seemed as if half the graduate students in the Physics Department were taken care of by Morrison, who spent hours talking to them, finding out which research ideas they could tackle.

Astrophysics

It was while Morrison was at Cornell that his interest turned from theoretical physics to astrophysics. "I was always rather interested in astrophysics," he recalls. "As a graduate student I published several small papers in nuclear astrophysical problems with Oppenheimer. At Cornell, though, I was actually trying to be a nuclear physicist until I took a sabbatical leave in 1952."

While on leave, Morrison determined to work on some of Bruno Rossi's problems; he knew Rossi's work from their days together at Los Alamos. "Along with many other scientists in the cosmic-ray domain, the early 1950s found me pushed into astronomy. The cosmic-ray people had always used this natural phenomenon as a source for high-energy particles—mesons were first discovered in cosmic rays—but in the early fifties machines became powerful enough to rival cosmic rays. Then, as machines improved, the cosmic rays were simply outcompeted. So cosmic rays were no longer of central interest from the point of view of their intrinsic physics; the interest was more in where they came from, first considering possible sources within the solar system, and then beyond. That interest gradually drew me and other scientists farther and farther into astronomy." He is pleased with the work he did on the origin of cosmic rays. "I do consider it as rather a high point. I regarded myself as a specialist in cosmic rays during the 1950s. At that time I proposed no single origin for them, but instead suggested they were highly hierarchical." Morrison argued that different places make different cosmic rays and that the highest energy concentrations might come from quasar-like objects such as the nearby radio galaxy M87.

At Cornell Morrison worked with Hans Bethe, a long-term friend and supporter. In 1956 they wrote a textbook together, *Elementary Nuclear Theory*. "It was a useful and happy collaboration," Bethe says today. "He has ideas which are not obvious. His genius is to connect many different parts of physics." As an example, Bethe cites Morrison's discussion of the radiogenic origin of the helium isotopes in rocks. Morrison argued that the ratio of helium-3 to helium-4 is much greater in the atmosphere than it is in rocks, because it rocks helium-4 comes mainly from radioactivity, whereas in the atmosphere there is relatively more helium-3 produced by the cosmic ray-mediated disintegration of nitrogen. "It is a typical insight of Philip's to connect two opposite things—such as cosmic rays and terrestrial radioactivity—to determine the composition of samples taken from such places as hot springs."

Morrison is known not only for his ability to connect disparate elements, but for his willingness to challenge assumptions. His interpretation of M82, once touted as an example of an exploding galaxy, is one instance of this characteristic. Morrison suggested that what we were seeing is not an explosion, but rather an intergalactic dust cloud through which the galaxy is passing, the interaction giving rise to features that one might interpret as an explosion. "Although M82 looks superficially as though it were exploding in a mini-quasarlike way," Morrison comments, "in fact it seems pretty clear it isn't at all." Instead of there being one point-like center—a tiny engine that does everything for the device—the central object is the whole core of the galaxy, thousands of light years across, in which hundreds, even thousands or millions of new stars are suddenly formed. "The rapid bursts of star formations can create in some ways the same kind of activity as if there were a quasar-like object. In this case, however, the energy is primarily nuclear instead of primarily gravitational."

Paul Joss, a theoretical astrophysicist at MIT, comments on Morrison's work: "Both with M82 and with his supernova model, Morrison proposed testable models that gave us something to attack, challenging us and forcing us to rethink." Morrison's supernova model is an attempt to account for the visible light that comes from supernovae "without worrying too much about the causes of the explosion." The central idea of his theory is that the observed light from the supernova consists of two portions: those photons that reach the observer directly along a straight line and those that interact at least once, travelling along a dogleg path. Because the original outburst is so brief, even the small delays that arise from the somewhat greater length of the dogleg path are significant. Simple geometrical arguments

J. Robert Oppenheimer (left) and Major W. A. Stevens in May 1944, selecting a site for the atomic-bomb test. (Photo by Kenneth Bainbridge, courtesy AIP Niels Bohr Library.)

show that the locus of the secondary emission points (places where light from the supernova is absorbed and then reemitted as fluorescence) form a sequence of expanding ellipsoids whose focal points are the point of the supernova outburst and the position of the observer. Because fluorescence efficiencies are typically a percent or less, the total energy of the explosion is from 100 to 1000 times more than can be detected on earth in the visible region.

Joss says, "Phil's work on supernovae is a very good example of his impact on astrophysics. He has a way of looking at fundamental assumptions and asking, 'Why do we believe this?' In supernovae, for instance, there was a standard picture, one that was probably right in a primitive sense, that is, supernovae result from violent explosions in massive stars, causing in turn both a very large expulsion of matter into interstellar space and a very large amount of electromagnetic radiation. But Phil noted that if you take a star the size of the sun and blow it up, you are not going to get a tremendous amount of visible light. The energy that comes out is 10^{10} or 10^{11} times the luminosity of the sun, and if it radiates as a blackbody, then that energy is not going to come out as visible light, it is going to come out as x rays. The expansion of the exploding material, increasing the size of the radiating surface, won't help either, because by the time the material has expanded as much as it has to—through several orders of magnitude times its original size—it will have undergone such adiabatic cooling it will hardly radiate at all. So the reason that one can see this visible light has to be more complicat-

ed. What Phil did was come up with a very specific model. It's been controversial, but that's not the point. It was a testable model that made specific predictions and challenged astrophysicists to reconsider some of their basic assumptions about the supernova phenomenon."

Teaching

Morrison has been at MIT since 1964, first as Francis Friedman Visiting Professor, and then as a permanent faculty member since 1965. Morrison's interest in educational theory influenced his move to MIT. "Gerald Zacharias invited me to the school. He had an intense interest in science education, an interest he knew I shared." MIT was a center of educational innovation, and Morrison was associated with the Physical Science Study Committee at its inception and coauthor of its secondary-school text *Physics*. Morrison, together with Don Holcomb of Cornell, also wrote a physics text for college students, *My Father's Watch*. Although not widely used, it had a special appeal to teachers introducing adults to physics, perhaps because of Morri-

son's care to relate scientific arguments to history, art and philosophy.

Throughout Morrison's career he has interpreted science for the public in popular articles, in science films, and in monthly book reviews for Scientific American. These book reviews in all fields of science are particularly well-known. One hundred years ago, Charles Darwin wrote[6] of the scientist Robert Brown: "He was rather given to sneering at anyone who wrote about what he did not fully understand. I remember praising Whewell's *History of the Inductive Sciences* to him, and he answered, 'Yes, I suppose that he has read the prefaces of very many books.'" Morrison is vulnerable to the same sneer, yet few would comment so of his incisive performances each month in Scientific American. Instead, one senses a polymath interested in every nook and cranny, as Morrison somehow makes his way through the 500 books he receives each month, choosing and then reviewing thoroughly the handful he selects as interesting and instructive.

"I judge my job to see what's inside, and then to unpack it. The nice part of the book-review column in Scientific American, and what makes it different from others, is that I don't need to review all the *important* books. I am not obliged to say, this is a lousy book but we have to review it because it is the work of an important author." Instead, Morrison tries to take a variety of books representing either a good popular approach or an approach at an introductory level. The reviews—serious, generous, often more entertaining than the original volumes—are a reflection of the intellectual energy that consumes Morrison; they are also the result of the peculiar ability he has to

Vatican Conference on nuclei of galaxies, 1970: Morrison, an unidentified priest, Donald Osterbrock, Martin Rees and Edwin Salpeter. (Courtesy of AIP Niels Bohr Library.)

read almost as rapidly as he can turn pages.

Throughout Morrison's book reviews, books and films, there is a stress on the evidence rather than on neatly packaged conclusions or indeed on the personality of the presenter. "The key thing in a science film is to show the evidence," Morrison says, "but the media believe more in testimony and atmosphere." Morrison tells an anecdote to illustrate this conflict. In his film "Whispers from Space," which Morrison considers his best, he spends half the program establishing and demonstrating experiments that are at least 100 years old. For instance, to illustrate one of the most important features of blackbody radiation, the viewers see a kiln loaded with dishes and piggybanks. These gradually heat up until all detail is lost: first the dishes disappear, then the piggybanks, until the viewer is left with a bland, smooth space. When the executive producer saw the clip, he exclaimed, "You're spending all this time and money on a thing you tell me was discovered 150 years ago. We can't do that old-hat stuff."

Morrison comments, "So long as science is seen largely as a personal view, so long as science films have a speaker who mainly ignores the evidence and presents the history of science as his own concoction of ideas and insights, it is possible to talk of Bermuda triangles and flying saucers. It's good enough if someone says it. If you invent myths and don't explain, people can't test the foundations of your beliefs, or be prepared to change when the foundation changes. Then another myth comes along and beats your myth. That's how the creationists can come along with their demands for equal time: as far as they are concerned, it is myth against myth."

Essentially Morrison was a radical as a youth and remains that way today.

His deep involvement in arms control extends from 1945 to the present. Two years ago he, his wife, and four Boston-area colleagues published *The Price of Our Defense: A New Strategy for Military Spending*. The book aimed at limiting the upward-spiraling arms trade and thus lightening what Morrison calls "the thermonuclear sword hanging over all mankind, sharper and heavier each decade." The authors take a look at how much the US needs to spend to maintain its national security, and propose a program for decreasing land, sea and air forces to give a "prudent military structure prepared for eventualities short of all-out nuclear attack." Against an all-out nuclear attack, the authors argue, there can be no defense; one must rely on deterrence alone.

How well has the book done? "The Pentagon was interested," Morrison comments. "It sold quite well in bookstores in Washington. It's also been popular with people in the peace movement. But we are in a period when the Russians are perceived as standing 10 feet tall. There are no signs that the government is considering the nuclear-arms cuts we proposed. In fact, it's quite the opposite." Morrison continues to act as a gadfly to the defense establishment with an energy characteristic of all his political struggles. One of his targets is the Air Force, which he says is on the edge of obsolescence. "Of course, it can't accept that, and so it tries harder. As the largest industrial organization in the world, it is up to all the sorts of things you would expect from a huge organization that cannot face its own obsolescence. The MX system is a perfect example; its chief value lies in its ability to keep the Air Force in the strategic-missile business."

Morrison continues to have a deep concern about nuclear weapons. "It is one of the great failings of the American political process," he says, "that there is a huge hue and cry against nuclear reactors, and nothing much about bombs. I think to some extent this had to do with displacement. People can't deal with bombs, and they displace their concern onto reactors, which turn out to be vulnerable objects. It's a most important phenomenon, the absence of attention to one, and the irrational attention to the other. But since the summer of 1981 I see a decisive change."

One of Morrison's most striking characteristics is the immense energy he has spent writing about science for the public. Why do this? "In part," he replies, "I think it is simply that I have a flair for it. But I imagine it's more than that. I feel very keenly an obligation to maintain the social nexus in which I've learned and become a scientist. The one obligation society makes on you is that you must explain your craft, because that is the cultural treasure you can pass on. People in the future will need the information."

References

1. A. K. Smith, *A Peril and a Hope: The Scientists' Movement in America, 1945–47*, U. of Chicago P., Chicago (1965).
2. F. Dyson, *Disturbing the Universe*, Harper & Row, New York (1979).
3. D. Lang, *From Hiroshima to the Moon: Chronicles of Life in the Atomic Age*, Simon & Schuster, New York (1959).
4. P. Morrison, Scientific American **213**, September 1965, page 257.
5. "Counterattack: Facts to Combat Communism," 6 March 1953, American Business Consultants, Inc., 55 West 42 Street, New York.
6. C. R. Darwin, *Autobiography of Charles Darwin, 1809–1882*, Norton, New York (1969). ☐

— Chapter 5 —
Personal Accounts

History begins with senior people telling younger ones what it was like back in the old days, and this remains by far the most popular kind of history. At meetings of their societies, physicists love to invite eminent people to speak for half an hour or so on how they made their famous discoveries. We find a peculiar fascination in every circumstance surrounding the discoveries, perhaps because the results have become dry facts embedded in textbooks—so neatly built into the structure of science that we could scarcely imagine what physics was like without this knowledge, if the discoverers did not remind us. Some of these recollections have found their way into PHYSICS TODAY; that is the origin of nearly all the articles in this section. From Einstein on relativity to Frisch and Wheeler on fission, from Goudsmit and Uhlenbeck on electron spin to Fermi on the early days of the neutron chain reaction, with Livingston and McMillan on cyclotrons in between, these first-person accounts take us into the hidden core of the scientists' work.

The AIP Center for History of Physics works to get tape recordings of such reminiscences, and asks anyone holding a memorial session to record the talks. The Center also prompts reminiscence directly by conducting tape-recorded oral history interviews, which are then transcribed, edited, and indexed. The Center has over five hundred such recordings in the archives of its Niels Bohr Library, and these are frequently used by researchers. A number of other programs and individual historians also conduct interviews, saving for posterity the recollections of physicists and astronomers.

Memory is notoriously fallible, and historians hesitate to accept anything as fact simply because someone remembers it as happening decades earlier. The AIP Center for History of Physics and other programs strive to secure documentary evidence, for example preserving laboratory notebooks in permanent repositories or making microfilms of unpublished correspondence, and historians use these materials assiduously. It may also be possible to check one person's memory against another's. In many cases the documents or additional interviews tend to support the original recollection; in other cases they do not. Often there is simply no way to check, and the historian must decide whether the story is internally consistent and the speaker generally credible. For example, Frisch's tale of how he and Meitner thought of fission while walking in the snow cannot be confirmed by other evidence, but it has been accepted and repeated (with the stipulation that it is simply a recollection) by many historians.

In any case there can be no doubt that reminiscence contains an inner truth: the experience as it was assimilated by a scientist who was on the spot. For psychological understanding this may actually be more accurate than the bare-bones documentary record. In some fashion that is hard to describe, the first-person account takes us closer than anything else to the living experience of discovery.

Contents

How I created the theory of relativity

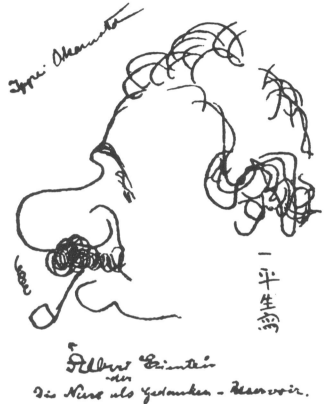

"**The nose** as a reservoir for thoughts" cartoon by Ippei Okamoto. (Courtesy AIP Niels Bohr Library.)

This translation of a lecture given in Kyoto on 14 December 1922 sheds light on Einstein's path to the theory of relativity and offers insights into many other aspects of his work on relativity.

Albert Einstein
Translated by Yoshimasa A. Ono

PHYSICS TODAY / AUGUST 1982

It is known that when Albert Einstein was awarded the Nobel Prize for Physics in 1922, he was unable to attend the ceremonies in Stockholm in December of that year because of an earlier commitment to visit Japan at the same time. In Japan, Einstein gave a speech entitled "How I Created the Theory of Relativity" at Kyoto University on 14 December 1922. This was an impromptu speech to students and faculty members, made in response to a request by K. Nishida, professor of philosophy at Kyoto University. Einstein himself made no written notes. The talk was delivered in German and a running translation was given to the audience on the spot by J. Ishiwara, who had studied under Arnold Sommerfeld and Einstein from 1912 to 1914 and was a professor of physics at Tohoku University. Ishiwara kept careful notes of the lecture, and published[1] his detailed notes (in Japanese) in the monthly Japanese periodical *Kaizo* in 1923; Ishiwara's notes are the only existing notes of Einstein's talk. More recently T. Ogawa published[2] a partial translation to English from the Japanese notes in *Japanese Studies in the History of Science*.

But Ogawa's translation, as well as the earlier notes by Ishiwara, are not easily accessible to the international physics community. However, the early account by Einstein himself of the origins of his ideas is clearly of great historical interest at the present time. And for this reason, I have prepared a translation of Einstein's entire speech from the Japanese notes by Ishiwara. It is clear that this account of Einstein's throws some light on the current controversy[3] as to whether or not he was aware of the Michelson–Morley experiment when he proposed the special theory of relativity in 1905; the account also offers insight into many other aspects of Einstein's work on relativity.

—*Y. A. Ono*

It is not easy to talk about how I reached the idea of the theory of relativity; there were so many hidden complexities to motivate my thought, and the impact of each thought was different at different stages in the development of the idea. I will not mention them all here. Nor will I count the papers I have written on this subject. Instead I will briefly describe the development of my thought directly connected with this problem.

It was more than seventeen years ago that I had an idea of developing the theory of relativity for the first time. While I cannot say exactly where that thought came from, I am certain that it was contained in the problem of the optical properties of moving bodies. Light propagates through the sea of ether, in which the Earth is moving. In other words, the ether is moving with respect to the Earth. I tried to find clear experimental evidence for the flow of the ether in the literature of physics, but in vain.

Then I myself wanted to verify the flow of the ether with respect to the Earth, in other words, the motion of the Earth. When I first thought about this problem, I did not doubt the existence of the ether or the motion of the Earth through it. I thought of the following experiment using two thermocouples: Set up mirrors so that the light from a single source is to be reflected in two different directions, one parallel to the motion of the Earth and the other antiparallel. If we assume that there is an energy difference between the two reflected beams, we can measure the difference in the generated heat using two thermocouples. Although the idea of this experiment is very similar to that of Michelson, I did not put this experiment to the test.

While I was thinking of this problem in my student years, I came to know the strange result of Michelson's experiment. Soon I came to the conclusion that our idea about the motion of the Earth with respect to the ether is incorrect, if we admit Michelson's null result as a fact. This was the first path which led me to the special theory of relativity. Since then I have come to believe that the motion of the Earth cannot be detected by any optical experiment, though the Earth is revolving around the Sun.

I had a chance to read Lorentz's monograph of 1895. He discussed and solved completely the problem of electrodynamics within the first [order of] approximation, namely neglecting terms of order higher than v/c, where v is the velocity of a moving body and c is the velocity of light. Then I tried to discuss the Fizeau experiment on the

Yoshimasa A. Ono is a member of the research staff of Hitachi Ltd. in Ibaraki, Japan.

Albert and Elsa Einstein embarking for the US on the S.S. Rotterdam, 1921, a year before their trip to Japan. (Courtesy AIP Niels Bohr Library.)

assumption that the Lorentz equations for electrons should hold in the frame of reference of the moving body as well as in the frame of reference of the vacuum as originally discussed by Lorentz. At that time I firmly believed that the electrodynamic equations of Maxwell and Lorentz were correct. Furthermore, the assumption that these equations should hold in the reference frame of the moving body leads to the concept of the invariance of the velocity of light, which, however, contradicts the addition rule of velocities used in mechanics.

Why do these two concepts contradict each other? I realized that this difficulty was really hard to resolve. I spent almost a year in vain trying to modify the idea of Lorentz in the hope of resolving this problem.

By chance a friend of mine in Bern (Michele Besso) helped me out. It was a beautiful day when I visited him with this problem. I started the conversation with him in the following way: "Recently I have been working on a difficult problem. Today I come here to battle against that problem with you." We discussed every aspect of this problem. Then suddenly I understood where the key to this problem lay. Next day I came back to him again and said to him, without even saying hello, "Thank you. I've completely solved the problem." An analysis of the concept of time was my solution. Time cannot be absolutely defined, and there is an inseparable relation between time and signal velocity. With this new concept, I could resolve all the difficulties completely for the first time.

Within five weeks the special theory of relativity was completed. I did not doubt that the new theory was reasonable from a philosophical point of view. I also found that the new theory was in agreement with Mach's argument. Contrary to the case of the general theory of relativity in which Mach's

argument was incorporated in the theory, Mach's analysis had [only] indirect implication in the special theory of relativity.

This is the way the special theory of relativity was created.

My first thought on the general theory of relativity was conceived two years later, in 1907. The idea occured suddenly. I was dissatisfied with the special theory of relativity, since the theory was restricted to frames of reference moving with constant velocity relative to each other and could not be applied to the general motion of a

A Japanese Tea Ceremony. The Einsteins' 1922 trip included the usual tourist attractions as well as scientific ones. (Einstein Archives, courtesy AIP Niels Bohr Library.)

reference frame. I struggled to remove this restriction and wanted to formulate the problem in the general case.

In 1907 Johannes Stark asked me to write a monograph on the special theory of relativity in the journal Jahrbuch der Radioaktivität. While I was writing this, I came to realize that all the natural laws except the law of gravity could be discussed within the framework of the special theory of relativity. I wanted to find out the reason for this, but I could not attain this goal easily.

The most unsatisfactory point was the following: Although the relationship between inertia and energy was explicitly given by the special theory of relativity, the relationship between inertia and weight, or the energy of the gravitational field, was not clearly elucidated. I felt that this problem could not be resolved within the framework of the special theory of relativity.

The breakthrough came suddenly one day. I was sitting on a chair in my patent office in Bern. Suddenly a thought struck me: If a man falls freely, he would not feel his weight. I was taken aback. This simple thought experiment made a deep impression on me. This led me to the theory of gravity. I continued my thought: A falling man is accelerated. Then what he feels and judges is happening in the accelerated frame of reference. I decided to extend the theory of relativity to the reference frame with acceleration. I felt that in doing so I could solve the problem of gravity at the same time. A falling man does not feel his weight because in his reference frame there is a new gravitational field which cancels the gravitational field due to the Earth. In the accelerated frame of reference, we need a new gravitational field.

I could not solve this problem completely at that time. It took me eight more years until I finally obtained the complete solution. During these years I obtained partial answers to this problem.

Ernst Mach was a person who insisted on the idea that systems that have acceleration with respect to each other are equivalent. This idea contradicts Euclidean geometry, since in the frame of reference with acceleration Euclidean geometry cannot be applied. Describing the physical laws without reference to geometry is similar to describing our thought without words. We need words in order to express ourselves. What should we look for to describe our problem? This problem was unsolved until 1912, when I hit upon the idea that the surface theory of Karl Friedrich Gauss might be the key to this mystery. I found that Gauss' surface coordinates were very meaningful for understanding this problem. Until then I did not know that Bernhard Riemann [who was a student of Gauss'] had discussed the foundation of geometry deeply. I happened to remember the lecture on geometry in my student years [in Zurich] by Carl Friedrich Geiser who discussed the Gauss theory. I found that the foundations of geometry had deep physical meaning in this problem.

When I came back to Zurich from Prague, my friend the mathematician Marcel Grossman was waiting for me. He had helped me before in supplying me with mathematical literature when I was working at the patent office in Bern and had some difficulties in obtaining mathematical articles. First he taught me the work of Curbastro Gregorio Ricci and later the work of Riemann. I discussed with him whether the problem could be solved using Riemann theory, in other words, by using the concept of the invariance of line elements. We wrote a paper on this subject in 1913, although we could not obtain the correct equations for gravity. I studied Riemann's equations further only to find many reasons why the desired results could not be attained in this way.

After two years of struggle, I found that I had made mistakes in my calculations. I went back to the original equation using the invariance theory and tried to construct the correct equations. In two weeks the correct equations appeared in front of me!

Concerning my work after 1915, I would like to mention only the problem of cosmology. This problem is related to the geometry of the universe and to time. The foundation of this problem comes from the boundary conditions of the general theory of relativity and the discussion of the problem of inertia by Mach. Although I did not exactly understand Mach's idea about inertia, his influence on my thought was enormous.

I solved the problem of cosmology by imposing invariance on the boundary condition for the gravitational equations. I finally eliminated the boundary by considering the Universe to be a closed system. As a result, inertia emerges as a property of interacting matter and it should vanish if there were no other matter to interact with. I believe that with this result the general theory of relativity can be satisfactorily understood epistemologically.

This is a short historical survey of my thoughts in creating the theory of relativity.

* * *

The translator is grateful to the late Professor R. S. Shankland for encouragement and for informing him of reference 2.

References

1. J. Ishiwara, *Einstein Ko-en Roku (The Record of Einstein's Addresses)*, Tokyo-Tosho, Tokyo (1971), page 78. (Originally published in the periodical Kaizo in 1923.)
2. T. Ogawa, Japanese Studies in the History of Science **18**, 73 (1979).
3. R. S. Shankland, Am. J. Phys. **31**, 47 (1963); **41**, 895 (1973); **43**, 464 (1974). G. Holton, Am. J. Phys. **37**, 968 (1972); Isis **60**, 133 (1969); or see *Thematic Origins of Scientific Thought*, Harvard U. P., Cambridge, Mass. (1973). T. Hiroshige, Historical Studies in the Physical Sciences, **7**, 3 (1976). A. I. Miller, *Albert Einstein's Special Theory of Relativity*, Addison-Wesley, Reading, Mass. (1981).   □

FIFTY YEARS OF SPIN

It might as well be spin

Compared to the competitive struggles of today's highly specialized physicists for recognition, the atmosphere in the "springtime of modern atomic physics" was like that of a "Peyton Place without sex."

Samuel A. Goudsmit

PHYSICS TODAY / JUNE 1976

It was a little over fifty years ago that George Uhlenbeck and I introduced the concept of spin. The United States, celebrating its bicentennial, is only four times as old as spin—not even an order of magnitude older. It is therefore not surprising that most young physicists do not know that spin had to be introduced. They think that it was revealed in Genesis or perhaps postulated by Sir Isaac Newton, which young physicists consider to be about simultaneous. There are many other fifty-year mileposts in physics, which also have been forgotten, such as the radio-pulse experiments of Merle Tuve and Gregory Breit that later led to radar.

Restless as a willow in a windstorm

When we reach the stage in life in which our future lies behind us, young people always ask us to look back. Most of us do not realize that we have reached that turning point until we are far beyond it; then looking back becomes a burden and often painful. We have many regrets— but never for what we did, always for what we failed to do. We realize that we failed to make use of many important opportunities, and so our looking back lacks objectivity. You must keep this in mind as you read this article, in which I propose to describe the contrast between today and the springtime of modern atomic physics, which spanned approximately the years from 1919 to the early 1930's and took place primarily in Europe.

Was it really springtime? In some respects, yes. Many little shrubs were planted that in fifty years grew into powerful trees full of fruit-bearing branches. Let me hasten to point out that at the time of planting, it was im-

possible to know which tree would flourish, although in hindsight it appears that some planters and observers made the right guesses, just as at the races and in the stock market.

Was it a romantic time? Were physicists better off and happier than they are today? Wasn't it exciting in those revolutionary years to have personally known Albert Einstein, H. A. Lorentz, Niels Bohr, Paul Ehrenfest, Arnold Sommerfeld, Pieter Zeeman and many others? The answer to all of these questions is, of course, that at that time the young people were not aware of or did not appreciate the circumstances in which they lived. In hindsight it must have been an unusually interesting period, even for minor participants. It is true that I was "restless as a willow in a windstorm" and often "starry-eyed and vaguely discontented" and perhaps suffered from spring fever. From an objective viewpoint however, it was merely different from today's physics; the concept of the "good old days" does not apply.

To describe the difference to those who did not personally experience that period, I must use analogies. The best analogy I have found so far is to say that the present community of physics represents life in a modern metropolis, exciting and full of frustration and dangers. In the 1920's, by comparison, we lived in a small village with its little feuds, a Peyton Place without sex. I am sure that the present generation would not have liked it, most of all because physics and physicists were unimportant to the outside world. The press did not care, the government did not care, the military did not care; isn't that awful? What is even worse, no one got his expenses paid for giving a paper at a meeting.

Marie Curie and Einstein were exceptions in that they had news value. Einstein knew how to capitalize on his fame.

Once, when Ehrenfest asked him why he had gone to Spain where there was no physics of interest to him, Einstein answered, "True, but the King gives such nice dinner parties." In general, publicity was frowned upon and many of Einstein's friends tried to persuade him to shun the press. The photograph of 1923 on the opposite page was not taken because Einstein was visiting in Leiden, but because Douglas Hartree happened to pass through. It shows also how small the number of physicists was: It represents the complete class of Ehrenfest. Only half of these students were physicists, the rest were astronomers and chemists.

Starry-eyed and vaguely discontented

To become a physics student in Europe was an anomaly in the 1920's. Physics was not a profession but a calling, like creative poetry, music composition or painting. I was considered a failure by my family. They expected me to become a businessman, as anybody who worked for a paycheck was considered a weakling. Almost all students I met came from academic families—their education came from the home, their training from the schools. As a physics student, I was considered a sort of misfit. This is quite different from what I found when I came to the US, and especially from the situation today. Physics is now a profession, like engineering or television repair, and physicists come from all walks of life. In Europe in the 1920's it was rather difficult to become a physicist. But once accepted as a serious research student, one had fairly easy access to the big shots in the field, easier than at present.

I never understood why relativity, such an abstract and difficult subject, caught the interest of the general public. World War I was followed by years of very severe economic problems, uncertainties and political upheavals in many areas. We

Samuel A. Goudsmit is visiting professor at the Department of Physics of the University of Nevada, Reno.

are living again in such a period of insecurity and again we see an increased public interest in the abstract, the occult, in extrasensory experiences and the Loch Ness monster. It scares me especially because this time the lunatic fringe includes some physicists. One of them was so taken in by the spellbinding, spoon-bending Uri Geller that he wrote a book about it. The world has lost confidence in scientific and rational reasoning; physics is now hard to sell.

It was in that old protected atmosphere that George Uhlenbeck and I came up with the concept of electron spin. The number of active physicists was small, and since I had already published several papers on spectra and atomic energy levels, I was personally acquainted with several of them. But I did not think spin important enough to send any of them preprints or to write to them about it. I did not worry at all about being scooped. This had happened to me a couple of

times with work on spectral lines, but there was so very much left to do that it was merely a disappointment, not a catastrophe.

Personally the spin solution gave me pleasure but not excitement. I did not appreciate its possible significance until Bohr showed such a great interest in it. There were other items in my physics work that gave me more of a thrill—for example the first experimental determination, together with Ernst Back in Tü-

Spin, Leiden University and all that. The class of Paul Ehrenfest (near the center) stands in front of the door of the Institute for Theoretical Physics at the University of Leiden, probably in 1923. Albert Einstein stands in the doorway, but the reason for the picture was a visit by Douglas Hartree (between Einstein and Ehrenfest). The author, Sam (then Sem) Goudsmit, is on the right. Jan Tinbergen (left, no hat) switched fields after obtaining his PhD in physics, and received a Nobel Prize in economics. The tall man next to Ehrenfest is Gerhard Dieke, who later became physics chairman at Johns Hopkins. The woman beside Einstein is Ini Roelofs; Jaap Voogt and Bernard Polak are fourth and fifth from left.

bingen, of a nuclear spin, that of bismuth. There was of course no such thing as a press conference when we discovered electron spin. For me there were no job offers either, not even as a high-school teacher.

Today, a new bump in a curve is enough to call Walter Sullivan of *The New York Times* out of bed to make sure that the work will get a headline in the paper. Today competition is fierce and often ruthless, because so much is at stake. Funding, promotion and a whole career may depend on publicity and on the *Citation Index*. In the 1920's, competition and animosities could be strong too, and sometimes affected careers, but funding was a minor consideration. There were very few academic openings and political considerations often determined who was chosen to fill them.

The published correspondence between Einstein and Max Born shows how difficult it was for Jews to get jobs in Germany long before Hitler came to power. The same was true over here, at many, but not all, universities and industrial laboratories. When my former student Robert F. Bacher was considered for a position at Cornell University in 1934, R. C. Gibbs asked me in confidence, on behalf of F. K. Richtmyer, whether Bacher was Jewish—if so, he would not have got the job. Some of these animosities had an international character, an aftermath of World War I.

The German physicist Sommerfeld published his great and influential book on atomic structure in 1921. It contained a chapter about radioactivity, which did not mention the Curies. The French were obviously offended. However, in those days the work of the Curies was considered not physics but chemistry and I very much doubt that Sommerfeld had deliberately omitted their name. But when the Dutch physicist Dirk Coster sent Sommerfeld a manuscript on x-ray energies for an opinion, he kept it so long that one of his own pupils, Walter Kossel, was able to scoop Coster. Similarly, on a visit to Holland, Sommerfeld learned that I was working on the spectrum of iron. He made sure that his pupil Otto Laporte got his results published first, making my efforts obsolete. A professor protected his pupils more than himself. These are just examples of common quibbles, minor compared to today's frantic races for a Nobel Prize.

It is sometimes pathetic to observe the present almost violent striving for publicity. The biochemist Erwin Chargaff describes it pointedly for his field and states: "That in our days such pygmies throw such giant shadows only shows how late in the day it has become.[1] What Chargaff overlooked is that pygmies also throw large shadows at dawn. This could be applied to me and several others in the 1920's, the dawn of the new physics. It is late in the day for physics too, but I am not going to predict the future—I leave

that to astrologers and computer addicts. Fortunately there are still and will always be a core of physicists who pursue their science for its great intrinsic value only. They love to teach and are not overanxious to burst into print and publicity with subliminal results and half-digested ideas. We can recognize their work because the adjective "beautiful" applies to it.

There are many colleagues who believe that we received the Nobel Prize for introducing electron spin. In fact, Lee DuBridge recently introduced me as an early Nobel Prize winner; I have also seen it in print. That is all very flattering but does not supplement my TIAA pension. About thirty years after introducing spin, we got the Research Corporation Award and shared $2500; the following recipients received $10 000. Again ten years later, we each received the Max Planck Medal from the German Physical Society. The Nobel Prize was not for us—there were too many physicists who made more important contributions at that time. For example, such spectacular advances as the explanation of radioactivity by George Gamow, Edward Condon and R. W. Gurney and that of chemical binding by Walter Heitler and Fritz London were not considered worthy of the award. Even the theory of relativity was ignored; Einstein was awarded the Nobel Prize for his explanation of the photoelectric effect. The discovery of spin was the main factor for our being offered, in 1926, jobs at the University of Michigan as instructors. That was for me a far more significant award than a Nobel Prize.

Busy as a spider spinning daydreams

This brings me to another difference between the 1920's and today: the rapidity of change. As a young student I was totally oblivious of possible changes. As in my theme song, "I was as busy as a spider spinning day dreams." When Sommerfeld's book appeared I literally believed that being cited in one of its footnotes meant immortality. I have forgotten whether I made it in a later edition! The book has been obsolete for decades. Another dream was some day to be the successor of Zeeman at the University of Amsterdam and continue experiments on spectra and the Zeeman effect. Years later when I was offered the position, that area of physics was dead. I did not take the job. Though changes did occur in the 1920's, one could follow them more easily than today, in almost all of physics. Today, extreme specialization is a necessity for a physicist who wants to make a meaningful contribution. The different branches of modern physics now speak different languages; each uses its own jargon, unintelligible to those working in other areas.

The present generation is hardly aware that we are living in a time of rapid change, of revolution. A physicist's work may be forgotten or considered as be-

PHOTO: A. PAIS, ROCKEFELLER UNIVERSITY

The originators of the concept of spin, George Uhlenbeck (center) and Samuel Goudsmit (right), are together, in 1926, with Oskar Klein, a Swedish physicist. Klein, who had spent the previous year at the University of Michigan, was responsible for their being invited to teach there.

longing in the public domain after two years instead of after fifty years. I did my best to adapt our journals to the present-day hectic activities, for example by creating *Physical Review Letters*. But that is not enough. The *Physical Review* has to change further also. That journal reminds me of an old mansion, still inhabited by remnants of a family that gradually has lost its fortune and its servants but clings to outer appearances. The physics community clings to the journal's format, which is too impressive a facade for contents no longer very impressive. Just read almost any article in the *Physical Review* of the 1920's and 1930's to see the difference.

Now a final remark. Many young people believe erroneously that wisdom comes with age. On the contrary, age brings fear of novelty and progress, fear of loss of status. Almost forty years ago I listened to the great Arthur Eddington lecturing about the fine-structure constant, 137. The little I understood was obviously farfetched nonsense. I asked my older friend, H. A. Kramers, whether all physicists went off on a tangent when they grew older. I was afraid. "No, Sam," answered Kramers. "You don't have to be scared. A genius like Eddington may perhaps go nuts, but a fellow like you just gets dumber and dumber."

* * *

This article is an adaptation of a talk presented 2 February at the joint New York meeting of The American Physical Society and the American Association of Physics Teachers as part of a symposium celebrating the 50th anniversary of the discovery of electron spin.

Reference

1 E. Chargaff, Science **172**, 637 (1971).

FIFTY YEARS OF SPIN

Personal reminiscences

How one student who was undecided whether to pursue a career in physics or history and another who had not taken his mechanics exam came to identify the fourth atomic quantum number with a rotation of the electron.

George E. Uhlenbeck

In a one-page Letter to the Editor of *Naturwissenschaften* dated 17 October 1925, Samuel A. Goudsmit and I proposed the idea that each electron rotates with an angular momentum $\hbar/2$ and carries, besides its charge e, a magnetic moment equal to one Bohr magneton, $e\hbar/2mc$. (Here, as usual, $\hbar$ is the modified Planck constant, m the mass of the electron and c the speed of light.) Sam, in his accompanying article, tells something of those times, fifty years ago. We have often talked about the circumstances that led to our idea, but it was mainly Goudsmit's recollections that have appeared in print before now—they are, however, not readily accessible in English.[1,2,3] Although I gave a short account[4] of the discovery of the spin as a part of my inaugural address for the Lorentz professorship in Leiden in 1955, it therefore appears to be my turn to reminisce.

I am a bit reluctant to do this; first, because my memories differ only in emphasis and in a few details from Sam's recollections, and second, because to describe the personal relationships and the circumstances properly requires, I think, almost an autobiography! However, since this is of course not meant to be a contribution to the history of the great consolidation of the quantum theory in the 1920's, I will just try to tell my side of the story, for what it is worth.

Note that I do not use the modish

George E. Uhlenbeck is professor emeritus of physics at The Rockefeller University, New York.

Sonderdruck aus Die Naturwissenschaften. 13. Jahrg., Heft 47
(*Verlag von Julius Springer, Berlin W 9*)

Ersetzung der Hypothese vom unmechanischen Zwang durch eine Forderung bezüglich des inneren Verhaltens jedes einzelnen Elektrons.

§ 1. Bekanntlich kann man die Struktur und das magnetische Verhalten der Spektren eingehend beschreiben mit Hilfe des LANDÉschen Vektormodelles R, K, J und m^1). Hierin bezeichnet R das Impulsmoment des Atomrestes — d. h. des Atoms ohne das Leuchtelektron — K das Impulsmoment des Leuchtelektrons, J ihre Resultante und m die Projektion von J auf die Richtung eines äußeren Magnetfeldes, alle in den gebräuchlichen Quanteneinheiten ausgedrückt. Man muß dann in diesem Modell annehmen:

a) daß für den Atomrest das Verhältnis des magnetischen Momentes zum mechanischen doppelt so groß ist, als man klassisch erwarten würde.

b) daß in den Formeln, wo R^2, K^2, J^2 auftritt, man diese durch $R^2 - \frac{1}{4}$, $K^2 - \frac{1}{4}$, $J^2 - \frac{1}{4}$ ersetzen muß. [Die HEISENBERGsche Mittelung²)].

Dieses Modell hat sich äußerst fruchtbar gezeigt und hat u. a. geführt zur Entwirrung der verwickeltesten Spektren.

§ 2. Man stößt aber auf Schwierigkeiten, sobald man versucht, das LANDÉsche Vektormodell anzuschließen an unsere Vorstellungen über den Aufbau des Atoms aus Elektronen. Z. B.:

a) PAULI³) hat schon gezeigt, daß bei den Alkaliatomen der Atomrest magnetisch unwirksam sein muß, da sonst der Einfluß der Relativitätskorrektion eine Abhängigkeit des ZEEMANeffektes von der Kernladung verursachen würde, welche in diesen Spektren nicht wahrgenommen ist.

b) Beim LANDÉschen Modell darf man das Impulsmoment des Atomrestes nicht mit demjenigen des positiven Ions identifizieren, sowie man es nach der Definition des Atomrestes erwarten würde. [Verzweigungssatz von LANDÉ-HEISENBERG⁴) — unmechanischer Zwang].

c) Bei einigen in der letzten Zeit mit Hilfe des LANDÉschen Schemas analysierten Spektren (z.B. Vanadium, Titan) stimmte das K des Grundtermes gar nicht mit dem Werte, welchen man aus dem BOHR-STONERschen periodischen Systems erwarten würde.

§ 3. Die obengenannten Schwierigkeiten zeigen alle in dieselbe Richtung, nämlich, daß die Bedeutung, welch[...] [...]NDÉschen Vektoren zukennt, wahr[...] [...] P[...]⁶) hat schon einen [...] an Schwier[...]

§ 4. In beiden Auffassungen bleibt jedoch das Auftreten des sog. relativistischen Doubletts in den Röntgen- und Alkalispektren ein Rätsel. Zur Erklärung dieser Tatsache kam man in letzter Zeit zur Annahme einer klassisch nicht beschreibbare Zweideutigkeit in den quantentheoretischen Eigenschaften des Elektrons¹).

§ 5. Uns scheint noch ein anderer Weg offen. PAULI bindet sich nicht an die Modellvorstellung. Die jedem Elektron zugeordneten 4 Quantenzahlen haben ihre ursprüngliche LANDÉsche Bedeutung verloren. Es liegt vor der Hand, nun jedem Elektron mit seinen 4 Quantenzahlen auch 4 Freiheitsgrade zu geben. Man kann dann den Quantenzahlen z.B. folgende Bedeutung geben:

n und k bleiben wie früher die Haupt- und azimuthale Quantenzahl des Elektrons in seiner Bahn.

R aber wird man eine eigene Rotation des Elektrons zuordnen³).

Die übrigen Quantenzahlen behalten ihre alte Bedeutung. Durch unsere Vorstellung sind formell die Auffassungen von LANDÉ und PAULI mit all ihren Vorteilen miteinander verschmolzen³). Das Elektron muß jetzt die noch unverstandene Eigenschaft (in § 1 unter a genannt), welche LANDÉ dem Atomrest zuschrieb, übernehmen. Die nähere quantitative Durchführung dieser Vorstellung wird wohl stark von der Wahl des Elektronenmodells abhängen. Um mit den Tatsachen in Übereinstimmung zu kommen, muß man also diesem Modell die folgenden Forderungen stellen:

a) Das Verhältnis des magnetischen Momentes des Elektrons zum mechanischen muß für die Eigenrotation doppelt so groß sein als für die Umlaufsbewegung⁴).

b) Die verschiedenen Orientierungen vom R zur Bahnebene (oder K) des Elektrons muß, vielleicht in Zusammenhang mit einer HEISENBERG-WENTZELschen Mittelungsvorschrift⁵), die Erklärung des Relativitätsdoubletts liefern können.

G. E. UHLENBECK und S. GOUDSMIT.

Leiden, den 17. Oktober 1925.
Institut voor Theoretische Natuurkunde.

¹) W. HEISENBERG, Zeitschr. f. Phys. **32**, 841. 1925.
²) Man beachte, daß man die hier auftretenden Quantenzahlen des Elektrons im Alkalispektren entnehmen muß. R hat also für jedes Elektron nur den Wert 1 (in LANDÉscher Normierung).
³) Z. B. wird nun auch die Bedeutung des HEISEN-BE[...] Schema III [...] worin m[...]

The spin hypothesis was proposed in this Letter, which might never have seen the light of day because of objections based on a rigid-electron model, but it was too late to withdraw it.

words "revolution" or "breakthrough." It was really a consolidation of many lines of thought, which admittedly occurred in the rather short period say from 1923 till 1928, but which required about twenty years of preparation. It will be a great but very difficult task to write a proper history about this period. Sam[1] is very skeptical about it and perhaps one must wait till more materials (such as the letters of Wolfgang Pauli) become available.

I will not go into the priority question. Sam has told all about this, especially in his *Delta* article,[3] and I agree with his conclusions. However, a short contribution by E. H. Kennard should be mentioned.[4] We were clearly not the first to propose a quantized rotation of the electron, and there is also no doubt that Ralph Kronig anticipated what certainly was the main part of our ideas in the spring of 1925, and that he was discouraged mainly by Pauli from publishing his results. In the memorial volume to Pauli, Kronig has written an article about the crucial period 1923–25, in which he also describes his personal experiences.[6] In the same volume there is a very useful survey by Bartel van der Waerden,[7] in which especially Pauli's contributions are discussed. Both articles are at most only mildly critical about Pauli's attitude about the spin hypothesis, and van der Waerden says explicitly that in his opinion Pauli and Werner Heisenberg can not be blamed for not having encouraged Kronig to publish his hypothesis. I do know, however, from a long conversation with Pauli in the 1950's during a summer school in Les Houches, that he blamed himself about the whole episode—"*Ich war so dumm wenn ich jung war!*" ("I was so stupid when I was young!") All I think one can say is that our proposal came just at the right time, that we had perhaps a better appreciation of its consequences—especially with respect to the fine structure of hydrogen—and, finally, that we had the luck and the privilege to be students of Paul Ehrenfest. His role in the story will become clear in the following.

Switching to paradise

Let me begin my story with some autobiographical notes. In September 1918 I started at the Technical University in Delft as a student in chemical engineering. I wanted to study physics and mathematics, but I did not have the classical education that the law required for admission to study at the University in Leiden. The work at Delft was very busy and disciplined. Every afternoon I worked in the chemical laboratory, which I especially disliked, probably because I was not very good at it.

In January 1919, the law was changed, thank God; the new so-called "Limburg law" (I will never forget the name!) allowed barbarians like me to study the sciences and medicine at the universities. I persuaded my parents to let me switch

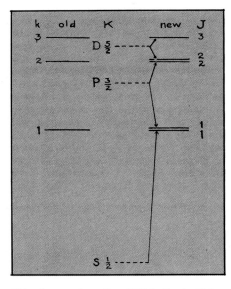

This diagram from the 1926 Letter to *Nature* illustrates how the spin hypothesis changes the explanation of the fine structure of hydrogen-like spectra. The principal quantum number is three; the dotted lines are the levels without spin. The new levels are at the same places as in the Sommerfeld theory, but the earlier disagreements with the correspondence principle have been resolved by using the concept of spin.

to Leiden, which was easy because no additional tuition was required—I only had to change my commuter ticket from Delft to Leiden. I lived at home with my parents in The Hague.

I found Leiden a kind of paradise. We had to take only five lectures a week and one afternoon of a rather standard physics laboratory. There was a wonderful physics and mathematics library, the so-called "Bosscha Reading Room" of which Ehrenfest was the director. In physics there were three professors. In addition to Ehrenfest there was H. Kamerlingh Onnes, the famous director of the low-temperature laboratory, and Johannes Kuenen, a very fine man, who gave the first-year courses. There were few students (in my year only four) and we all knew each other. And, to top it all, there were long vacations!

Since my high-school years I had been especially interested in the kinetic theory of gases, because to me it appeared to be a theory that really *explained* the observed phenomena. In all the free time I had, I therefore studied Boltzmann's *Vorlesungen über Gastheorie*. It was hard going; I had to learn analytical mechanics and several branches of mathematics just to be able to follow the argument. But I really did not understand what it was all about. I also dipped into Gibbs's *Statistical Mechanics*, with the same experience. It was therefore a revelation for me when I got hold of the famous *Encyclopedia* article of Paul and Tatiana Ehrenfest. Suddenly it became clear what the basic problems were and what had been achieved by the founders of statistical mechanics. There were a

whole series of open problems and questions, which showed the so-called "frontier" of the subject. Of course it did not occur to me to try to answer some of these questions—I did not have the chutzpah! I was a conscientious student and I thought that I had to study all the books before trying to do anything new.

In these years I hardly knew Goudsmit, who was two years younger and was therefore just coming over the horizon. I also had little contact with Ehrenfest. He knew that I existed, and once in a while he looked over my shoulder to see what I was studying. But I was too shy to ask him questions, which was almost a prerequisite for talking to him! All this changed completely after I had passed my so-called "candidate's examination" (roughly equivalent to the BS degree), which as a conscientious student I did in the required time (December 1920).

That year I also began to follow Ehrenfest's lectures, and was also allowed to come to the famous Wednesday colloquium. I have described Ehrenfest's methods of teaching elsewhere,[8] so not to go too far afield let me just say here that I remember those wonderful years especially because of the friendliness and feeling of community of the whole group. There was no competition. And this *all* came from Ehrenfest. He taught us that physics was not only fascinating but also fun, something we should share with each other. He had not a grain of pompousness, a trait that was (and still is!) rare among professors. We now know that, already in those years, he struggled with his feeling of inadequacy and with periods of depression, but he never showed it to us. I still remember his jokes and his laughter!

The years in Rome

For me the only trouble was that, to earn money, I accepted a job in my fourth year. I taught mathematics, ten hours a week, at the high school in Leiden. I did not mind the teaching, but I had trouble keeping order in my classes, and I begrudged the time it took. I did not get much sympathy from my father, who pointed out that, as I knew, even with a doctor's degree all I could expect was a job as a high-school or gymnasium teacher in some Dutch town. As he said, "*Tu l'as voulu, George Dandin*"! ("You wanted it, George Dandin.") When Ehrenfest asked in class, some time in the spring of 1922, whether anybody was interested in a tutoring job in Rome for a couple of years I immediately raised my hand. Thus began my Roman period, which almost changed the course of my life.

Since all this is meant to be an introduction to the wonderful summer of 1925 when Sam and I worked together, I will try to keep it short. My job in Rome, about ten hours a week, was to teach the youngest son of the Dutch ambassador, J. H. van Royen, all the subjects required in a Dutch gymnasium except the classical languages and history, for which there was a second tutor and which took the rest of the boy's time. Every summer the boy and I went back to Holland, where he was tested to see whether he was ready for the next grade in school. And so it went for three years.

I had never been outside Holland since my sixth year, so it was a real adventure for me. I got a princely salary, and except for studying the textbooks to keep ahead of my student, I did not have much to do. The first year I started to take Italian lessons, which I kept up in the following years. This was the most intelligent thing I did in those days, and I am still proud of it. The first year I also studied hard for my doctoral examination, which (again as a conscientious student) I passed in the required time (September 1923). After that time I became more and more interested and involved in the cultural history of Italy. I travelled a lot (I could afford it!) and I always took part in the activities of the Dutch Historical Institute in Rome. My first paper was a biographical sketch of the Dutch philosopher Johannes Heckius, who was one of the founders of the Academia dei Lincei in Rome.[9]

I still tried to study the old Bohr–Sommerfeld quantum theory, using the dissertation of Jan Burgers, and I kept in touch with Ehrenfest during the summer. In 1923 I also met and became good friends with Enrico Fermi, who was already at that time an accomplished physicist. Still, even his influence did not turn me back to physics. I suppose I went through what nowadays is called an identity crisis. Anyway, when I came back in June 1925, I thought that my real interest was in the study of cultural history, and that perhaps I should forget about physics. I had a long talk with my uncle, C. C. Uhlenbeck, the professor of linguistics at Leiden, who was the wise man in our family and who knew me very well. He was sympathetic and he shared my enthusiasm for the historian Johan Huizinga. But he reminded me that if I was serious I had better start to learn Latin and Greek, and he gave me the good advice first to try to finish my studies, especially since I had never done any work in physics by myself. Of course I also talked with Ehrenfest, who somehow still had enough confidence in me to propose that we work together on a study of the various solutions of the wave equation in n dimensions; this later appeared in a joint paper in the *Proceedings* of the Dutch Academy.[10] But he also told me that I had better start learning what the real problems in physics were, and that he would ask Sam Goudsmit to teach me what he knew and had done about the theory of atomic spectra.

The riddle of the gyromagnetic ratio

Thus began the remarkable summer of 1925. Two days a week I went to Leiden to work with Ehrenfest on the wave equation and on the other days Sam and I got together in The Hague to talk about the recent developments of atomic theory, which as I then slowly began to realize was at that time (1923–25) the "frontier" of physics.

At this point I think I should tell more about Sam Goudsmit, especially since in his accounts[1,2,3] he speaks rather deprecatingly about himself. It is true that Sam was not a very conscientious student and that he often had trouble passing the required examinations in the subjects that did not interest him. But on the other hand he was a very independent worker. Already in his first year (1921) he proposed a formula for the doublet splitting in atomic spectra and in the following years he wrote a number of papers on complex spectra and the vector model. This is not the place to try to describe this work,[1] so let me say only that in 1925 Sam was already a well known theoretical spectroscopist. He was the "house theoretician" in the Zeeman laboratory in Amsterdam, where he spent the first three days of the week—returning to Leiden in time for the Wednesday colloquium. Moreover, being from the Ehrenfest school, he was a good teacher!

So that summer Sam explained to me, in a nice orderly fashion, the work of Alfred Landé, Werner Heisenberg, Pauli and others (himself included) on the vector model of the atom, of which I was completely ignorant. Again I will not go into details, so let me only remind the reader that in this model (also sometimes called "*das Rumpf-Modell*") it was assumed, say for alkali atoms, that somehow the core (*der Rumpf*) had an angular momentum $\hbar/2$ and a magnetic moment of one Bohr magneton, so that the gyromagnetic ratio was twice the classical value, $e/2mc$, for the orbital motion of electrons. This was a riddle, but with this assumption one could understand very

H. A. Lorentz makes a point. The grand old man of Dutch physics was skeptical of spin; according to his calculations the velocity of the electron's surface was ten times the speed of light.

satisfactorily the coupling of the core with the outside electron, the influence of magnetic field (the Landé formula), and so on.

I remember that I was interested, but still detached. I asked many questions, and I made notes after each session. I remember that I was especially bothered by Goudsmit's statement that the model described all atoms except hydrogen, for which the old Sommerfeld theory was valid—as though that were somehow a horse of a different color! My skepticism infected Sam, and he then got the idea of looking into the way the level scheme of the fine structure of hydrogen would have to be if it were like an alkali atom. In our next session he already had it all worked out. It is the now accepted level scheme (except for the Lamb shift), which of course also follows from the Dirac theory of the electron. We realized that although the level splittings were the same as in the Sommerfeld theory, the selection rules were different; the theory thus explained a mysterious strong line in the spectrum of ionized helium that had been observed by Friedrich Paschen. This line was forbidden in the Sommerfeld theory and could also not be explained by the influence of electric fields, as I found out from H. A. Kramers's thesis.

It was our first success. We wrote a paper[11] in Dutch, which appeared in *Physica;* although it did not attract any attention until much later, I was quite proud of my first contribution to physics. However, I still had not yet completely made up my mind to continue. There was an opportunity for students at Leiden who wanted to switch to the humanities to take courses in the classical languages. So in the beginning of September I started to take Latin. Unfortunately this course was not like the Berlitz school in Rome, where I had started to learn Italian! It was very tough and, after a month or so, it became too much for me. During this time also my sessions with Sam became more and more absorbing. Ehrenfest was away, so we talked almost every day, trying to understand the ideas of Pauli.

Euphoria

Sam had earlier explained to me Pauli's criticism against *das Rumpf-Modell,* and he had told me about Pauli's proposal to ascribe *four* quantum numbers to each electron. He now continued with the discussion of the famous paper of January 1925 in which Pauli formulated the exclusion principle: No two electrons could have the same four quantum numbers. He explained to me how, by combining the four quantum numbers of the different electrons according to the rules of the vector model, one could understand the periodic system and the general multiplet structure of the atomic levels. Sam himself had simplified the argument by introducing the quantum numbers, n, l, m_l, and m_s (appropriate when a strong magnetic field is present) instead of those used by Pauli, and he noticed that then m_s was always $\pm\frac{1}{2}$.

I was impressed, but since the whole argument was purely formal, it seemed like abracadabra to me. There was no picture that at least qualitatively connected Pauli's formalism with the old Bohr atomic model. It was then that it occurred to me that, since (as I had learned) each quantum number corresponds to a degree of freedom of the electron, the fourth quantum number must mean that the electron had an additional degree of freedom—in other words, the electron must be rotating! Sam has written that he did not know at that time what a degree of freedom was. This may be so, as Sam had not done his exam in mechanics yet; in fact, he never passed this exam, and as a result he did not have the right to teach mechanics in the Dutch high schools even after he got his PhD. However, this did not prevent him later from teaching the graduate course in mechanics at the University of Michigan, which he did regularly because he liked the subject so much; furthermore, it was much appreciated by the students.

In spite of this he appreciated right away that if the angular momentum of the electron was $\hbar/2$, one had a picture of the alkali doublets as the two ways the electron could rotate with respect to its orbital motion. In fact, if one assumed that the gyromagnetic ratio for the rotation was twice the classical value, so that the magnetic moment was

$$2\,\frac{e}{2mc}\,\frac{\hbar}{2} = \text{one Bohr magneton}$$

then the properties formerly attributed to the core were now properties of the electron. The simple *"anschaulichen"* features of the original *Rumpf-Modell* were thus reconciled with Pauli's ideas.

I remember that when this became clear to us, we had a feeling of euphoria, but we also both agreed that one could not possibly publish such stuff. Since it had not been mentioned by any of the authorities (we did not know about Kronig, of course) it must for some reason be nonsense. But, of course, we told Ehrenfest, who was immediately interested. I am not sure precisely what happened next. Sam is wrong when he writes that he was satisfied and did not think any more about how our model could be justified. I remember that he wrote me a postcard from Amsterdam very soon afterward, in which he asked whether I

Bohr's letter to Ehrenfest. Although he misspelled both their names, Bohr was enthusiastic about the "electron-magnet gospel" of Uhlenbeck and Goudsmit. He thought it extremely likely that the quantum-mechanical calculation would reproduce all details correctly. (From the AIP Niels Bohr Library.)

was sure that the gyromagnetic ratio had to be $e/2mc$ classically—perhaps it was different for the rotation of an extended charged body. I showed this postcard to Ehrenfest, who then recalled a paper by Max Abraham[12] about the magnetic properties of rotating electrons. I studied this paper very hard and found there to my great satisfaction that if the electron has only surface charge the gyromagnetic ratio was $2\,e/2mc$, just as we had postulated! I think that when I showed this to Ehrenfest he thought (as he told us later) that our idea was either very important or nonsense, but that it should be published. The Abraham calculations were nonrelativistic and based on the old-fashioned rigid electron, so that they were at best only suggestive. Anyway, Ehrenfest told us to write a short, modest Letter to *Naturwissenschaften* and to give it to him. "*Und dann werden wir Herrn Lorentz fragen.*" ("And then we will ask Mr Lorentz.") A letter of 16 October to H. A. Lorentz in which he mentions this (among other things) was found and shown to me by Martin Klein.

Lorentz, who was of course the great old man of Dutch physics, was retired and lived in Haarlem but gave a lecture in Leiden every Monday at 11:00 am, in which he discussed the recent developments in physics. Everybody who could possibly make it came. So when school started in the middle of October (remember, we had long vacations) I had the opportunity to tell Lorentz about our ideas. Sam was not present because he had to resume his duties at the Zeeman laboratory. Lorentz was very kind and interested, although I also got the impression that he was rather skeptical. He said that he would think about it and that we should talk again the next Monday.

In fact, when we met that day he showed me a stack of papers full of calculations written in his beautiful handwriting, which he tried to explain to me. They were above my head but I understood enough to realize that there were serious difficulties. If the radius of the electron was

$$r_0 = e^2/mc^2$$

and if it rotated with an angular momentum $\hbar/2$, then the surface velocity would be about ten times the light velocity! If the electron had a magnetic moment $e\hbar/2mc$, its magnetic energy would have to be so big that, to keep the mass m, its radius would have to be at least ten times r_0.

It seemed to me that if one extended the Abraham calculations properly as Lorentz had apparently done, (and published in revised form[13]) then our picture of a quantized rotation of the electron could not possibly be reconciled with classical electrodynamics. I told this to Ehrenfest, of course, and said that his second alternative had turned out to be the right one. The whole thing was non-

sense, and it would be better that our Letter not be published. Then, to my surprise, Ehrenfest answered that he had already sent the Letter off quite a while ago, and that it would appear pretty soon. He added: "*Sie sind beide jung genug um sich eine Dummheit leisten zu können!*" ("You are both young enough to be able to afford a stupidity!")

This is not yet the end of the story. Our letter appeared in the middle of November, and soon afterwards (21 November) Goudsmit received a letter from Heisenberg, whom he knew quite well. In this letter (reproduced in reference 2) Heisenberg expressed his appreciation for Sam's courageous idea and agreed that it would remove all of the difficulties of the Pauli theory. He especially noted that it leads to the Landé–Sommerfeld formula for the alkali doublets except for a factor of two, and he asked how we had got rid of this factor. We had *not* derived this formula and therefore had no idea about the factor of two. In fact, I must say in retrospect that Sam and I in our euphoria had really not appreciated a basic difficulty—one with which Pauli and Bohr had been struggling for some time:

Clearly if one *formally* assigns the Landé quantum numbers of the atomic core to the electron as Pauli had done, then since there is *no model*, it is quite obscure how the "core" quantum number is coupled to the orbital quantum number of the electron. Bohr had speculated about a new force—the "*unmechanische Zwang*" (non-mechanical strain)—and Pauli spoke about an intrinsic two-valuedness of the motion of the electron. In our Letter we had maintained that such ideas could be replaced by a hypothesis about the structure of the electron. This explains the rather esoteric title of our Letter: "Replacement of the Hypothesis of the Non-mechanical Strain by an Assumption about the Internal Behavior of Every Single Electron."

Nevertheless, we had not actually explained how the basic difficulty would then be removed by the coupling of the rotational and orbital motion of the electron. Now we heard from Heisenberg that there was such a spin-orbit coupling and that it gave the right answer except for the mysterious factor of two. We still did not know how to derive the formula, but of course knowing the answer helps! Einstein, who visited Leiden every year for a month or so, gave us the essential hint. In the coordinate system in which the electron is at rest, the electric field $\mathbf{E}$ of the moving atomic core produces a magnetic field $[\mathbf{E} \times \mathbf{v}]/c$ (where $\mathbf{v}$ is the velocity of the electron) according to the transformation formula of relativity theory. This sounds learned (and in those days I liked that!), but it is of course just the magnetic field produced by the moving charged core. It is with respect to this magnetic field that the spin of the electron has its two orientations and the en-

ergy difference—the doublet splitting—could then be calculated by first-order perturbation theory. In this way we reproduced Heisenberg's formula with the same erroneous factor of two. By the way, there is no doubt that Kronig also had done this calculation (see "The Turning Point,"[6] page 20) and had shown it to Landé and Pauli. I find the reaction of Pauli mentioned there quite surprising, and it is certainly opposite to the sympathetic reaction of Ehrenfest to our ideas. Of course one must remember that Pauli was about of our age and was in the middle of the developments, while Ehrenfest was twenty years older and not deeply involved in what was sometimes called "spectral-term zoology."

This brings us to the beginning of December 1925 when Bohr came to Leiden to help celebrate the fiftieth anniversary of the doctorate of Lorentz, which was a great occasion. Bohr's visit was very lucky for us, since it gave us the opportunity to talk at length with him about our idea and the subsequent difficulties. Bohr had seen our Letter, but he still worried about how the coupling between the spin and the orbit could be understood. When we explained Einstein's argument he was completely convinced and became quite enthusiastic. He did not pay *any* attention to the calculations of Lorentz, which I mentioned to him. "They raise just classical difficulties," he said, "and they will disappear when the real quantum theory is found." The factor of two he took more seriously, but he somehow expected that a better calculation would also make it disappear.

He advised us to go back to the spectrum of hydrogen, especially when we told him about our earlier paper in *Physica*, with which he was not acquainted. Did the combination of the Sommerfeld relativistic effect with the spin-orbit coupling (forgetting about the factor of two) lead to the fine structure of the hydrogen levels as we had surmised in our *Physica* article? Sam could show this right away, and I think that together with the general Landé–Pauli unification, it completely convinced Bohr. On his way back to Copenhagen he made propaganda for our idea, as shown in the following part of a letter to Ehrenfest of 22 December:

"... I am convinced that it implies a large step forward for the theory of atomic structure. On my further travels I felt completely like a prophet for the electron-magnet gospel, and I believe that I have succeeded in convincing Heisenberg and Pauli that at least their present objections are not decisive and that it is very probable that a quantum-mechanical calculation will give all details correctly. I am looking forward to seeing the article of Goudsmit and Uhlenbeck ..."

This article was our second Letter to the Editor, this time that of *Nature*.[14] It was entitled: "Spinning Electrons and the

Structure of Spectra." It was dated December 1925 and appeared 20 February 1926. Bohr added an approving postscript. Since then our idea has been more or less accepted. The only holdout was Pauli, who had *not* been convinced by Bohr and who still spoke of it as the new "*Irrlehre*" ("erroneous teachings") (see van der Waerden's article, reference 7, page 215).

And there was of course still the mysterious factor of two! It is now well known that this difficulty was soon afterwards resolved[15] by L. H. Thomas, who showed that it was a forgotten relativistic effect. I remember that, when I first heard about it, it seemed unbelievable that a relativistic effect could give a factor of two instead of something of order v/c. I will not try to explain it, so let me only say that even the cognoscenti of the relativity theory (Einstein included!) were quite surprised. When Pauli understood it he finally withdrew his objections, as he mentioned later in his Nobel Prize lecture.[16]

This is the end of the story so far as Sam and I are concerned. I had become the assistant of Ehrenfest and in 1926 we worked together trying to digest the new quantum mechanics, and especially to understand the consequences for statistical mechanics. Sam continued his spectroscopic work, partially in Tübingen, where together with Ernst Back he worked out the theory of the hyperfine structure of the spectral lines when the atomic nucleus has a spin and magnetic moment.[17] In the spring of 1927 we both spent a few months in Copenhagen, where we wrote our dissertations. We received our doctor's degrees on the same day (7 July 1927) and then in the fall we went on the same boat to the US and to Ann Arbor, Michigan, where we had been appointed as instructors of physics.

With regard to the spin of the electron, it was of course Pauli who succeeded in incorporating the notion into Schrödinger wave mechanics.[18] It must have been a great satisfaction for him that it required a two-valued, or spinor, wave function. In a way it justified his old speculation on the two-valuedness of the electron motion. In his paper Pauli still had to assume the anomalous factor of two for the gyromagnetic ratio and he also had to take over the Thomas factor of two. The really complete explanation of these two factors of two, which had plagued the theory, did not come until 1928, when Paul Dirac developed the complete relativistic wave equation of the electron.[19]

*   *   *

I am indebted to Martin Klein for a copy of the letter from Bohr to Ehrenfest; the translation from the German is mine.

This article is an adaptation of a talk presented 2 February at the joint New York meeting of The American Physical Society and the American Association of Physics Teachers as part of a symposium celebrating the 50th anniversary of the discovery of electron spin.

References

1. S. A. Goudsmit, "The discovery of the electron spin," lecture given on the acceptance of the Max-Planck medal, in Proceedings of the Physikertagung, Frankfurt (1965); a German translation appeared in Physikalische Blätter, Heft 9/10 (1965).

2. S. A. Goudsmit, talk given at the 50th anniversary of the Dutch Physical Society in April 1971, Ned. Tydschrift voor Natuurkunde **37**, 386 (1971); in Dutch.

3. S. A. Goudsmit, Delta **15**, 77 (1972); excerpts from reference 2, in English.

4. G. Uhlenbeck, *Oude en Nieuwe Vragen der Natuurkunde*, North-Holland, Amsterdam (1955); partial English translation by B. L. van der Waerden, in *Theoretical Physics in the Twentieth Century*, Interscience, New York (1960).

5. E. H. Kennard, Phys. Rev. (2nd series) **19**, 420 (1922).

6. R. Kronig, "The Turning Point," in *Theoretical Physics in the Twentieth Century*, Interscience, New York (1960).

7. B. L. van der Waerden, "Exclusion Principle and Spin," in *Theoretical Physics in the Twentieth Century*, Interscience, New York (1960).

8. G. E. Uhlenbeck, "Reminiscences of Professor Paul Ehrenfest," Amer. J. Phys. **24**, 431 (1956).

9. G. E. Uhlenbeck, "Over Johannes Heckius," Comm. of the Dutch Historical Institute in Rome **4**, 217 (1924).

10. Collected Papers of P. Ehrenfest, North-Holland, Amsterdam, page 526 (1959).

11. S. A. Goudsmit, G. E. Uhlenbeck, Physica **5**, 266 (1925).

12. M. Abraham, Ann. der Physik **10**, 105 (1903).

13. H. A. Lorentz, *Collected Works*, Martinus Nyhoff, The Hague (1934), volume 7, page 179.

14. G. E. Uhlenbeck, S. A. Goudsmit, Nature **117**, 264 (1926).

15. L. H. Thomas, Nature **117**, 514 (1926).

16. W. Pauli, *Collected Scientific Papers*, volume 2, page 1080.

17. S. A. Goudsmit, PHYSICS TODAY, June 1961, page 18.

18. W. Pauli, Z. Physik **43**, 601 (1927).

19. P. A. M. Dirac, Proc. Roy. Soc. A **117**, 610 (1928); A **118**, 351 (1928); one should also not forget the contributions of H. A. Kramers: Quantentheorie des Elektrons und der Strahlung, in *Hand- und Jahrbuch der Chemischen Physik*, Akad. Verlagsges., Leipzig (1937); English translation, *Quantum Mechanics*, by D. ter Haar, North-Holland, Amsterdam (1957). □

Bosscha Reading Room, the physics and mathematics library at the University of Leiden. Students without a classical education, such as George Uhlenbeck, were barred from Leiden before 1919.

PART I

PHYSICS TODAY / OCTOBER 1959

History of the CYCLOTRON

On May 1, 1959, in memory of the late Ernest Orlando Lawrence, two invited lectures on the history of the cyclotron were presented as part of the American Physical Society's annual spring meeting in Washington, D. C. The present article is based on Prof. Livingston's talk on that occasion. The second speaker was E. M. McMillan, whose illustrated account also appears in this issue beginning on p. 24.

By M. Stanley Livingston

THE principle of the magnetic resonance accelerator, now known as the cyclotron, was proposed by Professor Ernest O. Lawrence of the University of California in 1930, in a short article in *Science* by Lawrence and N. E. Edlefsen.[1] It was suggested by the experiment of Wideröe [2] in 1928, in which ions of Na and K were accelerated to twice the applied voltage while traversing two tubular electrodes in line between which an oscillatory electric field was applied—an elementary linear accelerator. In 1953 Professor Lawrence described to the writer the origin of the idea, as he then remembered it.

The conception of the idea occurred in the library of the University of California in the early summer of 1929, when Lawrence was browsing through the current journals and read Wideröe's paper in the *Archiv für Elektrotechnik*. Lawrence speculated on possible variations of this resonance principle, including the use of a magnetic field to deflect particles in circular paths so they would return to the first electrode, and thus reuse the electric field in the gap. He discovered that the equations of motion predicted a constant period of revolution, so that particles could be accelerated indefinitely in resonance with an oscillatory electric field —the "cyclotron resonance" principle.

Lawrence seems to have discussed the idea with others during this early formative period. For example, Thomas H. Johnson has told the writer that Lawrence discussed it with himself and Jesse W. Beams during a conference at the Bartol Institute in Philadelphia during that summer, and that further details grew out of the discussion.

The first opportunity to test the idea came during the spring of 1930, when Lawrence asked Edlefsen, then a graduate student at Berkeley who had completed

his thesis and was awaiting the June degree date, to set up an experimental system. Edlefsen used an existing small magnet in the laboratory and built a glass vacuum chamber with two hollow internal electrodes to which radiofrequency voltage could be applied, with an unshielded probe electrode at the periphery. The current to the probe varied with magnetic field, and a broad resonance peak was observed which was interpreted as due to the resonant acceleration of hydrogen ions.

However, Lawrence and Edlefsen had not in fact observed true cyclotron resonance; this came a little later. Nevertheless, this first paper was the initial announcement of a principle of acceleration which was soon found to be valid and which became the basis for all future cyclotron development.

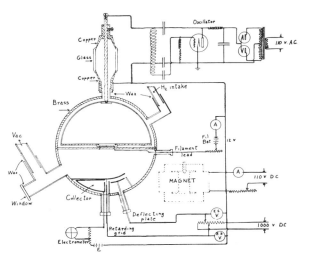

Fig. 1. Vacuum chamber of the first cyclotron. (PhD Thesis, M. S. Livingston, University of California, April 14, 1931)

M. Stanley Livingston, professor of physics at the Massachusetts Institute of Technology, is director of the Cambridge Electron Accelerator project at Harvard University, a program conducted under the joint auspices of Harvard and MIT.

Doctoral Thesis

IN the summer of 1930 Professor Lawrence suggested the problem of resonance acceleration to the author, then a graduate student at Berkeley, as an experimental research investigation. In my early efforts to confirm Edlefsen's results I found that the broad peak observed by him was probably due to single acceleration of N and O ions from the residual gas, which curved in the magnetic field and struck the unshielded electrode at the edge of the chamber.

It was my opportunity and responsibility to continue the study and to demonstrate true cyclotron resonance. A Doctoral Thesis [3] by the author dated April 14, 1931, reported the results of the study. It was not published but is on file at the University of California library. The electromagnet available was of 4-inch pole diameter. Fig. 1 is an illustration from this thesis, showing the arrangement of components which is still a basic feature of all cyclotrons. The vacuum chamber was made of brass and copper. Only one "D" was used, on this and several subsequent models; the need for a more efficient electrical circuit for the radiofrequency electrodes came later with the effort to increase energy. A vacuum tube oscillator provided up to 1000 volts on the electrode, at a frequency which could be varied by adjusting the number of turns in a resonant inductance. Hydrogen ions (H_2^+ and later H^+) were produced through ionization of hydrogen gas in the chamber, by electrons emitted from a tungsten-wire cathode at the center. Resonant ions which reached the edge of the chamber were observed in a shielded collector cup and had to traverse a deflecting electric field. Sharp peaks were observed in the collected current at the magnetic field for resonance with H_2^+ ions as shown in Fig. 2, a typical resonance curve taken from the thesis. Also present were 3/2 and 5/2 resonance peaks at proportionately lower magnetic fields,

due to harmonic resonances of H_2^+ ions. By varying the frequency of the applied electric field, resonance was observed over a wide range of frequency and magnetic field, as shown in Fig. 3, proving conclusively the validity of the resonance principle.

The small magnet used in these resonance studies had a maximum field of 5200 gauss, for which resonance with H_2^+ ions occurred at 76 meters wavelength or 4.0 megacycles frequency. In this small chamber the final ion energy was 13 000 electron volts, obtained with the application of a minimum of 160 volts peak on the D. This corresponds to about 40 turns or 80 accelerations. A stronger magnet was borrowed for a short time, capable of producing 13 000 gauss, with which it was possible to extend the resonance curve and to produce hydrogen ions of 80 000 ev energy. This goal was reached on January 2, 1931.

The First 1-Mev Cyclotron

LAWRENCE moved promptly to exploit this breakthrough. In the spring of 1931 he applied for and was awarded a grant by the National Research Council (about $1000) for a machine which could give useful energies for nuclear research. The writer was appointed as an instructor at the University of California on completion of the doctorate in order to continue the research. During the summer and fall of 1931, the writer, under the supervision of Lawrence, designed and built a 9-inch diameter magnet and brought it into operation, first with H_2^+ ions of 0.5-Mev energy. Then the poles were enlarged to 11 inches and protons were accelerated to 1.2 Mev. This was the first time in scientific history that artificially accelerated ions of this energy had been produced. The beam intensity available at a target was about 0.01 microampere. The progress and results were reported in a series of three

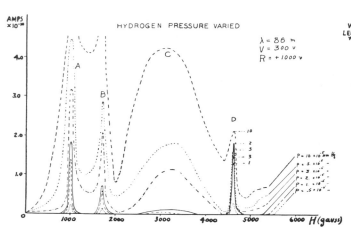

Fig. 2. Typical curves of current at the collector vs. magnetic field, showing resonant H_2^+ ions of 13 000 ev energy (peak D) and the variation of intensity with hydrogen gas pressure. (Thesis—Livingston)

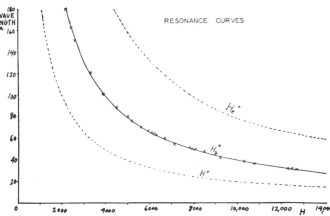

Fig. 3. Experimental values of cyclotron resonance for H_2^+ ions. (Thesis—Livingston)

Fig. 4. 1.2 Mev H⁺ cyclotron
at the University of California.[4]

Fig. 5. Vacuum chamber for 1.2 Mev
cyclotron with 11-inch pole faces.[4]

abstracts and papers by Lawrence and Livingston in *The Physical Review*.[4] Figs. 4 and 5 show the size and general arrangements of this first practical cyclotron.

Of course, Lawrence had other interests and other students in the laboratory. Milton White continued research with the first cyclotron. David Sloan developed a series of linear accelerators for heavy ions, limited by the radio power tubes and techniques available at that time, for Hg ions and later for Li ions. With Wesley Coates, Robert Thornton, and Bernard Kinsey, Sloan also invented and developed a resonance transformer using a radiofrequency coil in a vacuum chamber which developed 1 million volts. With Jack Livingood and Frank Exner he tried for a time to make this into an electron accelerator. I must again thank Dave Sloan for the many times that he assisted me in solving problems of the cyclotron oscillator.

The Race for High Voltage

TO understand the meaning of this achievement we must look at it from the perspective of the status of science throughout the world. When Rutherford demonstrated in 1919 that the nitrogen nucleus could be disintegrated by the naturally occurring alpha particles from radium and thorium, a new era was opened in physics. For the first time man was able to modify the structure of the atomic nucleus, but in submicroscopic quantities and only by borrowing the enormous energies (5 to 8 Mev) of radioactive matter. During the 1920's x-ray techniques were developed so machines could be built for 100 to 200 kilovolts. Development to still higher voltages was limited by corona discharge and insulation breakdown, and the multimillion volt range seemed out of reach.

Physicists recognized the potential value of artificial sources of accelerated particles. In a speech before the Royal Society in 1927 Rutherford expressed his hope that accelerators of sufficient energy to disintegrate nuclei could be built. Then in 1928 Gamow and also

Condon and Gurney showed how the new wave mechanics, which was to be so successful in atomic science, could be used to describe the penetration of nuclear potential barriers by charged particles. Their theories made it seem probable that energies of 500 kilovolts or less would be sufficient to cause the disintegration of light nuclei. This more modest goal seemed feasible. Experimentation started around 1929 in several laboratories to develop the necessary accelerating devices.

This race for high voltage started on several fronts. Cockcroft and Walton in the Cavendish Laboratory of Cambridge University, urged on by Rutherford, chose to extend the known engineering techniques of the voltage-multiplier, which had already been successful in some x-ray installations. Van de Graaff chose the long-known phenomena of electrostatics and developed a new type of belt-charged static generator to obtain high voltages. Others explored the Tesla coil transformer with an oil-insulated high-voltage coil, or the "surge-generator" in which capacitors are charged in parallel and discharged in series, and still others used transformers stacked in cascade on insulated platforms.

The first to succeed were Cockcroft and Walton.[5] They reported the disintegration of lithium by protons of about 400 kilovolts energy, in 1932. I like to consider this as the first significant date in accelerator history and the practical start of experimental nuclear physics.

All the schemes and techniques described above have the same basic limitation in energy; the breakdown of dielectrics or gases sets a practical limit to the voltages which can be successfully used. This limit has been raised by improved technology, especially in the pressure-insulated electrostatic generator, but it still remains as a technological limit. The cyclotron avoids this voltage-breakdown limitation by the principle of resonance acceleration. It provides a method of obtaining high particle energies without the use of high voltage.

Fig. 6. The "27-inch" cyclotron which produced 5 Mev D⁺ ions, with chamber rolled out.[7, 8]

Fig. 7. Vacuum chamber for the "27-inch" cyclotron.[7, 8]

The Cyclotron Splits its First Atoms

THE above digression into the story of the state of the art shows why the 1.2-Mev protons from the 11-inch Berkeley cyclotron were so important. This small and relatively inexpensive machine could split atoms! This was Lawrence's goal. This was why Lawrence literally danced with glee when, watching over my shoulder as I tuned the magnet through resonance, the galvanometer spot swung across the scale indicating that 1 000 000-volt ions were reaching the collector. The story quickly spread around the laboratory and we were busy all that day demonstrating million-volt protons to eager viewers.

We had barely confirmed our results and I was busy with revisions to increase beam intensity when we received the issue of the *Proceedings of the Royal Society* describing the results of Cockcroft and Walton in disintegrating lithium with protons of only 400 000 electron volts. We were unprepared at that time to observe disintegrations with adequate instruments. Lawrence sent an emergency call to his friend and former colleague, Donald Cooksey at Yale, who came out to Berkeley for the summer with Franz Kurie; they helped develop the necessary counters and instruments for disintegration measurements. Within a few months after hearing the news from Cambridge we were ready to try for ourselves. Targets of various elements were mounted on removable stems which could be swung into the beam of ions. The counters clicked, and we were observing disintegrations! These first early results were published on October 1, 1932, as confirmation of the work of Cockcroft and Walton, by Lawrence, Livingston, and White.[6]

The "27-inch" Cyclotron

LONG before I had completed the 11-inch machine as a working accelerator, Lawrence was planning the next step. His aims were ambitious, but supporting funds were small and slow in arriving. He was forced to use many economies and substitutes to reach his goals. He located a magnet core from an obsolete Poulsen arc magnet with a 45-inch core, which was donated by the Federal Telegraph Company. Two pole cores were used and machined to form the symmetrical, flat pole faces for a cyclotron. In the initial arrangement the pole faces were tapered to a 27½-inch diameter pole face; in later years this was expanded to 34 inches and still higher energies were obtained. The windings were layer-wound of strip copper and immersed in oil tanks for cooling. (The oil tanks leaked! We all wore paper hats when working between coils to keep oil out of our hair.) The magnet was installed in the "old radiation lab" in December 1931; this was an old frame warehouse building near the University of California Physics Building which was for years the center of cyclotron and other accelerator activities. Fig. 6 is a photograph of this magnet with the vacuum chamber rolled out for modifications.

Other dodges were necessary to meet the mounting bills for materials and parts. The Physics Department shops were kept filled with orders for machining. Willing graduate students worked with the mechanics installing the components. My appointment as instructor terminated, and for the following year Lawrence arranged for me an appointment as research assistant in which I not only continued development on the cyclotron but also supervised the design and installation of a 1-Mev resonance transformer x-ray installation of the Sloan design in the University Hospital in San Francisco.

The vacuum chamber for the 27-inch machine was a brass ring with many radial spouts, fitted with "lids" of iron plate on top and bottom which were extensions of the pole faces. This chamber is shown in Fig. 7. Sealing wax and a special soft mixture of beeswax and rosin were first used for vacuum seals, but were ultimately replaced by gasket seals. In the initial model only one insulated D-shaped electrode was used, facing a slotted bar at ground potential which was called a "dummy D". In the space behind the bar the collector could be mounted at any chosen radius. The beam was first observed at a small radius, and the magnet was "shimmed" and other adjustments made to give maximum beam intensity. Then the chamber was opened, the collector moved to a larger radius, and the tuning and shimming extended. Thus we learned, the hard way, of the necessity of a radially decreasing magnetic field for focusing. If our optimism persuaded us to install the collector at too large a radius, we made a "strategic retreat" to a smaller radius and recovered the beam. Eventually we reached a practical maximum radius of 10 inches and installed two symmetrical D's with which higher energies could be attained. Technical improvements and new gadgets were added day by day as we gained experience. The progress during this period of development from 1-Mev protons to 5-Mev deuterons was reported in *The Physical Review* by Livingston [7] in 1932 and by Lawrence and Livingston [8] in 1934.

I am indebted to Edwin M. McMillan for a brief chronological account of these early developments on the 27-inch cyclotron. (It seems that earlier laboratory notebooks were lost.) These records show, for example:

June 13, 1932. 16-cm radius, 28-meter wavelength, beam of 1.24-Mev H_2^+ ions.

August 20, 1932. 18-cm radius, 29 meters, 1.58-Mev H_2^+ ions.

August 24, 1932. Sylphon bellows put on filament for adjustment.

September 28, 1932. 25.4-cm radius, 25.8 meters, 2.6-Mev H_2^+ ions.

October 20, 1932. Installed two D's in tank, radius fixed at 10 in.

November 16, 1932. 4.8-Mev H_2^+ ions, ion current 10^{-9} amps.

December 2–5, 1932. Installed target chamber for studies of disintegrations with Geiger counter. Start of long series of experiments.

March 20, 1933. 5 Mev of H_2^+; 1.5 Mev of He^+; 2 Mev of $(HD)^+$. Deuterium ions acelerated for first time.

September 27, 1933. Observed neutrons from targets bombarded by D^+.

December 3, 1933. Automatic magnet current control circuit installed.

February 24, 1934. Observed induced radioactivity in C by deuteron bombardment. 3-Mev D^+ ions, beam current 0.1 microampere.

March 16, 1934. 1.6-Mev H^+ ions, beam current 0.8 microampere.

April–May, 1934. 5.0-Mev D^+ ions, beam current 0.3 microampere.

Those were busy and exciting times. Other young scientists joined the group, some to assist in the continuing development of the cyclotron and others to develop the instruments for research instrumentation. Malcolm Henderson came in 1933 and developed counting instruments and magnet control circuits, and also spent long hours repairing leaks and helping with the development of the cyclotron. Franz Kurie joined the team, and Jack Livingood and Dave Sloan continued with their linear accelerators and resonance transformers, but were always available to help with problems on the cyclotron. Edwin McMillan was a major thinker in the planning and design of research experiments. And we all had a fond regard for Commander Telesio Lucci, retired from the Italian Navy, who became our self-appointed laboratory assistant. As the experiments began to show results we depended heavily on Robert Oppenheimer for discussions and theoretical interpretation.

One of the exciting periods was our first use of deuterons in the cyclotron. Professor G. N. Lewis of the Chemistry Department had succeeded in concentrating "heavy water" with about 20% deuterium from battery acid residues, and we electrolyzed it to obtain gas for our ion source. Soon after we tuned in the first beam we observed alpha particles from a Li target with longer range and higher energy than any previously found in natural radioactivities—14.5-cm range, coming from the Li^6 (d,p) reaction. These results were reported in 1933 by Lewis, Livingston, and Lawrence,[9] and led to an extensive program of research in deuteron reactions. Neutrons were also observed, in much higher intensities when deuterons were used as bombarding particles, and were put to use in a variety of ways.

We had frustrations—repairing vacuum leaks in the wax seals of the chamber or "tank" was a continuing problem. The ion source filament was another weak point, and required continuous development. And sometimes Lawrence could be *very* enthusiastic. I recall working till midnight one night to replace a filament and to reseal the tank. The next morning I cautiously warmed up and tuned the cyclotron to a new beam intensity record. Lawrence was so pleased and excited when he came into the laboratory that morning that he jubilantly ran the filament current higher and higher, exclaiming each time at the new high beam intensity, until he pushed too high and burned out the filament!

We made mistakes too, due to inexperience in research and the general feeling of urgency in the laboratory. The neutron had been identified by Chadwick in 1932. By 1933 we were producing and observing neutrons from every target bombarded by deuterons.[10] They showed a striking similarity in energy, independent of the target, and each target also gave a proton group of constant energy. This led to the now forgotten mistake in which the neutron mass was calculated on the assumption that the deuteron was breaking up into a proton and a neutron in the nuclear field. The neutron mass was computed from the energy of the common proton group,[11] and was much lower than the value determined by Chadwick. Shortly afterward, Tuve, Hafstad, and Dahl in Washington, D. C., using the first electrostatic generator to be completed and used for research, showed that these protons and neutrons came from the $D(d,p)$ and $D(d,n)$ reactions,

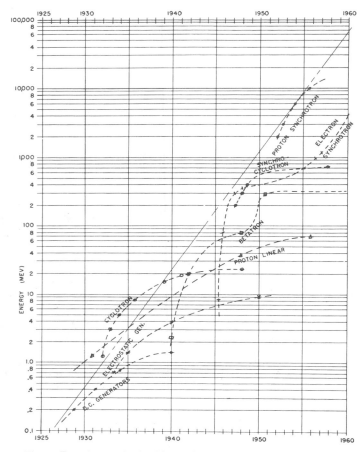

Fig. 8. Energies attained with accelerators as a function of time.

the target wheel to carbon, adjusted the counter circuits, and then bombarded the target for 5 minutes. When the oscillator switch was opened this time, the counter was turned on, and click-click--click---click----click. We were observing induced radioactivity within less than a half-hour after hearing of the Curie-Joliot results. This result was first reported by Henderson, Livingston, and Lawrence [12] in March, 1934.

I left the laboratory in July, 1934, to go to Cornell (and later to MIT) as the first missionary from the Lawrence cyclotron group. Edwin McMillan overlapped my term of apprenticeship by a few months, and stayed on to win the Nobel Prize and ultimately to succeed Professor Lawrence as director of the laboratory which he founded. McMillan can tell the rest of the story.

But it would be unfair to the spirit of Professor Lawrence if I failed to indicate some gleam of great things to come, some vision of the future. Recently I prepared a graph of the growth of particle energies obtained with accelerators with time, shown in Fig. 8. To keep this rapidly rising curve on the plot, the energies are plotted on a logarithmic scale. The curves show the growth of accelerator energy for each type of accelerator plotted at the dates when new voltage records were achieved. The cyclotron was the first resonance accelerator to be successful, and it led to the much more sophisticated synchronous accelerators which are still in the process of growth. The over-all envelope to the curve of log E vs time is almost linear, which means an exponential rise in energy, with a 10-fold increase occurring every 6 years and with a total increase in particle energy of over 10 000 since the days of the first practical accelerators. The end is not yet in sight. If you are tempted to extrapolate this curve to 1960, or even to 1970, then you are truly sensing the exponentially rising spirit of the Berkeley Radiation Laboratory in those early days, stimulated by our unique leader, Professor Lawrence.

in which the target was deuterium gas deposited in all targets by the beam. We were chagrined, and vowed to be more careful in the future.

We also had many successful and exciting moments. I recall the day early in 1934 (February 24) when Lawrence came racing into the lab waving a copy of the *Comptes Rendus* and excitedly told us of the discovery of induced radioactivity by Curie and Joliot in Paris, using natural alpha particles on boron and other light elements. They predicted that the same activities could be produced by deuterons on other targets, such as carbon. Now it just so happened that we had a wheel of targets inside the cyclotron which could be turned into the beam by a greased joint, and a thin mica window on a re-entrant seal through which we had been observing the long-range alpha particles from deuteron bombardment. We also had a Geiger point counter and counting circuits at hand. We had been making 1-minute runs on alpha particles, with the counter switch connected to one terminal of a double-pole knife-switch used to turn the oscillator on and off. We quickly disconnected this counter switch, turned

References

1. E. O. Lawrence and N. E. Edlefsen, Science **72**, 376 (1930).
2. R. Wideröe, Arch. Elektrotech. **21**, 387 (1928).
3. M. S. Livingston, "The Production of High-Velocity Hydrogen Ions without the Use of High Voltages". PhD thesis, University of California, April 14, 1931.
4. E. O. Lawrence and M. S. Livingston, Phys. Rev. **37**, 1707 (1931); Phys. Rev. **38**, 136 (1931); Phys. Rev. **40**, 19 (1932).
5. Sir John Cockcroft and E. T. S. Walton, Proc. Roy. Soc. **136A**, 619 (1932); Proc. Roy. Soc. **137A**, 229 (1932).
6. E. O. Lawrence, M. S. Livingston, and M. G. White, Phys. Rev. **42**, 150 (1932).
7. M. S. Livingston, Phys. Rev. **42**, 441 (1932).
8. E. O. Lawrence and M. S. Livingston, Phys. Rev. **45**, 608 (1934).
9. G. N. Lewis, M. S. Livingston, and E. O. Lawrence, Phys. Rev. **44**, 55 (1933); E. O. Lawrence, M. S. Livingston, and G. N. Lewis, Phys. Rev. **44**, 56 (1933).
10. M. S. Livingston, M. C. Henderson, and E. O. Lawrence, Phys. Rev. **44**, 782 (1933); E. O. Lawrence and M. S. Livingston, Phys. Rev. **45**, 220 (1934).
11. M. S. Livingston, M. C. Henderson, and E. O. Lawrence, Phys. Rev. **44**, 781 (1933); G. N. Lewis, M. S. Livingston, M. C. Henderson, and E. O. Lawrence, Phys. Rev. **45**, 242 (1934); Phys. Rev. **45**, 497 (1934); M. C. Henderson, M. S. Livingston, and E. O. Lawrence, Phys. Rev. **46**, 38 (1934).
12. M. C. Henderson, M. S. Livingston, and E. O. Lawrence, Phys. Rev. **45**, 428 (1934); M. S. Livingston and E. M. McMillan, Phys. Rev. **46**, 437 (1934); M. S. Livingston, M. C. Henderson, and E. O. Lawrence, Proc. Natl. Acad. Sci. US **20**, 470 (1934); E. M. McMillan and M. S. Livingston, Phys. Rev. **47**, 452 (1935).

PART II

PHYSICS TODAY / OCTOBER 1959

History of the CYCLOTRON

By Edwin M. McMillan

A S Dr. Livingston has told you, our activities over-lapped by a few months, so that between us we can give a continuous story of cyclotron development as carried out at Berkeley under the guidance of Professor Lawrence. My start in his laboratory was in April of 1934, but I was around Berkeley before that working in Le Conte Hall on a molecular beam problem. Therefore, I have two kinds of early memories of the Radiation Laboratory at that time. One is as a place that I visited occasionally before I was working there; the other is as a place where I came to work, which I remember better, although it still seems like a very long time ago. The whole way of working was rather different from what it is in most

Nobel Laureate Edwin M. McMillan is director of the Lawrence Radiation Laboratory at the University of California at Berkeley, having succeeded to that post following the death of the Laboratory's original director, E. O. Lawrence, in 1958. The article is based on the second of two talks presented before the American Physical Society last May in memory of Prof. Lawrence.

laboratories today. We did practically everything ourselves. We had no professional engineers, so we had to design our own apparatus; we made sketches for the shop, and did much of our own machine work; we took all of our own data, did all our own calculations, and wrote all our own papers. Things are now quite different from that, because everybody does just his share and the operations have become much larger and more professional. While the modern method produces more results, perhaps this older way may have been more fun.

What I have done in preparing a paper to give here is to let it be based mainly on a set of lantern slides, because I think pictures are more interesting than words. I would like to run through these pictures and try to recall what they illustrate and the various incidents, some amusing, some otherwise, that go along with them.

I'm going to start with another picture of the 27″ cyclotron. This shows the machine as it looked in 1934

Slide 2

when Stan and I were both there. (Slide 1.) Dr. Livingston is in the picture, and Professor Lawrence. The machine is the same as in the views shown by Stan, but here it is all assembled with the 27″ chamber in place. I have another view here of Professor Lawrence sitting at the control table, showing how one operated the machine. (Slide 2.) This was the major tool of nuclear research of that day and this was the control station. The switchboard in back had to do with magnet control, and the beam current was observed on the galvanometer scale.

As an illustration of the kind of experimental equipment one used, I have this drawing which was taken from a publication of about that period, early in 1935. (Slide 3.) This was an experiment to disintegrate aluminum with deuterons. You'll notice that in those days they were called deutons. The story was told that Ernest Rutherford objected to the name deuton; he didn't like the sound of it, but agreed that it would be all right if we put in his initials, E.R. (I don't think this story is really true, but at least the fact that it was told is true.) Well, these deutons came along inside the cyclotron vacuum chamber. This box is a cylinder soldered into the side of the brass wall of the cyclotron chamber. The beam that's inside passes through a thin target of aluminum foil. The secondary particles studied in this case were protons, making this an example of a (d,p) reaction. We didn't have that notation then, but that is what it would be called now. The secondary protons came out through a mica window, real old-fashioned mica, and into an ionization chamber counter and were counted. We measured the energy of these protons by simply sliding this counter back and forth inside of the tube, varying the range. We were measuring the range in air and plotting range curves in the

way that one did in those days. This was considered a piece of research in physics; this was published, but nowadays, of course, nobody would think of doing a thing quite that way.

Now, let us go on to the development of the cyclotron itself. The two principal parameters of the cyclotron, as far as its use is concerned, are the energy of the particles and the intensity. With that older vacuum tank that we saw, the one that was in place

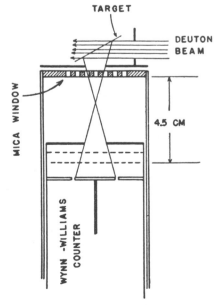

Slide 3: Arrangement of target, screens, and counter for bombarding in vacuum.

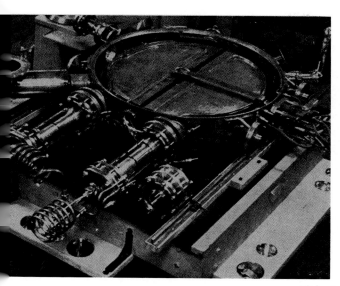

Slide 4

in Slide 1, the energy was up to about 3 Mev (this is the energy for deuterons). In 1936 a new chamber was built which is shown in the next slide. (Slide 4.) Comparing it with the chamber that Livingston showed, you'll see that there are many changes. For instance, the insulators for the two dees are made of Pyrex, with flanged ends which are clamped and bolted together rather than being waxed together, as the older ones were. The whole structure is more rugged, but there are still old-fashioned touches. You'll notice, coming into the center, a filament-type ion source that was still used then. Over in one corner you can see a glass liquid air trap, which was a very fragile and troublesome thing. People were always bumping into it and, of course, when it was bumped into, we'd have to pull the tank out, clean out the broken glass, and put the tank together all over again. With this new tank in place giving higher energies, up to 6 Mev for deuterons, and also larger currents, new types of experiments could be tried.

It was at about this time that an interest in biological work started in the laboratory, which has continued to the present. This was really started by John Lawrence, Ernest Lawrence's brother, who came out to the laboratory in 1935 to see what we were doing, and to see if there were any interest in the medical side. At this time biological experiments were started. I can recall the first time that a mouse was irradiated with neutrons. We put the mouse in a little cage and stuck him up on the side of the cyclotron tank and left him there for a while. Of course, nothing happened because there was not enough intensity. Then a serious attempt was made to see what neutrons did to mice. The first time this was done, it was done with an arrangement designed by Paul Aebersold in which the mouse could be put into the re-entrant tube shown in Slide 3, which was built into the cyclotron tank wall. In this way he could be close enough to the target to get some intensity. This mouse came out dead. This created a great impression at the time and I think perhaps was one reason why, in the Lawrence Radiation Laboratory,

people have always been careful with radiation even though it was soon discovered that somebody had forgotten to turn on the air supply which was supposed to provide ventilation for this mouse so that he died of anoxia. Anyhow, it was a very dramatic thing at the time.

Also at about this same time the first radioactive tracer experiments on human beings were tried. The first one that I recall, and I think the first use anywhere of an artificially produced radioisotope in human beings, was an early experiment of Joseph Hamilton in which he measured the circulation time of the blood by a very primitive method. The experimental subject takes some radioactive sodium dissolved in water in the form of sodium chloride, drinks it, and then has a Geiger counter which he holds in his hand, so that when the radioactive sodium reaches the hand, it starts to register. His hand is in a lead box so that the stuff that's just in his body doesn't affect the counter by gamma rays. I brought along a picture of this setup. (Slide 5.) This drawing, I believe, was made by Dr. Hamilton's wife, who is an artist. It shows the hand in the box, you see this cutaway lead box, holding a Geiger counter; the beaker with the radio sodium isn't shown but you might have shown him in the act of drinking it. After he does this, within just a few seconds, you begin to get some registration. After a few minutes, you begin to get equilibrium, and from these observations you get the circulation time of the blood. This, of course, is a very simple beginning, just like the simple beginning in physics that I showed with the primitive experiment of a (d,p) reaction. There were also simple beginnings of therapeutic use, coming a little bit later, in which neutron radiation was used, for instance, in the treatment of cancer. These things have gone on and built up so that there's now a whole field of radio medicine which had its beginning back in that time.

Another highlight from 1936 was the first time that anyone tried to make artificially a naturally occurring radionuclide (of course, we didn't have the word nuclide

Slide 5

Slide 6

then, but that is what it would now be called). This, I think, was a fairly classical experiment because there were then some people who didn't quite believe that the artificial radioactive materials were on the same status as the naturally occurring ones. Jack Livingood put some bismuth in the deuteron beam of the cyclotron, with an energy of about 6 Mev. This is high enough that one does get an appreciable yield of the (d,p) reaction forming radium E, a bismuth isotope, which then decays into polonium. The periods and energies were identical to those of natural radium E and polonium, so everybody was happy. This was the first time that one had gotten up that far in the periodic table with a charged-particle disintegration experiment.

Another thing that we were trying to do then was to bring the beam out of the tank. It seemed that there might some day be a use for a beam extractor. And so these experiments, which were spoken of as snouting experiments—getting the beam out of a snout—were done. Of course, in that re-entrant tube I showed you in Slide 3 you could get the beam in air by putting a little window on one side and letting the beam travel about two inches across the diameter of that brass tube. It was in air but it wasn't really outside the tank, because it plunged back into the wall of the tube. To get the beam the rest of the way out, we had to increase the strength of the deflecting field and move the deflector plate out some, so as to get enough radial displacement that the beam would come out to the edge of the magnetic field. The next slide I'm going to show is the first time that a beam was brought outside the tank in this sense. I remember this occasion very well because when we first tried, the beam didn't

quite clear the edge of the tank; it was coming almost tangentially and the thickness of the tank wall stopped it, so I spent about half a day with a file. curled up alongside the cyclotron, filing a groove in the thickness of the tank wall so that the beam could come out. This beam is shown in the next picture. (Slide 6.) There's a copper fitting, which is truly a snout, since it is a nose-shaped affair, which is fastened to the side of the tank, and the beam comes out through it, with the meter stick indicating the range. A little later, about two months after this, the beam was carried farther around—about a quarter of the way around the magnet. (Slide 7.) This shows where it came out of the window, way outside the cyclotron field. This, one might say, is the ancestor of modern beam extraction which has become a very sophisticated art in comparison to what it was in those days.

Slide 7

Slide 8

Everything up to now has been about the so-called 27-inch cyclotron. By the way, one thing I should apologize for at some point is my concentration on work at Berkeley. This is supposed to be the history of the cyclotron. But, in the first place, for some time this was the only place where there was a cyclotron, so that's where cyclotron history was being made. Secondly, this talk is in honor of Professor Lawrence, and that's where he was doing his work. Nevertheless, when we get to about 1936 or 1937, there did begin to be feedback of cyclotron lore from other parts of the world. At the end of 1936 there were about twenty other cyclotrons in the world; so the art had spread and things were coming back—improved ion sources, improved arrangements of radiofrequency systems, magnet control circuits, and all kinds of things. And from then on, of course, development of the cyclotron really became an international matter. Nevertheless, I shall continue to show pictures taken at Berkeley.

This is the 37-inch cyclotron, which used the same magnet as the 27-inch. (Slide 8.) All one had to do was to take out the old pole pieces, which had a reduced diameter, and put in larger diameter poles and the new tank shown on this slide. This was in late 1937 and begins to show signs of professionalism. You'll notice a gasket groove around the top, you'll notice nicely machined surfaces and things welded together, bolted together, and gasketed together, showing improved standards of design and construction. Still, you see a few old-fashioned touches; I think that the tank coil on the top side looks a bit primitive. We were still using a simple resonant circuit and two dees, plus an inductance forming the resonant circuit, which was loosely coupled to an oscillator. With this larger diameter and

better designed tank, the deuteron energy was now up to 8 Mev. The energy was climbing; currents were getting up to 100 microamperes which were tremendous currents at that time. Experiments were beginning to get sophisticated. It was in 1938 that Dr. Alvarez first introduced the method of time of flight for neutrons. By keying the cyclotron beam and then having a gated detector, one could use the time of flight to measure the velocity and to select out given energy ranges. That was the birth of that method.

Also in this period the first artificial element, technetium, was discovered by Segrè and Perrier, using a piece of the cyclotron. As you know, where the beam emerges from the dee there is a deflecting plate, and just next to the deflecting plate the boundary of the dee is made of a thin sheet of metal which has to decide whether a given turn of the beam is inside the dee or outside. Because the front edge of this metal sheet gets a lot of bombardment it is always made of a refractory metal. In this case it was made of molybdenum, and when the old tank was dismantled and thrown away and the new tank went in (the one I just showed you), Segrè said he wanted the old molybdenum strip, so we gave it to him. He was then in Italy and, with the help of Perrier, was able to get a definite proof that it contained the new element technetium made by deuteron bombardment of the molybdenum. If it hadn't been for the fact that this particular spot —this particular item—in the anatomy of the cyclotron gets a lot of bombardment, this new discovery would have been considerably delayed.

Another thing that started in this period is that the theorists were getting interested in the cyclotron. Before, you see, it was an experimental art, and the people that worked on the cyclotron sort of knew what they were doing, but they weren't very sophisticated about it. They didn't stop to think much about how and why it worked; they knew that it worked and that was enough. But it was at this time that Bethe and Rose first pointed out the relativistic limit on cyclotron energies and, a little after that, that L. H. Thomas devised an answer to the relativistic limit. This answer turned out to be a little hard for the experimenters to understand, so it lay fallow for many years. Now, of course, everybody wants to build Thomas-type cyclotrons or FFAG machines (which are, in a sense, extreme examples of Thomas cyclotrons), so it is now a great thing; but it lay dormant for quite a while because nobody took it very seriously at first. Also, at that time in 1937, cyclotron energies were limited by other factors such as sizes, budgets, and things like that, and not by the relativistic effect, which was thought of before it became a practical limit.

Shortly after, in my history, comes the 60-inch cyclotron, which was the first really professionally designed cyclotron that was built in Berkeley. There were some elsewhere in the world, but this was the first in Berkeley. Before I get to that, as a sort of transition, I want to show a picture, taken around 1938, that

illustrates several things. (Slide 9.) Now, let's see, what does this illustrate? First, it illustrates that people had started worrying about shielding against radiation around the cyclotron. Those were 5-gallon cans that were filled with water and simply stacked around and above the cyclotron to give shielding. As a matter of fact, the cans in this picture were originally on top of the cyclotron. They developed leaks, and the people that worked underneath would get tired of having water drip on them, and then they would take the leaky ones down and kick big dents in them so that nobody would be tempted to put them back.

The second thing that this slide illustrates is the type of building this work was done in, the Old Radiation Laboratory. I might inject a slightly sad touch, in that as I left Berkeley to come to this meeting, the last boards of the Old Radiation Laboratory were being battered down by a great big clam shell. We managed to save a few pieces as historical relics; otherwise it is all gone now. The third thing illustrated is that the man pictured here is Bill Brobeck, who was our first professional engineer hired at the Laboratory, showing the coming in of the more professional approach to the design and building of accelerators.

Now I will say a little about the 60-inch cyclotron, starting with a picture that was taken in 1938, showing

Slide 10 (Left to right and top to bottom): A. S. Langsdorf, S. J. Simmons, J. G. Hamilton, D. H. Sloan, J. R. Oppenheimer, W. M. Brobeck, R. Cornog, R. R. Wilson, E. Viez, J. J. Livingood, J. Backus, W. B. Mann, P. C. Aebersold, E. M. McMillan, E. M. Lyman, M. D. Kamen, D. C. Kalbfell, W. W. Salisbury, J. H. Lawrence, R. Serber, F. N. D. Kurie, R. T. Birge, E. O. Lawrence, D. Cooksey, A. H. Snell, L. W. Alvarez, P. H. Abelson.

the magnet, which had just been installed, and (approximately) the scientific staff of the Radiation Laboratory as of that time. (Slide 10.) You can see Professor Lawrence in the center, with Professor Birge, who was then chairman of the Physics Department, at his right, and Dr. Cooksey at his left. There are probably quite a few people here who can recognize themselves in that picture. It is always a little shocking to look at these old pictures and realize what time has done to us all!

This is the 60-inch cyclotron shortly after it was put together. (Slide 11.) A good many modifications in design were embodied in this machine and one of the most important ones is one of the things that fed back from outside; that is, the idea of getting away from glass insulators altogether, and having the dees plus their stems form a resonant system which is entirely inside the vacuum. The two tanks at the right hold the dee stems. This system has no insulators except in the lead-in for radiofrequency power. The power lead-ins come down the slanting copper cylinders at the right. The round tank on top of the magnetic yoke contains the deflector voltage supply, a rectified voltage supply under oil. And I think you can recognize the people in there: Don Cooksey, Dale Corson, Ernest Lawrence, Robert Thornton, John Backus, Winfield Salisbury, Luis Alvarez on the magnet coil, and myself on a dee-stem tank.

Now, just to show that physicists are not always serious, I have made a slide of the following pose: Laslett, Thornton, and Backus posing in the dee-stem tank of the 60-inch cyclotron before it was assembled. (Slide 12.) The next slide shows the control station of the 60-inch; now we have a real control desk, designed and not thrown together. (Slide 13.) At the desk are

Slide 12

Professor Lawrence and his brother, John Lawrence, who initiated the medical work and is still continuing it at the Lawrence Radiation Laboratory.

We are now up to 1939. Fission has been discovered. I should point out that the old 37-inch cyclotron was still running, since the 60-inch had a new magnet and a new building, the Crocker Laboratory. So some of these things I mention now were done on the old 37-inch, which ran, with some interruptions, right up to the time when it was used for the first model test on the principle of the synchrocyclotron in 1946. But when fission was discovered, everybody in the Laboratory immediately jumped on the band wagon the way people do, and tried to think of an experiment having to do with fission. They did things with cloud chambers and counters and and made recoil experiments and various things of that kind.

Slide 13

Slide 14

Slide 15

In 1940 came the first production of a transuranium element, which was done with the 60-inch cyclotron, although some of the experiments that led up to it had been done with the 37-inch. Carbon 14, which is perhaps the most important of all the tracer isotopes, came in this period. Kamen and Ruben finally pinned that down. Carbon 14 was something people had been trying to discover for a long time. I tried once myself but didn't quite get it. The mass 3 isotopes, hydrogen 3 and helium 3, were discovered then, helium 3 being found by an unusual use of a cyclotron. It was used as a mass spectrometer rather than as a cyclotron; that is, it was set for a resonance point for particles with charge 2 and mass 3, and when something came through at that resonance it had to be helium 3. This was done by Alvarez.

Perhaps the crowning event of that time was the award of the Nobel Prize to Professor Lawrence. Somebody, I think Cooksey, had the foresight to take a photograph of what appeared on the blackboard then. (Slide 14.) You see there is a two-stage announcement: first it says ASSOCIATED PRESS—UNCONFIRMED and then it says CONFIRMED with an arrow. The column down the left is a schedule of dates when people in the Laboratory received blood counts. I see Kruger, Corson, Alvarez, Aebersold, Livingston, Wright, Backus, Helmholz, Salisbury, and Cooksey. That's the other Livingston, Bob Livingston.

Now Ernest Lawrence was never a man who wanted to rest on achievement; he always wanted to go a step farther. I think it was this forward-looking spirit, and his ability to communicate it to others, that was his true greatness. So, even though the 60-inch cyclotron was a beautiful machine, was running fine, and was doing a great deal of important work, he had this dream of 100 million volts. I've looked at some of his old correspondence and it's always referred to as "100 million volts"; and he believed this could be achieved with the cyclotron. When he got the Nobel Prize, this helped things by focusing attention on this whole concept, and he set out on a campaign to see if he could raise the money to build a 100-million-volt cyclotron. Of course, in those days, money was essentially private money. There was no Manhattan District; there was no Atomic Energy Commission; and so he was trying to get this money by private funds.

In the course of this effort a good many things were written, plans and calculations were made, and one rather interesting picture was drawn which I will show you now. This was an artist's concept of a cyclotron for 100 million volts. (Slide 15.) This is what is now called the 184-inch cyclotron. You can see that this concept is rather different from the way the machine really looks. The magnet yoke is the same, but you see two tremendous tanks projecting on either side. Those were the dee-stem tanks; the beam was supposed to be deflected at one dee, make a complete turn inside, pass through a slit in one dee stem, and emerge as shown in the picture. But the important point this illustrates is that one was designing this as a conventional cyclotron, and one could easily estimate what dee voltages would be required to reach a given particle voltage, following the ideas of Rose and Bethe. We estimated that to reach 100 million electron volts for deuterons with this sort of design we would have wanted about

Slide 16

1.4 million volts between dees, or 700 000 volts to ground on each dee. We were planning to go ahead with floods of rf power to reach this voltage, and perhaps we would have, who knows?

The next picture shows a conference in the Old Radiation Laboratory, the building that has just been torn down, between Ernest Lawrence, Arthur Compton, Vannevar Bush, James Conant, Karl Compton, and Alfred Loomis. (Slide 16.) They were discussing ways of getting support for the project, and were obviously in a happy mood. Dr. Cooksey, who took the picture, tells me that someone had just told a joke, but the happiness may have had a deeper justification, for a few days later, on April 8, 1940, the Rockefeller Foundation

decided to give 1.15 million dollars for the cyclotron. This grant, with help from the Regents of the University and others, made it possible for the project to go ahead.

But then the war came along and the whole effort of the Laboratory was diverted to other things. The magnet for this cyclotron was used for research on the electromagnetic isotope separation process, and it wasn't until quite a while later that it came back to use as a cyclotron. By that time other ideas had come out—the idea of the use of phase stability and frequency modulation—and so when the machine finally was built as a cyclotron, it didn't look like that picture on Slide 15 but looked like this one. (Slide 17.) Here

Slide 17

is what the 184-inch cyclotron looked like when it was first assembled. You can get some idea of the size, since there's a man there for scale. Of course, by now this is a synchrocyclotron. When I think of the history of the cyclotron in the sense of this talk, I think of it as the history of the fixed frequency cyclotron, so I won't say much more about this machine except that it does work. I'll show you a picture of about the way it looks today, encased in concrete blocks for shielding, which is a better solution to the shielding problem than 5-gallon cans of water. (Slide 18.) If you look hard, you can see a man in this picture, too.

I shall close this talk with an aerial view of the present establishment in Berkeley of the Lawrence Radiation Laboratory. (Slide 19.) In the foreground, in the circular building, is the Bevatron, which is of course a descendant of the cyclotron since it does use the magnetic resonance principle. A little farther back is another circular building which houses the 184-inch cyclotron, the machine I just showed you. The other buildings house other accelerators, research laboratories, shops, and all the things which make up the laboratory which really, one can say in all truth, is the outgrowth of the ideas and the faith and the strength of Professor Lawrence, in whose memory we have spoken today.

Slide 18

Slide 19

The Discovery of Fission

*Initial formulations of nuclear fission are colored
with the successes, failures and just plain bad luck of several scientists
from different nations. The winning combination of good
fortune and careful thought made this exciting concept a reality.*

by Otto R. Frisch and John A. Wheeler

PHYSICS TODAY / NOVEMBER 1967

How It All Began

by Otto R. Frisch

THE NEUTRON was discovered in 1932. Why, then, did it take seven years before nuclear fission was found? Fission is obviously a striking phenomenon; it results in a large amount of radioactivity of all kinds and produces fragments that have more than ten times the total ionization of anything previously known. So why did it take so long? The question might be answered best by reviewing the situation in Europe from an experimentalist's point of view.

Research in Europe

In Europe there were few laboratories in which nuclear-physics research was conducted, and I think the word "team" had not yet been introduced into scientific jargon. Science was still pursued by individual scientists who worked with only one or two students and assistants.

Paris harbored some of the most active research laboratories in Europe. It is the city in which radioactivity had been discovered and where Madame Curie was working until her death in 1934. She still dominated the situation: Techniques were quite similar to those used at the turn of the century; that is, ionization chambers and electrometers. This state of affairs is good enough for performing accurate measurements on natural radioactive elements, but it is not really adequate for much of the work on nuclear disintegration. Madame Curie

had little respect for theory. Once, when one of her students suggested an experiment, adding that the theoretical physicists next door thought it hopeful, she replied, "Well, we might try it all the same." Their disregard of theory may have cost them the discovery of the neutron.

Cambridge is the second place worthy of discussion. Ernest Rutherford, whose towering personality dominated Cambridge research, had split atomic nuclei in 1919; since 1909 he had, in fact, been keenly concerned with the observation and counting of individual nuclear particles. He first introduced the scintillation method and stuck firmly to it. His great preference was for simple, unsophisticated methods, and he possessed a strong distrust of any complicated instrumentation. Even in 1932, when John Cockcroft and Ernest Walton first disintegrated nuclei by artificially-accelerated protons, they used scintillations to detect the process. By that time Rutherford had realized that electronic methods of particle counting must be developed. The reason was that the scintillation method clearly had its shortcomings. It did not work for very low or high counting rates and was not really reliable. This deficiency was highlighted by the results that came from the third laboratory I want to mention—Vienna.

Vienna is where I began my career and it was in those days a sort of en-

Otto R. Frisch, professor of natural philosophy (physics) at Cambridge University, England, did research in Berlin (1927–30), Hamburg (1930–33), London (1933–34), Copenhagen (1934–39) and Birmingham (1939–40). During the war he worked on the A-bomb at Los Alamos. He was first to observe energy liberated in the fission of a single uranium nucleus.

John A. Wheeler, one of the first American scientists to concentrate on nuclear fission, worked at the U. of Copenhagen in 1934 as a National Research Fellow with Niels Bohr. Wheeler received his PhD in physics at the Johns Hopkins University prior to his research in Copenhagen. In 1938 he joined Princeton's physics department, where he remains active.

fant terrible of nuclear physics. Several physicists were claiming that not only nitrogen and one or two others of the light nuclei could be disintegrated by alpha particles but that practically all of them could and did give many more protons than anybody else could observe. I still do not know how they found these wrong results. Apparently they employed students to do the counting without telling them what to expect. On the face of it, that operation appears to be a very objective method because the student would have no bias; yet the students quickly developed a bias towards high numbers because they felt that they would be given approval if they found lots of particles. Quite likely this situation caused the wrong results along with a generally uncritical attitude and considerable enthusiasm over beating the English at their own game.

I still remember when I left Vienna at just about that time (after having escaped the duty of counting scintillations). My supervisor, Karl Przibram, told me with sadness in his voice, "You will tell the people in Berlin, won't you, that we are not quite as bad as they think?" I failed to persuade them.

Germany had nuclear-physics research in several places. The team of Otto Hahn and Lise Meitner, which had been one of the first groups to study radioactive elements, had at that time separated to carry out indepen-

dent research. Hahn was working on various applications of radioactivity for the study of chemical reactions, structures of precipitates and similar subjects, whereas Lise Meitner was using radioactive materials chiefly to elucidate the processes of beta and gamma emission and the interaction of gamma rays with matter.

In addition, Hans Geiger was in Germany. He had been with Rutherford from 1909 onwards, in the early days before the nucleus was discovered. Rutherford felt uncertain about the scintillation method and asked Geiger to develop an electric counter to check on it. But as soon as Rutherford saw that the two gave the same results, Rutherford returned to the scintillation method, which appeared to be simpler and more reliable when used with proper precaution. Geiger went back to Germany and perfected his electric counters, and in 1928, together with a student named W. Müller, he developed an improved counter that could count beta rays. Earlier counters were inadequate for this purpose, and scintillation methods were also incapable of detecting beta rays. However the new counters were still very slow because the discharge between the central wire and the cylindrical envelope was quenched by a large resistor of many megohms placed in the circuit; consequently the counting rate was limited to numbers not much greater than with the scintilla-

tion method. Even at a few hundred particles a minute there were quite large corrections to be applied.

Walther Bothe was the first to use the coincidence method, both in an attempt to do something about cosmic rays and also for measuring the energy of gamma rays by the range of the secondary electrons they produced. This was really the first reliable method for measuring the energy of weak gamma radiations.

Until 1932, the only source of particles for doing atomic nuclear disintegration was natural alpha particles: either polonium, which was difficult to come by (in fact one practically had to go to Paris) or sources of one of the short-lived decay products of radium, which were very clean but were short-lived and usually had lots of gamma radiation.

The year of discovery

But in 1932, that *annus mirabilis*, not only the neutron was discovered but two other developments took place. In the US Ernest O. Lawrence made the first cyclotron that showed promise of being useful, and in England Cockcroft and Walton built the first accelerator for protons capable of producing nuclear disintegrations. I need not state that this was the beginning of an enormous development; most of nuclear physics as we know it would have never come about without at least one of those two instruments. But the interesting thing is that they played practically no role in that narrow thread that led to the discovery of nuclear fission.

I do not want to dwell on the discovery of the neutron very much because it was discussed in several interesting lectures in 1962 at the History of Science Congress held in Ithaca, New York. The published proceedings contained interesting contributions by Norman Feather and Sir James Chadwick, who showed that the neutron was discovered in Cambridge, not simply by chance with everybody else having done the groundwork, but because a search for the neutron had been going on in Cambridge (admittedly with wrong ideas). The people at Cambridge were keyed up for this discovery. They had made one observation that was important and that tends to be overlooked: H. C. Web-

COMPUTATIONS, indicating chains of radioactive elements, were published in a 1938 *Die Naturwissenschaften* article by Hahn, Meitner and Strassmann. —FIG. 1

ster showed that those queer penetrating rays that beryllium emitted when alpha particles fell on it were more intense in the forward direction than in the backward direction. This result was quite incomprehensible if the radiation were gamma rays as everybody believed. Even the French physicists Curie and Joliot shared that belief in the teeth of all theoretical predictions. Then Chadwick's experiment showed clearly that the mysterious radiation consisted of particles having approximately the mass of the proton. There was a bit of confusion at the time because the word "neutron" had been used by Enrico Fermi and Wolfgang Pauli to indicate the particle that later came to be called the "neutrino."

After the neutron was discovered, there was of course a certain rush of activity, but nobody knew quite what to do. Neutrons were rather few in number. They were, after all, secondary products of nuclear disintegration. With only natural alpha sources available at first, neutron production was low.

Moreover the main instrument for detection was essentially the cloud chamber. With cloud chambers only a limited number of tracks due to neutrons could be found. And it was slow work to make any sense out of the few detected tracks of recoil nuclei. Leo Szilard once joked that if a man suddenly does something unexpected there is usually a woman behind it, but if an atomic nucleus suddenly does something unexpected, there is probably a neutron behind it.

Electronic counting methods had only just been developed; largely as a reaction to the wrong results coming out of Vienna that nobody else could confirm, it had been decided that it really was necessary to build electronic amplifiers and counters. Actually the Viennese themselves started that kind of work but were not very successful. The work was also started in Switzerland with some success by Hermann Greinacher. Yet I think the main thread that led to the development of decent counters took place in England, where Charles Wynn-Williams used proper screening and tubes with low noise level etc. to produce electronic counters. Nevertheless those counters, although Chadwick

GREAT AND GOOD FRIENDS. Lord and Lady Rutherford (left) with Niels and Margrethe Bohr in Rutherford's garden. The photograph was taken about 1930.

had used them with good effect to pin down the neutron, were still too noisy to be of much use.

Artificial radioactivity

Things really got moving when, in 1934, artificial radioactivity was found by Curie and Joliot. I think they must have been very happy to have made up for their failure to spot the neutron two years previously. Almost to the day two years previously both discoveries came out in the middle of January. They had known for many months before that aluminum bombarded with alpha particles emits positrons, but it had never occurred to them that this might be a delayed process. They had only observed the positron during bombardment. Lawrence and his cyclotron people in California had made the same mistake. In fact they had noticed that the counters misbehaved after the cyclotron was switched off. I am told that they built in special gadgetry so that the counters were automatically switched off together with the cyclotron! Otherwise they would have found artifi-

cial radioactivity before the French.

It is astonishing that nobody appears to have thought beforehand that the result of a nuclear disintegration might be an unstable nucleus although the existence of unstable nuclei had, of course, been known for thirty years or more. I have been told that, after the discovery, Rutherford wrote to Joliot and congratulated him on his discovery saying that he himself had thought that some of the resulting nuclei might be unstable, but had always looked for alpha particles only because he was not really interested in beta particles.

As soon as this work became known in January 1934 a lot of people rushed to repeat and extend the experiment. But most of them rushed in a straight line indicated by Curie-Joliot, bombarding other elements with alpha particles. (So did I in Blackett's laboratory in London.)

But in Rome Fermi at that time had already decided that nuclear physics was an important and interesting line, and he had started to set up some instrumentation. So when this discovery came along, he began working quite fast to see whether neutrons would form radioactive nuclei.

I remember that my reaction and probably that of many others was that Fermi's was a silly experiment because neutrons were much fewer than alpha particles. What that simple argument overlooked of course was that they are very much more effective. Neutrons are not slowed down by electrons, and they are not repelled by the Coulomb field of nuclei. Indeed, within about four weeks of the discovery by Curie and Joliot, Fermi published the first results proving that various elements did become radioactive when bombarded with neutrons. Only another month later he announced that bombarding uranium produced some new radioactivity that he felt must be due to transuranic elements. Because both on theoretical grounds (Coulomb barrier and all that) and as far as the experiments confirmed it, all heavier elements were known to absorb neutrons without splitting anything off. And so it was felt that must also be the case with uranium.

This work was of course considerably interesting to radiochemists. Several took it up, but once again, oddly enough, one false result started things really moving—a note by Aristid von Grosse, a German-born chemist working in the US, who thought one of these elements behaved like protactinium. He had done some of the early work with Hahn on protactinium soon after it was discovered in 1917; so his suggestion put Hahn and Meitner on their mettle. They felt protactinium was their own baby and they were going to check it. Lise Meitner persuaded Hahn to join forces again. They soon showed that von Grosse was wrong: It was not protactinium. On the other hand there were so many odd things there that they were captured by this phenomenon and had to go on. The results were most peculiar.

Figure 1 shows one of the tabulations indicating the chains of radioactive elements that Hahn and Meitner had thought identified them. They did not give new names to the transuranic elements that they thought they had identified, but they used the prefix "eka" to indicate that they were higher homologues of rhenium, osmium, etc. up to ekagold. Obvious-

LINKS IN THE CHAIN. Cockcroft (top) and Walton contributed to the new ideas when they disintegrated nuclei by artificially-accelerated protons.

ly, Hahn was excited to have a whole new lot of chemical elements to play with and to study their properties. Today, of course, these elements after uranium are known as neptunium, plutonium, americium etc., and are known to be chemically quite different from those that Hahn was studying.

Parallel chains

The results were astonishing for two reasons. In the first place, it appeared that there were three parallel series. And from the yields obtained they must all derive from uranium 238 or possibly one of them from 235 (which is already much rarer). So it looked as if there were at least two parallel chains of isomeric elements. This isomeric property had to be propagated all along the chain of beta disintegrations.

Nuclear isomerism was still fairly new in 1938, and its interpretation was not altogether clear. It had been suggested (as we now accept) that it was due to high angular momentum, but there were also proposals that it might be due to the existence of rigid structures inside nuclei. One could imagine that such a rigid structure might survive a beta decay and might influence the half-life of the subsequent product.

But then there was still the mystery of the great length of those chains. Uranium, after all, was not beta unstable itself. The other elements in that region never had more than two beta decays in succession; yet here four or five had been found. So Hahn the chemist was delighted by so many new elements, but Hahn the radiophysicist or radiochemist was rather worried about the mechanism that could account for them.

All this work was made difficult by the political situation in Germany. Hitler was in power and the institute had to play a delicate game of politics to prevent racial persecution from removing some of its personnel. In 1938, when Austria was occupied by the Nazis, Lise Meitner felt very insecure; rumors began to float around that she might lose her post and be prevented thereafter from leaving Germany because of her knowhow. A certain amount of panic resulted. Dutch colleagues offered to smuggle her to Holland without a visa. Thus she left Germany in the early summer of 1938, went from Holland for a brief stay in Denmark, and was offered hospitality by Manne Siegbahn at the Nobel Institute in Stockholm.

Near misses

After that, the team that had already brought Strassmann in with Hahn as a second chemist had to carry on without her. In the meantime some work had been started in Paris. It is interesting that they had a different angle. They were at first not so interested in the transuranic elements; but they realized that if thorium is bombarded with neutrons, one ought to find the beginning of the new and missing radioactive chain with the atomic weight $4n + 1$. One realizes that the others, $4n$, $4n + 2$, $4n + 3$, are all represented by the natural radioactive series. But the $4n + 1$ was missing, and so Irene Curie, the daughter of Madame Curie, together with Hans von Halban, an Austrian, and Peter Preiswerk, a Swiss, set out to search for that series and published some work on it.

Later that team broke up because Halban came to Copenhagen and, for a time, worked with me on the study of slow neutrons. Irene Curie found a new collaborator in Pavel Savitch, a Yugoslav. They tried to disentangle the transuranic elements. Having realized that there was a great variety of different materials, Irene Curie had the good idea of selecting one of them simply by the high penetration of its beta rays. They covered their samples with a fairly thick sheet of brass and only studied the substance whose radiation penetrated. They did not realize that even that method might not select a single substance although the substance appeared to have a reasonably unique lifetime of 3.5 hours. From the chemical behavior they first thought it looked like thorium.

This work was checked by Hahn, who concluded that it was not thorium and wrote so to Paris. Curie and Savitch continued the work and in a later paper in the summer of 1938 acknowledged that the 3.5-hour substance was not thorium but behaved a bit more like actinium and even more like lanthanum. She had come very close indeed to the concept of nuclear fission but unfortunately did not state it clearly. She said that it was definitely not actinium and that it was quite similar to lanthanum, "from which it could be separated only by fractionation." But she did think it could be separated. The reason was probably that she still had a mixture of two substances; in that case of course one does effect a partial separation. Then this work was in turn checked by Hahn and Strassmann who discovered radioactive products that behaved partly like actinium, partly a bit like radium.

RUTHERFORD was the first to use scintillation methods to detect particles.

There was another near miss at about the same time: Gottfried von Droste, a physicist working with Lise Meitner, looked for long-range alpha rays from uranium during neutron bombardment. If he had supressed the ordinary alpha rays by applying a bias to the amplifier, he would not have failed to find fission. Unfortunately instead of using a bias he used a foil, and that foil was thick enough to stop not only uranium alpha rays but also the fission fragments; nor did he find any long-range alpha rays, which had to be there if radium or actinium isotopes were formed.

Then Hahn and Strassmann checked the chemical properties of this "radium" with care and found that they were identical with those of barium.

A propitious visit

This is where I came in because Lise Meitner was lonely in Sweden and, as her faithful nephew, I went to visit her at Christmas. There, in a small hotel in Kungälv near Göteborg I found her at breakfast brooding over a letter from Hahn. I was skeptical about the contents—that barium was formed from uranium by neutrons—but she kept on with it. We walked up and down in the snow, I on skis and she on foot (she said and proved that she could get along just as fast that way), and gradually the idea took shape that this was no chipping or cracking of the nucleus but rather a process to be explained by Bohr's idea that the nucleus was like a liquid drop; such a drop might elongate and divide itself. Then I worked out the way the electric charge of the nucleus would

diminish the surface tension and found that it would be down to zero around $Z = 100$ and probably quite small for uranium. Lise Meitner worked out the energies that would be available from the mass defect in such a breakup. She had the mass defect curve pretty well in her head. It turned out that the electric repulsion of the fragments would give them about 200 MeV of energy and that the mass defect would indeed deliver that energy so the process could take place on a purely classical basis without having to invoke the crossing of a potential barrier, which of course could never have worked.

We only spent two or three days together that Christmas. Then I went back to Copenhagen and just managed to tell Bohr about the idea as he was catching his boat to the US. I remember how he struck his head after I had barely started to speak and said: "Oh, what fools we have been! We ought to have seen that before." But he had not—nobody had.

Lise Meitner and I composed a paper over the long-distance telephone between Copenhagen and Stockholm. I told the whole story to George Placzek, who was in Copenhagen, before it even occurred to me to do an experiment. At first Placzek did not believe the story that these heavy nuclei, already known to suffer from alpha instability, should also be suffering from this extra affliction. "It sounds a bit," he said, "like the man who is run over by a motor car and whose autopsy shows that he had a fatal tumor and would have died within a few days anyway." Then he

THE JOLIOT-CURIES discovered artificial radioactivity.

WIDE WORLD PHOTOS

CENTRAL FIGURES in the discovery were Otto Hahn and Lise Meitner, here shown in front of the institute that bears their names

said, "Why don't you use a cloud chamber to test it?" I did not have a cloud chamber handy and thought it would be difficult anyway. But I used an ionization chamber and it was a very easy experiment to observe the large pulses caused by ion fragments.

I do not think chronology means very much and certainly cannot claim any particular intelligence or originality. I was just lucky to be with Lise Meitner when she received advance notice of Hahn's and Strassmann's discovery. Then I had to be nudged before I did the crucial experiment on 13 January. By that time our joint paper was nearly written. I held it back for another three days to write up the other paper, and then they were both sent to *Nature* on 16 January but published a week apart. In the first paper I used the word "fission" suggested to me by the American biologist, William A. Arnold, whom I asked what one calls the phenomenon of cell division.

The second paper also contained a suggestion from Lise Meitner that fission fragments emerging from a bombarded uranium layer could be collected on a surface and their activity measured. The same thought independently occurred to Joliot, and he successfully did this experiment on 26 January. About that same time the news reached the US; what happened then is discussed by Wheeler.

Serendipitous searches

To come back to my initial question: Why did it take so long before fission was recognized? Indeed, why wasn't the neutron found earlier? Rutherford thought about it and foretold some of its properties as early as his Bakerian lecture in 1920; but Joliot did not read it, expecting a public lecture to contain nothing new! When Curie and Joliot found that the "beryllium radiation" ejected protons from paraffin, they put it down to a kind of Compton effect of a very hard gamma radiation (some 50 MeV), ignoring the objections of theoretical physicists. The neutron was finally observed in Cambridge, where such a particle was expected and had been sought.

At the time the neutron was found in 1932 pulse amplifiers and ionization chambers were available for a facile detection of fission pulses. But that would have been too big a jump to expect. The liquid-drop model of the nucleus was born late; the compound-nucleus idea was conceived by Bohr only late in 1936. It would have been a stroke of genius to think of fission then, and nobody did.

The discovery of artificial radioactivity in 1934 was again a chance discovery; no one had looked for it except Rutherford, who looked in vain for alpha decay. And indeed the

Berkeley team turned a blind eye when their counters "misbehaved." After the discovery there was a sheep-like rush to repeat the experiment with only the most obvious variation (I was one of the sheep). Only Fermi had the intelligence to strike out in a different and tremendously fruitful direction.

But then Fermi got on the wrong track: He felt sure that uranium, like other heavy nuclei, would obediently swallow any slow neutron that fell on it. He did make sure that the radioactive substances that were formed from it were different from any of the known elements near uranium. Ida Noddack, a German chemist, quite rightly pointed out that they might be lighter elements; but her comments (published in a journal not much read by chemists and hardly at all by physicists) were regarded as mere pedantry. She did not indicate how such light elements could be formed; her paper had probably no effect whatever on later work.

In the end it was good solid chemistry that got things on the right track. Irene Curie and Pavel Savitch came very close to it; only the presence of two substances with maliciously similar properties prevented them from establishing uranium fission before Hahn and Strassman finally accomplished it. □

Mechanism of Fission

by John A. Wheeler

IN EARLY JANUARY 1939 the Swedish-American liner, MS Drottningholm carried a short message across the stormy sea from Copenhagen to New York. This message symbolized the steady transfer of nuclear discoveries from Europe to the US that had been going on during the Hitler years.

Although these transfers were fateful for the US and the rest of the world, the act of relaying this particular message was simple: words of Otto Frisch to Niels Bohr and Leon Rosenfeld at Copenhagen before departure and words spoken by Rosenfeld to me that Monday afternoon, January 16, when I met them at the pier, and by Bohr to me when he and I started working on the issue the next day at Princeton. As a junior participator in the events that occurred then and in subsequent months, I shall relate the activities that led to the publication of a *Physical Review* paper by Bohr and me. In this paper we summarized the thoughts expressed in the message: the liquid-drop model that Frisch had applied to the mechanism of fission and the determinations of packing fraction that Lise Meitner considered when arriving at the first estimate of energy release in fission.

No one looking at such a novel process at that time could fail to call on everything he knew about nuclear physics to seek an interpretation. Fortunately the key ideas for unraveling the puzzle had already been developed. It may be appropriate to recall what had been learned about nuclear physics in the preceding half a dozen years.

Clues to the answer

1933 was a fruitful year for someone like me, who was just earning his doctor's degree. It was the year of the discovery of the neutron and Werner Heisenberg's great paper on the structure of nuclei built out of neutrons and protons. These discoveries made one feel that he might soon know as much about the nucleus as he already knew about the atom.

Encouraged by the vision that inspired so many young men, me included, at that time, I spent 1933–34 working with Gregory Breit, to whose insights I owe so much. He and the group of which I soon found myself a member accepted almost unconsciously the model of the nucleus of that day: neutrons and protons moving in a common self-consistent potential, closely analogous to the electric potential of the atom. "Unconscious" our acceptance of the model was, yes; but also shadowy. None of us took it too literally, especially not Breit, with his caution and insight. Thus he was always willing to consider alpha particles in the nucleus as well as neutrons and protons when that point of view made sense in considering a particular reaction. Breit also directed especial attention to areas of investigation as nearly free as possible of model-dependent issues. Thus much work was done on the penetration of charged particles into nucleli and how the cross section for a nuclear reaction depends on energy. The analysis of scattering processes in terms of phase shifts also received much attention.

With Breit's warm endorsement I spent the following year at Niels Bohr's institute in Copenhagen. Here I was initiated into the study of many new ideas, but nothing was more impressive in nuclear physics than the message that Møller brought back during the spring of 1935 from a short Easter visit to Rome: It told of Fermi's slow-neutron experiments and the astonishing resonances that he had discovered. Every estimate ever made before then indicated that a particle passing through a nucleus would have an extremely small probability of losing its energy by radiation and undergoing capture if the current nuclear model was credible. Yet, directly in opposition to the predictions of this model, Fermi's experiments displayed huge cross sections and resonances that were quite beyond explanation.

Of course a number of weeks went by before the most significant results of this discovery could be sorted out. Everyone was actively concerned, but no one more so than Bohr, who paced up and down in the colloquium and took a central part in discussions.

Liquid drops

The story of the development of the liquid-drop model and the compound-nucleus picture is a familiar one. What is not so clear and was certainly not evident at the time is the distinction between these ideas: (1) The compound-nucleus model shows, in essence, that the fate of a nucleus is independent of the mechanism by which it has been formed, and (2) the liquid-drop model is, so to speak, a special case of the compound-nucleus model, a particular way of making such a model of nuclear structure reasonable. Bohr proposed that the mean free path of nucleon is short in relation to nuclear dimensions instead of being long, as assumed in all previous estimates. This new idea made something like a liquid-drop model exceedingly attractive.

No one looking back on the situation from today's vantage point can fail to be amazed at "the great accident of nuclear physics"—the circumstance that the mean free path of particles in the nucleus is neither extremely short compared with nuclear dimensions (as assumed in the liquid-drop picture) nor extremely long (as assumed in the earlier model) but of an intermediate value. Moreover, all the marvelous detail of nuclear physics turns out to depend in such a critical way on the value of this parameter. As Aage Bohr and Ben Mottelson have taught us in recent years, no one could have predicted the precise one among many alternative regimes in which the phenomenology would actually lie from any advance estimate of the mean free path. Only observation could suffice! Knowing as little as one did in 1935 about the value of this decise parameter, still less about its cirticality, one had no option but to explore with all vigor the idea that the mean free path is very short.

The development of the liquid-drop model, which was applied to a variety of processes, took place in the hands of Fritz Kalckar and Niels Bohr in 1935–37. They applied it to a variety of processes. At the center of every such application stood the idealization of the compound nucleus, that is, the concept that a nuclear reaction occurs

STROLLING THINKERS, Fermi (left) and Bohr, are well known for their important applications and expansions of early ideas of nuclear fission.

in two well separated stages: First, the particle arrives in the nucleus and imparts an excitation; then in some way the nucleus uses that energy for radiation, neutron or alpha-particle emission or any other competing process.

Bohr brings the news

The message that Frisch gave Bohr as Bohr left Copenhagen opened up a new domain of application for this concept of the compound nucleus. By the time Bohr had arrived in New York he had already recognized that fission is one more process in competition with neutron reëmission and gamma-ray emission. Four days after his arrival he and Rosenfeld finished a paper sum-

marizing this general picture of fission in terms of formation and breakup of the compound nucleus.

Rosenfeld had originally accompanied Bohr to Princeton for several months of work on the problem of measurement in quantum electrodynamics. During Rosenfeld's Princeton sojourn Bohr gave less than half a dozen lectures on that issue. Nevertheless, that and many other questions conspired to take much of his time. No one could go into his office without seeing the long list of duties and people he had to give time to. That list made it easy to appreciate the pleasure with which he came into my office to discuss the work that we had under way. We were trying to understand in detail the

mechanism of fission and, not least, analyze the barrier against fission and the considerations that determine its height.

First of all, of course, we had to formulate the very idea of a threshold or barrier. How can there even be any barrier according to the liquid-drop picture? Is not an ideal fluid infinitely subdivisible? And therefore cannot the activation energy required to go from the original configuration to a pair of fragments be made as small as one pleases? We obtained guidance on this question out of the theory of the calculus of variations in the large, maxima and minima, and critical points. This subject we absorbed by osmosis from our environment, so thoroughly charged over the years by the ideas and results of Marston Morse. It became clear that we could find a configuration space to describe the deformation of the nucleus. In this deformation space we could find a variety of paths leading from the normal, nearly spherical configuration over a barrier to a separated configuration. On each path the energy of deformation reaches a highest value. This peak value differs from one path to another. Among all these maxima the minimum measures the height of the saddle point or fission threshold or activation energy for fission.

While we were estimating barrier heights and the energy release in various modes of fission, the time came for the fifth annual theoretical physics conference held in Washington on 26 Jan. Bohr felt a responsibility toward Frisch and Meitner and thought that word of their work-in-progress and their concepts should not be released until they had the proper opportunity to publish, as is the custom throughout science. Even though this was the situation, at the outset Rosenfeld did not appreciate all the complications and demands of Bohr's position. On the day of Bohr's arrival in the US Rosenfeld went down to Princeton on the train. (Bohr had an appointment later that day in New York.) Rosenfeld reported the new discovery at the journal club—the regular Monday night journal club—and of course everybody was very excited. Isidor I. Rabi, who was at the journal club, carried the news back to Columbia, where John Dunning started to plan an experiment.

Nevertheless, even on 26 Jan., Bohr was reluctant to speak about Frisch's and Meitner's findings until he received word that they had actually been published. Fortunately that afternoon an issue of *Die Naturwissenschaften,* which contained work by Hahn and Fritz Strassmann, was handed to him; thus he could tell about it. Of course everybody started his experiments. The first direct physical proof that fission takes place appeared in the newspapers of the twenty-ninth.

Shaping the theory

The analysis of fission led to the theory of a liquid drop and this in turn led back to a favorite love of Bohr, who, for his first student research work, experimented on the instability of a jet of water against breakup into smaller drops. He was quite familiar with the work of John W. Strutt, the third Lord Rayleigh. This work furnished a starting point for our analysis. However, we had to go to terms of higher order than Rayleigh's favorite second-order calculations to pass beyond the purely parabolic part of the nuclear potential, that is, the part of the potential that increases quadratically with deformation. We determined the third-order terms to see the turning down of the potential. They enabled us to evaluate the height of the barrier, or at least the height of the barrier for a nucleus whose charge was sufficiently close to the critical limit for immediate breakup.

Here we found that we could reduce the whole problem to finding a function f of a single dimensionless variable x. This "fissility parameter" measures the ratio of the square of the charge to the nuclear mass. This parameter has the value 1 for a nucleus that is already unstable against fission in its spherical form. For values of x close to 1, by the power-series development mentioned above one could estimate the height of the barrier and actually give quite a detailed calculation of the first two terms in the power series for barrier height, or f, in powers of $(1 - x)$. The opposite limiting case also lent itself to analysis. In this limit the nucleus has such a small charge that the barrier is governed almost entirely by surface tension. The Coulomb forces give almost negligible assistance in pushing the material apart.

ROSENFELD, with Bohr, summarized the idea of fission.

Between this case (the power series about $x = 0$) and the other case (the power series about $x = 1$) there was an enormous gap. We saw that it would take a great amount of work to calculate the properties of the fission barrier at points in between. Consequently we limited ourselves to interpolation between these points. In the 28 years since that time many workers have done an enormous amount of computation on the topography of the deformation energy as depicted over configuration space as a "base" for the topographic plot. We are still far from completing the analysis. Beautiful work by Wladyslaw J. Swiatecki and his collaborators at Berkeley has taught us much more than we ever knew before about the structure of this fission barrier and has revealed many unsuspected features for values of x that are remote from the two simple, original limits.

From fission barrier we turned to fission rate. All of us have always recognized that nuclear physics consists of two parts: (a) the energy of a process and (b) the rate at which the process will go on. The compound-nucleus model told us that the rate should be measured by the partial width of the nuclear state in question for breakup by the specified process.

Toward a simpler theory

How could we estimate this width? Happily, in earlier days, several persons in the Princeton community— among them Henry Eyring and Eugene Wigner—had been occupied by the theory of the rates of chemical reac-

tions. Also we derived some useful information from cosmic-ray physics. Who does not recall the many detailed calculations Størmer and his associates made on the orbits of cosmic-ray particles in the earth's magnetic field? Fortunately Manuel Sandoval Vallarta and later workers were able to spare themselves almost all of these details. They had only to employ Liouville's theorem. It said that the density of systems in phase space remains constant in time.

The same considerations of phase space were equally useful for evaluating the rate of fission. It turned out that we could express the probability of going over the barrier as the ratio of two numbers. One of these numbers is related to the amount of phase space available in the transition-state configuration as the nucleus goes over the top of the barrier. We were forced to think of all the degrees of freedom of the nucleus other than the particular one leading to fission. All these other degrees of freedom are summarized in effect in the internal excitations of the nucleus as it passes over the fission barrier. In classical terms this concept is well defined. It is a volume in phase space completely determined by the amount of energy.

The other quantity, appearing in the denominator of the rate-of-fission expression, is linked with the volume of phase space accessible to the compound system. In all the complex motion short of actual passage over the barrier the ensemble of systems under consideration remains confined to the narrow band of energies, ΔE, defined by the energy of the incident neutron. What counts is this energy interval multiplied with the rate of change of volume in phase space with energy for the undissociated nucleus. The beauty of this derivation is the fact that these classical ideas lend themselves to direct transcription into quantum-mechanical terms. Thus the Wentzel-Kramers-Brillouin approximation taught us that volume in phase space determines the number of energy levels. So we concluded that the width—the desired width measuring the probability for fission—is given by a ratio in which the numerator is the number of states accessible to the transition-state nucleus as it is going over the barrier, that is, the number of

PLACZEK was helpful in formulating theories of fission.

states of excitation other than motion in the direction of fission. In the denominator appears the spacing between nuclear energy levels, divided by 2π. Thus we had attained the most direct tie with experimentally interesting quantities. The formula that was obtained in this way for the reaction rate, or the level width, applied to a wide class of reactions as well as to fission, and was more general than any that had previously been available in reaction-rate theory. The new formula gave considerable insight into the rate of passage over the fission barrier.

At this particular point it is interesting to note the caution with which Bohr adopted the formula. He would come in every other day or so, and we would go at it for perhaps a half a day, trying out first this approach and then that approach. But his supreme caution was most evident when we wanted to interpret the number of levels accessible in the transition state. Today that number is called "the number of channels," and we use it as a formula to describe the channel-analysis theory of fission rate. Also we apply similar channel-analysis considerations to other nuclear reactions. But at that time the idea that each one of these individual channels has in principle a definite experimentally observable significance was, for us, of dubious certainty. Still less did we appreciate, until the later work of Aage Bohr, the possibility that each individual channel would have its individual angular

distribution from which one could determine the K values of that channel. The cautious phrase that was used in reference to that channel number appears in the following quotation: "It should be remarked that the specific quantum-mechanical effects which set in at and below the critical fission energy may even show their influence to a certain extent above this energy and produce slight oscillations in the beginning of the yield curve, allowing, possibly, a direct determination of the number of channels." Of course we know how later on in the 1950's these variations were observed by Lamphere and Green and others and how they led to direct measurement of the channel number.

Bohr's epiphany

The most important part of this Princeton period happened when I was not in direct touch with Bohr. One snowy morning he was walking from the Nassau Club to his office in Fine Hall. As a consequence of a breakfast discussion with George Placzek, who was deeply skeptical of these fission ideas, Bohr began struggling with the problem of explaining the remarkable dependence of fission cross section on neutron energy. In the course of the walk he concluded that slow-neutron fission is caused by U^{235} and fast-neutron fission by U^{238}. By the time he had arrived at Fine Hall and he and I had gathered together with Placzek and Rosenfeld, he was ready to sketch out the whole idea on the blackboard. There he displayed the concept that U^{238} is not susceptible to division by neutrons of thermal energy, nor is it susceptible to neutrons of intermediate energy but only to neutrons with energies of a million electron volts or more. Further, the fission observed at lower energies occurs because U^{235} is present and has a $1/v$ cross section for capture. We already knew experimentally that neutrons of intermediate energy undergo resonance capture. And, with the help of simple considerations, we could show that the resonance reaction of neutrons with uranium could not be due to U^{235}. We concluded this because we knew that the resonance cross section would exceed the theoretical limit given by the square of the wavelength if U^{235} were

responsible for the resonance effect. So the resonance had to be due to U^{238}, and the very fact that the resonance neutrons did not bring about fission proved that U^{238} was not susceptible to fission by neutrons of such low energy. Thus if it was not susceptible at that energy, it would certainly not be susceptible at lower energies; consequently low-energy fission must be due to U^{235}.

Around this time, Szilard, Placzek, Wigner, Rosenfeld, Bohr, myself and others discussed whether one could ever hope to make a nuclear explosive. It was so preposterous then to think of separating U^{235} that I cannot forget the words that Bohr used in speaking about it: "It would take the entire efforts of a country to make a bomb." He did not foresee that, in truth, the efforts of thousands of workers drawn from three countries would be needed to achieve that goal.

The theory of fission made it possible to predict in general terms how the cross section for fission would depend upon energy. In Palmer Physical Laboratory Rudolf Ladenberg, James Kanner, Heinz H. Barschall and Van Voorhies, just at the time we were working on the theory, actually measured the cross section of uranium in the region from two million to three million volts—and also the cross section for thorium, all of which fitted in with predictions. The same considerations of course made it possible to predict that plutonium 239 would be fissile. For this application of the theory we are especially indebted to Louis A. Turner. One started on the way that ultimately led to the giant plutonium project having only this theoretical estimate to light and encourage the first steps.

Spontaneous fission offered a most attractive application of these ideas in conjunction with the concept of barrier penetration. Another application dealt with the difference between prompt neutrons and delayed neutrons. In conclusion, nuclear fission brought us a process distinguished from all the other processes with which we ever dealt before in nuclear physics, in that we have for the first time in fission a nuclear transformation inescapably *collective* in character. In this sense fission opened the door to the development of the collective model of the nucleus in the postwar years. □

ENRICO FERMI

PHYSICS at

By Enrico Fermi

The following is a verbatim transcript of Enrico Fermi's last address before the American Physical Society, delivered informally and without notes at Columbia University's McMillin Theater on Saturday morning, January 30, 1954. His retiring presidential address was delivered one day earlier. The present speech, transcribed from a tape recording, is left deliberately in an unpolished and unedited form. Such informality would no doubt have been frowned upon by Fermi, who was very particular about his published writings. For those who knew Fermi or heard him speak, however, the verbatim transcript may serve (as no formal document could ever serve) to bring back for a moment the very sound of his voice. The paper was presented as part of the session "Physics at Columbia University" during the Society's 1954 annual meeting.

Mr. Chairman, Dean Pegram, fellow members, ladies and gentlemen:

IT seems fitting to remember, on this 200th anniversary of Columbia University, the key role that the University played in the early experimentation and the organization of the early work that led to the development of atomic energy.

I had the good fortune to be associated with the Pupin Laboratories through the period of time when at least the first phase of this development took place. I had had some difficulties in Italy and I will always be very grateful to Columbia University for having offered me a position in the Department of Physics at the most opportune moment. And in addition this offer gave me, as I said, the rare opportunity of witnessing the series of events to which I have referred.

In fact I remember very vividly the first month, January, 1939, that I started working at the Pupin Laboratories because things began happening very fast. In that period, Niels Bohr was on a lecture engagement in Princeton and I remember one afternoon Willis Lamb came back very excited and said that Bohr had leaked out great news. The great news that had leaked out was the discovery of fission and at least an outline of its interpretation; the discovery as you well remember goes back to the work of Hahn and Strassmann and at least the first idea for the interpretation came through the work of Lise Meitner and Frisch who were at that time in Sweden.

COLUMBIA UNIVERSITY

PHYSICS TODAY / NOVEMBER 1955

the Genesis of the

Nuclear

Energy

Project

Then, somewhat later that same month, there was a meeting in Washington organized by the Carnegie Institution in conjunction with George Washington University where I took part with a number of people from Columbia University and where the possible importance of the new-discovered phenomenon of fission was first discussed in semi-jocular earnest as a possible source of nuclear power. Because it was conjectured, if there is fission with a very serious upset of the nuclear structure, it is not improbable that some neutrons will be evaporated. And if some neutrons are evaporated, then they might be more than one; let's say, for the sake of argument, two. And if they are more than one, it may be that the two of them, for example, may each one cause a fission and from that one sees of course a beginning of the chain reaction machinery.

So that was one of the things that was discussed at that conference and started a small ripple of excitement about the possibility of releasing nuclear energy. At the same time experimentation was started feverishly in many laboratories, including Pupin, and I remember before leaving Washington I had a telegram from Dunning announcing the success of an experiment directed to the discovery of the fission fragments. The same experiment apparently was at the same time carried out in half a dozen places in this country and in three or four, in fact I think slightly before, in three or four places in Europe.

Now a rather long and laborious work was started at Columbia University in order to firm up these vague suggestions that had been made as to the possibilities that neutrons were emitted and try to see whether neutrons were in fact emitted when fission took place and

if so how many they would be, because clearly a matter of numbers is in this case extremely important because a little bit greater or a little bit lesser probability might have made all the difference between possibility and impossibility of a chain reaction.

Now this work was carried on at Columbia simultaneously by Zinn and Szilard on one hand and by Anderson and myself on the other hand. We worked independently and with different methods, but of course we kept close contact and we kept each other informed of the results. At the same time the same work was being carried out in France by a group headed by Joliot and Von Halban. And all the three groups arrived at the same conclusion—I believe Joliot may be a few weeks earlier than we did at Columbia—namely that neutrons are emitted and they were rather abundant, although the quantitative measurement was still very uncertain and not too reliable.

A curious circumstance related to this phase of the work was that here for the first time secrecy that has been plaguing us for a number of years started and, contrary to perhaps what is the most common belief about secrecy, secrecy was not started by generals, was not started by security officers, but was started by physicists. And the man who is mostly responsible for this certainly extremely novel idea for physicists was Szilard.

I don't know how many of you know Szilard; no doubt very many of you do. He is certainly a very peculiar man, extremely intelligent (LAUGHTER). I see that is an understatement (LAUGHTER). He is extremely brilliant and he seems somewhat to enjoy, at least that is the impression that he gives to me, he seems to enjoy startling people.

So he proceeded to startle physicists by proposing to them that given the circumstances of the period—you see it was early 1939 and war was very much in the air—given the circumstances of that period, given the danger that atomic energy and possibly atomic weapons could become the chief tool for the Nazis to enslave the world, it was the duty of the physicists to depart from what had been the tradition of publishing significant results as soon as the *Physical Review* or other scientific journals might turn them out, and that instead one had to go easy, keep back some results until it was clear whether these results were potentially dangerous or potentially helpful to our side.

So Szilard talked to a number of people and convinced them that they had to join some sort of—I don't know whether it would be called a secret society, or what it would be called. Anyway to get together and circulate this information privately among a rather restricted group and not to publish it immediately. He sent in this vein a number of cables to Joliot in France, but he did not get a favorable response from him and Joliot published his results more or less like results in physics had been published until that day. So that the fact that neutrons are emitted in fission in some abundance—the order of magnitude of one or two or three—became a matter of general knowledge. And, of course, that made the possibility of a chain reaction appear to most physicists as a vastly more real possibility than it had until that time.

Another important phase of the work that took place at Columbia University is connected with the suggestion on purely theoretical arguments, by Bohr and Wheeler, that of the two isotopes of uranium it was not the most abundant uranium 238 but it was the least abundant uranium 235, present as you know in the natural uranium mixture to the tune of 0.7 of a per cent, that was responsible at least for most of the thermal fission. The argument had to do with an even number of neutrons in uranium 238 and an odd number of neutrons in uranium 235 which, according to a discussion of the binding energies that was carried out by Bohr and Wheeler, made plausible that uranium 235 should be more fissionable.

Now it clearly was very important to know the facts also experimentally and work was started in conjunction by Dunning and Booth at Columbia University and by Nier. Nier took the mass spectrographic part of this work, attempting to separate a minute but as large as possible amount of uranium 235, and Dunning and Booth at Columbia took over the part of using this minute amount in order to test whether or not it would undergo fission with a much greater cross section than ordinary uranium.

Well, you know of course by now that this experiment confirmed the theoretical suggestion of Bohr and Wheeler, indicating that the key isotope of uranium, from the point of view of any attempt of—for example —constructing a machine that would develop nuclear energy, was in fact uranium 235. Now you see the mat-

ter is important primarily for the following reasons that at the time were appreciated perhaps less definitely than at the present moment.

The fundamental point in fabricating a chain reacting machine is of course to see to it that each fission produces a certain number of neutrons and some of these neutrons will again produce fission. If an original fission causes more than one subsequent fission then of course the reaction goes. If an original fission causes less than one subsequent fission then the reaction does not go.

Now, if you take the isolated pure isotope U-235, you may expect that the unavoidable losses of neutrons will be minor, and therefore if in the fission somewhat more than one neutron is emitted then it will be merely a matter of piling up enough uranium 235 to obtain a chain reacting structure. But if to each gram of uranium 235 you add some 140 grams of uranium 238 that come naturally with it, then the competition will be greater, because there will be all this ballast ready to snatch away the not too abundant neutrons that come out in the fission and therefore it was clear at the time that one of the ways to make possible the production of a chain reaction was to isolate the isotope U-235 from the much more abundant isotope U-238.

Now, at present we have in our laboratories a row of bottles labeled, more or less, isotope—what shall I say—iron 56, for example, or uranium 235 or uranium 238 and these bottles are not quite as common as would be a row of bottles of chemical elements, but they are perfectly easily obtainable by putting due pressure on the Oak Ridge Laboratory (LAUGHTER). But at that time isotopes were considered almost magically inseparable. There was to be sure one exception, namely deuterium, which was already at that time available in bottles. But of course deuterium is an isotope in which the two isotopes hydrogen one and hydrogen two have a ratio of mass one to two, which is a very great ratio. But in the case of uranium the ratio of mass is merely 235 to 238, so the difference is barely over one per cent. And that, of course, makes the differences of these two objects so tiny that it was not very clear that the job of separating large amounts of uranium 235 was one that could be taken seriously.

Well, therefore, in those early years near the end of 1939 two lines of attack to the problem of atomic energy started to emerge. One was as follows. The first step should be to separate in large amounts, amounts of kilograms or maybe amounts of tens of kilograms or maybe of hundreds of kilograms, nobody really knew how much would be needed, but something perhaps in that order of magnitude, separate such at that time fantastically large-looking amounts of uranium 235 and then operate with them without the ballast of the associated much larger amounts of uranium 238. The other school of thought was predicated on the hope that perhaps the neutrons would be a little bit more and that perhaps using some little amount of ingenuity one might use them efficiently and one might perhaps be able to achieve a chain reaction without having to separate the isotopes, a task as I say that at that time looked almost beyond human possibilities.

Now I personally had worked many years with neutrons, and especially slow neutrons, so I associated myself with the second team that wanted to use non-separated uranium and try to do the best with it. Early attempts and studies, discussions, on how to separate the isotopes of uranium were started by Dunning and Booth in close consultation with Professor Urey. On the other hand, Szilard, Zinn, Anderson, and myself started experimentation on the other line whose first step involved lots of measurements.

Now, I have never yet quite understood why our measurements in those days were so poor. I'm noticing now that the measurements that we are doing on pion physics are very poor, presumably just because we have not learned the tricks. And, of course, the facilities that we had at that time were not as powerful as they are now. It's much easier to carry out experimentation with neutrons using a pile as a source of neutrons than it was in those days using radium-beryllium sources when geometry was the essential item to control or using the cyclotron when intensity was the desired feature rather than good geometry.

Well, we soon reached the conclusion that in order to have any chance of success with natural uranium we had to use slow neutrons. So there had to be a moderator. And this moderator could have been first water or other substances. Water was soon discarded; it's very effective in slowing down neutrons, but still absorbs a little bit too many of them and we could not afford that. Then it was thought that graphite might be perhaps the better bet. It's not as efficient as water in slowing down neutrons; on the other hand little enough was known of its absorption properties that the hope that the absorption might be very low was quite tenable.

This brings us to the fall of 1939 when Einstein wrote his now famous letter to President Roosevelt advising him of what was the situation in physics—what was brewing and that he thought that the government had the duty to take an interest and to help along this development. And in fact help came along to the tune of $6000 a few months after and the $6000 were used in order to buy huge amounts—or what seemed at that time when the eye of physicists had not yet been distorted—(LAUGHTER) what seemed at that time a huge amount of graphite.

So physicists on the seventh floor of Pupin Laboratories started looking like coal miners (LAUGHTER) and the wives to whom these physicists came back tired at night were wondering what was happening. We know that there is smoke in the air, but after all (LAUGHTER).

Well, what was happening was that in those days we were trying to learn something about the absorption properties of graphite, because perhaps graphite was no good. So, we built columns of graphite, maybe four feet on the side or something like that, maybe ten feet high. It was the first time when apparatus in physics, and these graphite columns were apparatus, was so big

that you could climb on top of it—and you had to climb on top of it. Well, cyclotrons were the same way too, but anyway the first time when I started climbing on top of my equipment because it was just too tall—I'm not a tall man (LAUGHTER).

And then sources of neutrons were inserted at the bottom and we were studying how these neutrons were first slowed down and then diffused up the column and of course if there had been a strong absorption they would not have diffused very high. But because it turned out that the absorption was in fact small, they could diffuse quite readily up this column and by making a little bit of mathematical analysis of the situation it became possible to make the first guesses as to what was the absorption cross section of graphite, a key element in deciding the possibility or not of fabricating a chain reacting unit with graphite and natural uranium.

Well, I will not go into detail of this experimentation. That lasted really quite a number of years and required really quite many hours and many days and many weeks of extremely hard work. I may mention that very early our efforts were brought in connection with similar efforts that were taking place at Princeton University where a group with Wigner, Creutz and Bob Wilson set to work making some measurements that we had no possibility of carrying out at Columbia University.

Well, as time went on, we began to identify what had to be measured and how accurately these things that I shall call "eta", f, and p—I don't think I have time to define them for you—these three quantities "eta", f, and p had to be measured to establish what could be done and what could not be done. And, in fact, if I may say so, the product of "eta", f, and p had to be greater than one. It turns out, we now know, that if one does just about the best this product can be 1.1.

So, if we had been able to measure these three quantities to the accuracy of one per cent we might have found that the product was for example 1.08 plus or minus 0.03 and if that had been the case we would have said let's go ahead, or if the product had turned out to be 0.95 plus or minus 0.03 perhaps we would have said just that this line of approach is not very promising, and we had better look for something else. However I've already commented on the extremely low quality of the measurements in neutron physics that could be done at the time—where the accuracy of measuring separately either "eta", or f, or p was perhaps with a plus or minus of 20 per cent (LAUGHTER). If you compound, by the well-known rules of statistics, three errors of 20 per cent you will find something around 35 per cent. So if you should find, for example, 0.9 plus or minus 0.3—what do you know? Hardly anything at all (LAUGHTER). If you find 1.1 plus or minus 0.3—again, you don't know anything much. So that was the trouble and in fact if you look in our early work—what were the detailed values given by this or that experimenter to, for example, "eta" you find that it was off 20 per cent and sometimes greater amounts. In fact I think it was strongly influenced by the temperament of the physicist. Shall we say optimistic physicists felt it unavoidable to push these quantities high and pessimistic physicists like myself tried to keep them somewhat on the low side (LAUGHTER).

Anyway, nobody really knew and we decided therefore that one had to do something else. One had to devise some kind of experiment that would give a complete over-all measurement directly of the product "eta", f, p without having to measure separately the three, because then perhaps the error would sort of drop down and permit us to reach conclusions.

Well, we went to Dean Pegram, who was then the man who could carry out magic around the University, and we explained to him that we needed a big room. And when we say big we meant a really big room, perhaps he made a crack about a church not being the most suited place for a physics laboratory in his talk, but I think a church would have been just precisely what we wanted (LAUGHTER). Well, he scouted around the campus and we went with him to dark corridors and under various heating pipes and so on to visit possible sites for this experiment and eventually a big room, not a church, but something that might have been compared in size with a church was discovered in Schermerhorn.

And there we started to construct this structure that at that time looked again in order of magnitude larger than anything that we had seen before. Actually if anybody would look at that structure now he would probably extract his magnifying glass (LAUGHTER) and go close to see it. But for the ideas of the time it looked really big. It was a structure of graphite bricks and spread through these graphite bricks in some sort of pattern were big cans, cubic cans, containing uranium oxide.

Now, graphite is a black substance, as you probably know. So is uranium oxide. And to handle many tons of both makes people very black. In fact it requires even strong people. And so, well we were reasonably strong, but I mean we were, after all, thinkers (LAUGHTER). So Dean Pegram again looked around and said that seems to be a job a little bit beyond your feeble strength, but there is a football squad at Columbia (LAUGHTER) that contains a dozen or so of very husky boys who take jobs by the hour just to carry them through College. Why don't you hire them?

And it was a marvelous idea; it was really a pleasure for once to direct the work of these husky boys, canning uranium—just shoving it in—handling packs of 50 or 100 pounds with the same ease as another person would have handled three or four pounds. In passing these cans fumes of all sorts of colors, mostly black, would go in the air (LAUGHTER).

Well, so grew what was called at the time the exponential pile. It was an exponential pile, because in the theory an exponential function enters—which is not surprising. And it was a structure that was designed to test in an integral way, without going down to fine details, whether the reactivity of the pile, the reproduction factor, would be greater or less than one. Well,

it turned out to be 0.87. Now that is by 0.13 less than one and it was bad. However, at the moment we had a firm point to start from, and we had essentially to see whether we could squeeze the extra 0.13 or preferably a little bit more. Now there were many obvious things that could be done. First of all, I told you these big cans were canned in tin cans, so what has the iron to do? Iron can do only harm, can absorb neutrons, and we don't want that. So, out go the cans. Then, what about the purity of the materials? We took samples of uranium, and with our physicists' lack of skill in chemical analysis, we sort of tried to find out the impurities and certainly there were impurities. We would not know what they were, but they looked impressive, at least in bulk (LAUGHTER). So, now, what do these impurities do?—clearly they can do only harm. Maybe they make harm to the tune of 13 per cent. Finally, the graphite was quite pure for the standards of that time, when graphite manufacturers were not concerned with avoiding those special impurities that absorb neutrons. But still there was some considerable gain to be made out there, and especially Szilard at that time took extremely decisive and strong steps to try to organize the early phases of production of pure materials. Now, he did a marvelous job which later on was taken over by a more powerful organization than was Szilard himself. Although to match Szilard it takes a few able-bodied customers (LAUGHTER).

Well, this brings us to Pearl Harbor. At that time, in fact I believe a few days before by accident, the interest in carrying through the uranium work was spreading; work somewhat similar to what was going on at Columbia had been initiated in a number of different Universities throughout the country. And the government started taking decisive action in order to organize the work. and, of course, Pearl Harbor gave the final and very decisive impetus to this organization. And it was decided in the high councils of the government that the work on the chain reaction produced by nonseparated isotopes of uranium should go to Chicago.

That is the time when I left Columbia University, and after a few months of commuting between Chicago and New York eventually moved to Chicago to keep up the work there, and from then on, with a few notable exceptions, the work at Columbia was concentrated on the isotope-separation phase of the atomic energy project.

As I've indicated this work was initiated by Booth, Dunning, and Urey about 1940, 1939, and 1940, and with this reorganization a large laboratory was started at Columbia under the direction of Professor Urey. The work there was extremely successful and rapidly expanded into the build-up of a huge research laboratory which cooperated with the Union Carbide Company in establishing some of the separation plants at Oak Ridge. This was one of the three horses on which the directors of the atomic energy project had placed their bets, and as you know the three horses arrived almost simultaneously to the goal in the summer of 1945. I thank you. (APPLAUSE)

Chapter 6
Particles and Quanta

The physics of atoms, quanta, and elementary particles has caught the attention of more historians than any other field of modern science. At no time were more than a minority of physicists working in this area, but the area has had an unmatched impact on all of science and even on philosophy. Quantum mechanics in particular, by way of chemical physics, condensed matter physics, and so forth, has been at the center not only of a revolution in thought but of a new industrial era. Study of the history of the area has been accordingly extensive, indeed so extensive that we can give it, alone among the subfields of physics, a separate section of this book. As usual the articles are arranged roughly in chronological order.

The section begins with the discovery of the electron, which was the first true elementary particle to be identified, and also the key to atomic physics, and also ultimately the key to understanding quanta. Contrary to a myth that is still widespread, few physicists prior to the time of that discovery saw their field as moribund; most put their faith in exciting and ambitious programs to get at the mysterious heart of matter and energy. Many physicists were motivated by a confusion of electromagnetic theories which seem peculiar today, but which convinced them that great secrets lay near the surface, and it was this intellectual ferment that produced the discoveries of Röntgen, Becquerel, Thomson, and Planck. But the discoveries were even more astonishing than they had anticipated. The next several decades were a turmoil of misunderstandings and incredible ideas. After the late 1930s with the coming of nuclear physics, field theory, and a distinct physics of elementary particles, things settled down into a new intellectual configuration, whose basic outlines have stayed intact down to the present.

Among the articles here, those by Thomson, Condon, Bloch, Schmitt, and Weisskopf are written by scientists who witnessed or came close to witnessing the events they describe. These articles thus have some of the advantages, and pitfalls, of first-person accounts, as discussed in the introduction to the previous section. However, the physicists in this section are not so much describing their own work as the work that was going on around them. Such descriptions, whether relatively informal accounts of the sort reprinted here, or highly stylized review articles, are the traditional starting-point for work by professional historians. The remaining articles in this section are written by some of our foremost professional historians of physics. Note that these historians approach the subject from a quite different viewpoint than the physicists, typically less personal and more analytic. It is such a mixture of two viewpoints, the original scientists' experiences and the subsequent interpretation by historical scholarship, that gives history of modern physics its tremendous vigor and appeal.

Contents

J. J. THOMSON
and the discovery of the Electron

PHYSICS TODAY / AUGUST 1956

By **George P. Thomson**

Conference members (from left to right) L. B. Leder, L. Marton, H. S. W. Massey, G. P. Thomson, F. L. Hereford, A. Klein, R. D. Birkhoff, A. W. Kenney, H. A. Tolhoek.

Sir George P. Thomson, FRS (right), is Master of Corpus Christi College, Cambridge, England. The son of Sir. J. J. Thomson, Sir George shared the 1937 Nobel Prize in physics with C. J. Davisson for the discovery of the diffraction of electrons by crystals. The present paper is the text of his after-dinner address at the Electron Physics Conference Banquet.

MAY I say that I am particularly glad and happy that my father's hundredth anniversary of his birth should be celebrated here in this way. My father had a great affection for America at all times, and of all the places in America, he best knew and loved Baltimore; and that it should be the University of Maryland which is honouring him in this way is to me a very great pleasure, as it would have been to him.

Now, I have to speak not only of J. J. but also of the discovery of the electron; and the electron was, as most people know, named before it was discovered, and the anomaly that this implies has its counterpart in an uncertainty of meaning which I think to some extent still subsists. The two parent strains from which the electron sprang have not even now completely fused, and indeed it was the blending of two different trains of thought which constituted its discovery. They are, first, the idea of a natural unit of electric charge, and second, the existence of very light electrified particles fundamental in the structure of matter. The first is

Sir Joseph J. Thomson

inherent in the work of Faraday on electrolysis, and it is to me one of the most curious features of the history of physics that it should have taken people so long to realise that Faraday's laws of electrolysis are only intelligible if you suppose that there is a fundamental unit of charge involved. But in fact it seems to have taken a long time for people to do so. Johnstone Stoney in 1874 called attention to the importance of the charge carried by the hydrogen ion in electrolysis, and in the early 80's he estimated its value; I may say he got it 16 times too small, but he named it the electron in 1891.

There were electron theories in the course of the 19th century, but they were not very important. It was not until the great theory of Lorentz in the early 90's which predicted the effect that Zeeman found in '96, that electrons became important in theoretical physics. But since I am speaking in commemoration of the 100th anniversary of the birth of J. J., I will leave that side of the picture and turn to the side with which he was personally concerned. But I should like first to say a few words about his background.

He was, as you have been told, born 100 years ago. He was a son of a Manchester book-seller and publisher who died when he was 16, and there is little recorded about his boyhood. He was given at some suitable age a microscope, and I would like to say of course that

Sir George P. Thomson

that turned his thoughts to science and was the beginning of his career. I do not think it was, but there is a story about it which I might as well tell you.

He got the microscope and examined things in it. A friend of his father's came along one day and to show it off he put a hair, a human hair, presumably one of his own because he had no sisters to pull hairs out of, on the stage, focussed it duly, and asked the father's friend to look. He looked and seemed puzzled. J. J. said "Can't you see it?"

"Oh, yes, I can see it all right", said the friend, "but where is the number?"

"Number?" asked J. J.

"Yes, you know, it's in the Bible. All the hairs of your head are numbered."

My father's original ambition was to be an engineer and I think it is probable that if his father had lived he would have been. In those days in order to become an engineer you had to be taken on by a firm in a kind of quasi-apprenticeship and pay a very substantial premium for the privilege. The family was not rich enough to afford this, but J. J. went, at a very early age, to what was then called Owens College and is now the University of Manchester. He was trained there as a mathematician and at 19 he went to Cambridge with a scholarship to Trinity College, of which he was master for the last 22 years of his life. After he had taken his mathematical degree he worked in the Cavendish Laboratory under Lord Rayleigh who was professor for a not very long time, in succession to Maxwell. His early theoretical work was really inspired by Maxwell's theory. It is perhaps difficult for us who have had to face so many worse things, to realise that in these early days Maxwell's theory was regarded as the limit of obscurity. Actually, if you read Maxwell, you will see that there was something to be said for this view. He was not very skilful, perhaps, in putting forward what he was thinking about and it is expressed in a rather curious fashion. The displacement, for example. which figures so largely in it: it is not really clear precisely how Maxwell did in fact envisage it, and it must have been a very difficult idea for his contemporaries.

My father published an edition of Maxwell's theory and supplemented it by a volume called "Recent Researches" which were really a kind of commentary on it and the working out of a number of problems suggested by it. He discovered, if "discovered" is the right word (predicted the existence of is perhaps better), electromagnetic mass, the first example of the connection of mass with energy. That was in the year 1881. About this time he was attracted by a theory which I suppose few people have even heard of, and that is Helmholtz's vortex theory of atoms, a theory based on vortex rings presumed to exist in the ether, and capable of the most delightful mathematical complexities as they interlocked and performed curious gyrations round one another. It is interesting in its way as a theory of matter which is almost purely kinematic: dynamics hardly comes into it; it is the motion produced by each vortex in the other vortices, you see, which

governs the whole thing. J. J. published a long prize essay on it, but the most important point, from our point of view tonight, is that it first called his attention to the gaseous discharge. He thought that it was possible that what happened in the gaseous discharge was that molecules of two atoms each were pulled apart, and by the theory it was reasonable to guess that this might produce electrification. Anyhow he started experiments on the gaseous discharge in 1886 and for about 50 years afterwards he was rarely, if ever, without some work on gaseous discharge in one form or another on hand. He was always fascinated with it and indeed I think those who have worked in that field, as most of us here perhaps have done, recognize the fascination. But in fact no great advance was made until Roentgen's discovery of x-rays in 1895. This was certainly one of the major things in physics, quite comparable with Galvani's discovery of the twitching frog's leg, and it gave the same kind of thrill to its age that the discovery of nuclear fission gave to ours.

I would like just to tell a very short story that is not very relevant but it is of interest to physicists as to how, if you wish, you can avoid making discoveries.

At the time that Roentgen was working with the discharge tube, other people were also. And one of them, who shall be nameless, he was not a very famous physicist, discovered that photographic plates, when near this tube, got fogged. Well, he was intent on the job he was doing, and he was a man of common sense, so he took the plates further away.

Rutherford had only recently come to Cambridge from New Zealand. His work there had been on the electromagnetic waves, and he was a pioneer in what we in England call "wireless" and you call radio. He continued this work for a little while at Cambridge. But this discovery of x-rays called him to atomic physics, and he moved on to create nuclear physics.

J. J. turned to the line which led to the discovery of the electron, namely cathode rays. Cathode rays had been known for about 50 years, but there was at the time a great controversy as to their nature, a controversy which was of an international character for the two sides were, in a sense, divided by the Rhine. The Germans held the view that these cathode rays were in their nature waves; the French and the British for the most part held that they were particles. Lenard had shown that they could emerge from the exhausted tube into the air and appear as a visible streak of luminosity. They could emerge through thin but appreciable thicknesses of solid metal. Now to people in those days it seemed quite inconceivable that any material particle could get through metal and go on in something like the same straight line. It was a very strong argument in favour of some kind of wave motion. Perrin, on the other hand, had shown that when received in a Faraday cylinder they gave it an electric charge. I'm not so sure whether that would be quite such a strong argument now-a-days—we should think rather of the possibilities of secondaries. He also showed that when the cathode rays were deflected away by a magnet the charge ceased.

One of my father's first experiments in this field was to carry Perrin's one stage further to show that when the Faraday cylinder was not in line with the original rays it did not receive a charge, but that the charge appeared when they were bent by a magnet so that their path as shown by their luminosity made them hit the cylinder.

But a more fundamental attack was the attempt to measure the ratio of the charge to the mass. Many attempts at measuring this and so comparing it with the same ratio for the ions in electrolysis were made in the year 1897. Now, of course, from the knowledge of the stiffness in a magnetic field, it was easy to find e/mv. But you had to find something else, either v or perhaps $\frac{1}{2}mv^2$. A number of these attempts were made. Some, I think, were based on unjustifiable assumptions, but one, at least (that due to Wiechert which was published in January 1897), was quite sound in principle. It attempted to measure v by comparing the time that the cathode rays took to go a certain distance with the time of oscillation of an electrical circuit. But, in fact, in his early experiments Wiechert did not make a measurement; he only got a lower limit and he only did it for one gas. J. J. always stressed the importance of the constancy of e/m for different gases and different electrodes; and in February of that same year he showed that mv/e, that is to say the stiffness of the rays, was the same for all gases provided the voltage was constant. My father's first measurement of e/m appeared in a lecture at the Royal Institution of London on the 30th of April 1897 and was published a fortnight later in *The Electrician*. In this first measurement he used a thermopile and a Faraday cylinder to measure the total energy in a given time and also the charge; that is, he measured effectively e/mv^2 for the individual particle, and then, having e/mv from the magnetic deflection, of course both e/m and v followed. This led to the result that e/m was about a thousand times that for the ion of hydrogen.

Then came the better known method using the electrostatic deflection, the origin of the cathode ray oscillograph. And this perhaps was most important, not so much for its actual measurement as for its accounting for what was the strongest argument against the particle nature of the cathode rays—a piece of work by Hertz who had tried to deflect them by electrostatic action and had failed. Well, of course, we now know that the reason that he failed was because the vacuum wasn't good enough. The gas between the plates between which the field was supposed to be applied was ionized. The ions would flow towards the two plates and virtually neutralize the field between them. By slightly improving the vacuum (I don't think it was more than slightly improving, seeing what vacua were like in those days and indeed much later) my father was able to get it good enough so that he got the deflection, which indeed was an essential part of the measurement.

In his earlier papers J. J. emphasized not merely the large value of e/m but the fact that these corpuscles, as he liked to call them, were a universal constituent of matter, that they were the same whatever the gas in the tube—he tried 4 I think, and whatever the metal of which the electrodes were made, of which he tried 3. Then followed a measurement of the charge e for the x-ray ions which he had discovered a couple of years before. It followed the now well-known method, due to C. T. R. Wilson's work on the condensation of clouds on ions, work which incidentally was a consequence of the theory my father had developed earlier, on the connection between energy, and chemical and physical reactions. The charge was about the same as for a monovalent ion. indicating that m for a cathode ray was a thousand times less than for a hydrogen atom.

This, in a sense, was not conclusive because there was no absolute reason to suppose that the x-ray ions had got anything very particular to do with the cathode rays; one could not assert dogmatically that the charge on the cathode ray was the same as the charge on the negative x-ray ion, though most people thought it was. Nevertheless, the proof was completed in 1899 by the measurement simultaneously of e/m and e for the photoelectric particles, which are now called photoelectric electrons. This showed that he had found something very much smaller, at least a thousand times smaller, than the mass of a hydrogen atom, and something, and this was much more important, something which was a universal constituent of all matter. After that no reasonable person could really refuse belief that there were particles smaller than atoms, or lighter than atoms at least, and that these particles played a fundamental part in the constitution of matter.

I have no time in an after-dinner speech to speak of his work on the electron theory of metals, or his theory of atoms in which the electrons were supposed to be imbedded in a uniform distribution of positive electricity, except perhaps to give the reason why he held what now seems to be this rather odd idea. He did so because he knew from the Newtonian theory that the other obvious explanation, something like a solar system, would be unstable. A solar system in which the planets repel one another can be shown to be inherently unstable. And unfortunately my father knew that, for he was a very good mathematician. Nor will I speak of his estimates by three methods of the number of electrons in the atoms of various elements which led to the conclusion that this number was not very different from half the atomic weight. His very important work on positive rays does not, I suppose, really come under this subject, but it did of course lead to the discovery of nonradioactive isotopes and was indeed, and this is perhaps not always realised, the first experimental proof that atoms of any one substance are equal in mass apart from isotopes, and that the atomic weight is not merely a statistical property which might really represent a continuous spread over a very wide range. Before those experiments there was no real evidence to show that that was not the case.

I should like to conclude, if I may, by saying a few words about my father as a thinker and a man. His character was in some ways rather anomalous. He was a mathematician, a very good mathematician, who yet liked his theories concrete. All his life he was attracted by the idea of tubes of force, Faraday's tubes of force, and always tried to ascribe to them some kind of actual physical reality. He liked something he could picture and he entirely distrusted metaphysics. He preferred the wave atom, the wave atom with the wave electron, to the Bohr atom, at least as long as the waves could be allowed to remain pictorial. He was a great experimentalist who was liable to break any apparatus he got near. He was singularly clumsy with his hands and my mother, who was good at that kind of thing, never dreamed of allowing him to knock a nail in.

He had most of the actual preparing of the experiments done by his personal assistant Everett; my father just took the readings, which very often took the form of examining a photographic record, for example of positive rays, which he would measure. But he had an uncanny power of diagnosing the reasons why apparatus, his own or other people's, would not work, and suggesting what had to be done to make it work. He was a man who was normally silent, but he was a witty and amusing host at any sort of party, including the daily teas held in his room in the Cavendish, which he introduced. He loved flowers, wild and cultivated, and knew a very great deal about them, but he seldom gardened. He was fond of watching cricket, tennis, and football, and could recall the names and achievements of most of the leading people at Cambridge for the last 30 or 40 years in those sports. But in fact he had played little himself. He was a man of exceptionally wide sympathies. He could enjoy talking to almost anybody, and had the knack of making other people talk well about their own particular subject. He founded, and these sympathies helped him to found, the first school of physics, in a modern sense, at least outside Germany, and at one time his pupils, Cavendish men, held a very large fraction of the professorships throughout the world. Though he had a strong sense of humour, physics was too important to be funny, certainly too important to be laughed at. For him the two great qualities of a physicist, the two that really mattered, were originality and enthusiasm; and though he rated originality extremely high, it was enthusiasm which stood at the top.

Thermodynamics and Quanta in Planck's Work

Planck's search for a deeper understanding of the second law of thermodynamics led him to a strange and unexpected result—the concept of energy quanta. His conservative attitude toward this revolutionary discovery expressed itself in his attempts to reconcile the quantum with classical electrodynamics.

PHYSICS TODAY / NOVEMBER 1966

by Martin J. Klein

IN JANUARY 1910 Max Planck sent a paper to *Annalen dar Physik* on the theory of black-body radiation.[1] It was his first paper on this subject since the epoch-making work in which he had introduced the concept of energy quanta almost a decade earlier. Planck had no new results to report, but he felt that it was time he expressed his views on what had been going on in the intervening years. Not that there was so very much to discuss: neither the problems of radiation nor Planck's startling idea that energy could sometimes vary only in discrete steps had yet seriously caught the attention of most of his colleagues. Planck himself, of course, had thought a great deal about these things, as he remarked in a letter to Walther Nernst a few months later:[2] "I can say without exaggeration that for ten years, without interruption, nothing in physics has so stimulated me, agitated me, and excited me as these

quanta of action." But his approach to the problems did not coincide with those of the relatively few others who had concerned themselves with the theory of radiation, and Planck wanted to point out the path that he considered most sensible and most promising for future success.

In his paper, Planck arranged the current views on radiation into a spectrum, placing his own in the solid central position. The extreme right wing, represented by James Jeans, was still trying to maintain the soundness of Hamilton's equations and the equipartition theorem. The fact that the equipartition theorem could not account for the existence of the equilibrium distribution of black-body radiation, much less for its observed form, had to be explained, according to Jeans, by the absence of true thermodynamic equilibrium in the radiation. At the opposite end of the spectrum of opinion were the radicals who

interpreted the failure of the equipartition theorem as a sign that nineteenth-century physics, for all its great successes, now needed sweeping changes. The most daring of their proposals suggested that radiation be considered as a collection of independent particles of energy—light quanta—rather than as continuous electromagnetic waves. This position was advanced most forcefully by Albert Einstein, who supported it with a variety of arguments, drawing upon his un-

Martin J. Klein, who is acting head of the Physics Department at Case Institute of Technology, received his Ph.D. at MIT in 1948. He is editor of *Collected Scientific Papers* of Paul Ehrenfest and has written extensively on the early history of the quantum theory.

PLANCK

matched insight into statistical mechanics.

Planck could not accept either of these extreme viewpoints. Jeans' attempts to salvage the equipartition theorem left him unconvinced. Something in classical physics had to be given up. To that extent he could agree with the radicals, but only to that extent. For he was concerned that they wanted to throw out too much. He would not grant the cogency of the arguments for a new corpuscular theory of light, even though Einstein claimed that his light quanta were a *necessary* consequence of the observed form of the black-body radiation law. Planck was not ready to give up the whole development from Huygens to Maxwell and Hertz which had established the electromagnetic wave theory of light, "all those achievements which belong to the proudest successes of physics, of all science," for the sake of what he called a few highly controversial arguments.

He was, however, ready to sacrifice the equations of mechanics, and stated his assurance that Hamilton's equations could no longer be taken as generally valid. In that way the equipartition theorem and its unfortunate consequences could be avoided.

Planck was sure of something else: The discontinuity expressed by his quantum of action was real and would have to be reckoned with. He foresaw a future theory that would somehow reconcile the existence of the quantum of action with electrodynamics, but in the meantime he advocated caution: "One should proceed as conservatively as possible in introducing the quantum of action into the theory, making only those changes in existing theory that have proved to be absolutely necessary."

Planck's stand amounted to this: He had no doubts about the fundamental importance of the quantum of action itself, but he saw no need for a real quantum *theory* of radiation and matter of the kind that already seemed inevitable to Einstein. I think that this statement of Planck's views helps one to understand his work during the next few years, in which he seemed to retreat steadily from his own radical step in 1900. I shall discuss some of this work later on in this paper, but I want first to go back and try to point out the way in which the development of Planck's ideas had led him to adopt this attitude towards the quantum and the quantum theory.

Second law as absolute

In his later years Planck often expressed his deep conviction that "the search for the absolute" was "the loftiest goal of all scientific activity."[3] The context of his remarks clearly indicated that he saw the two laws of thermodynamics as a prototype of that "loftiest goal." For Planck had formed himself as a physicist by his self-study of the writings of Rudolf Clausius, that lucid but rather argumentative man who first distinguished and formulated the two laws of thermodynamics, and it was thermodynamics as seen by Clausius that set the pattern of Planck's scientific career. He devoted the first fifteen years or so of that career to clarifying, expounding and applying the second law of thermodynamics and especially the concept of irreversibility. Planck's solid and successful work in this field did not bring him all the satisfaction he might properly have expected. One reason was that he learned, too late, that some of his results had been anticipated a few

years earlier in the memoirs of Willard Gibbs. More disturbing was the rise of a powerful school of thought, the "Energeticists," led by Wilhelm Ostwald and Georg Helm, which rejected the clear distinctions made by Clausius, and offered a new master-theory that would have replaced the elegant mathematical structure of thermodynamics by a confused and inconsistent tangle.[4] Planck later described his failure to persuade the Energeticists of the errors of their ways as "one of the most painful experiences of my entire scientific life."

As a disciple of Clausius, Planck looked upon the second law of thermodynamics as having absolute validity: Processes in which the total entropy decreased were to be strictly excluded from the natural world. He did not care to follow Clausius in pursuing "the nature of motion which we call heat," or in searching for a mechanical explanation of the second law of thermodynamics.[5] And he most certainly did not follow Ludwig Boltzmann in his reformulation of the second law of thermodynamics as a statistical law. Boltzmann's statistical mechanics made the increase of entropy into a highly probable rather than an absolutely certain feature of natural processes, and this was not in keeping with Planck's own commitments. The statistical interpretation of entropy is conspicuously absent from the papers Planck wrote in the early 1890's under such titles as "General Remarks on Modern Developments in the Theory of Heat"[6] and "The Essence of the Second Law of Thermodynamics."[7]

One should not think, however, that Planck was content to keep thermodynamics a completely independent subject, separate from the rest of physics. He preferred the rigorous arguments of pure thermodynamics to the difficult but approximate treatment of molecular models in kinetic theory, but he also felt strongly the need to relate the irreversibility described by the second law to the other fundamental laws governing the basic conservative processes. He rejected Boltzmann's approach because it rested on statistical assumptions, and Planck wanted to avoid these. He hoped that the principle of increasing entropy could be preserved intact as a rigorous theorem in some more comprehensive and more fundamental theory.

Second law and Wien distribution

In March 1895 Planck presented a paper to the Academy of Sciences at Berlin that seemed to represent a basic shift in his interests.[8] He had just put aside his usual thermodynamic concerns to discuss the problem of the resonant scattering of plane electromagnetic waves by an oscillating dipole of dimensions small compared to the wave length. A careful reader would have noticed, however, that at the end of the paper Planck admitted that this study was only undertaken as a preliminary to tackling the problem of black-body radiation. The scattering process offered a way of understanding how the equilibrium state of the radiation in an enclosure at fixed temperature could be maintained. The thermodynamics of radiation was the underlying problem, and Planck's attention may have been drawn to it by Wien's paper of 1894 which presented the displacement law.[9]

The following February Planck had further results to report to the Academy.[10] He had extended his studies to the radiation damping of his charged oscillators, and he was impressed by the difference between radiation damping and damping by means of the ordinary resistance of the oscillator. Radiation damping was a completely conservative mechanism that did not require one to invoke the transformation of energy into heat, or to supply another characteristic constant of the oscillator in order to describe its damping. Planck thought this could have far-reaching implications for this fundamental question of irreversibility and the second law. As he put it, "The study of conservative damping seems to me to be of great importance, since it opens up the prospect of a possible general explanation of irreversible processes by means of conservative forces—a problem that confronts research in theoretical physics more urgently every day."

One year later, in February 1897, he communicated the first of what would become a series of five papers, extending over a period of more than two years, on irreversible phenomena

HERTZ

MAXWELL

in radiation.[11] The extended introduction itself indicated that Planck was planning a major work. He began by asserting that no one had yet successfully explained how a system governed by conservative interactions could proceed irreversibly to a final state of thermodynamic equilibrium. He explicitly discounted Boltzmann's H-theorem as an unsuccessful attempt in this direction, citing the criticisms recently raised by E. Zermelo, Planck's own student, against Boltzmann's analysis.[12] Planck then announced his own program for deriving the second law of thermodynamics for a system consisting of radiation and charged oscillators in an enclosure with reflecting walls. He would introduce no damping other than radiation damping, but would take the basic mechanism for irreversibility to be the alteration of the form of an electromagnetic wave by the scattering process—its apparently irreversible conversion from incident plane to outgoing spherical wave. The ultimate goal of this program would be the explanation of irreversibility for conservative systems and, as a valuable by-product, the determination of the spectral distribution of black-body radiation.

Planck had high hopes: His goal was precisely right for a disciple of Clausius. It would have been a splendid conclusion to his work in thermodynamics, and it would have put an end, once and for all, to claims that the second law was merely a matter of probability. How was Planck to

know that he was headed in a very different direction, that he had started on what he would later call "the long and multiply twisted path" to the quantum theory?[13]

There was, unfortunately, a fundamental flaw in Planck's proposal and it was promptly pointed out by Boltzmann.[14] The equations of electrodynamics could not produce a monotonic approach to equilibrium any more than the equations of mechanics; both needed to be supplemented by appropriate statistical assumptions. Nothing in the equations of electrodynamics would, for example, forbid the inverse of Planck's scattering process. (It is reasonable to suppose that Boltzmann was, at the least, not deterred from pointing out this error by Planck's negative comments on his own work. Planck's support of Zermelo did not help matters either, since Boltzmann had found Zermelo's criticism particularly irksome; Boltzmann commented that Zermelo's paper showed that if, after a quarter of a century, his work had still not been understood, at least it had finally been noticed in Germany!)[15]

Planck finally granted that a statistical assumption was necessary, and introduced what he called the hypothesis of "natural radiation," [16] the appropriate analogue of Boltzmann's hypothesis of "molecular chaos," the hypothesis underlying the H-theorem.[17] With the help of this hypothesis Planck was able to complete his program, in a sense, and he reported his

HUYGENS

BOLTZMANN

LORENTZ

work in the last paper of the series in June 1899.[18] He proved first that the spectral distribution of the equilibrium radiation at temperature T, $\rho(\nu,T)$ (the energy per minute frequency interval at ν in a unit volume), was related to the average energy, $E(\nu,T)$, of an oscillator of frequency ν by the equation,

$$\rho(\nu,T) = (8\pi\nu^2/c^3)\ E(\nu,T) \qquad (1)$$

This average energy could be determined once he fixed the dependence of the entropy S of the oscillator on its energy E, but he had no independent method for determining the function $S(E)$. He knew, however, that the spectral distribution had to satisfy Wien's displacement law,

$$\rho(\nu,T) = \nu^3\ f(\nu/T) \qquad (2)$$

where f is a function of the ratio (ν/T) only, and that Wien had proposed a particular form of the distribution that accounted for all available experimental measurements.[19] Wien's distribution had the form,

$$\rho(\nu,T) = \alpha\ \nu^3 \exp\ (-\beta\nu/T) \qquad (3)$$

and, with the help of equation 1 and the thermodynamic definition of the temperature, this would fix the form of the entropy function $S(E)$.

Planck proceeded to *define* $S(E)$ by $S(E) = -(E/\beta\nu)\ \{\ln E/\alpha\nu\} - 1\}$ (4) the form fixed by equation 3, where $a = (\alpha c^3/8\pi)$. He convinced himself that this definition was the only possible one in the sense that if and only if the entropy had this form could he prove that the total entropy of the system increased monotonically to an equilibrium value. This is what 1 meant when I said that Planck com-

pleted his program "in a sense." He had shifted his ground so that he actually *used* the second law to fix the entropy function and thereby the spectral distribution of the black-body radiation.

Planck formulated his result in these words: "I believe that it must therefore be concluded that the definition given for the entropy of radiation, and also the Wien distribution law for radiation that goes with it, are necessary consequences of applying the principle of entropy increase to the electromagnetic theory of radiation, and that the limits of this law, should there be any, therefore coincide with those of the second law of thermodynamics. For this reason further experimental tests of this law naturally acquires so much the more interest."

The absolute system of units

This last statement is remarkable enough in the clear light of our hindsight, especially since this paper was also published, with only minor revisions, in the *Annalen der Physik* early in 1900, only months before the introduction of the quantum.[20] But Planck ended his paper with an even more remarkable section. His expression for the entropy of an oscillator (4) contained two constants, a and β, which also appear in the Wien distribution law, two universal constants as Planck called them when he introduced them. He evaluated these constants numerically from the available experimental data on black-body radiation and found for β the value

0.4818 $\times\ 10^{-10}$ sec °K and for a the value 6.885 $\times\ 10^{-27}$ erg sec. Planck observed that these two constants together with the velocity of light c and the gravitational constant G could be used to define new units of mass, length, time and temperature and that these units properly deserved the title of "natural units".

All systems of units previously employed owed their origins to the accidents of human life on this earth, wrote Planck. The usual units of length and time derived from the size of the earth and the period of its orbit, those of mass and temperature from the special properties of water, the earth's most characteristic feature. Even the standardization of length using some spectral line would be quite as arbitrary, as anthropomorphic, since the particular line, say the sodium D line, would be chosen to suit the convenience of the physicist. The new units that he was proposing would be truly "independent of particular bodies or substances, would necessarily retain their significance for all times and for all cultures, including extraterrestrial and non-human ones," and therefore deserved the name of "natural units." That they were of awkward sizes (10^{-33} cm, 10^{-42} sec. etc) was obviously of no importance. "These quantities preserve their natural significance so long as the laws of gravitation and the propagation of light in vacuum, and the two laws of thermodynamics retain their validity."[21]

I have referred earlier to Planck's conviction that the search for the ab-

NERNST

CLAUSIUS

solute was the physicist's proper goal. The universal constants as well as the most general physical laws belonged to that category of the absolute for him. As he put it in an essay written in his ninetieth year, "The endeavor to discover [the absolute constants] and to trace all physical and chemical processes back to them is the very thing that may be called the ultimate goal of scientific research and study."[22] He had obviously felt the same way half a century earlier.

It will not have escaped your notice that the constant he called *a* in 1899 was soon to be renamed and reinterpreted. The "further experimental tests" that Planck had called for were promptly made, and as the measurements were extended to longer wavelengths it became apparent to Planck that either the second law of thermodynamics did not have universal validity or there was an error in his arguments.[23] For the Wien distribution law could not represent the new data in the infrared. I do not have space here to recount in detail the exciting events of 1900, but by October Planck was ready to offer a new distribution law which did account for the experimental results obtained by his colleagues Rubens and Kurlbaum, as well as for all subsequent results on the black-body radiation spectrum.[24] The new law had the now familiar form,

$$\rho(\nu,T) = \alpha\nu^3[\exp(\beta\nu/T)-1]^{-1} \quad (5)$$

Planck's earlier analysis of the way that entropy increased with time had suggested this as the next simplest

possibility after Wien's law. The problem was to create a suitable theoretical foundation for the new distribution law.

Planck had to take a difficult and probably painful step. He had to put aside his opposition to statistical mechanics and his years of occasional controversy with Boltzmann and try to adapt Boltzmann's methods to his problem.[25] All other resources had failed him. The crux of the matter was still the energy-entropy relation for an oscillator; perhaps Boltzmann's equation for the entropy in terms of the number of complexions could fix this one missing relationship. Planck had the great advantage of knowing what the answer had to be, since his new distribution law, equation 5, determined the form of the entropy of an oscillator as a function of its energy. It too had the kind of logarithmic structure that Boltzmann's equation would suggest. Using Boltzmann's great memoir[26] of 1877 as his guide Planck plunged in, and "after a few weeks of the most strenuous work of my life," as he put it, "the darkness lifted and an unexpected vista began to appear."

"An act of desperation"

In order to calculate the "thermodynamic probability" of a state in which a certain energy was shared among many oscillators of the same frequency, that is to say, the number of ways in which this sharing could be accomplished, it was essential that Planck imagine the energy to be composed of

a finite number of identical units, each of magnitude ϵ. This by itself would not have been a novel step: Boltzmann had often done it as a computational device, particularly in the 1877 memoir that Planck used as his guide. But Planck had to refrain from taking the accepted next step, namely going to the limit where ϵ vanishes.[27] He had to refrain, that is, if he were to arrive at the entropy formula required by the distribution law that he knew to be the correct one. He was willing to take this step, to restrict the energy of one of his oscillators to multiples of the energy unit or quantum ϵ, radical though he must have known it to be.

Thirty years later, in a letter to R. W. Wood,[28] Planck described what he had done as "an act of desperation," undertaken against his naturally peaceful and unadventurous disposition. "But," he went on, "I had already been struggling with the problem of the equilibrium of matter and radiation for six years (since 1894) without success; I knew that the problem is of fundamental significance for physics; I knew the formula that reproduces the energy distribution in the normal spectrum; a theoretical interpretation *had* to be found at any cost, no matter how high." He described himself as ready to sacrifice any of his previous convictions except the two laws of thermodynamics. When he found that the hypothesis of energy quanta would save the day he considered it "a purely formal assumption, and I did not give it much

THE 1911 SOLVAY CONGRESS brought together many of those who were interested in quantum theory. Planck is standing second from left.

thought except for this: that I had to obtain a positive result, under any circumstances and at whatever cost."

Planck actually did give his assumption of quanta a good deal of thought along one particular line. His theory, which I must omit here, once again contained two universal constants: the constant k, the proportionality constant that related entropy to the logarithm of the "thermodynamic probability," and the constant h, brought into existence by the requirements of the displacement law which made the energy quantum ϵ proportional to the frequency of the oscillator, so that ϵ could be expressed as $h\nu$. These constants were equivalent to those that Planck had emphasized a year earlier: h was the former a and k was the ratio of the former a and β. But now Planck could discuss their detailed physical importance as well as their absolute significance. The constant k, in particular, had to be equal to the ratio of the gas constant R to Avogadro's number N_0, the number of atoms in a gram atomic weight. And Planck's determination of k and h from the measurements on black-body radiation, with the help of his distribution law in the form

$$\rho(\nu, T) = (8\pi\nu^2/c^3)\,(h\nu)\,\{\exp\,(h\nu/kT) -1\}^{-1} \qquad (6)$$

gave him an accurate value of Avogadro's number and with it the mass of the individual atom.[29]

This was a major achievement. Planck's value for Avogadro's number was far more accurate than any of the existing indirect estimates based on the kinetic theory of gases, and he used it not only to get the mass of the atom but also, together with the Faraday

constant, to determine the charge on the recently discovered electron, the natural unit of electric charge. His value of e was 4.69×10^{-10} e.s.u.—at a time when the early attempts at direct measurement gave results from 1.3 to 6.5 in the same units. Unfortunately, Planck's contemporaries did not properly appreciate these results; the handbooks went on printing crude determinations of Avogadro's number, ignoring Planck's value.[30] The first experimentalist to quote Planck's value of e seems to have been Rutherford, in 1908, probably because he and Geiger had obtained essentially the same value, 4.65×10^{-10} e.s.u. from the charge on the alpha particle and were glad to have a confirmation of a result 50% higher than J. J. Thomson's current best determination.[31]

Planck himself laid heavy emphasis on these concrete results of his theory, both in his papers and in his *Lectures on the Theory of Heat Radiation*[32] published in 1906. I am convinced that, with Planck's particular sensitivity to the importance of the natural constants, it was these results that assured him that quanta were more than an ad hoc hypothesis, useful only for arriving at the radiation law. Of course h, the second constant in his equation, the essentially new constant in the theory, was yet unexplored. He remarked in his *Lectures* at several points that h must have some direct electrodynamic meaning, that this meaning must be found before the theory of radiation could be considered fully satisfactory, but that a lot more research would be needed before this meaning was revealed.

The kind of electrodynamic meaning that Planck had in mind for h

was suggested in a letter he wrote to Paul Ehrenfest[33] in July 1905. Ehrenfest was engaged in an analysis of Planck's assumptions and had written to Planck asking several questions about them. In his answer Planck pointed out that the existence of a discrete unit of electric charge imposed certain limitations on the electromagnetic field. He went on to write: "Now it seems to me not completely impossible that there is a bridge from this assumption (of the existence of an elementary quantum of electric charge e) to the existence of an elementary quantum of energy h, especially since h has the same dimensions and also the same order of magnitude as (e^2/c). But I am not in a position to express any definite conjecture about this." Planck never published this remark, so far as I can tell. Almost the same thought, however, was expressed by Einstein in 1909 in the course of a dimensional analysis of the displacement law.[34] He too pointed out the dimensional equivalence of h and (e^2/c). But I am not in noted, correctly, that their magnitudes differed by a factor of about a thousand. "The most important thing in this derivation," Einstein went on, "is that it reduces the constant for light quanta h to the elementary unit of electricity e. Now one must remember that the elementary charge e is a stranger in the Maxwell-Lorentz electrodynamics. . . . It seems to me to follow from the relationship, $h = e^2/c$, that the same modification of the theory which contains the elementary charge as one of its consequences will also contain the quantum structure of radiation."

Retreat from energy quantization

I have been trying to give the background for my earlier statement that Planck was fully committed to the quantum, but not necessarily to a quantum theory in Einstein's sense. Planck's work in the years after 1910, when he resumed publication in this field shows him holding fast to the quantum of action but retreating steadily from his earlier strict quantization of the oscillator. In a paper[35] read to the German Physical Society in February 1911 he explained that he was revising his original theory because of the valid criticism to which it had been subjected, particularly by H. A. Lorentz.[36] The objection was basically that the intensity of the radiation at high frequencies was very low, whereas at these frequencies the energy quantum was very large. As a consequence the time it would take an oscillator to absorb one quantum would have to be unreasonably long, and the oscillator might not even be able to absorb one full quantum if the radiation should be cut off. This criticism naturally presupposed that radiation was properly described by electromagnetic waves, and it is interesting to note that Lorentz had used this argument to show how difficult it was to explain phenomena like the photoelectric effect without having recourse to Einstein's light quanta instead of the wave description. Planck, however, did not take it that way.

He proposed instead to give up his hypothesis that the energy of an oscillator had to be an integral multiple of h_ν and could therefore absorb or emit energy only in discrete units. In his new theory the oscillator would absorb energy continuously, just as it did classically, so that Lorentz's criticism could be set aside. The emission process, however, was still quantized. This procedure would eliminate another difficulty, an internal contradiction in the original theory pointed out by Einstein.[37] In that theory Planck had used the classically derived relationship between the radiation density and the oscillator's energy, but that classical derivation was, of course, incompatible with the assumption of quantum states for the oscillator.

Planck gave several versions of his new theory of quantized emission in 1911 and 1912, finally settling on one in which the oscillator, absorbing energy continuously, could emit only when its energy was a multiple of h_ν.[38] If it emitted at all it had to emit all the energy it possessed, however many quanta that might be. Whether or not it emitted as its energy reached nh_ν, for any n, was governed by a probability η. This probability was fixed by the assumption that the ratio of the probability of no emission to the probability of emission, $(1-\eta/\eta)$, should be propor-

tional to the intensity of the incident radiation. The proportionality constant, in turn, was determined by the requirement of classical behavior in the limit of high intensity radiation. (This is surely one of the first uses of the correspondence principle. There is reason to believe that this paper of Planck's had considerable influence on Bohr's first papers on atomic structure.[39])

This second quantum theory of Planck's led to the same law for black-body radiation as had the first (this must have been an unexpressed boundary condition on the work). But it made an interesting change in the expression for the average energy of an oscillator,

$$\bar{E} = h_\nu \{\exp(h_\nu/kT)-1\}^{-1} + h_\nu/2 \qquad (7)$$

The additional term meant that the energy of an oscillator would not vanish at the absolute zero of temperature but would be just $(h_\nu/2)$; hence its usual name of zero-point energy. Planck saw a variety of phenomena that might be interpreted as favoring his concept of quantum emission, and also some that supported the reality of the zero-point energy. He suggested, for example, that this might be the source of the energy of particles emitted by radioactive atoms, and that the sharply defined energy of these particles was an example of quantum emission.

The novel idea of zero-point energy attracted a good deal of attention, first of all from Einstein, as one might have expected. Early in 1913 he and Otto Stern discussed its possible relevance for understanding Eucken's new measurements of the heat capacity of hydrogen gas at low temperatures.[40] A number of physicists then tried to apply the zero-point energy to phenomena as diverse as deviations from Curie's law in paramagnetism[41] and the equation of state of gases.[42] The most significant application was made by Debye in his theory of the effect of thermal vibrations on x-ray scattering from crystals.[43] Debye showed that the presence or absence of the zero-point energy could be brought to experimental test by a study of the intensities of x-ray diffraction spots. This was eventually done, and the existence of zero-point

energy was confirmed, but by that time it had lost its connection with Planck's largely forgotten second quantum theory.[44]

For Planck the zero-point energy was an interesting by-product of his work, but the important thing was that he had arrived at the radiation law without having to restrict the energy of the oscillator to quantized energies. Actually he was ready to give up even the quantized emission of radiation, and did so in a paper he wrote in 1914, where the crucial h governed only the interaction between oscillators and free particles, and the absorption and emission of radiation followed the classical laws.[45] Planck was always arguing *to* the radiation law and tried to restrict the use of the quantum to the minimum sufficient for deriving that law.

Nernst's law, entropy and quanta

Planck's book on radiation included one important new step in the search for an understanding of h. He constructed an argument showing that h could be interpreted directly as a quantum of action in the sense that h measured the areas of the regions of equal statistical weight in the phase space of the oscillator.[46] The concept of a cell in phase space had already played an important part in Boltzmann's statistical mechanics, but as Planck emphasized in his parallel discussion of the ideal gas, its magnitude was apparently of no significance there since it appeared only in the additive constant in the entropy.

At this stage he did not yet see that there was anything general about the use of h to fix the size of a cell in phase space.

The lectures on heat radiation on which Planck's book were based were delivered during the winter semester of 1905-1906, and while they were going on, Planck's colleague at Berlin, Nernst, reported a significant advance in thermodynamics.[47] This was Nernst's famous heat theorem which, although he did not formulate it that way, amounted to the statement that the entropy differences between all states of a system disappear at absolute zero. It is clear that a new result in thermodynamics of such general import would have been of interest to Planck, but it is not so clear, in view of Planck's background as I have described it, that he should have been the one to probe its statistical significance as well.

He discussed his views in a lecture entitled "On Recent Thermodynamic Theories: Nernst's Heat Theorem and the Hypothesis of Quanta," delivered before the German Chemical Society in December 1911.[48] Planck described the importance of Nernst's theorem, which was really a new and independent postulate, by pointing out the incompleteness of the classical thermodynamics based on the first and second laws. Classical thermodynamics could not lead to a full specification of the conditions for equilibrium (phase equilibrium or chemical equilibrium) precisely because it provided no way of fixing the undetermined constant in the entropy equation. Just this gap was filled by Nernst's law, and Planck stated it in what he considered its simplest and most far-reaching form: the entropy of a chemically pure substance in a condensed phase vanishes at absolute zero. Nernst's law, in other words, allowed one to fix the absolute value of the entropy and therefore represented a major addition to thermodynamics.

Planck then went on to ask for "the real, the more profound physico-chemical meaning" of the law, that is, its meaning on the atomic scale, "not only because this promises

greater intuitive insight, but also because only it can help one to discover regularities and relationships . . . which pure thermodynamics cannot touch." And this atomistic interpretation of a law involving the entropy would have to be found, he said, by using Boltzmann's fundamental relationship between entropy and probability. Planck had come a long way in his thinking in the decade or so since he had reconciled himself to trying Boltzmann's methods!

If one wanted to calculate the entropy of a system with the help of Boltzmann's relationship, the whole procedure was fully determined except for one point: there was no *a priori* criterion for choosing the size of the elementary cells in phase space. This lack of definiteness was the exact counterpart of, and could be considered the reason for, the indeterminateness of the entropy constant (as mentioned earlier). Conversely, then, if Nernst's law fixed the entropy constant, this must imply that its "deeper meaning" must be that the sizes of the cells in phase space are not arbitrary but must have definite values. This statement would have been hard to accept, Planck went on, if not for the totally unexpected support it received from the theory of black-body radiation, that is from his own interpretation of h as precisely the size of the phase cell for oscillators of any frequency. Further analysis of the "meaningful and attractive problem" of

determining these quite definite elementary cells for calculating the thermodynamic probability was called for, since Planck now saw this as the essential content of the hypothesis of quanta.

He put it this way some months later in the preface to the second edition of his book on heat radiation.[49] "For the hypothesis of quanta as well as the heat theorem of Nernst may be reduced to the simple proposition that the thermodynamic probability of a physical state in a definite integral number, or what amounts to the same thing, that the entropy of a state has a quite definite, positive, value, which, as a minimum, becomes zero, while in contrast therewith the entropy, may, according to the classical thermodynamics, decrease without limit to minus infinity. For the present, I would consider this proposition as the very quintessence of the hypothesis of quanta." Planck must have been thoroughly gratified to have found this way of relating his two favorite concepts—entropy and the quantum of action. He devoted much thought to the general problem of determining the size and shape of the elementary cells in phase space over the next decade,[50] but I cannot discuss that work here.

"A far more significant part"

In the *Scientific Autobiography* that he wrote near the end of his long life Planck frankly discussed the attitude prevalent among many physi-

OSTWALD

EINSTEIN

WIEN

cists about his work after 1901.[51] "My futile attempts to fit the elementary quantum of action somehow into the classical theory continued for a number of years, and they cost me a great deal of effort. Many of my colleagues saw in this something bordering on a tragedy. But I feel differently about it. For the thorough enlightenment I thus received was all the more valuable. I now knew for a fact that the elementary quantum of action played a far more significant part in physics than I had originally been inclined to suspect."

It was in this same spirit that he had prophetically closed his lecture to the German Chemical Society in 1911. "To be sure, most of the work remains to be done; . . . but the beginning is made: the hypothesis of quanta will never vanish from the world. . . . And I do not believe I am going too far if I express the opinion that with this hypothesis the foundation is laid for the construction of a theory which is someday destined to permeate the swift and delicate events of the molecular world with a new light." ☐

All quotations from Planck's unpublished letters are made with the kind permission of Frau Dr. Nelly Planck, to whom I should like to express my thanks.

For an analysis coming to rather different conclusions see Thomas S. Kuhn, Black-Body Theory and the Quantum Discontinuity, 1894–1912 (New York, 1978). See also Allan A. Needell, Irreversibility and the Failure of Classical Dynamics: Max Planck's Work on the Quantum Theory 1900–1915 (Yale Univ. PhD. Diss., 1980).

References

Planck's scientific papers are collected in three volumes: *Physikalische Abhandlungen und Vorträge* (Friedrich Vieweg & Sohn, Braunschweig, 1958). Referred to below as *Papers*.
1. M. Planck, Ann. Phys. (4) **31**, 758 (1910); *Papers* II, 237.
2. M. Planck to W. Nernst 11 June 1910. This letter is quoted in full in an unpublished manuscript by Jean Pelseneer entitled "Historique des Instituts Internationaux de Physique et de Chimie Solvay." The manuscript is part of the archive "Sources for the History of Quantum Physics," at the Library of the American Philosophical Society in Philadelphia.
3. M. Planck, *Scientific Autobiography and Other Papers*, translated by F. Gaynor (Philosophical Library, New York, 1949) p. 35.
4. M. Planck, Ann. Phys. (3) **57**, 72 (1896); *Papers* I, 459.
5. M. Planck, *Treatise on Thermodynamics*, (1897) translated by A. Ogg. (Longmans, Green, and Co., London, 1903), Preface.
6. M. Planck, Z. phys. Chem. **8**, 647 (1891); *Papers* I, 372.
7. M. Planck, Z. f. phys. und chem. Unterricht **6**, 217 (1893); *Papers* I, 437.
8. M. Planck, Ann. Phys. (3) **57**, 1 (1896); *Papers* I, 445.
9. W. Wien, Ann. Phys. (3) **52**, 132 (1894).
10. M. Planck, Ann. Phys. (3) **60**, 577 (1897); *Papers* I, 466.
11. M. Planck, S.-B. Preuss. Akad. Wiss. (1897), p. 57; *Papers* I, 493.
12. E. Zermelo, Ann. Phys. (3) **57**, 485 (1896), **59**, 793 (1896); Also see R. Dugas, *La théorie physique au sens de Boltzmann* (Editions Griffon, Neuchâtel, Suisse, 1959) pp. 206-219.
13. M. Planck, Nobel Prize Address in *A Survey of Physical Theory* reprinted (Dover, New York, 1960), p. 102; *Papers* III, 121.
14. L. Boltzmann, S.-B. Preuss. Akad. Wiss. (1897) pp. 660, 1016, (1898) p. 182.
15. L. Boltzmann, Ann. Phys. (3) **57**, 773 (1896). Also his *Populäre Schriften* (Barth, Leipzig; 1905) p. 406.
16. M. Planck, S.-B. Preuss. Akad. Wiss. (1898), p. 449; *Papers* I, 532.
17. See, for example, P. and T. Ehrenfest, *The Conceptual Foundations of the Statistical Approach in Mechanics*, translated by M. J. Moravcsik (Cornell University Press, Ithaca, N.Y., 1959) p. 41.
18. M. Planck, S.-B. Preuss. Akad. Wiss. (1899), p.440.; *Papers* I, 560.
19. W. Wien, Ann. Phys. (3) **58**, 662 (1896).
20. M. Planck, Ann. Phys. (4) **1**, 69 (1900); *Papers* I, 614.
21. See references 18 and 20. I would like once again to thank Dr. Joseph Agassi for calling my attention to Planck's pre-quantum determination of h.
22. Op. cit. reference 3, p. 78.
23. M. Planck, Ann. Phys. (4) **1**, 719 (1900); *Papers* I, 668.
24. M. Planck, Verh. d. Deutsch. Phys. Ges. **2**, 202 (1900); *Papers* I, 687.
25. See M. J. Klein, Archive for History of Exact Sciences **1**, 459 (1962), and The Natural Philosopher (Blaisdell Publishing Company, New York) **1**, 81 (1963). Also see K. A. G. Mendelssohn in *A Physics Anthology*, edited by N. Clarke (Chapman and Hall, London, 1960), p. 62 and L. Rosenfeld, Osiris **2**, 149 (1936).
26. L. Boltzmann, Wien. Ber. **76**, 373 (1877).
27. M. Planck, Verh. d. Deutsch. Phys. Ges. **2**, 237 (1900); *Papers* I, 698; Ann. Phys. (4) **4**, 553 (1901); *Papers* I, 717; Also the papers of reference 25.
28. M. Planck to R. W. Wood, 7 October 1931. This letter is part of the collection in the Archives of the Center for the History and Philosophy of Physics of the American Institute of Physics in New York City.
29. See the first article in reference 27 and also M. Planck, Ann. Phys. (4) **4**, 564 (1901); *Papers* I, 728.
30. G. Hertz in *Max Planck zum Gedenken* (Akademie-Verlag, Berlin, 1959) pp. 33-35.
31. E. Rutherford and H. Geiger, Proc. Roy. Soc. A **81**, 162 (1908).
32. M. Planck, *Vorlesugen über die Theorie der Wärmestrahlung* (Barth, Leipzig, 1906) p. 162.
33. M. Planck to P. Ehrenfest, 6 July 1905. This letter is part of the Ehrenfest collection at the National Museum for the History of Science in Leyden.
34. A. Einstein, Phys. Z. **10**, 192 (1909).
35. M. Planck, Verh. d. Deutsch. Phys. Ges. **13**, 138 (1911); *Papers* II, 249.
36. H. A. Lorentz, Phys. Z. **11**, 1248 (1910). This is actually a report by Max Born of Lorentz's Wolfskehl lectures, "Alte und neue Fragen der Physik."
37. A. Einstein, Ann. Phys. (4) **20**, 199 (1906).
38. M. Planck, Ann. Phys. (4) **37**, 642 (1912); *Papers* II, 287.
39. See T. Hirosige and S. Nisio, "Formation of Bohr's Theory of Atomic Constitution," Japanese Studies in the History of Science No. 3, p. 6 (1964).
40. A. Einstein and O. Stern, Ann. Phys. (4) **40**, 551 (1913).
41. E. Oosterhuis, Phys. Z. **14**, 862 (1913).
42. W. H. Keensom, Phys. Z. **14**, 665 (1913).
43. P. Debye, Ann. Phys. (4) **43**, 49 (1914).
44. R. W. James, I. Waller, and D. R. Hartree, Proc. Roy. Soc. A**118**, 334 (1928). See also *Fifty Years of X-Ray Diffraction*, P. P. Ewald, ed. (Oosthoek, Utrecht 1962), pp. 126, 230.
45. M. Planck, S.-B. Preuss. Akad. Wiss. (1914) p. 918; *Papers* II, 330.
46. Op. cit. reference 32, pp. 154-156.
47. W. Nernst, Gött. Nachr. (1906), p. 1. See also F. Simon's Guthrie Lecture in Yearbook of the Physical Society of London 1956, p. 1.
48. M. Planck, Phys. Z. **13**, 165 (1912); *Papers* III, 54.
49. M. Planck, *The Theory of Heat Radiation*, translated by M. Masius 2nd Ed. (1913), reprinted (Dover, New York, 1959), p. vii.
50. See his Wolfskehl Lecture of 1913 in *Vorträge über die kinetische Theorie der Materie und der Elektrizität* (B. G. Teubner, Leipzig, 1914) p. 3; *Papers* II, 316. Also see L. Rosenfeld in *Max-Planck-Festschrift 1958* (Deutscher Verlag der Wissenchaften, Berlin 1959), p.203.
51. Op. cit. reference 3, p. 44.

J. J. Thomson and the Bohr atom

Far from being merely "scientific curiosities," J. J. Thomson's seemingly naive models actually contained some of the fundamental ideas of Niels Bohr's revolutionary quantum theory of the atom.

John L. Heilbron

PHYSICS TODAY / APRIL 1977

In 1911 Niels Bohr went to Cambridge, hoping to talk physics with J. J. Thomson; the discoverer of the electron was friendly but uninterested. Two years later Bohr published his epochal three-part paper on the constitution of atoms and molecules, which challenged the program and goal of the Cambridge school. Bohr's new views soon won out; Thomson's quaint atomic models were declared worthless—old lumber fit only, as Ernest Rutherford put it, "for a museum of scientific curiosities." For his part Thomson rejected the advances made by Bohr as meretricious superficialities obtained without, or at the price of, an understanding of the mechanism of atoms.

As in many other instances in the history of science, Bohr's revolutionary theory became such a success that its origins in the views it superseded were all but forgotten. In particular, Thomson's opposition and the quick replacement of his research program by Bohr's obscured the connection between the theory of the quantized atom and the deceptively simple and apparently naive models of the Cambridge school. So has the odd circumstance that the three installments of Bohr's first paper on atomic structure inverted the order of his discoveries. The first installment, the only one now remembered, gives the theory of the Balmer spectrum, which Bohr worked out in a few weeks in February 1913; the other two record Bohr's attempts, beginning in June 1912, to bring Rutherford's nuclear model—itself a product of Thomson's research program—to bear on the chief problems of atomic theory as Thomson had identified them.

John L. Heilbron is professor of history and director of the Office for History of Science and Technology, University of California, Berkeley.

To Thomson the key problem in atomic theory was the explanation of the variation in the periodic properties of the chemical elements represented in Mendeleev's table. Already in 1897, when announcing the discovery of the electron, he intimated that the new particles might well provide this periodicity when they are bound into an atom. Not then knowing how this might be accomplished, he resorted to the sort of analogy characteristic of the Cambridge school of mathematical physics during Thomson's time.

Magnets and a plum pudding

As an analogue to the arrangement of electrons in an atom, Thomson offered Alfred Mayer's floating magnets, which distribute themselves into concentric circles under the influence of a large stationary magnet, as shown in figure 1. In 1903, having secured the electron, measured its charge and mass, and laid the foundation of the electron theory of metals, Thomson took up the question how his favorite corpuscle could play the part of Mayer's magnets.

The first problem was to choose a representation for the positive portion of the atom. The arrangement that is perhaps the most obvious, the nuclear model, had already been proposed and discarded on the ground of mechanical instability: In any Saturnian atom—one with several electrons arranged in a plane ring or rings—there exists at least one unstable mode of oscillation about the equilibrium orbits. The amplitudes of these unstable modes grow until the system flies apart. However, a stable variant can be obtained by allowing the positive charge to fill the entire volume of the atom; the electrons then circulate within the positive charge, subject to a restoring force varying directly as the distance rather than as its inverse square. This so-called "plum-pudding model" is the one Thomson adopted.

Note that the instability that led to the initial rejection of the nuclear model was a mechanical one: It did not derive from that drain of energy by radiation that plays so important a role in the standard historical accounts. Indeed, as Thomson showed, the total radiation from a ring of p symmetrically placed electrons describing the same circular orbit decreases very rapidly as p increases; for moderate values of p the ring—and hence the atom—has almost eternal life.

Even the eventual mortality of atoms was no inconvenience to Thomson: He merely associated radioactivity with ancient atoms, the internal motions of which had decayed to the point of instability and explosion. At this time (1904) he thought that the atom contained a great many electrons, perhaps—as the richness of spectral lines and the ratio of the masses of the electron and the hydrogen ion suggest—as many as a thousand times the atomic weight. He did not lack particles to populate his rings and plug the radiation drain.

The urge of individual electrons in an atom to radiate can therefore be curbed by the social pressure of their neighbors. But this pressure can not be driven too far: Electrons are not friendly; they repel one another. When enough of them are assembled in a ring to extinguish their radiation, there may be too many for mechanical stability; a little disturbance to any one of them might cause the ring to fly apart. Thomson conceived the idea that the condition of mechanical stability might be the clue to the periodicity in the electronic arrangements of the atoms. The electrons' need for elbow room might fix their population distribution. In 1904 he put this idea to the test.

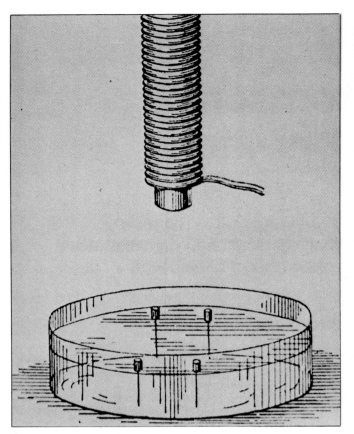

Mayer's magnets—magnetized needles floated on corks, under a large stationary magnet—provided J. J. Thomson with an analogy to the arrangements of electrons in atoms. These diagrams, made by pressing paper against the inked tops of the magnets, displayed stable configurations with a periodicity suggestive of Mendeleev's table. From A. M. Mayer, Am. J. Science **116**, 248 (1878). Figure 1

Rutherford's first calculations on the passage of alpha particles through atoms. In his "theory of structure of atoms," Rutherford used a nuclear atom that was a variant of Thomson's model, of electrons in a sphere of positive charge: It had a positive central nucleus of charge *ne* surrounded by a diffuse sphere of negative electricity. From the Rutherford Papers, Cambridge University Library. Figure 2

The heart of Thomson's analysis was the calculation of the frequencies of the perturbed oscillations of the electrons in a single-ring atom as a function of their number p. He hoped to learn from the frequencies how large p might be before mechanical instability set in: The number turned out to be six. To accommodate more electrons in a single ring, the rate at which the restoring force varied with distance had to be greater than that afforded by the diffuse charge alone.

Rings of electrons

Nothing could be simpler than increasing this rate: One needed merely to put one or more electrons (q in all, say) at the atom's center. Thomson calculated the values of q that would result in a stable outer ring of p electrons. It turned out that the inner electrons themselves must be distributed in rings, and that for each value of the total electron population, $n = p + q$, the distribution is unique.

This distribution represents an electronic parallel to Mayer's magnets, but one that is far more suggestive of the physics of atoms. Thomson shows that if $p = 20$, q must lie between 39 and 47, inclusive; his results are presented in Table 1. If q is close to the minimum, the

atom could increase its stability by losing one of its 20 outer electrons; such an atom would act electropositively. If q is near a maximum, the atom would tend to gain an electron, and therefore act electronegatively. The models characterized by p = 20 consequently offered a striking analogy to the elements of the second and of the third periods of Mendeleev's table.

It was this elucidation of the periodic table, expanded and translated into German, that brought continental physicists an inkling that something might come from the Cambridge theory of atomic structure. In 1909 Max Born thought Thomson's model sufficiently promising to take it as the subject of his inaugural lecture as *Privatdozent*, and in 1911 Arnold Sommerfeld's physics colloquium studied it with the help of floating magnets.

"If it resembled a little, it was so"

Three points about Thomson's analogy deserve attention:
▶ He has introduced the fundamental idea that atoms of successive elements in the periodic table differ from one another by the addition of a single electron.
▶ He has, from a modern point of view, interchanged the roles of core and valence

electrons. The atoms of each period are characterized by the same number of *external* electrons, and differ only in the populations of their *inner* rings. Chemical and optical properties consequently derive primarily from the deeper-lying electrons; the members of a chemical family have only internal structures in common. Likewise all the electrons in the atom, and not just the deepest, are implicated in radioactivity, and it is therefore difficult to find room in Thomson's scheme for structures with identical chemical and different radioactive properties. The existence of isotopes, as Bohr later emphasized, could not be explained plausibly on the basis of the diffuse-sphere atom.
▶ Lastly, despite the mathematical labor that secured it, Thomson's analogy was essentially qualitative. Here we reach a perplexing and perennial characteristic of Thomson's physics. At the very beginning of his career, in 1882, he had won the prestigious Adams Prize at Cambridge for a lengthy essay on Kelvin's vortex atoms. To describe encounters between such atoms, which resembled smoke rings in air, required severe and rigorous calculations, the application of which to physical or chemical phenomena proved all but impossible. Already then Thomson had

Margrethe Norlund and Niels Bohr announce their engagement in 1911. That year Bohr defended his thesis at the University of Copenhagen and left for the Cavendish Laboratory.　　Figure 3

spectra too complicated to reveal anything useful about atomic structure—and in this opinion too he was followed by Bohr.

Bohr's approach

Bohr came to the problem of atomic structure almost by chance. His subject had been the electron theory of metals, on which he had written a thesis defended at the University of Copenhagen in the spring of 1911. He then went to the Cavendish Laboratory, intending to rework his thesis for English publication. And why the Cavendish?

"I considered first of all Cambridge as the center of physics," Bohr later said of his decision to study there, "and Thomson as a most wonderful man . . ., a genius who showed the way for everybody." Thomson received him politely and promised to read the rough translation of his thesis that Bohr had brought him.

"I have just talked to J. J. Thomson," Bohr wrote his brother after his first interview, "and I explained to him as well as I could my views on radiation, magnetism, etc. You should know what it was for me to talk to such a man. He was so very kind to me; we talked about so many things; and I think he thought there was something in what I said. He has promised to read my thesis, and he invited me to have dinner with him next Sunday at Trinity College, when he will talk to me about it . . ."

The exchange of views Bohr desired did not take place. Thomson, who had long before given up active cultivation of the electron theory, probably never read Bohr's thesis; in any case he did not enjoy having his ancient errors rehearsed by a tenacious foreigner whose English he could scarcely understand. But even had language and divergent interests not been barriers, one doubts that the intellectual communion that Bohr sought could have developed.

For one thing, the imprecise and contradictory analogies Thomson fancied were inadequate for Bohr, who sought coherent, consistent models from which quantitative predictions about experimental results might be drawn. For another, Thomson, though friendly and receptive to questions, worked alone; he seldom solicited his students' views on scientific questions, nor did he develop his own through extended conversations with others. Bohr's life-long practice, on the other hand, was to refine his ideas in lengthy discussions, which often became monologues, with informed individuals. Whether his colloquist was a full collaborator, a sounding board or an amanuensis, he required some human contact at almost every stage of his work.

It is perhaps not too fanciful to see a reflection of their styles in their photographs. Figure 3 shows Bohr about 1911, aged 26, boyish, callow, soft-featured and gentle. With him is one of his aman-

to content himself with the sort of qualitative and suggestive connections he was later to make with his electronic atom. He never identified particular chemical atoms with definite models, whether vortical or electronic. "Things needed not to be very exact for Thomson," Bohr used to say, "and if it resembled a little, it was so."

The most important undetermined parameter in Thomson's model was the total electron number n. On its magnitude depended not only the security of the atom against radiation collapse, but also inferences about the nature of positive charge and the process of spectral emission. Thomson worked on the problem for five or six years, bringing to bear his powerful mathematics and the experimental resources of the Cavendish Laboratory. He was the first to explore the atom by shooting charged particles through it, and the first to work out formulas, including probability considerations where appropriate, for the scattering of x rays and beta rays.

The chief result of comparing the experiments to the formulas was that n was about equal to the atomic weight A. The outcome of Rutherford's variant of Thomson's scattering theory—alpha scattering elucidated by the nuclear atom—was an n of about $A/2$. That the nuclear atom was an outgrowth of Thomson's research program appears plainly from the first page of Rutherford's first calculations on the "theory of the structure of atoms," reproduced in figure 2. Note the depiction of the scatterer as a tiny positive nucleus of charge ne, surrounded by a *diffuse sphere of negative electricity of fixed radius.*

The thousandfold reduction of the atomic population brought Thomson and his co-workers very close to the doctrine of atomic number. It also made acute the problem of the radiation collapse of light atoms. It was quite characteristic of Thomson to acknowledge this unpleasantness and move on; he considered

uenses, his future wife Margrethe, who wrote out his first papers on atomic structure. Figure 4 portrays Thomson at about the same time, aged 53. He had not changed much since his discovery of the electron.

"You ask whether J.J. is an old man," Rutherford had written his fiancée in 1896. "He is just 40 and looks quite young, small, rather straggling moustache, short, wears his hair (black) rather long, but has a very clever-looking face, and a very fine forehead and a most radiating smile, or grin as some call it, when he is scoring off anyone."

A little piece of reality

Thomson's indifference by no means deflected Bohr from the pursuit of the electron theory. It was the chief subject of his research throughout the eight months he spent at Cambridge, and it remained so during the first three months of his stay at Manchester, where he moved in March 1912, to learn something of the experimental side of radioactivity. It is important to recognize that Bohr did not go to Manchester, Rutherford's citadel, to help develop the consequences of the nuclear atom. He went to take a six-week course on experimental technique, a standard service of the laboratory for beginners in radioactivity, after which they usually began a small research task proposed by Rutherford. Figure 5 shows a page of Bohr's carefully kept laboratory notebook.

It was not that Bohr wished to become an experimentalist: His object was to capitalize on his time in England, and to make contact with Rutherford, evidently the coming power in English physics. After finishing the laboratory work for the day he would return to the electron theory of metals.

Bohr came to atomic physics in a casual way. The research topic Rutherford had assigned him was interrupted for want of radium emanation (radon). While waiting for more to grow he studied a paper on the absorption of alpha particles that had just been published by C. G. Darwin, the only mathematical physicist besides himself in Rutherford's group. Bohr found that Darwin's treatment rested on an unsatisfactory assumption about the interaction between alpha particles and atomic electrons: Darwin had ignored the binding forces. Bohr, following a technique used by Thomson, proposed to take the forces into account by treating the interaction as a resonance phenomenon depending on the ratio of $1/\nu'$, the natural period of the electrons' vibrations about equilibrium, to the time required by an alpha particle to pass the atom.

Bohr expected to make an easy calculation, which would quickly furnish a short note for the *Philosophical Magazine;* that was in early June, 1912. By the middle of the month he had abandoned the laboratory, shelved the electron

J. J. Thomson in 1909. In 1896 Rutherford had written of his "most radiating smile . . . when scoring off anyone." (Photo in G. P. Thomson, *J. J. Thomson,* Doubleday, 1965)   Figure 4

theory and given himself up entirely to the design of atomic models. A letter from Bohr to his brother Harald, dated 12 June 1912, gives a clue to what happened:

"It could be that I've perhaps found out a little bit about the structure of atoms. You must not tell anyone anything about it; otherwise I certainly could not write you this soon. If I'm right, it would not be an indication of the nature of a possibility (like J. J. Thomson's theory) but perhaps a little piece of reality. It has all grown out of a little piece of information I obtained from the absorption of alpha particles . . . You can imagine how anxious I am to finish quickly and I've stopped going to the laboratory for a couple of days to do so (that's also a secret)."

And what was the "little piece of information"? It may well have been the discovery that the nuclear atom is mechanically unstable.

Thomson and the Cambridge school had rejected the nuclear model on account of its mechanical instability; Bohr welcomed it precisely because it needed a nonmechanical force to exist. Already in his Copenhagen dissertation he had pointed to certain phenomena—heat radiation and paramagnetism in particular—that eluded the electron theory and appeared to require the ascription of a nonmechanical rigidity to the paths of atomic electrons. He was drawn to the nuclear model as a possible representation or reification of the sorts of difficulties he had encountered in his earlier studies.

Bohr's fiat

To make further progress possible he exempted, by fiat, electrons that describe closed orbits satisfying the condition

$$T = K\nu' \qquad (1)$$

(where T is the electron's kinetic energy, ν' its orbital frequency and K a constant) from the ordinary necessities of their existence: They did not radiate energy and they did not respond to small perturbations. Electrons so characterized, electrons in their ground or permanent state,

Table 1.   A Thomson atom with twenty external electrons

Total number of atomic electrons n	59	60	61	62	63	64	65	66	67
Number in outermost ring p	20	20	20	20	20	20	20	20	20
Number of electrons in successive rings q; innermost ring at the bottom	16	16	16	17	17	17	17	17	17
	13	13	13	13	13	13	14	14	15
	8	8	9	9	10	10	10	10	10
	2	3	3	3	3	4	4	5	5

Adapted from J. J. Thomson, Phil. Mag. **7**, 237 (1904).

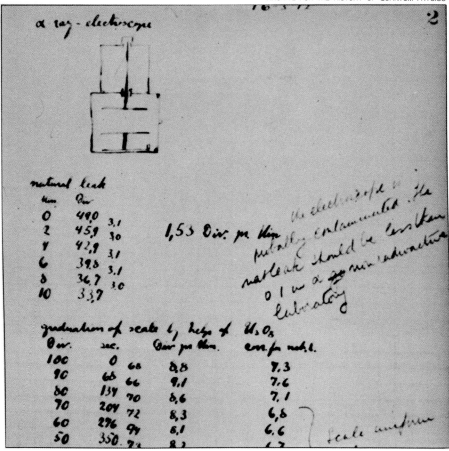

Bohr's laboratory notebook at Manchester 1912. During one experiment he ran out of radon and read a paper that launched him into the problem of atomic structure. Figure 5

Bohr's calculation of the energy of an *n*-electron ring, from his "Manchester Memorandum." It contains an error in the potential energy, and hence also in the total energy. Figure 6

are more like beads on a wire than like freely orbiting particles.

There is no doubt that Bohr's introduction of the stability condition marked a fundamental departure from Thomson's program. The form of the condition was chosen in analogy to Max Planck's quantum theory, and with the expectation that K might be a submultiple of Planck's constant h. It turned out that $K = h/2$, a fact Bohr discovered in February 1913, when at last he came to examine Balmer's formula. The resulting account of the Balmer lines and the concept of stationary states forced him to conclude that the frequencies of spectral lines are not the mechanical frequencies of the atoms that emit them.

As we now know, there followed a progressive relaxation of the dominion of mechanics in the microphysical world, culminating in the invention of quantum mechanics and the principles of uncertainty and complementarity. Nothing so radical was in Bohr's mind in June 1912, however. Having taken a step that was to have revolutionary consequences, he immediately turned back to the problems of the Thomsonian atomist.

In June or July of 1912 Bohr drew up the notes now known as the "Manchester Memorandum" for discussion with Rutherford. The Memorandum opens with a definition of the nuclear atom and an acknowledgment of its mechanical instability, which can be demonstrated, as Bohr put it, "by an analysis similar to the one used by Sir J. J. Thomson in his theory about the constitution of an atom." How then can one account for periodicity? This was a pressing problem: No atomic model unable to elucidate Mendeleev's table could decisively defeat Thomson's.

Bohr thought he had a simple solution. He computed the total energy W of each electron in a ring of n electrons, and discovered that W was negative for $n \leq 7$, but positive for $n > 7$. Evidently for $n > 7$ the electrons leave the atom; for $n \leq 7$ they may be bound securely if their motions satisfy condition 1. For an atom with more than seven electrons, several rings will be required; but, in marked contrast to Thomson's model, the additional rings will be formed outside the first, and the population of the outermost will determine the valence of the atom.

"This," said Bohr, "seems to offer a very strong indication of a possible explanation of the periodic law of the chemical properties of the elements."

An error

What is particularly interesting about this analysis—other than the fact that it addresses, as its first order of business, Thomson's central problem—is that it is altogether wrong. Figure 6 shows Bohr's calculations. From the equation of motion, which is correct, it follows that T, the kinetic energy of each electron, is $Q/2r$, where $Q = e^2(n - A_n/4)$ and $A_n = \Sigma_{i=1}^{n-1}\csc(\pi i/n)$. (The balance of forces makes $mv^2/r = Q/r^2$.) Bohr's computation of the potential energy U is, however, incorrect; because U belongs to both of the interacting particles, the sum in Bohr's expression for U should be divided by two. Then we have $U = -Q/r$ and

$$W = U + T = -Q/2r = -T$$

The total energy is the negative of the kinetic energy. Bohr's error is the more remarkable because his value for the potential energy conflicts both with his expression for the equation of motion and with a theorem proved later in the Memorandum, namely that any particle bound into an orbit by an inverse-square force has a potential energy twice the negative of its kinetic energy. Bohr's slip may betray his anxiety to solve Thomson's problem of periodicity. (The sign of W does change, but at a value of $n > 500$, not 7 or 8 as Bohr wanted.)

For the rest, the Memorandum concerns the structure of simple molecules such as those illustrated in figure 7. Bohr aimed to show, among other things, why the H_2 molecule occurs and He_2 does not, and to demonstrate that no charge is transferred in the combination of identical atoms. He probably took the problem of charge distribution in symmetric diatomic molecules from Thomson's *Corpuscular Theory of Matter,* which gave perhaps the earliest useful explanation of chemical bonding *via* electron exchange.

Thomson had decided that charge transfer occurs in the formation of H_2 and O_2 because identical plum-pudding atoms can not remain in stable equilibrium. For say they are symmetrically combined, by interpenetration of their positive spheres; any subsequent jostling would create a flow of electrons from one sphere to the other, and a permanent polar bond. Thomson made this conclusion plausible by a characteristic analogy. This system, one of identical water-filled jars suspended from identical springs and connected with a siphon, is unstable; for any relative vertical displacement of the jars will grow with the flow of water through the siphon. Thomson thought the evidence favored asymmetric H_2 and O_2; Bohr thought the case for symmetry stronger; hence the considerable attention given to the structure of simple molecules in the Memorandum.

The second and third parts of Bohr's paper of 1913 remain within the set of problems posed by the Memorandum. Part II concerns the problem of the distribution of electrons into rings. Bohr takes for granted the chief result of Thomson's program, the doctrine of atomic number. He then lays down two principles:

▶ In the ground state of an atom every electron, regardless of its distance from the nucleus, has just one quantum of angular momentum.

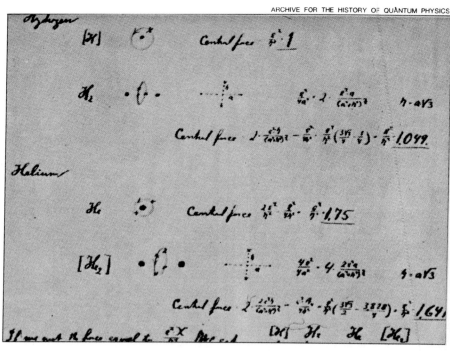

The structures of simple molecules, according to Bohr's Memorandum. The earliest useful explanation of chemical bonding by electron exchange was probably that of Thomson.   Figure 7

▶ The ground-state configuration is the one with the lowest possible potential energy consistent with the principle of angular momentum.

Alas! these directions do not suffice, for they point to structures—such as a single-ring lithium atom—in obvious disagreement with atomic volumes and chemical data. So Bohr assigned distributions more by intuition than by principle, with the curious result given in Table 2. Note particularly the confluence of inner rings at neon ($Z = 10$) and argon ($Z = 18$), brought about, Bohr thought, by the demands of the usual laws of mechanics. Bohr's care and trouble in constructing Table 2 may be indicated by the alternative distributions of figure 8.

Part II of Bohr's paper of 1913 also resolves—or rather shelves—the problem of radioactivity by tucking it into the nucleus. As for Part III, it argues the merits of Bohr's hydrogen molecules.

Thomson's response

Thomson did not salute Bohr's work as the capstone of his own. To him, setting down an arbitrary condition like $T = K\nu'$ and pretending it had dynamical signifi-

cance, was not doing physics; it was a screen of ignorance, a cowardly substitute for "a knowledge of the structure of the atom." Nothing could be easier, or so Thomson told the British Association for the Advancement of Science in September 1913, than to obtain quantum theoretical results in an orthodox mechanical manner.

Take Einstein's formula for the photoelectric effect, $mv^2/2 = h\nu$, for example. (For simplicity Thomson omitted the work function.) Assume, he said, that the usual Coulomb attraction A/r^2 operates only in a few separated, pie-shaped regions in the atom and that, in addition, an inverse-cube repulsion B/r^3 exists everywhere. An electron can sit in stable equilibrium within the pie-shaped regions at a distance a from the atom's center, where $a = B/A$. The frequency of small vibrations about this position is

$$\nu' = \frac{1}{2\pi}\frac{B}{a^2}\left(\frac{1}{mB}\right)^{1/2}$$

Assume that a passing light wave of frequency $\nu = \nu'$ strikes the electron, and gives it enough energy to cross from the pie-shaped region into one of uncom-

Table 2. Bohr's electronic distribution, 1913

1(1)	7(4,3)	13(8,2,3)	19(8,8,2,1)
2(2)	8(4,2,2)	14(8,2,4)	20(8,8,2,2)
3(2,1)	9(4,4,1)	15(8,4,3)	21(8,8,2,3)
4(2,2)	10(8,2)	16(8,4,2,2)	22(8,8,2,4)
5(2,3)	11(8,2,1)	17(8,4,4,1)	23(8,8,4,3)
6(2,4)	12(8,2,2)	18(8,8,2)	24(8,8,4,2,2)

From Phil. Mag. **26**, 476 (1913). The symbol $N(n_1, n_2, \ldots)$ indicates the total number of electrons and their distribution counting outward from the nucleus.

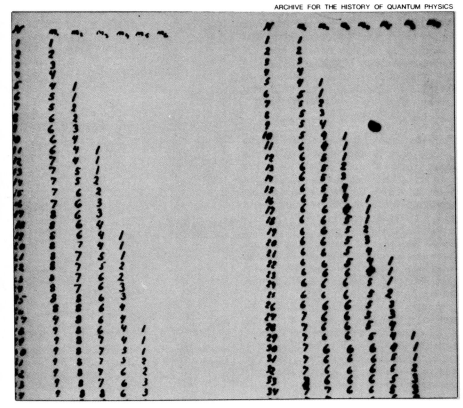

Tentative electron distributions, 1912–13. This is a part of Bohr's manuscript with two sets of ring populations (the n's) for atoms with electron number N up to 40.   Figure 8

pensated repulsion. It will be pushed out into the world with kinetic energy

$$\frac{1}{2}mv^2 = \int_a^\infty \frac{Bdr}{r^3} = \frac{B}{2a^2}$$
$$= \pi(mB)^{1/2}\nu' = \pi(mB)^{1/2}\nu$$

Now set $\pi(mB)^{1/2} = h$: Einstein's formula emerges, and h discloses its true nature, a shorthand for the product of certain electronic parameters.

This *tour de force* was widely applauded by Thomson's school. *Nature* called it a "brilliant attempt" not soon to be forgotten. Other sympathizers rushed to reinterpret Bohr's fundamental contribution, the elucidation of the Balmer lines. One likened the plum pudding to a rotating, pulsating sphere of gas, and imagined that the Balmer lines were emitted by electrons running around on nodal surfaces. Another made what he called a "spherical counterpart" to Thomson's sectioned atom, a baroque structure with many niches of stable equilibrium about which an electron could vibrate at one or another of the Balmer frequencies.

Thomson himself contributed to this curious literature. "If [the Bohr theory] is true," he said, "it must be the result of forces whose existence has not been demonstrated." He set out to find these forces, and to represent them in "the working of a model"; and so, for a time, he occupied himself in reinterpreting Bohr —as Bohr had been reinterpreting him. He ended by appealing to a force varying sinusoidally with the distance between the radiating electron and what he coyly called the "positive center" of the atom.

These rearguard actions did nothing to divert the progress of the quantum theory of the atom. When academic physics resumed after World War I, Thomson recognized that he was out of date and resigned the Cavendish professorship in favor of Rutherford. Not that he gave up physics; but he could never be persuaded that quantum theory was a fundamental one.

In his *Recollections and Reflections*, an autobiography published in 1937, Thomson allowed that Bohr's papers had "changed chaos into order" in certain branches of spectroscopy. And that, he thought, was "the most valuable contribution which the quantum theory has ever made to physical science."

Further reading

- For Thomson: *Dictionary of Scientific Biography, XIII,* Scribners, New York (1976), page 362; "The Scattering of α and β Particles and Rutherford's Atom," Archive for History of Exact Science **4**, 247 (1968).
- For Bohr: J. L. Heilbron, T. S. Kuhn, "The Genesis of the Bohr Atom," Historical Studies in the Physical Sciences **1**, 211 (1969).
- For the Archive for History of Quantum Physics: T. S. Kuhn, J. L. Heilbron, P. Forman, L. Allen, *Sources for History of Quantum Physics,* American Philosophical Society, Philadelphia (1967).   □

60 YEARS of QUANTUM PHYSICS

By Edward U. Condon

PHYSICS TODAY / OCTOBER 1962

I WAS invited to speak on the occasion of the 1500th Regular Meeting of the Society, and of course am delighted to be able to come and do it. But those who conveyed the invitation could not refrain from reminding me that I owed the Society a retiring presidential address. I was president in 1951, and it was in the fall of that year that I departed hastily to go to Corning Glass Works to be director of research. That was a very interesting experience, and I am still connected with the glass business, though I am also doing professing. I started my career in experimental physics and lasted one day. When I started work on a doctoral thesis at the University of California in 1925 I had to set up a vacuum system. All experimental physicists in those days had to get a Cenco pump on the floor and glass tubing up to something that was on the table. I started out like all the rest but broke so much glass the first day that they suggested I go into theoretical physics. I told this story at Corning after I became their director of research. Mr. Amory Houghton, chairman of the board, who is now our ambassador to France, said, "Isn't it good that at last you are in a place where you can't possibly break enough glass to make any difference."

Looking back over the various possibilities of things that might be suitable to talk about this evening, I thought it would be interesting to review the historical development of what I now would like to call quantum physics, rather than quantum mechanics, because it has grown and expanded in such a way that it permeates all of modern physics. In fact it is extremely difficult to think of any actively cultivated part of physics that is not directly involved with Planck's quantum constant h. The basic discovery by Planck was made within a week or two of exactly sixty years ago, so I thought it might be interesting to discuss this subject.

THE subject of quantum physics started with the statistical theory of the distribution of energy in the black-body spectrum. The spectrum of radiated energy in equilibrium with matter in an enclosure is commonly called black-body radiation because it is the kind of radiation that would be emitted by a perfect absorber. The active problem in 1900 was the explanation of the distribution of energy in the spectrum.

It is interesting to realize that the subject has quite an ancient history. The first application of thermodynamics to black-body radiation goes back to 1859, when Kirchhoff first developed the ideas of radiative exchanges, and the connections between emission and absorption, rules according to which a good emitter is a good absorber, and a poor emitter is a poor absorber. In 1884 the discovery had been made of what we now call the Stefan-Boltzmann law, that the total radiation goes up as the fourth power of the absolute temperature. It was discovered by Stefan experimentally and interpreted theoretically by Boltzmann, making it one of the earliest applications of thermodynamics to radiation after those first ideas of Kirchhoff's.

In 1894 came the discovery by Wien of the displacement law, which tells how the distribution of energy over various wavelengths changes with the absolute temperature. The big problem at that time was to try to understand the reason for this distribution. Contrary to the general belief, which has become true in the last thirty years or so, that all physics is really

E. U. Condon is Wayman Crow Professor of Physics and head of the Department of Physics at Washington University in St. Louis, Mo. He presented the address upon which this article is based on the occasion of the 1500th regular meeting of the Philosophical Society of Washington, which was held on December 2, 1960, at the Natural History Museum Auditorium of the Smithsonian Institution in Washington, D. C. His address is included in Volume 16 of the archival *Bulletin* of the Society.

Planck

Rayleigh

done by young men in their twenties, the discovery of Planck was made when he was at the advanced age of 42. In 1900 he had already put a part of his career of research work behind him and was a professor in the University of Berlin, so that his work on quantum physics was done twenty-one years after he had received his doctorate for a thesis on the second law of thermodynamics. His thesis, it is interesting to note, was done under Kirchhoff and Helmholtz at Berlin. In his autobiography he says that he is quite confident neither of them ever read it.

Thermodynamics was Planck's first love, his principal love throughout physics. In fact there are many indications that he was rather annoyed with his discovery of the Planck constant of action and did his best for about fifteen years or so, on up to about 1915, to find ways of evading his own discovery and reconciling the theory that he had discovered with classical theory. This resembles somewhat the story that I used to hear from Professor Ladenburg at Princeton, about Roentgen. Everybody knows about the great consequences of Roentgen's discovery of the Roentgen rays, or x rays. Ladenburg was a student of Roentgen. He said that Roentgen was annoyed with his x rays because he did not understand what they were and much preferred classical subjects. So the upshot of it was that Ladenburg did a doctoral thesis under Roentgen just a few years after Roentgen had discovered x rays, on the subject of the correction to Stokes' law for a body falling through a viscous medium in a cylindrical tube, allowing for the finite diameter of the tube and the wall effect. They had a long pipe filled with castor oil, which is the traditional viscous material. It reached from the top floor of the laboratory to the basement. He said nothing ever gave Roentgen quite as much pleasure as to see the steel ball arrive down at the basement just when the calculation said it ought to. You can tell by a great deal of Planck's writings and readings that he felt much the same way about classical physics in relation to the modern developments.

Lord Rayleigh had published a theory, based on the equipartition-of-energy doctrine that goes back to Maxwell, Waterston, and Boltzmann, whereby every degree of freedom in the radiation field should have had the energy kT. He knew it did not, because that would have given an infinite or divergent result. But nevertheless that was where the theoretical thinking of his time led, which served to point up the importance of the quantum modifications that had to be made.

One of the things that I found interesting in looking back in the history of this theory is that it has always been referred to as the Rayleigh-Jeans law, and I had supposed that Rayleigh and Jeans had worked together on it. In point of fact, Rayleigh derived it and made a mistake by a factor of 8, which Jeans corrected in a letter to *Nature,* so that dividing the original Rayleigh formula by 8 was Jeans' contribution.

It was an essential contribution because it is a mistake that we all might make very readily. In counting up the degrees of freedom in the radiation field that are associated with frequencies between ν and $\nu + d\nu$, one has to calculate how many integers there are whose squares add up to a certain value, and it is natural to take the volume of a sphere of a certain radius. But in fact one takes only an octant out of this sphere because the integers, all three of them, have to be positive, and that is where Rayleigh went wrong.

The radiation measurements that served to inspire Planck were being made at the Physikalisch-Technische Reichsanstalt by some of the great names of early days of radiation-measurement work: Lummer, Pringsheim, and Rubens. The problem of distribution of energy in the spectrum was thus very much in the foreground and very good measurements were being made.

It was on October 19, 1900, that Planck presented his radiation formula to the German Physical Society at a meeting in Berlin, strictly as an empirical interpolation formula between the Rayleigh-Jeans law, which is valid at long wavelengths, and the Wien law, which is valid at short wavelengths. By interpolating in between, he had been able to find a simple formula that extended across the whole region, but at that time he had no theoretical basis for it whatever.

That night Rubens took the data to which he had access and made a very careful comparison with Planck's formula—a more careful one than Planck himself had made at that time—and found that it represented the data with extraordinary accuracy, much better than an empirical formula usually does. He called on Planck the next morning with a strong conviction that there was some real fundamental truth in the formula and not just an accidental agreement. Planck then set to work to find a theoretical basis for this formula and worked very hard for quite a while. In his autobiography he speaks of this as the most difficult period of his whole life.

Then, within less than two months, on December 14, 1900—so we are just twelve days ahead of the 60th Anniversary—he presented a paper to the Physical Society of Berlin in which he took the decisive step. By applying the Boltzmann principle for the connection of entropy with probability, which up to that time had hardly been used at all, he was able to work out the spectral distribution of energy that would be in equilibrium with a system of electrical oscillators.

In order to get the desired result, he had to suppose that the energy of each oscillator was built up in finite steps of energy, whereas, in all of physics hitherto, energy had been a continuous variable. To agree with the Wien displacement law he then had to assume that the finite size of these steps was proportional to the frequency, and so the energy quanta were $h\nu$. In that way he arrived at the famous formula, $u_\nu = [8\pi/c^3][h\nu^3/(e^{h\nu/kT} - 1)]$, for the density of the energy in the spectral frequency range between ν and $\nu + d\nu$ in black-body radiation at absolute temperature T. As is readily seen, in the limit of $h\nu/kT$ small compared with 1, this formula transforms into the Jeans formula; in the limit of $h\nu$ large compared with kT, it becomes the Wien formula and represents the data with great accuracy in between. Additional measurements of the same sort, which were later made with great precision at the National Bureau of Standards by W. W. Coblentz, greatly improved our knowledge of the subject.

One of the most extraordinary aspects of this work of Planck's is the accuracy with which he was able to define these fundamental constants. At that time there was no good value available for Avogadro's number, or for the charge on the electron, and the values that Planck was able to derive were much closer than is usually appreciated. When he first represented the data, in order to obtain a fit with the old black-body data, he had to assume that h was 6.885×10^{-27} erg·sec, and that k, which we now call the Boltzmann constant, was 1.429×10^{-16} erg/°K. The present best value for first number is 6.6252×10^{-27}, instead of 6.885×10^{-27}; and for the second number, it is 1.3804×10^{-16}, instead of 1.429×10^{-16}. At that very first time Planck

got Planck's constant only about 4.4 percent too high, and Boltzmann's constant about 3.5 percent too high, relative to the best modern values.

This was actually the first time that the Boltzmann constant had been evaluated. Let me just remind you of its relation to the other basic constants that have so much importance.

The gas constant, R, as we ordinarily know it, per gram mole, is equal to the Avogadro number, N, times k; and the Faraday, F, the amount of charge needed to plate out a gram mole of univalent ions, is equal to Avogadro's constant times the charge of the electron. That is, $R = Nk$, and $F = Ne$.

These molar quantities, R and F, were well known, and good values for them were available in those days, but what was not known was the Avogadro number N. However, if you know any one of these quantities you can get the other, so, as it turns out, obtaining the Boltzmann constant, k, enables one to get N by the first equation, and then, by using that N in combination with the knowledge of the Faraday, F, one is able to get the charge on the electron.

The electron had only been recognized about three years earlier by J. J. Thomson, and while the ratio of its charge to mass was known, its charge by itself was not well known. You will find, in the literature of that time, values published for e, the charge on the electron, ranging all the way from 1.29×10^{-10} electrostatic units, on up to 6.5×10^{-10} electrostatic units, which was given by J. J. Thomson, and a little while later revised back down to 3.4×10^{-10}. In other words at that time one only knew the charge on the electron to a factor of about 5 or 6.

On the other hand if you take the value of the Faraday and the value of k and solve for N from the gas constant and then solve for e, you find, surprisingly enough, that e equals 4.69×10^{-10} electrostatic units, which is only 2.3 percent below the currently recognized value.

Thus, in the space of just a month or two, Planck first found an empirical formula which to this day gives the most accurate representation of the spectral distribution of the radiant energy; second, he found a derivation of that formula. In order to get the derivation he had to introduce the extraordinary idea of energy quantization into physics. Third, he obtained an excellent value for the charge on the electron, which everybody was trying to do at that time.

You might expect that this would cause a great deal of excitement among physicists at that time, but it did not. If you search through the journals you find practically nothing is said about Planck in the years 1900 through 1904. I was very much intrigued, therefore, when just before this meeting Mr. Marton recalled that a search of the records of this Society indicated that in 1902 Arthur L. Day gave a report on Planck's work. Thus, The Philosophical Society of Washington was one of the earliest to pay attention to it.

The first real extension of Planck's work came with

Einstein's famous paper of 1905, the paper for which he got the Nobel Prize. (It is important to realize that Einstein did not get the Nobel Prize for the theory of relativity. They might give it to him now if he were around, but they did not in those days.) Planck wrote only one other paper on the subject in that period between 1900 and 1905 and this was mainly an expository paper. There is one brief mention by Burbury, another paper by van der Waals, Jr., and that is all. In those days Planck was almost completely ignored.

In Planck's own autobiography he tells of his own attitude toward the Planck constant, and I thought it would be interesting to read his own words on that, of course translated into English. He said:

> While the significance of the quantum of action for the interrelation between entropy and probability was thus conclusively established, the great part played by this new constant in the uniform regular occurrence of physical processes still remained an open question. I therefore tried immediately to weld the elementary quantum of action, *h,* somehow into the framework of classical theory. But in the face of all such attempts the constant showed itself to be obdurate.
>
> So long as it could be regarded as infinitesimally small, i.e., when dealing with higher energies and longer periods of time, everything was in perfect order. But in the general case difficulties would arise at one point or another, difficulties which became more noticeable as higher frequencies were taken into consideration. The failure of every attempt to bridge that obstacle soon made it evident that the elementary quantum of action plays a fundamental part in atomic physics and that its introduction opened up a new era in natural science, for it heralded the advent of something entirely unprecedented and was destined to remodel basically the physical outlook and thinking of man which, ever since Leibniz and Newton laid the ground work for infinitesimal calculus, were founded on the assumption that all causal interactions are continuous.

He goes on in a more personal vein to say:

> My futile attempts to fit the elementary quantum of action somehow into the classical theory continued for a number of years [actually until 1915] and they cost me a great deal of effort. Many of my colleagues saw in this something bordering on a tragedy. But I feel differently about it, for the thorough enlightenment I thus received was all the more valuable. I now knew for a fact that the elementary quantum of action played a far more significant part in physics than I had originally been inclined to suspect, and this recognition made me see clearly the need for the introduction of totally new methods of analysis and reasoning in the treatment of atomic problems.

In spite of Jeans' intimate association with this problem, you find no reference whatever to the Planck black-body law in the first edition of his *Dynamical Theory of Gases,* which was published in 1904, four years after Planck's work. In the Landolt-Bornstein Tables, published in 1905, we find an extraordinary thing, namely, that it gives widely different values for what is often called the Loschmidt number, the num-

ber of molecules in one cubic centimeter of various gases under standard conditions. Of course this value should be the same for all gases. But they solemnly give you a table with 2.1×10^{19} for air, 4.2×10^{19} for nitrogen, 7.3×10^{19} for hydrogen, and so on. Apparently Landolt and Bornstein did not believe in the Avogadro number. Planck got 2.76×10^{19} for this number, which as we have seen is a good value.

Josiah Willard Gibbs was America's first great theoretical physicist. He was elected, I find, to membership in the Washington Academy of Sciences in 1900. He died in 1903 at the age of 64. There is no indication in any of his publications or notes that he left behind that he paid any attention to Planck's work. He had puzzled over the problem of the specific heat of polyatomic gases, which everybody was puzzled about at that time, because it has too low a value to correspond with the equipartition law. There is some indication that he found these difficulties with the equipartition law revealed in the specific heat of gases somewhat depressing, and I find an indication of that, perhaps, in an interesting paragraph from the preface to his famous work on statistical mechanics, published in 1902, a year before Gibbs' death:

> In the present state of science it seems hardly possible to frame a dynamic theory of molecular action which shall embrace the phenomena of thermodynamics, of radiation and of the electrical manifestations which accompany the union of atoms. Yet any theory is obviously inadequate that does not take account of all these phenomena. [Then comes a wonderful sentence at the end of this paragraph which I think we all ought to realize was written by Gibbs in 1902:] Certainly one is building on an insecure foundation who rests his work on hypotheses concerning the constitution of matter.

Lord Kelvin's Baltimore lectures, which were delivered at The Johns Hopkins University in 1884 but were not published until 1904, had undergone a great deal of revision up to the latter date. The preface to these lectures is very interesting to those who have anything to do with editing or getting things through the press. He admits that he had been working on the revision for all of the nineteen years. I can well imagine that he was a popular fellow around the print shop.

That work includes as its appendix B, the famous lecture to which I am sure you have all heard allusions, "Nineteenth Century Clouds over the Dynamical Theory of Heat and Light." That lecture was delivered in April 1900, some months before Planck's work, and then it was published originally in the *Philosophical Magazine* of July 1901. It makes no reference to the black-body radiation or to Planck's work, although cloud 2—he had his clouds numbered—was this same concern about the failure of equipartition of energy as evidenced by the specific heats of gases, the same problem that was troubling Gibbs.

Lord Rayleigh's publication of what we now call the Rayleigh-Jeans law was in the *Philosophical Magazine* in 1900. He did not return to the subject again until 1905 when he wrote several notes in *Nature* in which he concedes or agrees with the comment that Jeans had made about being wrong by a factor of 8. It is interesting in that he says that he "has not succeeded in following Planck's reasoning". That is how Planck's work was received by Lord Rayleigh, one of the greatest British physicists. Rayleigh actually published papers actively through 1919, but he seems to have had no more to say on black-body radiation than what he said in that one 1905 paper.

Search through his published papers reveals two more items relating to modern quantum physics. In the 1906 *Philosophical Magazine* he comments on the classical radiative properties of the atom models that resemble J. J. Thomson's. However, he goes beyond them in that he regards the negative charge as distributed more like a continuous fluid and studies it as a normal mode-of-vibration problem. In an editorial note added to his collected papers, written in 1911, he refers back to some old work of an 1897 paper. An interesting thing to me is his comment that all kinds of models of normal modes of vibrations of continuous systems always lead to formulas in which the square of the frequency is written additively as the sum of contributions coming from the different degrees of freedom —from what we would now call the different quantum numbers. Rayleigh was wedded to a classical vibration-theory model where the squares of the frequencies get in because of the second derivative with regard to the time, based on Newton's law of mechanics. Nowadays, when we lecture on quantum mechanics, we just quietly *make* the Schrödinger equation contain the first time derivative so we will not have this trouble, which is the advantage of making up your equations as you go along as compared with getting them from someone like Newton.

Steeped in acoustics as he was, Rayleigh does say: "A partial escape from these difficulties might be found in regarding the actual spectrum lines as due to difference tones from primaries of much higher pitch." That is a well-known device giving physicists license to pass from a square term to a linear term; that is, a small change in the square is linear.

There is still something else in the 1906 paper which intrigues me. Rayleigh devotes a paragraph to the problem of the sharpness of spectral lines despite the random character of the conditions of excitation, and concludes with a paragraph that sounds very modern. I quote:

> It is possible, however, that the conditions of stability or of exemption from radiation may, after all, demand this definiteness, notwithstanding that in the comparatively simple cases treated by Thomson, the angular velocity is open to variation. According to this view, the frequencies observed in the spectrum may not be frequencies of disturbance or of oscillation in the ordinary sense at all, but rather form an essential part of the original constitution of the atom as determined by conditions of stability.

Maybe one reads into the statement one's present knowledge of the later developments of quantum theory, but I found it very interesting as a foreshadowing of the way we look at it now.

Even as late as 1911, we find Lord Rayleigh worrying about Kelvin's cloud 2, the specific-heat difficulty, although Einstein had really put that difficulty to rest in 1907. In 1911, Rayleigh wrote to Walter Nernst to express his concern:

> If we begin by supposing an elastic body to be rather stiff, the vibrations have their full share of kinetic energy [that is the equipartition law] and this share cannot be diminished by increasing the stiffness. . . .

We all know that increasing the stiffness makes the interval between the vibration quantum levels greater, so that they do not take part practically in the equipartition law simply because they cannot get enough energy even to be excited to the first state.

However, Rayleigh goes on:

> Perhaps this failure might be invoked in support of the views of Planck and his school that the laws of dynamics as hitherto understood cannot be applied to the smallest part of the bodies. But I must confess that I do not like this solution of the puzzle . . . I have a difficulty in accepting it as a picture of what actually takes place.
>
> We do well I think to concentrate attention on the diatomic gaseous molecule. Under the influence of collisions the molecule freely and rapidly acquires rotation. [He knows this from the specific heat.] Why does it not also acquire vibration along the line joining the two atoms?
>
> If I rightly understand the answer of Planck is that in consideration of the stiffness of the union, the amount of energy that should be acquired at each collision falls below the minimum possible and that therefore none at all is acquired [this is of course exactly what we know] an argument which certainly sounds paradoxical.

This is the end of it for Rayleigh.

So we can see that the acceptance of these ideas was something that came very, very slowly. The examples I have chosen illustrate that very little was stated about the subject at all from 1900 to 1905, and even after that you find the great men of the period hesitant and unwilling to build it into their thinking.

LET us now turn to Einstein's famous 1905 paper, which I must confess I had not read until I got to thinking over the preparation for this lecture. It is one of the papers we all hear about in school and worship, but do not read. One of the odd things about this paper is that "*h*" is not in it, believe it or not. In the paper Einstein denotes by the letter β what we now would call h/k, and then he writes R/N for what we call k, and thus you find in that paper that the energy of a light quantum is not $h\nu$ at all. It is $R\beta\nu/N$, which certainly takes a bit of getting used to.

The title of his paper is an interesting one: "Heuristic Viewpoint Concerning the Emission and Transformation of Light", indicating, I think, that he meant that there is something in the paper, but he does not quite know what. At least that is what I mean when I say "heuristic". Einstein might, of course, have meant something else. He says:

> The energy of a ponderable body cannot be divided into indefinitely many indefinitely small parts, whereas the energy emitted by a point light source is regarded on the Maxwell theory or more generally according to every wave theory as continuously spread over a continuously increasing volume.
>
> Such wave theories of light have given a good representation of purely optical phenomena and will surely not be replaced by any other theory. [He was right in that. They have not been replaced.]

He continues:

> It is to be remembered, however, that the optical observations referred to time mean values, not to instantaneous values, and it is quite conceivable that, in spite of complete success in dealing with diffraction, reflection, refraction, dispersion, et cetera, such a theory of continuous fields could lead to contradictions with experience when applied to phenomena of light emission and absorption.

After a little more discussion he makes the key declaration that played such a decisive role in all subsequent developments in which, in one sentence, he says:

> According to the supposition here considered, the energy in the light propagated from rays from a point is not smeared out continuously over larger and larger volumes, but rather consists of a finite number of energy quanta localized at space points, which move without breaking up and which can be absorbed or emitted only as wholes.

Oddly enough, though nowadays this paper is quoted purely for the photoelectric effect in the discussion, the photoelectric effect is only one section, paragraph 8, of the whole paper. The first six paragraphs are concerned entirely with another way of looking at the details of the statistical distribution of the black-body radiation law, and the entropy of radiation along the lines of the quantized wave theory.

Finally, paragraph 7 is an interpretation of the Stokes' rules for photoluminescence. In ordinary fluorescence and phosphorescence, Stokes' law, which goes way back to 1860, says that the wavelength of the fluorescent light is always greater than the wavelength of the exciting light, or nearly so. There is some radiation, called anti-Stokes radiation, for which the wavelength is a little shorter.

Why there was so little stress in Einstein's paper on the photoelectric effect, compared with these other things, puzzled me. Then I came to section 8, which deals with the photoelectric effect, and I asked Professor A. L. Hughes, one of the pioneers of photoelectric work, who is at our place—he is emeritus professor in Washington University in St. Louis—how

that could be. He told me how very primitive the knowledge of the photoelectric effect was at that time. No vacuum work on photoelectricity had been done, and, even if it had, it would have been done with very poor vacuums, under very poor conditions. In point of fact, no effort had been made to determine the retarding potential required to stop the photocurrent in a definite circuit.

What had been observed was that, if you insulate a metal object and shine light on it, it will build up to a certain potential and then stay at that potential. That is, it builds up its own retarding potential and finally prevents the escape of further electrons into the air. Metals differ with regard to the potential built up by certain kinds of light, and it was found that as one went to more and more violet light one got a higher potential. But these were only very crude measurements indeed, so crude that one would hardly think that they might offer any possibility of fundamental understanding.

That, perhaps, is the reason why the photoelectric effect was so little stressed in Einstein's paper. The Stokes'-law argument was much more directly experimental, and, conversely, it seems rather odd to me, as I think about it, that Stokes' law is not more stressed today in teaching the subject.

It was in 1907 that the specific-heat work of Einstein clarified the problem of low-temperature specific heat.

It is fascinating to look up some of the historical information that is available in the literature. I do not mean that one has to go to ancient history, just the history of the last century. For example, in 1904, when the great St. Louis World's Fair was held, various distinguished visitors presented lectures. Lord Kelvin gave a speech suggestive of a sort of inverse neutrino theory. The thing bothering him at that time was—this is a little off the subject of quantum theory, but I think it is interesting—the measurement that had just been made by the Curies of the amount of energy given off by radium per unit time. They had not measured the half life, and the energy given off did not show any signs of weakening, and you know that physicists are great on extrapolation. They said that radium gives off energy *perpetually*—that was the word, perpetually.

So the question was, how could anything radiate perpetually at this tremendous rate?—a rate unheard of when expressed in terms of energies of usual chemical reactions.

Kelvin had an idea which he propounded at this talk; perhaps, he suggested, there was some kind of energy that one could not detect—like the neutrinos—floating around in space, and perhaps radium had the property of absorbing it in that form and then reconverting it, like a fountain, and shooting it out, and that was what was observed. Even in those days, people were perfectly willing to balance the books on conservation of energy in such a manner.

J. J. Thomson

the young Bohr

THE next major historical event was the development of the Bohr atom model in 1913. At this point, since we are just talking a little bit of anecdotal material about the history of our subject, I will tell a story that I learned from George Gamow. The young Bohr—he was about 26 at that time—came to England from Copenhagen to work in the Cavendish Laboratory. The great J. J. Thomson was at the height of his powers. Bohr came to the great center to study fundamental atomic physics, but within a few months he left the Cavendish Laboratory and went up to Manchester to work under a relatively unknown fellow named Rutherford. The question is, why did he do that? According to Gamow, Bohr had gotten into trouble with "J. J." because he was a little critical of the Thomson atom model, and "J. J." had politely indicated to him that it might be nice if he left Cambridge and went to work with Rutherford. That is how Bohr went to work for Rutherford, which was advantageous, I think, for all. It was not so good for the Thomson model but it was fine for the future development of physics.

To bring this story up to date, Gamow told me that in 1928, when he worked on the alpha-particle tunneling paper, the basic work which Gurney and I did simultaneously in Princeton, Rutherford sent Gamow to see Bohr and to tell him about this exciting new development. He also wrote him a letter—of which Gamow said he still has a copy—saying, "Please pay attention to this fellow; there is something in it. It isn't cockeyed. You remember how it was with you when you went to 'J. J.' and he wouldn't listen; so now you listen to Gamow." I do not know whether there

is any truth in that or not, but at any rate Bohr did listen to Gamow.

Of course the most exciting immediate experimental consequence of Bohr's work, besides the direct interpretation of the spectrum of hydrogen which was well known at that time (I mean the facts of the Balmer series, which went way back into the 19th century), was the interpretation of spectroscopic-term values as being energy levels with the associated implication that controlled electron impact would produce controlled excitation of atoms and molecules. This was the work that was immediately carried further by James Franck and Gustav Hertz, and for which they received the Nobel Prize in 1926.

That work was very quickly taken up here in Washington, in the pioneer work of Paul Foote and F. L. Mohler. At the Bureau of Standards the accountants were rather stuffy, and had rather sharp lines about appropriations and budgets, so all the work on critical potentials for which the Bureau became famous was carried on under a budget number which had something to do with improving pyrometric methods. I am not sure it helped much in advancing pyrometry, but it certainly was a great addition to the development of science.

The period of the second decade of our subject was also characterized by the very first extension of the idea of quantized energy levels to the interpretation of band spectra, rotation and vibration spectra, and infrared.

A curious thing about the atom-model work of Bohr, prior to 1923 or 1924, was that if you look at the then-current papers you get the impression that everybody

Schrödinger

de Broglie

Born

Heisenberg

in the world was terrifically excited about the Bohr model and believed in it hook, line, and sinker, including the electron orbits as they are used in the ads for the atomic age nowadays. Bohr, on the other hand, was constantly making remarks, speeches, and admonitions to the effect that this is temporary and we ought to be looking for a way to do it right.

THE great breakthrough, as the modern saying goes, came about in 1924, 1925, and 1926, when the idea of waves accompanying electrons was first published by de Broglie as a doctor's thesis—and was also ignored. I do not know anybody who read that paper until a year or two later. Schrödinger then founded the great discoveries of wave mechanics on de Broglie's work in the series published in the spring of 1926.

Just before Schrödinger's work in late 1925, Born, Jordan, and Heisenberg had developed the matrix-mechanics methods. For about a year they were thought of as two rival and distinct theories, until Schrödinger and Carl Eckart, then a young physicist in Chicago, who is now in La Jolla, recognized the mathematical identity of the two theories.

I had the good fortune to get my doctorate in the summer of 1926 when all these things were at their highest peak of excitement, and went to Göttingen to

work with Born. There was a young graduate student there named Robert Oppenheimer with whom I got acquainted at that time.

It was an extremely difficult period because the rate of advance was so great, and the whole subject was so obscure to all of us, that it was hard to keep up with the state of affairs. I remember that David Hilbert was lecturing on quantum theory that fall, although he was in very poor health at the time. (He had anemia, and liver extract was then unavailable, so he was eating a vast quantity of liver every day and saying he would rather not live than eat that much liver. His life was saved by the fact that liver extract was discovered just about that time.) But that is not the point of my story. What I was going to say is that Hilbert was having a great laugh on Born and Heisenberg and the Göttingen theoretical physicists because when they first discovered matrix mechanics they were having, of course, the same kind of trouble that everybody else had in trying to solve problems and to manipulate and really do things with matrices. So they went to Hilbert for help, and Hilbert said the only times that he had ever had anything to do with matrices was when they came up as a sort of by-product of the eigenvalues of the boundary-value problem of a differential equation. So if you look for the differential equation which has these matrices you can probably do more with that. They had thought it was a goofy idea and that Hilbert did not know what he was talking about, so he was having a lot of fun pointing out to them that they could have discovered Schrödinger's wave mechanics six months earlier if they had paid a little more attention to him.

I mention some of the occurrences of those years because I do not believe that anybody who did not live through that period can fully appreciate what a tremendous number of things happened then that are still very basic, and have blossomed out into whole areas of physics which now are subjects for courses in themselves.

In 1926 we had the whole wave mechanics, as we know it, and the whole matrix mechanics formulated, And just a little before that, we had the discovery of the electron spin and of the Pauli exclusion principle. In 1927 came the whole theory of the chemical valence bond as a perturbation problem in quantum mechanics, with correlations over electron pairs with their spins antiparallel. Almost simultaneously, there occurred the whole development of Fermi-Dirac statistics and its clarification of the problems of metal theory. A few months later, the Dirac papers on the quantization of the electromagnetic field explained, at last, the difference between spontaneous and induced emission and put the two together in a unified theory. Soon after that came the whole Dirac relativistic theory of the electron, which later led to the prediction of the positron.

Shortly thereafter, in 1928, the interpretation of natural alpha radioactivity came as a consequence of the barrier-leakage idea, also an essential element of quantum mechanics and an essential element of its statistical or probability interpretation. I think it is fair to say that the barrier-leakage idea was the opening of the modern period of the application of quantum mechanics to nuclear physics. Nuclear physics, in terms of real specific models, has never had a classical past. Nobody tried in those days to develop specific models of the structure of a nucleus.

Another big year for discoveries was 1932, the year in which Urey discovered heavy hydrogen, which from a nuclear point of view means the deuteron, the year in which the first production of an artificial nuclear reaction was accomplished by Cockcroft and Walton, the year in which the positron was discovered—the antiparticle associated with the electron, as we call it nowadays, and the year in which the neutron was discovered.

In that same decade, a few years later, 1936 saw the development of the Fermi theory of beta decay based on the neutrino hypothesis that had been introduced by Pauli, in almost a joking way, a year or two earlier.

I remember that in the summer of 1937, when we had a conference on beta-decay theory at Cornell University, and a lot of us were having trouble worrying about it, Fermi was in the audience sitting in the back row just smiling and smiling as he usually did. People tried to get him to comment, and he said, "I have always been surprised that people take that theory so seriously." But, as we know, it has turned out to be remarkably correct—that is, the basic formalism which Fermi developed then for accounting for the four-fermion interactions, even in spite of the great crisis it went through in 1957 with the discovery of nonconservation of parity. The basic formalism, as Fermi first introduced it, has beautifully stood the test of time.

The year 1936 is also important to us here because of work done by prominent people in Washington. I refer to the work of Hafstad, Heidenberg, and Tuve in the first real studies of proton-proton scattering, which gave direct evidence of forces between protons other than the Coulomb forces, that is, short-range nuclear forces between protons. The theoretical interpretations of those results were largely done in Princeton by Gregory Breit and myself, in association with Richard Present, who is now at the University of Tennessee. That provided the first evidence of what is now called the charge independence of nuclear forces, because the additional short-range force that was revealed in this way turned out quantitatively to be very close to the force between a proton and a neutron which is revealed in the normal state of the deuteron.

From about 1932 on, the whole field of nuclear physics came into being in a big way, with deuterons available and with machines available, both cyclotrons and Van de Graaff machines. In the latter part of that decade, we began to have the first theories of Bethe and Marshak on the application of specific models of nuclear reactions to the problem of finding satisfactory sources of stellar energy.

I THINK perhaps I must give up at this point, because the last two decades have seen such an overwhelmingly rapid and vast amount of progress, spreading out into so great many different fields, that one could not possibly, in the short time remaining, do more than just mention it.

We had, in the decade from 1940 to 1950, the whole development of the modern point of view on quantum electrodynamics. It came rather late in the decade, with the discovery of the Lamb shift and the experimental confirmation of the abnormal magnetic moment of the electron, which was somewhat off from the original Dirac theory. We had at last the clarification of the puzzling features of the mesons in cosmic rays, whereby it turned out that there were the two kinds, the pi mesons and the mu mesons, the pi mesons decaying into the mu mesons. The latter part of the decade represented the beginning of public knowledge of fission, and the engineering and political uses of fission.

At the same time, going off in quite another direction, what has turned out to be of equal importance has been the whole wide development of the application of Fermi statistics to electrons in solids, first resulting in the major classification of properties of metals, then of semiconductors, and then finally of really modern tailored effects that led to transistors and other devices.

The decade just passed has corresponded to an enormous further development along these same lines. We have the study of nuclear reactions going on up to higher energies of some hundreds of millions of volts, with predominant interest in the study of polarization effects in nuclear reactions as another way of getting at points of detail; the recognition of the nonconservation of parity; the experimental discovery of the neutrino; the recognition that the Fermi interaction that applies in weak interactions is more general than simply the beta decay, applying also to muon decay and other related processes; and the discovery of the strange particles.

And then finally, as a roundup of mentioning things that we do not have time to talk about, there are the extraordinarily fine extensions that have been made in the last five years of the theory of broad, modern, good perturbation-theory methods for dealing with the many-body problem. They involved not only the better calculation of nuclear models but also, at last, after many years of effort, they are beginning to provide a real understanding of superfluids.

I want to close by remarking that all this started, as I said, almost exactly 60 years ago—barring two weeks—on December 14, 1900, when Planck's constant was first introduced into physics. In the 60 years that have intervened it is now almost impossible to find many papers in physics which do not deal directly or indirectly with phenomena that are fully and basically conditioned by the existence of that one universal constant.

REMINISCENCES OF

Heisenberg and the early days of quantum mechanics

Recollections of the days, 50 years ago, when a handful of students
in the "entirely useless" field of physics heard of a strange new mechanics
invented by Maurice de Broglie, Werner Heisenberg and Erwin Schrödinger.

Felix Bloch

PHYSICS TODAY / DECEMBER 1976

It is appropriate in this year, when we celebrate the 50th anniversary of quantum mechanics, and during which we have been saddened by the death of one of its leading founders, Werner Heisenberg, to reminisce about the formative years of the new mechanics. At the time when the foundations of physics were being replaced with totally new concepts I was a student of physics. I sat in the colloquium audience when Peter Debye made the suggestions to Erwin Schrödinger that started him on the study of de Broglie waves and the search for their wave equation. It was from Heisenberg, as his first doctorate student, that I caught the spirit of research, and that I received the encouragement to make my own contributions.

First inklings

Let me begin by going back to 1924, when I entered the Swiss Federal Institute of Technology in my home town of Zurich. I began as a student of engineering but after a year and good deal of soul searching I decided, against all good sense, to switch over to the "entirely useless" field of physics. The E. T. H., as it is known from its German name, was an institution of great international repute and in my newly chosen field of studies I had heard of such famous men as Peter Debye and Hermann Weyl. In fact, the first introductory course of physics I took was taught by Debye and, without knowing much about his scientific work, I realized from the high quality of his lectures at the Institute that here was a great master of his field.

There was a good deal less to be enthusiastic about in the other courses one

could take, and there was nothing like the complete menu that is presented to the students nowadays. Once in a while, a professor would offer a special course on a subject he just happened to be interested in, completely disregarding the tremendous gaps in our knowledge left by this system. Anyway, there was only a handful of us foolish enough to study physics and it was evidently not thought worthwhile to bother much about these "odd fellows." The only thing we could do about it was to go to the library and read some books, although nobody would advise us which ones to choose.

Among the first I hit upon was Arnold Sommerfeld's *Atomic Structure and Spectral Lines,* which I found fascinating; the only trouble was that I could not understand most of it because I knew far too little of mechanics and electrodynamics. So at first I had to learn about these subjects from other books, to truly appreciate what Sommerfeld said; but then it conveyed the good feeling that everything about atoms was completely known and understood. The fact that one really could handle only periodic systems and only those that allowed a separation of variables did not seem a great cause for concern. Therefore, when I saw a paper in which somebody tried to squeeze the theory of the Compton Effect into that scheme, I was more impressed than discouraged by the complicated mathematics spent in the effort.

The news that the foundations of a new mechanics had already been laid by Maurice de Broglie and Heisenberg had hardly leaked to Zurich yet and certainly had not penetrated to our lower strata. The first inklings of such a thing came to me in early 1926; I had by then started to attend the physics colloquium regularly, although most of what I heard there was far above my head. The colloquium, run

with firm authority by Debye, might have had an audience of as much as a couple of dozen—on a good day.

Physics was also taught at the University of Zurich by a smaller and rather less illustrious faculty than that at the E. T. H. Theory there was in the hands of a certain Austrian of the name of Schrödinger, and the colloquium was alternately held at both institutions. I apologize to my friends who already have heard from me what I am going to tell you now. My account may not conform to the strictest standards of history, which accord validity only to written documents, nor will I be able to render the exact words I heard on those occasions, but I can vouchsafe that, in content, I shall report the truth and only the truth.

A wave equation is found

Once at the end of a colloquium I heard Debye saying something like: "Schrödinger, you are not working right now on very important problems anyway. Why don't you tell us some time about that thesis of de Broglie, which seems to have attracted some attention."

So, in one of the next colloquia, Schrödinger gave a beautifully clear account of how de Broglie associated a wave with a particle and how he could obtain the quantization rules of Niels Bohr and Sommerfeld by demanding that an integer number of waves should be fitted along a stationary orbit. When he had finished, Debye casually remarked that he thought this way of talking was rather childish. As a student of Sommerfeld he had learned that, to deal properly with waves, one had to have a wave equation. It sounded quite trivial and did not seem to make a great impression, but Schrödinger evidently thought a bit more about the idea afterwards.

Just a few weeks later he gave another

Felix Bloch, winner (with E. M. Purcell) of the 1952 Nobel Prize in physics, is professor emeritus of physics at Stanford University.

HEISENBERG

talk in the colloquium which he started by saying: "My colleague Debye suggested that one should have a wave equation; well, I have found one!"

And then he told us essentially what he was about to publish under the title "Quantization as Eigenvalue Problem" as a first paper of a series in the *Annalen der Physik*. I was still too green to really appreciate the significance of this talk, but from the general reaction of the audience I realized that something rather important had happened, and I need not tell you what the name of Schrödinger has meant from then on. Many years later, I reminded Debye of his remark about the wave equation; interestingly enough he claimed that he had forgotten about it and I am not quite sure whether this was not the subconscious suppression of his regret that he had not done it himself. In any event, he turned to me with a broad smile and said: "Well, wasn't I right?"

Of course, there was afterwards a lot of talk among the physicists of Zurich, including even the students, about that mysterious "psi" of Schrödinger. In the summer of 1926, a fine little conference was held there and at the end everyone joined a boat trip to dinner in a restaurant on the lake. As a young *Privatdozent*, Erich Hückel worked at that time on what is now well known as the Debye–Hückel theory of strong electrolytes, and on the occasion he incited and helped us to compose some verses, which did not show too much respect for the great professors. As an example, I want to quote the one on Erwin Schrödinger in its original German:

"Gar Manches rechnet Erwin schon
Mit seiner Wellenfunktion.
Nur wissen möcht' man gerne wohl
Was man sich dabei vorstell'n soll."

In free translation:

Erwin with his psi can do
Calculations quite a few.
But one thing has not been seen:
Just what does psi really mean?

Well, the trouble was that Schrödinger did not know it himself. Max Born's interpretation as probability amplitude came only later and, along with no less a company than Max Planck, Albert Einstein and de Broglie, he remained skeptical about it to the end of his life. Much later, I was once in a seminar where someone drew certain quite extended conclusions from the Schrödinger equation, and Schrödinger expressed his grave doubts that it could be taken that seriously; whereupon Gregor Wentzel, who was also there, said to him: "Schrödinger, it is most fortunate that other people believe more in your equation than you do!"

Schrödinger thought for a time that a wave packet would represent the actual shape of an electron, but it naturally bothered him that the thing had a tendency to spread out in time as if the electron would gradually get fatter and fatter.

As I said before, I was too green then to understand these things and still struggled with the older theories. In reading Debye's paper of 1923 on the Compton effect, it occurred to me that, instead of

his assumption of the electron being originally at rest, one should take into account its motion on a stationary orbit in the atom. I thought this was such a good idea that I even had the incredible courage to go to Debye's office and tell it to him. It really wasn't all that wrong but he only said: "That's no way any more to talk about atoms; you better go and study Schrödinger's new wave mechanics."

Well, you would not disobey the authorities and, of course, he was again quite right. So this is what I did; Schrödinger's next papers on wave mechanics appeared shortly, one after the other. I did not learn about the matrix formulation of quantum mechanics by Heisenberg, Born and Pascual Jordan until I read that paper of Schrödinger's in which he showed the two formulations to lead to the same results. It did not take me too long to absorb these new methods, and I wish I could confer to the younger physicists who read this article the marvellous feeling we students experienced at that time in the sudden tremendous widening of our horizon. Since we were not burdened with much previous knowledge, the process was quite painless for us, and we were blissfully unaware of the deep underlying change of fundamental concepts that the more experienced older physicists had to struggle with.

Although I had already begun an experiment in spectroscopy, I was now entirely captured by theory and I felt the legal entrance into the guild to be confirmed through my acquaintance with Walter Heitler and Fritz London. They had just obtained their PhD's and had come to Schrödinger's Institute, where together they worked on their theory of covalent bonds. I must have met them in a seminar, and it was a great thing for me that they asked me to join them in some of their walks through the forests around Zurich. For us students the professors lived somewhere in the clouds, and that two real theorists at the ripe age of almost 25 should even bother about a greenhorn like me was ample cause for my gratitude to them.

Leipzig

This great period in Zurich came to a sudden end in the fall of 1927 when some of the most important men there simultaneously succumbed to the pull of the large magnet in the North, represented by the flourishing science in Germany. Weyl had accepted a position in Göttingen, Schrödinger in Berlin and Debye in Leipzig, and it was clear to me that I had to join the exodus if I did not want my time as a student to drag on much longer. The question was only where to go; I was tempted to follow either London's example and go with Schrödinger to Berlin, or Heitler's, and go to Göttingen.

Before deciding, however, I went to ask Debye for his opinion, and he advised me to do neither but instead to come to

Leipzig. There I would work with Heisenberg whom he, as the new director of the Institute of Physics of the University, had persuaded to accept the professorship for theoretical physics. Debye's power of persuasion was quite formidable and I could not resist it either, particularly because I had previous evidence of his sound judgment.

So, in October 1927 before the beginning of the winter semester, I left my nice home town for the first time, to arrive on a cold gray morning in that rather ugly city of Leipzig. The little room I found for rent from a family overlooked a railroad yard; the noise and smoke did not help much to cheer me up! As soon as I had completed the simple formality of registering as a student of the University in the center of the city I went to the Physics Institute, which was located near the outskirts.

It was an old building opposite a cemetery on one side and adjoining the garden of a mental institution on the other, but occupied by people who were far from being either dead or crazy. Heisenberg had not arrived yet and the theorist in charge was Wentzel who, a year later, was to become Schrödinger's successor in Zurich. I did not find him in his office and was told by an assistant that I could see him in his apartment on the third floor of the building.

It was quite customary at that time for professors to have official living quarters in or adjacent to their institutes; Debye had the Director's villa in a side wing, and for young bachelors like Wentzel and also Heisenberg upon his arrival there were small but comfortable apartments under the roof.

I was not at all sure whether it was really all right to go up there and knock at his door but I dared to do it anyhow, and almost from the moment he opened it I realized that I had come to a new and much warmer academic climate. Used to the great distance that separated the students and professors in freedom-loving Switzerland, I had expected the proverbial discipline of the Germans to call for an even stricter caste system. Instead, Wentzel received me with the informal cordiality of a colleague, which made it almost difficult for me to address him with the normal *"Herr Professor"* but very easy to show him a little paper I had written before I came to Leipzig.

My paper had been motivated by Schrödinger's old dislike of electron wavepackets' disagreeable habit of spreading, and I had had the naive idea that they might be cured from it at least partially by radiation damping. To try it out, I had done a serious calculation for the harmonic oscillator, with the result that a suitable gaussian wavepacket, without spreading, would perform a nice damped oscillation that led asymptotically to the wavefunction of the ground state. Wentzel made some kind com-

DEBYE

ments but modestly disclaimed sufficient expert knowledge to pass judgment; he said I should ask Heisenberg, who was expected in a few days.

My first paper

Although his great achievements dated back no more than about two years, Heisenberg was already very famous as the founder of the new form of mechanics, which accounted for quantum phenomena by abandoning such fundamental ideas as motion in an orbit and replacing them by concepts referring to the actual observation of atomic processes. I think I lost my breath for a moment when Wentzel introduced me to this great physicist in the person of a slender young man. Maybe Debye had already mentioned to him that he knew me from Zurich; in any case, as soon as he shook hands and started to talk to me in his simple natural way, I had the feeling that I was "accepted."

Just as with Wentzel, there was no indication whatever of a barrier to separate us on the grounds of Heisenberg's vastly superior standing, and this was the experience I had with many of the other prominent scientists I later met in Germany. While it surprised me at first, it had quite a simple reason: These men were so entirely devoted to their science and their work spoke so clearly for itself that there was really no room or reason for any pretense, be it in the form of grand manners or of false modesty. With Heisenberg there was the additional factor of his youth; as a professor at the age of 26 he was only about four years older,

although in the time scale of theorists this already put him something like two generations ahead of me.

As to my hopes for keeping wavepackets together by radiation damping, he only smiled and said that, if anything, it could of course only make them spread even more. Nevertheless he thought my calculations on the harmonic oscillator were a good start, and that I should go on to work them out for the general case. With the help of P. A. M. Dirac's paper on radiation effects and a few more tricks, I managed to do that rather quickly, confirming Heisenberg's prediction, and it became my first published paper. It appeared in the *Physikalische Zeitschrift* as a precursor to the well known paper of Victor Weisskopf and Eugene Wigner on radiation damping and natural line widths.

Before the Christmas vacations, Heisenberg said that I should think about a topic for my doctor's thesis: This I did mostly while skiing in Switzerland after I had gone home. I knew the importance of Paul Ehrenfest's adiabatic theorem in the older quantum theory, and when I went back to Leipzig after New Year I proposed for my thesis its reformulation in quantum mechanics.

"Yes," said Heisenberg, "one might do that, but I think you had better leave such things to the learned gentlemen of Göttingen."

What he meant was the school of Born, which had the reputation of being particularly skilled in, and rather fond of, elaborate mathematical formalisms. Instead, he suggested something more

SCHRÖDINGER

down to earth such as, for example, ferromagnetism or the conductivity of metals.

As to ferromagnetism, he thought that it had to be explained by an exchange integral between electrons, with the opposite sign from that in helium so as to favor a parallel rather than opposite orientation of their spins. He had shown before that the difference between the ortho and para states of the helium atom were due to the dependence of the exchange energy on their symmetry properties and had also recognized that the analogous phenomenon for the protons in the hydrogen molecule led to the two forms, ortho and para, of hydrogen. Well, his idea sounded so convincing that I felt there was no point of my going into it. It was obvious to me that Heisenberg already knew the essentials; indeed, he soon wrote the paper on the subject that laid the groundwork for the modern theory of ferromagnetism. It was not until two years later that I somewhat embellished his treatment by the introduction of spinwaves.

Electrons in crystals

There was a greater challenge in his other suggestion, to do something more about the properties of metals. Going beyond the earlier work of Paul Drude and H. A. Lorentz, Wolfgang Pauli had already given a first new impetus to the field by invoking Fermi statistics to explain the temperature-independent paramagnetism of conduction electrons; Sommerfeld had gone further by discussing the consequences for the specific heat and the relation between the thermal and the electric conductivity of metals. Both, however, had treated the conduction electrons as an ideal gas of free electrons, which didn't appear in the least plausible to me.

When I started to think about it, I felt that the main problem was to explain how the electrons could sneak by all the ions

in a metal so as to avoid a mean free path of the order of atomic distances. Such a distance was much too short to explain the observed resistances, which even demanded that the mean free path become longer and longer with decreasing temperature. But Heitler and London had already shown how electrons could jump between two atoms in a molecule to form a covalent bond, and the main difference between a molecule and a crystal was only that there were many more atoms in a periodic arrangement. To make my life easy, I began by considering wavefunctions in a one-dimensional periodic potential. By straight Fourier analysis I found to my delight that the wave differed from the plane wave of free electrons only by a periodic modulation.

This was so simple that I didn't think it could be much of a discovery, but when I showed it to Heisenberg he said right away: "That's it!" Well, that wasn't quite it yet, and my calculations were only completed in the summer when I wrote my thesis on "The Quantum Mechanics of Electrons in Crystal Lattices."

I then left Leipzig to become for a year the assistant of Pauli in Zurich and to spend another year as Lorentz Fellow in Holland. It was not until the fall of 1930 that I returned to Leipzig, this time as Heisenberg's assistant, and by then the early days of quantum mechanics were really over, although many of its important consequences were yet to come—and are still coming.

I don't think many of us realized that we had just gone through quite a unique era; we thought that this was just the way physics was normally to be done and only

wondered why clever people had not seen that earlier. Almost any problem that had been tossed around years before could now be reopened and made amenable to a consistent treatment. To be sure, there were a few minor difficulties left, such as the infinite self-energy of the electron and the question of how it could exist in the nucleus before beta decay; and nobody had yet derived the numerical value of the fine-structure constant. But we were sure that the solutions were just around the corner and that any new ideas that might be called for in the process would be easily supplied in the unlikely event that this should be necessary. Well, the last fifty years have taught us at least to be a little more modest in our expectations.

Heisenberg the teacher and scientist

From what I have told about the year when I had the good fortune to be Heisenberg's first student it may already be evident that he stands in the center of my memories of this most formative period in my life as a physicist. It is not only that he suggested the theme of my thesis, but I owe it to him that I caught the real spirit of research and that I dared to take the first steps in learning how to walk. If I should single out one of his great qualities as a teacher, it would be his immensely positive attitude towards any progress and the encouragement he thereby conferred.

This does not mean that one always received praise from him and that, on occasions, he could not be quite severe. Once during my thesis work I became stuck on a rather awkward difficulty and hoped that he would help me out. But

WENTZEL

PAULI

after I had explained it to him he only said: "Now that you have analyzed the source of the trouble it can't be all that hard to see what to do about it."

Of course, I felt rather depressed, but just to get out of it I pushed once more and in some cumbersome way finally managed indeed to get over the obstacle. It was not the mathematical method but only physical content that ever mattered to Heisenberg. As to elegance he might have agreed with Ludwig Boltzmann's opinion that it was "best left to tailors and bootmakers."

Besides my year as Heisenberg's student, I spent the two more years, 1930–31 and 1932–33, in Leipzig until Hitler succeeded in forming a new Germany in his own frightful image. What followed is too well known for me to dwell upon, but I cannot refrain from one sad comment on human nature. The very devotion to their work and their detachment from the dark irrational passions spreading around them caught most of even the finest German scientists unprepared for the oncoming flood. Those who did not leave were with few exceptions swept along and were left, each in his own way, to struggle with their inner conflicts.

But my memories of Heisenberg belong to the happier time before those events. Many of them relate to entirely informal and anything-but-professional conversations on walks, in his ski hut in the Bavarian Alps or under other relaxed circumstances. These remain no less precious to me than our talks on physics, and I want to tell in conclusion about two of them that I remember most vividly.

Once I came back after dinner to my room in the Institute to finish some work. While I sat at my desk I heard Heisenberg, who was an excellent pianist, playing in his apartment under the roof of the building. It was already late at night when he came down to my room and said he just wanted to talk a little before going to bed after he had practiced a few bars of a Schumann concerto for three hours. And then he told me that Franz Liszt, when he was already a famous pianist, found that his scales of thirds and fifths were not smooth enough. So he cancelled all engagements, and for a year practiced nothing but these scales before he started to perform again. The reason I remember this so well is that I felt that Heisenberg, without intention, had told me something important about himself. The audience of Liszt after that year must have thought it a wonder how easily he was able to play those difficult scales. But the real wonder was of course that he had had the strength and the gift of concentration to keep on perfecting them incessantly for a whole year.

Now, one of the most marvellous traits of Heisenberg was the almost infallible intuition that he showed in his approach to a problem of physics and the phenomental way in which the solutions came to him as if out of the blue sky. I have asked myself whether that wasn't a form of the "Liszt phenomenon," and for that the more admirable. Not that Heisenberg would ever have cancelled all other activity for a year to master a special technique. But we all knew the dreamy expression on his face, even in his complete attention to other matters and in his fullest enjoyment of jokes or play, which indicated that in the inner recesses of the brain he continued his all-important thoughts on physics.

There is another remark he once made that I consider even more characteristic. We were on a walk and somehow began to talk about space. I had just read Weyl's book *Space, Time and Matter,* and under its influence was proud to declare that space was simply the field of linear operations.

"Nonsense," said Heisenberg, "space is blue and birds fly through it."

This may sound naive, but I knew him well enough by that time to fully understand the rebuke. What he meant was that it was dangerous for a physicist to describe Nature in terms of idealized abstractions too far removed from the evidence of actual observation. In fact, it was just by avoiding this danger in the previous description of atomic phenomena that he was able to arrive at his great creation of quantum mechanics. In celebrating the fiftieth anniversary of this achievement, we are vastly indebted to the men who brought it about: not only for having provided us with a most powerful tool but also, and even more significant, for a deeper insight into our conception of reality.

* * *

This article is an adaptation of a talk given 26 April 1976 at the Washington, DC meeting of The American Physical Society.

Electron diffraction: fifty years ago

A look back at the experiment that established the wave nature of the electron, at the events that led up to the discovery, and at the principal investigators, Clinton Davisson and Lester Germer.

Richard K. Gehrenbeck

PHYSICS TODAY / JANUARY 1978

An article that appeared in the December 1927 issue of *Physical Review,* "Diffraction of Electrons by a Crystal of Nickel," has been referred to in countless articles, monographs and textbooks as having established the wave nature of the electron—in principle, of all matter.[1] Now, fifty years later, it is fitting to look back at the events that led up to this historical discovery and at the discoverers, Clinton Davisson and Lester Germer. Figure 1 shows them in their lab in 1927, together with their assistant Chester Calbick.

A shy midwesterner

Clinton Joseph Davisson, the senior investigator, was born in Bloomington, Illinois, on 22 October 1881, the first of two children. His father, Joseph, who had settled in Bloomington after serving in the Civil War, was a contract painter and paperhanger by trade. His mother, Mary, occasionally taught in the Bloomington school system. Their home was, as Davisson's sister, Carrie, characterized it, "a happy congenial one—plenty of love but short on money."

Davisson, slight of frame and frail throughout his life, graduated from high school at age 20. For his proficiency in mathematics and physics he received a one-year scholarship to the University of Chicago; his six-year career there was interrupted several times for lack of funds. He acquired his love and respect for physics from Robert Millikan; Davisson was "delighted to find that physics was the concise, orderly science [he] had imagined it to be, and that a physicist [Millikan] could be so openly and earnestly concerned about such matters as colliding bodies."

Richard K. Gehrenbeck is an associate professor of physics and astronomy at Rhode Island College, Providence, Rhode Island.

Before finishing his undergraduate degree at Chicago, he became a part-time instructor in physics at Princeton University, where he came under the influence of the British physicist Owen Richardson, who was directing electronic research there. Davisson's PhD thesis at Princeton, in 1911, extended Richardson's research on the positive ions emitted from salts of alkaline metals. Davisson later credited his own success to having caught "the physicist's point of view—his habit of mind—his way of looking at things" from such men as Millikan and Richardson.

After completing his degree, Davisson married Richardson's sister, Charlotte, who had come from England to visit her brother. After a honeymoon in Maine Davisson joined the Carnegie Institute of Technology in Pittsburgh as an instructor in physics. The 18-hour-per-week teaching load left little time for research, and in six years there he published only three short theoretical notes. One notable break during this period was the summer of 1913, when Davisson worked with J. J. Thomson at the Cavendish laboratory in England.

In 1917, after he was refused enlistment in the military service because of his frailty, Davisson obtained a leave of absence from Carnegie Tech to do war-related research at the Western Electric Company, the manufacturing arm of the American Telephone and Telegraph Company, in New York City. His work was to develop and test oxide-coated nickel filaments to serve as substitutes for the oxide-coated platinum filaments then in use. At the end of World War I he turned down an offered promotion at Carnegie Tech to accept a permanent position at Western Electric. It was at this time that he began the sequence of investigations that ultimately led to the

discovery of electron diffraction; it was also at this time that he was joined by a young colleague, Lester Halbert Germer, just discharged from active service.

An adventurous New Yorker

Germer was born on 10 October 1896, the first of two children of Hermann Gustav and Marcia Halbert Germer, in Chicago, where Dr Germer was practicing medicine. In 1898 the family moved to Canastota in upper New York state, the childhood home of Mrs Germer. Germer's father became a prominent citizen in the little town on the Erie canal, serving as mayor, president of the board of education and elder in the Presbyterian church.

Germer attended school in Canastota and won a four-year scholarship to Cornell University, graduating from there in the spring of 1917, six weeks early because of the outbreak of the war. The local newspaper, after applauding 18-year-old Lester for working as a laborer for the local paving contractors during his summer vacation, proceeded to ridicule his lazier contemporaries for sitting "day after day in the lounging places of the village," saying there is "nothin' doin'" and that "a young feller has no chanst in this durn town." (Lester, must have taken a bit of ribbing from the "idle boys" after this appeared!) Germer's studies at Cornell were partly self-directed; in their junior year he and two classmates, finding themselves "unsatisfied with the course in electricity and magnetism given . . . bought a more advanced text and met regularly in the vacant class room . . . and really learned something."

Upon graduation from Cornell, Germer obtained a research position at Western Electric, which he held for about two months before volunteering for the Army (aviation section of the signal corps). He apparently made no contact with Davisson then. Lieutenant Germer, among those piloting the first group of airplanes on the Western Front, was officially credited with having brought down four German warplanes. Discharged on 5 February 1919, Germer was treated in New York City for severe headache, nervousness, restlessness and loss of sleep, conditions attributed to his military campaigns, but he refused to file for compensation because "others were worse off." After three weeks of rest, he was re-hired by Western Electric—and had as his first assignment the preparation of an annotated bibliography for a new project being directed by his new supervisor, Davisson.

That fall Germer married his Cornell sweetheart, Ruth Woodard of Glens Falls, New York.

Electron emission—in court

The assignment that engaged Davisson and Germer in their first joint effort reflects one of the chief interests of the parent company, AT&T, at this time: to conduct a fundamental investigation into the role of positive-ion bombardment in electron emission from oxide-coated cathodes. Although Germer later remembered this project as having been directly related to the famous Arnold–Langmuir patent suit, that occupied Western Electric (Harold Arnold) and General Electric (Irving Langmuir) from 1916 until it was finally settled[2] by the US Supreme Court (in favor of Western Electric) in 1931, a careful examination of the documents makes it clear that Davisson and Germer's project could have related to it only in a very indirect way. The patent case concerned improvements to the earliest deForest triode tubes with metallic (tungsten or tantalum) cathodes; it dealt with evidence obtained in the

years 1913 to 1916, before Davisson and Germer appeared on the scene. Nevertheless, because AT&T was deeply concerned about the efficiency and effectiveness of its triode amplifiers—key components in its recently constructed transcontinental telephone lines—Arnold assigned Davisson and Germer the task of conducting tests on oxide-coated cathodes. They published their results in the

Physical Review in 1920, concluding that positive-ion bombardment has a negligible effect on the electron emission from oxide-coated cathodes.[3]

With this problem settled, a related question came up: What is the nature of secondary electron emission from grids and plates subjected to electron bombardment? Davisson was assigned this new task and given an assistant, Charles

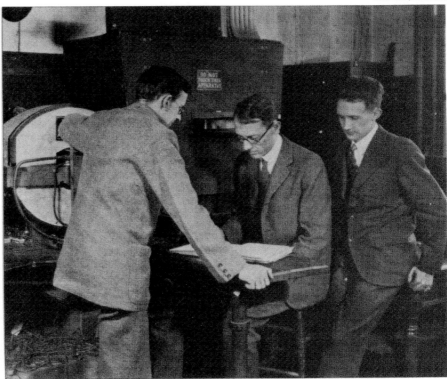

Davisson, Germer and Calbick in 1927, the year they demonstrated electron diffraction. In their New York City laboratory are Clinton Davisson, age 46; Lester Germer, age 31, and their assistant Chester Calbick, age 23. Germer, seated at the observer's desk, appears ready to read and record electron current from the galvanometer (seen beside his head); the banks of dry cells behind Davisson supplied the current for the experiments.   Figure 1

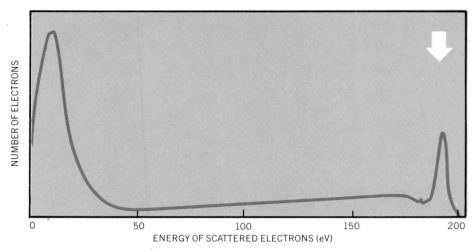

Electron-scattering peak. The energy of the scattered electrons varies from almost zero to that of the incident beam (indicated by the arrow). This is a reconstruction of the type of observation that led Davisson and Charles Kunsman to conclude that some electrons were being scattered elastically. Davisson saw these as possible probes of the electronic structure of the atom, in analogy to Rutherford's use of alpha particles to explore the nucleus.   Figure 2

H. Kunsman, a new PhD from the University of California. For this work they were able to convert the positive-ion apparatus to an electron-beam apparatus. Meanwhile Germer was shifted to a project on the measurement of the thermionic properties of tungsten, a topic he pursued for about four years, both under Davisson's direction and as part of a graduate program he undertook at nearby Columbia University part time.

A startling observation

Soon after Davisson and Kunsman began their secondary electron emission studies, they observed an unexpected phenomenon that was to have crucial importance for their future experimental program: A small percentage (about 1%) of the incident electron beam was being scattered back toward the electron gun with virtually no loss of energy—the electrons were being scattered elastically. Figure 2 reconstructs this phenomenon. Previous observers had noticed this effect for low-energy electrons (about 10 eV), but none had reported it for electrons of energies over 100 eV.

Although this discovery undoubtedly had no immediate impact on the stockholders of AT&T, it affected Davisson profoundly. To him these elastically scattered electrons appeared as ideal probes with which to examine the extranuclear structure of the atom. Ernest Rutherford announced his nuclear model of the atom in 1911, the year Davisson completed his PhD; Hans Geiger and Ernest Marsden completed their definitive experimental tests of Rutherford's theory and Niels Bohr announced his planetary model of the atom in 1913, when Davisson worked with Thomson at Cambridge. So it is not surprising that Davisson was enthusiastic about the prospect of using these electrons for basic research on the structure of the atom. In

Davisson's own words,

"The mechanism of scattering, as we pictured it, was similar to that of alpha ray scattering. There was a certain probability that an incident electron would be caught in the field of the atom, turned through a large angle, and sent on its way without loss of energy. If this were the nature of electron scattering it would be possible, we thought, to deduce from a statistical study of the deflections some information in regard to the field of the deflecting atom . . . What we were attempting . . . were atomic explorations similar to those of Sir Ernest Rutherford . . . in which the probe should be an electron instead of an alpha particle."

In fact, Davisson was so enthusiastic about a full-scale assault on the atom that he was able to convince his superiors to let him and Kunsman devote a large fraction of their time to it, and to give them the necessary shop backup.

The basic piece of apparatus, built to order by a talented machinist and glassblower, Geroge Reitter, was a vacuum tube with an electron gun, a nickel target inclined at an angle of 45° to the incident electron beam and a Faraday-box collector, which could move through the entire 135° range of possible scattered electron paths; it is diagrammed in figure 3. The Faraday box was set at a voltage to accept electrons that were within 10% of the incident electron energy.

After two months of experimentation, Davisson and Kunsman submitted a two-column paper to *Science*, in which they sketched the main features of their scattering program, presented a typical curve of their data, proposed a shell model of the atom for interpreting these results, and offered a formula for the quantitative prediction of the implications of the model.[4] Unfortunately their attempts to

link together their data, the model and the predictions were anything but definite—quite out of keeping with the Rutherford–Geiger–Marsden tradition.

Although Davisson (and Kunsman) must have been somewhat disappointed at the limited success of their initial venture, they pressed on with additional experiments. In the next two years they built several new tubes, tried five other metals (in addition to nickel) as targets, developed rather sophisticated experimental techniques at high vacuum ("the pressure became less than could be measured, i.e., less than 10^{-8} mm Hg,") and made valiant theoretical attempts to account for the observed scattering intensities. The results were uniformly unimpressive; several of the studies were not even published. In fact, the generally disheartened atmosphere that seems to have prevailed by the end of 1923 is indicated by the fact that Kunsman left the company and Davisson abandoned the scattering project.

A year later, however, Davisson was ready to have another try at electron scattering. Was this change of heart prompted by Davisson's strong attraction to the project? Was it his eagerness to obtain additional information about the extranuclear structure of the atom? In any case, in October 1924 Germer was put back on the scattering project in place of the departed Kunsman. Germer, who had already completed several thermionic-emission studies, was returning to Western Electric after a 15-month illness. Regarding his development as a physicist by this time, Germer later recollected:

"I learned relatively little at Columbia . . . but was nevertheless fortunate in working . . . with Dr C. J. Davisson. I learned a simply enormous amount from him. This included how to do experiments, how to think about them, how to write them up, how even to learn what other people had previously done in the field . . . I am quite certain that I do really owe to Dr Davisson much the best part of my education, and I am not really convinced that it is so inferior to that obtained in more conventional ways. It is certainly different."

A "lucky break" and a new model

So the scattering experiments were finally resumed. One can easily imagine, then, the feelings of disappointment and frustration that Davisson and Germer must have shared when, soon after the project had been restarted, they discovered a cracked trap and badly oxidized target on the afternoon of 5 February 1925, as the notebook entry in figure 4 shows. What it meant in simple terms was that the experiments with the specially polished nickel target, discontinued for almost a year, were to be delayed again. Apparently Germer's attempts to revitalize the tube after its long period of disuse by repumping and baking (out-gassing) were to be for nought; an additional delay for repairs was necessary.

This was not the only time that a tube had broken during a scattering experiment, nor was it to be the last. Nor was the method of repair unique, for the method of reducing the oxide on the nickel target by prolonged heating in vacuum and hydrogen had been used once before (unsuccessfully; that time it had led to the formation of a "black precipitate" and "no apparent cleaning up of the nickel"). This particular break and the subsequent method of repair, however, had a crucial role to play in the later discovery of electron diffraction.

By 6 April 1925 the repairs had been completed and the tube put back into operation. During the following weeks, as the tube was run through the usual series of tests, results very similar to those obtained four years earlier were obtained. Then suddenly, in the middle of May, unprecedented results began to appear, as shown in figure 5. These so puzzled Davisson and Germer that they halted the experiments a few days later, cut open the tube, and examined the target (with the assistance of the microscopist F. F. Lucas) to see if they could detect the cause of the new observations.

What they found was this: The polycrystalline form of the nickel target had been changed by the extreme heating until it had formed about ten crystal facets in the area from which the incident electron beam was scattered. Davisson and Germer surmised that the new scattering pattern must have been caused by the new crystal arrangement of the target. In other words, they concluded that it was the arrangement of the atoms in the crystals, not the structure of the atoms, that was responsible for the new intensity pattern of the scattered electrons.

Thinking that the new scattering patterns were too complicated to yield any useful information about crystal structure, Davisson and Germer decided that a large single crystal oriented in a known direction would make a more suitable target than a collection of some ten small facets randomly arranged. Because neither Davisson nor Germer knew much about crystals, they, assisted by Richard Bozorth, spent several months examining the damaged target and various other nickel surfaces until they were thoroughly familiar with the x-ray diffraction patterns (note!) obtained from nickel crystals in various states of preparation and orientation.

By April 1926 they had obtained a suitable single crystal from the company's metallurgist, Howard Reeve, and cut, etched and mounted it in a new tube that allowed for an additional degree of freedom of measurement; the collector could now rotate in azimuth (the 360° angle circling the beam axis) as well as in colatitude. The design of the new tube re-

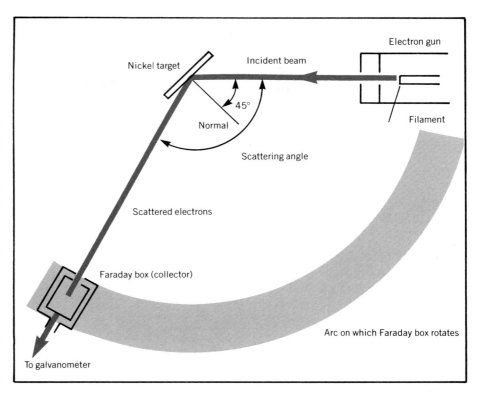

Scheme of the first scattering tube, which served as a prototype for the group's later models. Davisson and Germer later included mechanisms for rotating the target azimuthally 360° about the beam axis and for changing the angle of the incident beam with respect to the normal to the target. In their 1926–27 work the incident beam was perpendicular to the target face, and the scattering angle was called the "colatitude angle." Figure 3

flected their expectation of finding certain "transparent directions" in the crystal along which the electrons would move with least resistance. They expected these special directions to coincide with the unoccupied lattice directions.

More than a "second honeymoon"

Having suffered disappointment with the results of the original scattering experiments performed with Kunsman, Davisson must have been doubly disheartened by the meager returns he and Germer obtained with the new tube. After an entire year spent in preparation, and with a new tube and a new theory in hand, they obtained experimental results that were even less interesting than those from the earliest experiments. The new colatitude curves showed essentially nothing, and even the new azimuth curves gave at best only a weak indication of the expected three-fold symmetry of the nickel crystal about the incident beam.

Davisson must have been quite pleased with the prospect of getting away for a few months during the summer of 1926, when he and his wife had planned a vacation trip to relax and visit relatives in England. Mrs Davisson recalled that this summer had been chosen for the trip because her sister, May, and brother-in-law, Oswald Veblen of Princeton University, were available to stay with the Davisson children at that time. As Davisson wrote to his wife, then at the Maine cottage mak-

ing arrangements for the children: "It seems impossible that we will be in Oxford a month from today—doesn't it? We should have a lovely time—Lottie darling—It will be a second honeymoon—and should be sweeter even than the first." Something was to happen on this particular trip, however, to turn it into more than the "second honeymoon" Davisson envisioned.

Theoretical physics was undergoing fundamental changes at this time. In the early months of 1926 Erwin Schrödinger's remarkable series of papers on wave mechanics appeared, following Louis de Broglie's papers of 1923–24 and Albert Einstein's quantum-gas paper of 1925. These papers, along with the new matrix mechanics of Werner Heisenberg, Max Born and Pascual Jordan, were the subject of lively discussions at the Oxford meeting of the British Association for the Advancement of Science. Davisson, who generally kept abreast of recent developments in his field but appears to have been largely unaware of these recent developments in quantum mechanics, attended this meeting. Imagine his surprise, then, when he heard a lecture by Born in which his own and Kunsman's (platinum-target) curves of 1923 were cited as confirmatory evidence for de Broglie's electron waves![5]

After the meeting Davisson met with some of the participants, including Born and possibly P.M.S. Blackett, James

The notebook entry for 5 February 1925 records, in Germer's handwriting, the discovery of the broken tube that interrupted the scattering experiments once again. It was this break, however, which initiated a chain of events that eventually led to the preparation of a single crystal of nickel as the target, and to a shift of Davisson's interest from atomic structure to crystal structure. Reproduced by courtesy of Bell Laboratories.   Figure 4

Franck and Douglas Hartree, and showed them some of the recent results that he and Germer had obtained with the single crystal. There was, according to Davisson, "much discussion of them." All this attention might seem strange in light of the relatively feeble peaks Davisson and Germer had obtained, but even these may have been exciting to physicists already convinced of the basic correctness of the new quantum theory. It may also reflect the fact that several European physicists, Walter Elsasser (Göttingen), E.G. Dymond (Cambridge, formerly Göttingen and Princeton), and Blackett, James Chadwick and Charles Ellis of Cambridge[6] had attempted similar experiments and abandoned them because of the difficulties of producing the required high vacuum and detecting the low-intensity electron beams. Apparently they were encouraged by these results, which appeared so unimpressive to Davisson. At any rate, Davisson spent "the whole of the westward transatlantic voyage . . . trying to understand Schrödinger's papers, as he then had an inkling . . . that the explanation might reside in them"—no doubt to the detriment of the "second honeymoon" in progress.

Back at Bell Labs (as the engineering arm of Western Electric has been called since 1925), Davisson and Germer examined several new curves that Germer had obtained during Davisson's absence. They found a discrepancy of several degrees between the observed electron intensity peaks and the angles they expected from the de Broglie–Schrödinger theory. To pursue this matter further they cut the tube open and carefully examined the target and its mounting. After finding that most of the discrepancy could be accounted for by an accidental displacement of the collector-box opening, they "laid out a program of thorough search" to pursue the quest of diffracted electron beams. In typical Davisson fashion, however, this quest was preceded by a period of careful preparation, including an important change in the experimental tube. As Davisson wrote to Richardson in November,

"I am still working at Schrödinger and others and believe that I am beginning to get some idea of what it is all about. In particular I think that I know the sort of experiment we should make with our scattering apparatus to test the theory."

Found—a "quantum bump"

It was three weeks before the "thorough search" was begun. The importance that Davisson (and Bell Labs) had come to attach to this project can be surmised from the addition to it of a new assistant, Chester Calbick, a recently graduated electrical engineer. After about a month of experimenting, during which time Calbick took charge of operating the experiment, they gave the newly prepared tube a thorough set of consistency tests. During one attempt by Germer to reactivate the tube in late November the tube broke, but with little damage. (Strangely, little damage can be considered "lucky" in this case, whereas it would have been "unlucky" in the case of the 1925 break!)

The first experiments with the new tube yielded no significant results; the colatitude and azimuth curves looked much as before, and the new experiments added by Davisson "to test the theory" were uninformative as well. These tests consisted of varying the accelerating voltage, and hence electron energy E, for fixed colatitude and azimuth settings, and were designed to see if any effect could be discerned for a changed electron wavelength λ, according to the de Broglie relationship, $\lambda = h/(2\,mE)^{1/2}$.

A concerted search for "quantum peaks" (voltage-dependent scattered electron beams) was launched by late December. These attempts revealed only "very feeble" peaks. The situation changed dramatically on 6 January 1927, however; the data for that day are accompanied by the remark, in Calbick's neat handwriting: "Attempt to show 'quantum bump' at an intermediate [colatitude] angle. Bump develops at 65 V, compared with calculated value for 'quantum bump' of V = 78 V." Then, stretched across the bottom of the page in Germer's unmistakable bold strokes, is the additional remark: "First Appearance of Electron Beam." A portion of the notebook page is reproduced in figure 6.

The data for this curve are extremely interesting. Noting from the figure that the readings were taken in one-volt intervals on either side of 79 volts, whereas the steps are 2, 5 and then 10 volts elsewhere, we see that a peak was expected at about 78 volts. But the experiment yielded a single large current at 65 volts. The experimenters took immediate notice of this spike, making a second run in one-volt steps around 65 volts, which on a graph shows a clear peak centered on 65 volts. It is easy to imagine the excitement that must have accompanied this sudden turn of events, moving Germer to sprawl his glad tidings across the bottom of the page!

With this single critical result in hand, the experimental situation changed suddenly. The next day, 7 January, they ran several additional voltage curves, one for each of four different colatitude positions. A voltage peak appeared at a colatitude angle of 45° that was even greater than that at 40°, where the collector had been set the previous day. On the eighth, a new colatitude curve was run at a voltage of 65 volts, and the first true and unmistakable colatitude peak was observed—this was what Davisson had been looking for since 1920! Skipping Sunday, they next ran an azimuth curve at 65 volts and a colatitude of 45°. This time the threefold azimuthal symmetry was immediately apparent. Figure 7 shows these curves.

The experiments that were carried out during the next two months show that Davisson, Germer and Calbick, having finally found and positively identified one set of electron beams, could now find and identify others quickly. This block of experiments continued through 3 March, when Calbick left for a month on family business. Comparing this with earlier periods of Davisson's long contact with electron scattering, we see that not since the early days of the original Davisson–Kunsman experiments had there been such intense and concentrated effort in a single well defined direction. The presence of a clear, unambiguous goal certainly must have been a major factor in the two cases, an ingredient lacking at other times.

Another factor undoubtedly urging Davisson on to rapid (but careful) experimentation and possible early publication was his feeling that others might be pursuing similar investigations at that time. Recalling his conversations at Oxford and the comments that had been made about the interest of others in this matter, he sent off an article to Richardson in March with the accompanying note:

"I hope you will be willing, if you think it at all desirable, to get in touch with the editor of *Nature* with the idea of securing early publication. We know of three other attempts that have been made to do this same job, and naturally we are somewhat fearful that someone may cut in ahead of us." As it turned out these efforts had long been abandoned, but he had no way of knowing that. Nevertheless, another investigator, unknown to Davisson at that time, was indeed making progress at revealing the phenomena of electron diffraction with high-voltage electrons and thin metal foils. This was J.J.'s son, G.P. Thomson; his and Andrew Reid's first note was published in *Nature* just one month after Davisson and Germer's.[7]

A conservative note and a bold one

Davisson and Germer's *Nature* article was an extremely conservative expression of the new experimental evidence for electron diffraction.[8] Its title, "The Scattering of Electrons by a Single Crystal of Nickel," bears a closer connection to the early work of Davisson and Kunsman than it does to the new wave mechanics. Although the paper included a table linking the scattered electron peaks to the corresponding de Broglie wavelengths, it was not until the last two paragraphs that a tentative suggestion was made about the important implications of the work: The results were "highly suggestive . . . of the ideas underlying the theory of wave mechanics."

This cautious attitude may have been due to the problem that Davisson and Germer had in making the proper correlation between their data points and the theory; they found it necessary to hypothesize an *ad hoc* "contraction factor" of about 0.7 for the nickel-crystal spacing to get approximate correspondence between the de Broglie wavelengths and their data. Even at that, only eight of the thirteen beams described were clearly amenable to this analysis.

This cautious attitude appears to have been abandoned in a concurrent article by Davisson alone for an in-house publication, the *Bell Labs Record*.[9] The very title, "Are Electrons Waves?" suggests this difference. After reviewing the evidence that led Max von Laue to think of x rays as being wave-like, he cited his and Germer's recent work with electrons, urging a similar conclusion in this case. Although this article gave its readers no actual data on the experimental evidence for electron waves, it clearly indicates that

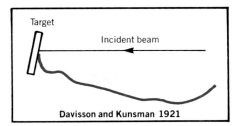
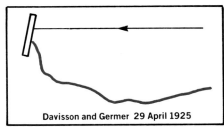
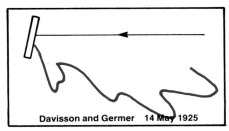

Target
Incident beam
Davisson and Kunsman 1921

Davisson and Germer 29 April 1925

Davisson and Germer 14 May 1925

Before and after the accident of 5 February 1925. Although the first scattering curves after the repair of the broken tube (middle curve) resembled the 1921 results of Davisson and Kunsman (top curve), striking peaks soon made a sudden appearance (bottom). This development led Davisson and Germer to make a major change in their program. **Figure 5**

Davisson's thoughts (and certainly Germer's as well) on the subject were not nearly as reserved as the *Nature* article suggests.

One other public announcement of the recent discoveries was made at this time. In a paper presented at the Washington meeting of The American Physical Society on 22–23 April 1927 and abstracted in the *Physical Review* in June,[10] Davisson and Germer basically repeated what they had stated in their *Nature* article, and then added an intriguing final paragraph. Referring to the three anomalous beams that could not be fitted into the analysis in the *Nature* article, they suggested that these "offer strong evidence that there exists in this crystal a structure which has not been hitherto observed for nickel." This statement implies Davisson and Germer had already gone beyond the point of using the "known" structure of the nickel crystal to find out about the possibility of the wave properties of the electron; they were now using the "known" electron waves to learn new facts about the nickel crystal. Between March, when the *Nature* article was submitted, and April, when the *Phys. Rev.* abstract was prepared, results that had been embarrassing to the theory had become a potential new application of that very theory!

True to form, however, Davisson and Germer did not sit back and rest on a "job well done"; they recognized the considerable work necessary to resolve a number of questions still outstanding. Among these were:
▸ the problem of the "anomalous" beams mentioned above,
▸ the *ad hoc* "contraction factor" that they had found necessary to attribute to the nickel crystal and
▸ extension of their electron energies over a greater range, and sharpening and refining their diffraction peaks.

Instant acclaim

Toward this end they initiated an extensive experimental and theoretical attack that lasted from 6 April (when Calbick returned from his month's absence) until 4 August. At that time the tube was cut open for a final careful examination of the target and the other tube components. As it turned out, this intention was foiled when, in the process of being brought back to room temperature, the tube "blew up and [was] partially ruined . . . the leads being broken, filament also, and a large part of the nickel oxidized." A broken tube had served to initiate the decisive experiments on 5 February 1925, and a broken tube ended them on 4 August 1927, two and a half years later.

The most interesting of this last group of experiments was a series designed to investigate "the anomalous peaks after bombardment," which appeared for a restricted period of time after the target had been heated by bombardment. The experiments showed that the nature—even the existence—of certain beams was not static but varied with temperature and time (and hence conditions of the target in terms of occluded gases). The notebook entries include a great variety of different terms, diagrams and calculations designed to try to make sense out of these data. Davisson and Germer found a "gas crystal" model, in which "gas atoms fit into the crystal," to be the most effective.

The task of welding data and interpretation into a comprehensive report for publication was begun in mid June, well before the experiments were completed. It appears that Davisson was responsible for most, if not all, of the writing; in a letter to his family at the summer cottage he wrote:

"I'm busy these days writing up our experiment—It's an awful job for me. I didn't get much done yesterday as Prof. Epstein from Pasadena turned up and had to be entertained and shown things—and today I'm too sleepy [after having spent last evening at the theater with Karl Darrow]. However, I must keep at it."
More than three weeks later (23 July) he

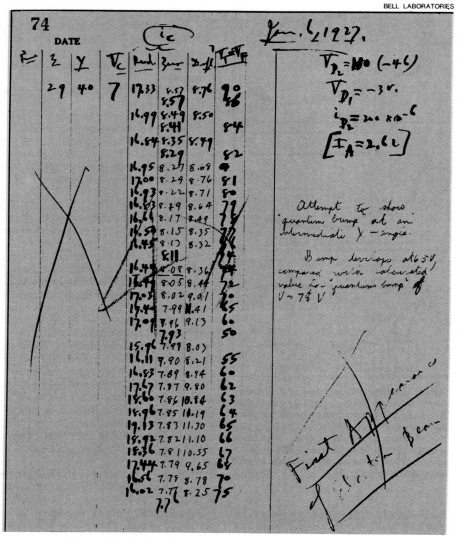

The sixth of January 1927 might well be regarded as the birthday of electron waves, for it was the day that data directly supporting the de Broglie hypothesis of electron waves were first observed. Note the peak deflection at 65 volts, and the detailed study of the region directly below. Calbick's handwriting is neat and cautious; Germer's is bold and expansive. Davisson made no entries in any of the research notebooks kept in the Bell Labs files.   Figure 6

was at last able to exclaim,

"I finished the first draft of our paper this morning. It is going to take a lot of going over and revising ... I will leave [the drawings] to Lester—and also the thing is full of blanks in which he will have to stick in the right numbers."

A week later he made his final changes before departing for Maine. Germer, too, needed a break, and after finishing his tasks he left on 14 August for a canoe trip with several friends. The final copy was sent to the *Physical Review* in August and the article appeared in December.

The paper itself was a detailed, comprehensive report on experiments performed, conclusions reached and questions left unanswered. One of the significant features of the paper was its thoughtful examination of the possible ways of interpreting the systematic differences between observed and calculated electron wavelengths (either the sug-

gested "contraction factor" or an "index of refraction" proposed by Carl Eckart, A.L. Patterson, and Fritz Zwicky in independent responses to the *Nature* article).[11] Summarizing the evidence, the paper concluded that of the 30 beams that had been observed, 29 were adequately accounted for by attributing wave properties to free electrons. It acknowledged, however, that the wave assumption implied the existence of eight additional beams, which had not been observed.

The discrepancies between theory and experiment, apparently fairly minor, that Davisson and Germer recorded, evidently did not reduce their fundamental belief that free electrons behave like waves. The physics community appears to have concurred, for I have not found a single voice raised in opposition. This may well have been due as much to the success of the earlier theory of wave mechanics and the acceptance of a wave–particle duality for light as to the force of the evidence

inherent in the paper itself.

This may be illustrated by some remarks made by prominent physicists prior to the publication of the *Phys. Rev.* article. In the reports and discussions of the fifth Solvay Conference held in Brussels in October 1927, Niels Bohr, de Broglie, Born, Heisenberg, Langmuir and Schrödinger all hailed the experiments of Davisson and Germer (as described in the *Nature* article) as being, in the words of de Broglie, "very important results which [appear] to confirm the general provisions and even the formulas of wave mechanics."[12] Bohr, speaking before the International Congress of Physics assembled in Como, Italy, on 16 September 1927, drew upon these experiments in establishing his views on complementarity:

"... the discovery of the selective reflection of electrons from metal crystals ... requires the use of the wave theory superposition principle ... Just as in the case of light ... we are not dealing with contradictory but with *complementary* pictures of phenomena."[13]

Planck, addressing the Franklin Institute on 18 May 1927, even *before* he had heard of the Davisson–Germer results, stated about the electron: "[Its] motion [in the atom] resembles ... the vibrations of a standing wave ... [Thanks] to the ideas introduced into science by L. de Broglie and E. Schrödinger, these principles have already established a solid foundation."[14] Yet in the same address Planck stated that he was *still* (in 1927, four years *after* the decisive Compton experiments) reluctant to accept the corpuscular implications for electromagnetic radiation inherent in his own quantum hypothesis! It appears that physicists were willing to accept the experimental evidence for electron waves almost before those experiments were performed!

The world that is physics

Davisson and Germer succeeded where others had failed. In fact, the others mentioned above (Elsasser, Dymond, Blackett, Chadwick and Ellis), who had the *idea* of electron diffraction considerably *ahead* of Davisson and Germer, were *not able* to produce the desired experimental evidence for it. G.P. Thomson, who did find that evidence by a very different method, testified to the magnitude of the technical achievement as follows:

"[Davisson and Germer's work] was indeed a triumph of experimental skill. The relatively slow electrons [they] used are most difficult to handle. If the results are to be of any value the vacuum has to be quite outstandingly good. Even now [1961] ... it would be a very difficult experiment. In those days it was a veritable triumph. It is a tribute to Davisson's experimental skill that only two or three other workers have used slow

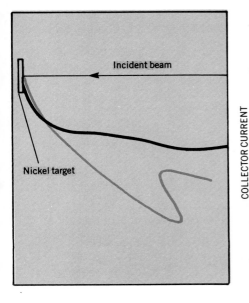

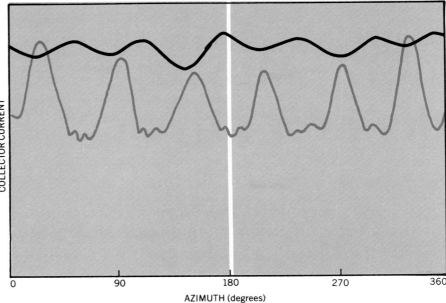

New colatitude and azimuth curves. The black lines show the appearance of the colatitude (left) and azimuth (right) distributions of the scattered electrons when Davisson took the curves to England in 1926. The colored curves are from data taken after 6 January 1927, when the first "quantum bump" was observed. The azimuth curves also confirm the threefold symmetry of the nickel crystal.

Figure 7

electrons successfully for this purpose."[15]

Davisson and Thomson shared in the Nobel Prize for physics in 1937 for their accomplishments. Germer and Reid, as junior partners to Davisson and Thomson, did not share in the prize. Reid was tragically killed in a motorcycle accident shortly after his and Thomson's definitive papers appeared in 1928.

Davisson and Germer actively pursued the topic of electron diffraction for about three years after 1927, publishing, together and separately, about twenty more papers on the subject; reference 16 gives three of the most important. By the early 1930's, both Davisson and Germer had turned to new fields: Davisson to electron optics (including early television); Germer to high-energy electron diffraction and later still to electrical contacts. Davisson retired from Bell Labs in 1946 and spent the remaining twelve years of his life in Charlottesville, Virginia, summering as usual in Maine. Germer regained his interest in low-energy electron diffraction in 1959–60, at which time he and several co-workers at Bell Labs perfected a technique, eventually referred to as the "post-acceleration" technique,[17] which had been devised in 1934[18] and then abandoned, by Wilhelm Ehrenberg. With this work Germer was able to follow up with great success the study of surfaces, to which he had been attracted in his original work with Davisson; the field of low-energy electron diffraction (LEED) is now widespread and very active. Germer retired from Bell Labs in 1961 and remained active in this "new" field and in his favorite recreation, mountain climbing, until his death in 1971.

In trying to answer the question of

"Why Davisson and Germer, and not someone else?" one's thoughts leap to such things as the "luck" of the broken tube in 1925 and the trip to England in 1926. Davisson and Germer themselves freely admitted the key importance of these events. But to dwell on them exclusively would be a mistake. Neither of these events would even have been remembered had they not been followed by thorough, careful and creative experiment and reflection. Perhaps of equal importance is the habit of attention to technical detail established by Davisson in his student days and extended in the long series of Davisson–Kunsman and earlier Davisson–Germer experiments. Another important factor is the time for pure research provided by Western Electric–Bell Labs, and the technical support in areas such as high vacua and electrical detection techniques available at that industrial laboratory.

All in all, this case history on the discovery of electron diffraction appears to illustrate the complex nature of the world that is physics, the difficulty of singling out any one factor as being responsible for a great discovery, and the importance of establishing and nurturing the ties that bind together the generations of physicists, as well as the physicists of each generation.

References

References to correspondence, personal remarks and other archival material are documented in the author's PhD dissertation, "C.J. Davisson, L. H. Germer, and the Discovery of Electron Diffraction," University of Minnesota, 1973, available from Xerox University Microfilms, 3000 North Zeeb Road, Ann Arbor, Michigan 48106, Order No. 74–10 505.

1. C. J. Davisson, L. H. Germer, Phys. Rev. **30**, 705 (1927).
2. US Reports **283**, 665 (1931).
3. C. J. Davisson, L. H. Germer, Phys. Rev. **15**, 330 (1920).
4. C. J. Davisson, C. H. Kunsman, Science **54**, 523 (1921).
5. M. Born, Nature **119**, 354 (1927).
6. W. Elsasser, Naturwissenschaften **13**, 711 (1925); E. G. Dymond, Nature **118**, 336 (1926).
7. G. P. Thomson, A. Reid, Nature **119**, 890 (1927).
8. C. J. Davisson, L. H. Germer, Nature **119**, 558 (1927).
9. C. J. Davisson, Bell Lab. Record **4**, 257 (1927).
10. C. J. Davisson, L. H. Germer, Phys. Rev. **29**, 908 (1927).
11. C. Eckart, Proc. Nat. Acad. Sci. **13**, 460 (1927); A. L. Patterson, Nature **120**, 46 (1927); F. Zwicky, Proc. Nat. Acad. Sci. **13**, 518 (1927).
12. L'Institut International de Physique Solvay, *Electrons et Protons: Rapports et Discussions du Cinquieme Conseil de Physique*, Gauthier-Villars, Paris (1928), pages 92, 127, 165, 173, 274, 288.
13. N. Bohr, *Atomic Theory and the Description of Nature*, Macmillan, New York (1934), page 56; italics supplied.
14. M. Planck, J. Franklin Inst. **204**, 13 (1927).
15. G. P. Thomson, *The Inspiration of Science*, Oxford U.P., London (1961); reprinted by Doubleday, Garden City, New York (1968), page 163.
16. C. J. Davisson, L. H. Germer, Proc. Nat. Acad. Sci. **14**, 317 (1928); Proc. Nat. Acad. Sci. **14**, 619 (1928); Phys. Rev. **33**, 760 (1929).
17. A. U. MacRae, Science **139**, 379 (1963).
18. W. Ehrenberg, Philosoph. Mag. **18**, 878 (1934).

□

1932—Moving into the new physics

The exciting events of the early 1930's raised high hopes
for progress in nuclear physics and, before the end of the decade,
had changed its pace, scale, cost and social applications.

Charles Weiner

PHYSICS TODAY / MAY 1972

In 1972 we celebrate the fortieth anniversary of the "annus mirabilis" of nuclear and particle physics. Seen from the perspective of the present, the cluster of major conceptual and technical developments of 1932 mark that "marvelous" year as a very special one. It began with Harold Urey's announcement in January that he had discovered a heavy isotope of hydrogen, which he called "deuterium." In February James Chadwick demonstrated the existence of a new nuclear constituent, the neutron. In April John Cockcroft and E. T. S. Walton achieved the first disintegration of nuclei by bombarding light elements with artificially accelerated protons. In August Carl Anderson's photographs of cosmic-ray tracks revealed the existence of another new particle, the positively charged electron, soon to be called the "positron." And later that summer Ernest Lawrence, Stanley Livingston and Milton White disintegrated nuclei with the cyclotron, an instrument that would generate almost 5-million electron volts by the end of that eventful year.

New particles, new constituents of the nucleus and powerful new techniques for probing its structure—they all provided a wealth of fresh challenges and opportunities for theory and experiment. Physicists who remember the excitement of those days sometimes sound as if they were relishing an excellent wine when they smile and comment: "It was a great year."

What were the circumstances and the immediate impact of these events? Was their significance recognized at the time? And what effect did they have in the decade that followed? Historians ask such questions in the hope that the answers may reveal more about the nature of scientific activity and the processes and consequences of scientific change than is evident in a mere listing of key discoveries. Particularly interesting are the social structures and processes that helped create the environment for doing nuclear physics and influenced its relationships to the scientific community and to the larger society in which it functions. The events of 1932 helped open new fields of research and led to important changes in the pace, scale, cost, organization and style of experimental physics research. In addition, the rapid growth of nuclear physics gave rise in the 1930's to public expectations of applications, expectations that were to be fulfilled in unanticipated ways before the end of the decade.

These developments are being illuminated through historical documentation and research studies underway at the American Institute of Physics Center for History and Philosophy of Physics. Here I shall draw on some of the results to provide glimpses of the circumstances of the 1932 discoveries and the immediate effect on some of the discoverers and their colleagues. Wherever possible, these individuals will speak for themselves, in excerpts from the letters they exchanged or from interviews I have more recently conducted with several of them. I shall also sketch the effect of the 1932 events on the growth of nuclear-physics research in the US in the 1930's and indicate briefly the special role of Lawrence's Berkeley laboratory: We shall see that one of the most striking effects of the "annus mirabilis" was its impact on the social organization and support of academic research.

These glimpses cannot provide a full or balanced picture, nor even a chronological listing of the many interconnected conceptual, technical and social factors involved. But they do offer some insight into the spirit of the times.

News from the US

The stage was set at the very beginning of 1932, and the action that was soon to unfold into the dramatic developments of that year was already underway. Some of the highlights of the developing situation in the US are seen in a letter written on 8 January by Joseph Boyce of Princeton to John Cockcroft, his friend and former colleague at the Cavendish Laboratory in Cambridge:

"I have just been on a very brief visit in California and thought you might be interested in a brief report on high voltage work there and in the eastern US as well. At Pasadena [Charles] Lauritsen continues work with his 700 000 volt x-ray tube. . . . He is now waiting for the GE to furnish him other transformers to go to still higher voltages. [Robert] Millikan and Anderson are working a Wilson chamber between the poles of a very large magnet and are obtaining cosmic ray recoil and disintegration tracks whose curvatures can be measured. . . . Everyone seems most enthusiastic about [the results], even people outside Pasadena. Some of the photographs show simultaneous ejection of (+) and (−) particles of high speed, as if

Charles Weiner is professor of History of Science and Technology at MIT.

29 March 1934
Copenhagen.

Dear Cockcroft !

Thank you very much for your letter with last news from nuclear world. It is always a great excitation in de Institute when your letters come and a spatial commision of english-speaking pleople and spetialists on egiptian and babilonian scripture discuss four hours the questions wether „Wevser" means „however" or „hidrogen" and what is to be understood under de notation: „Wexler". After these difficult philological questions are settled de tekst comes into hands of physicists.

Well, I am very interested to hear about your mesurements of energy-distribution in

present ! I should also expect that some discrepancies in de explanation of nuclear reactions may be explained by hypothesis that the isomeric nuclei are build up in different cases. What do you dink about it?
With best wishes to Lord Rutherford and his children
Yours trully G. Gamow.

P.S. On the way to USA we stop in London for two days (11 & 12 June) and I hope to pay you a visit in one of dese days.

both a proton and an electron were knocked out from a nucleus by the cosmic ray.... With that and the high voltage developments everywhere it looks as if cosmic ray work will become a laboratory problem for a while rather than a mountain-climbing excuse."[1]

Boyce's news about the Caltech cosmic-ray photographs and the possible nuclear reactions involved had in fact already been brought to Cambridge by Millikan, who was visiting at the Cavendish Laboratory in November 1931 when he received copies of the photographs in a letter from Anderson. After describing the puzzling tracks—for example, "a *positive* particle moving *downward* or an *electron* moving *upward*"—Anderson concluded: "A hundred questions concerning the details of these effects immediately come to mind.... It promises to be a fruitful field and no doubt much information of a very fundamental character will come out of it...."[2]

The fundamental information did "come out" in August 1932, when Anderson identified the curious tracks in some new photographs as evidence of a "positively-charged particle comparable in mass and in magnitude of charge with an electron."[3]

Boyce's January 1932 letter went on:

"But the place on the coast where things are really going on is Berkeley. Lawrence is just moving into an old wooden building back of the physics building where he hopes to have six different high-speed particle outfits. One is to move over the present device by which he whirls protons in a magnetic field and in a very high frequency tuned electric field and so is able to give them

John Cockcroft and George Gamow (right) work on a nuclear-physics problem in December 1933. Gamow's theoretical ideas of 1928 spurred on the building of the Cockcroft–Walton accelerator, which in 1932 achieved the first nuclear disintegration by artificially accelerated particles. In a letter written 29 March 1934, Gamow exchanges the latest news with Cockcroft. Cockcroft's minuscule handwriting and Gamow's unique English spelling were notorious but apparently did not interfere with communication between the two men. (Photo, by K. T. Bainbridge, from Niels Bohr Library; original letter in Churchill College Library, Cambridge.)

velocities a little in excess of a million volts. With this he has already had proton currents of the order of 10^{-9} amps.... Then there is the Hg ion outfit.... This has already given Hg ions in excess of a million volts, by the use of about 50 000 volts high frequency.... Several more units can be added to it, all driven by a master oscillator. Then a similar device with higher applied voltages and longer electrodes to use with protons. The fourth is a whirling device for protons in a magnet with pole pieces 45 inches in diameter, with which he hopes for at least 3 million volts, perhaps more.... Then a small tesla-coil x-ray outfit is already installed, and the remaining room is reserved for a Van de Graaff electrostatic generator. On paper this sounds like a wild damn fool program, but Lawrence is a very able director, has many graduate students, adequate financial backing, and in his work so far with protons and mercury ions has achieved sufficient success to justify great confidence in his future...

Back in the east [Merle] Tuve at Washington [Carnegie Institution] is working on the development of tubes to stand high voltages, and has ordered a six foot sphere to build a one-ball Van de Graaff outfit for about 3 million [volts]. I think I sent you clippings about Van's [Robert Van de Graaff's] own results and plans...

On the way west I stopped at New Orleans for the Physical Society meeting. The most interesting paper was Urey's on the hydrogen isotope. The spectroscopic evidence alone, as reported in the abstract is quite convincing, but [Walker] Bleakney in our [Princeton] laboratory has been able to confirm it with a mass spectrograph..."[1]

Discovery of the neutron

All of these developments described by Boyce were of great interest to the physicists in the Cavendish Laboratory, where work aimed at probing the nature and structure of the nucleus had been pursued under Ernest Rutherford for more than a decade. These efforts began to pay off dramatically early in 1932.[4] James Chadwick had been searching for the neutron ever since Rutherford had suggested in his Bakerian Lecture in 1920 that such a particle might exist. He followed up observations made in 1930 by two German scientists, Walther Bothe and H. Becker, which were subsequently extended at the end of 1931 in Paris by Frédéric and Irène Joliot-Curie. Chadwick's own recollections of the circumstances provide some of the flavor of the event:

"One morning I read the communication of the Curie-Joliots in the *Comptes Rendus,* in which they reported a still more surprising property of the radiation from beryllium, a most startling property. Not many minutes afterwards [Norman] Feather came to my room to tell me about this report, as astonished as I was. A little later that morning I told Rutherford. It was a custom of long standing that I should visit him about 11 a.m. to tell him any news of interest and to discuss the work in progress in the laboratory. As I told him about the Curie-Joliot observa-tion and their views on it, I saw his growing amazement; and finally he burst out 'I don't believe it.' Such an impatient remark was utterly out of character, and in all my long association with him I recall no similar occasion. I mention it to emphasize the electrifying effect of the Curie-Joliot report. Of course, Rutherford agreed that one must believe the observations; the explanation was quite another matter.

It so happened that I was just ready to begin experiment, for I had prepared a beautiful source of polonium from the Baltimore material [used radon tubes brought back by Feather]. I started with an open mind, though naturally my thoughts were on the neutron. I was reasonably sure that the Curie-Joliot observations could not be ascribed to a kind of Compton effect, for I had looked for this more than once. I was convinced that there was something quite new as well as strange. A few days of strenuous work were sufficient to show that these strange effects were due to a neutral particle and to enable me to measure its mass: the neutron postulated by Rutherford in 1920 had at last revealed itself."[5]

Chadwick's letter announcing the discovery was to appear in *Nature* on 27 February, 1932[6] and on 24 February he sent proofs of the letter to Niels Bohr in Copenhagen. Bohr then invited Chadwick to come and discuss his work at the small informal conference that had been planned for the second week of April at the Copenhagen institute.[7] These annual week-long conferences had been started in 1929 and

Participants in the April 1932 conference at Niels Bohr's Institute of Theoretical Physics in Copenhagen. Seated in the first row are Léon Brillouin (left), Lise Meitner and Paul Ehrenfest. Seated behind and to the right of Ehrenfest is H. A. Kramers. The first six people, from the left, standing along the wall are Werner Heisenberg, Piet Hein, Niels Bohr, Léon Rosenfeld, Max Delbrück and Felix Bloch. Seated second from the right in the last row is P. A. M. Dirac, with R. H. Fowler on his right. Other visitors to Copenhagen in the group include Walter Heitler, Karl von Weiszäcker, Guido Beck and C. G. Darwin. (Photo: Niels Bohr Institute, Copenhagen.)

they brought together physicists from many different countries to discuss, as Bohr put it, "actual atomic problems."

Chadwick was unable to attend the meeting, but R. H. Fowler of Cambridge was present and provided an up-to-the-minute account of the experimental work underway by Chadwick, Feather and P. I. Dee in their follow-up of Chadwick's discovery. The conference was truly international: The 22 foreign physicists were from 17 institutions in nine countries. Among the participants were C. G. Darwin, Max Delbrück, Paul Ehrenfest, P. A. M. Dirac, R. H. Fowler, Werner Heisenberg, Walter Heitler, H. A. Kramers and Lise Meitner.[8] Bohr's personal style of thinking out loud set the tone for the Copenhagen conferences and stimulated a lively exchange of information, ideas and interpretations. The neutron, like the other topics discussed at the meeting, found a place in the parody of Faust written and performed there by some of the participants:

"Now a reality,/Once but a vision.
What classicality,/Grace and precision!
Hailed with cordiality,/Honored in song,
Eternal neutrality/Pulls us along!"[9]

In June, only two months after the Copenhagen conference, Heisenberg submitted the first in a three-part series of papers that incorporated the neutron in a theory of the nucleus to demonstrate that quantum mechanics could be applied to many existing nuclear problems.[10] That summer he was a lecturer at the University of Michigan's annual summer schools in theoretical physics, which attracted physicists from all over the US and Europe. In November Samuel Goudsmit wrote to Bohr from Michigan, commenting on Heisenberg's lectures:

"We followed with great interest his new ideas about the nucleus but everyone feels that there still are great difficulties. It is strange and regrettable that the discovery of the neutron did not give some more fertile clues for progress. In many respects the situation has not changed much from what it was at the Rome meeting a year ago, except that the difficulties can now be formulated more sharply. I have been playing around with nuclear magnetic moments, but none of my speculations yielded any results certain enough to communicate."[11]

Bohr replied:

"Not least in connection with the [difficulties of relativistic quantum mechanics] we have all been very interested [in] the problem of nuclear constitution and the possible clue to this problem offered by the discovery of the neutron. Still I quite agree with you as regards the very preliminary character of any attempt hitherto made to attack the problem on such lines."[12]

An acceptable theory of the nucleus was still beset with difficulties by the end of 1932, but the neutron did attract theorists to nuclear problems because it provided fresh challenges and possibilities for theory. One senior nuclear theorist recently explained: "I went into nuclear physics only after 1932 ... after the discovery of the neutron in 1932, it was in a general way clear what had to be done ... I cannot invent something out of nothing..." Another recalled: "For me [nuclear physics] started with Heisenberg's paper ... [he] pointed out that now that the neutron has been discovered, one can think of starting a theory of the nucleus. This impressed me very much."[13]

Accelerators attack the nucleus

Other news from the Cavendish followed on the heels of the discovery of the neutron. On 21 April 1932, about a week after the neutron was discussed at the Copenhagen meeting, Rutherford wrote to Bohr:

"I was very glad to hear about you all from Fowler when he returned to Cambridge and to know what an excellent meeting of old friends you had. I was interested to hear about your theory of the Neutron....

It never rains but it pours, and I have another interesting development to tell you about of which a short account should appear in *Nature* next week. You know that we have a High Tension Laboratory where steady D.C. voltages can be readily obtained up to 600 000 volts or more. They have recently been examining the effects of a bombardment of light elements by protons...."

Rutherford went on to describe the work of Cockcroft and Walton in which they achieved the first artificial nuclear disintegrations with the high-voltage accelerator that they had been developing at the Cavendish since 1929. He concluded:

"I am very pleased that the energy and expense in getting high potentials has been rewarded by definite and interesting results.... You can easily appreciate that these results may open up a wide line of research in transmutation generally."[14]

Bohr's response reveals that he fully shared Rutherford's evaluation of the significance of this latest development:

"By your kind letter with the information about the wonderful new results arrived at in your laboratory you made me a very great pleasure indeed. Progress in the field of nuclear constitution is at the moment really so rapid, that one wonders what the next post will bring, and the enthusiasm of which every line in your letter tells will surely be common to all physicists. One sees a broad new avenue opened, and it should soon be possible to predict the behavior of any nucleus under given circumstances."[15]

Thirty-five years later, Cockcroft warmly recounted the atmosphere in the Cavendish when they achieved their results:

"It was extremely exciting to see the alpha particles in this transmutation. The first thing we did was to call up Rutherford on the laboratory exchange and invite him to come down and have a look at the scintillations, which he did. He, of course, was very excited about it."[16]

The story was soon carried in newspapers throughout the world, reviving alchemical dreams and hopes for new energy sources. For example, *The New York Times* carried articles on the Cockcroft–Walton work, with the following headlines: 1 May, "Atom Torn Apart with Energy Rise;" 3 May, "Hail New Approach to Energy of Atom;" 3 May, "Value Put in Energy Gain;" 4 May, "Atomic Energy;" and 8 May, "Atom Bombarders."

What of the reaction within the physics community? Cockcroft recalled the rapid response that "came from Berkeley and from Tuve's lab in Washington, where they had all been working on development of high-voltage equipment, such as the cyclotron or the Van de Graaff machine, toward just this kind of experiment."[16] On 20 August 1932, Lawrence wrote to

Cockcroft and Walton:

"I want to thank you very much for the reprints of your epoch making experiments on the disintegration of the elements by high velocity protons, and I hope you will continue to send me accounts of your work in the future. Under separate cover I am sending you reprints of the work of myself and my coworkers on methods for the acceleration of ions, which you may find of some interest.

At the present time we are attempting to corroborate your experiments using protons accelerated to high speeds by our method of multiple acceleration. We have some evidence already of disintegration, though as yet we can not be certain. Unfortunately our beam of protons is not nearly as intense as yours—although of higher voltage. Whenever we obtain some reliable results, of course, we will let you know promptly."[17]

Stanley Livingston recently recalled the Berkeley response:

"With the 11-inch [cyclotron] we had resonant particles of full energy in a collector cup at the edge of the pole. Our first publication was sent in to the *Physical Review* on February 20th, 1932 reporting 1 200 000 volts. Cockcroft and Walton's paper came out later that spring and showed that they had disintegrated nuclei with even lower energies.

Well, we weren't ready for experiments yet. We didn't have the instruments for detection. I had built the machine but had not included any devices for studying disintegrations. So we had to rebuild it. Now, Milton White was a student at that time, following right along behind me. He joined with me that spring in helping to rebuild the machine, and Lawrence also put in an emergency call to his friend Don Cooksey at Yale, who came out. Franz Kurie, a graduate student, also came out with Cooksey for the summer. Meanwhile we re-equipped the chamber with a target mounted inside where it would be hit by the beam, and a thin-foiled window on the side where we could mount counters. I think the first devices used for detecting the product particles were Geiger point counters. We set the threshold low so that they wouldn't trigger with x-rays or ultra-violet and they would count with particles. It wasn't long before we started to observe disintegrations, too. . . ."[18]

The exciting developments of 1932 stirred new interest in nuclear physics and the pace of activity began to quicken as the new techniques and concepts were put into action. The need for personal visits to the laboratories involved was obvious, if one was to keep abreast of the new work. At the beginning of 1933, Cockcroft planned to visit the US to study the work of Tuve, Lauritsen and Lawrence and to discuss with them the future of their various methods of nuclear disintegration. Applying for a travel grant for Cockcroft's trip, Rutherford wrote to the Rockefeller Foundation: "During the last year we have had visits from a number of workers interested in this field, and have given them as much information as we possess on our own methods."[19] He stressed that now it was equally valuable for the Cavendish workers to have similar first hand knowledge of work underway in US Laboratories.

Before Cockcroft's June trip regular letters kept physicists at the various institutions informed of one another's techniques and results. Lawrence's enthusiasm was evident when he wrote to Cockcroft at the beginning of June 1933: "We have been having a most exciting month in the laboratory. We have obtained so many disintegration effects that it is impossible for me to keep them all in mind. I am almost bewildered by the results. I will mention only a few as I will be seeing you soon. . . ."[20]

Cockcroft later recalled his impressions of that visit to the Berkeley laboratory:

"It was really interesting to see it actually in operation after having read so much about it in the journals. I was very much impressed by the way of working; to see the sealing-wax and string way of working on the cyclotron, which functioned for very short periods of time. They had a two-shift system, one shift doing the experiments, the other shift keeping the cyclotron going. And as soon as a leak developed, the maintenance shift would dash in and the experiment shift would retire backwards. A highly organized system."[16]

Just before Cockcroft left for the US, and a little more than a year after Rutherford had written Bohr that the Cockcroft-Walton results "may open up a wide line of research," Rutherford wrote to Gilbert N. Lewis, the renowned physical chemist at Berkeley, who had supplied him with heavy hydrogen for use as a projectile in the Cavendish accelerators. The aging dean of nuclear physics was enthusiastic about the new prospects for research in the field in which he had pioneered for many decades:

"I was delighted to receive your concentrated sample of the new hydrogen isotope in good shape, and we shall certainly take an early opportunity of examining its effects in our low voltage apparatus which Dr. Oliphant and I have been using the past year.

I have been enormously interested in your work of concentration of the new isotope with almost unbelievable success. I congratulate you and your staff on this splendid performance. I can appreciate the extraordinary value of this new element in opening up a new type of chemistry. If I were a younger man I think I would leave everything else to examine the effects produced by the substitution of H^2 for H^1 in all reactions.

Next, I should like to congratulate Lawrence and his colleagues for the prompt use they have made of this new club to attack the nuclear enemy. Cockcroft showed me the letter of Lawrence giving his preliminary results which are very exciting. These developments make me feel quite young again as in the early days of radioactivity when new discoveries came along almost every week, for it is a double scoop not only to prepare this new material but also to have the powerful method of Lawrence to examine its effects on nuclei. I wish them every success in their work and as soon as we can arrange it, I will try out the effects we can observe at our low voltages."[21]

In October 1933 the Solvay Congress in Brussels brought together most of the major participants in the burgeoning field, and a year later, in London another international conference on nuclear physics was held. By that time there were many more important new developments to discuss, including Enrico Fermi's theory of beta decay, the discovery of artificially induced radioactivity by the Joliot-Curies in Paris, and the technique of neutron bombardment to produce artificial radioactivity, which was systematically applied and developed by Fermi's group in Rome.

An American who attended the London conference was Frank Spedding, a former student of Lewis. His comments, in a letter to Lewis in December 1934, characterize the rapid pace of nuclear physics in the aftermath of the 1932 events, and show the reaction of a nonspecialist:

"There was also a symposium on nuclear physics. This field is moving so rapidly that one becomes dizzy contemplating it. With talk of the experimental properties of H^3, He^3, He^5, the new artificial radioactive elements, the neutron and positron, and the predicted properties of the neutrino and proton of minus charge, one who has been brought up on the old naïve picture of protons and electrons in the nucleus feels bewildered. I managed to attend a few of these sessions and found them

E. T. S. Walton and John Cockcroft (right) flank Ernest Rutherford in a 1932 photograph taken outside the Cavendish, after their accelerator had disintegrated nuclei by bombardment with protons. (Photo: UK Atomic Energy Authority.)

James Chadwick, working at the Cavendish Laboratory, had been searching for evidence of the neutron ever since Ernest Rutherford's suggestion, in 1920, that such a particle might exist. In 1932, about the time of this photograph, his efforts became successful. (Photo: Meggers Collection, Niels Bohr Library.)

extraordinarily interesting. There was one rather amusing incident that occurred here. Prof. Born had prepared a rather involved paper on the quantum theory of the nucleus. (An extension of Dirac's theory of the electron.) He wrote the paper longhand labelling it "For the Conference on Nuclear Physics." He made his "n"'s and "u"'s much alike so that his stenographer in copying it wrote "For the Conference on Unclear Physics."[22]

Indications of growth

One thing *was* clear about nuclear physics in the early 1930's: It was growing more rapidly than any other field of physics. This was especially true in the US, which provided a particularly fertile environment for the new and growing field to take root. The effect of the *annus mirabilis* can be clearly seen in the jump in nuclear-physics publications in *The Physical Review* between 1932 and 1933. The results of a study by Henry Small at the AIP Center for History and Philosophy of Physics show that in *The Physical Review,* nuclear-physics papers, letters and abstracts increased from 8% of the publications in 1932 to 18% in 1933 and reached 32% by 1937. A further examination of the dramatic increase in the number of nuclear-physics publications in *The Physical Review* between 1932 and 1933 shows that 42% of the increase was due to publications involving the neutron and 18% to those involving disintegration by protons. Publications from Berkeley alone accounted for 38% of the total increase in nuclear-physics papers between 1932 and 1933.[23]

While the total number and proportion of nuclear-physics papers was rising, the number of nuclear-physics papers that acknowledged funding was increasing even faster. In 1930, before the *annus mirabilis,* nuclear-physics papers constituted a very minor percentage of the papers in *The Physical Review* and a similarly small percentage of the funded papers. By 1935, however, when nuclear physics accounted for 22% of all papers in *The Physical Review,* fully 46% of the total funded papers were nuclear. And by 1940, when 34% of *The Physical Review* papers were in nuclear physics, they accounted for 55% of the funded papers. Clearly, nuclear physics was not only growing but also becoming a relatively heavily funded research subject. In fact, by 1939 fully one third of the nuclear-physics papers were being funded.

Another indication of the growth of a field is the number of new physics PhD's whose dissertation research is on a topic within the field. Here again nuclear physics showed an increase in the US from 2 new PhD's in 1930 to 41 in 1939. It was the only field of physics to increase steadily through the decade, and from 1937 on more new PhD's specialized in nuclear physics than in any other single field.

Of course, the growth of nuclear physics in the 1930's was not due solely to the discoveries of 1932. But these discoveries did help to focus the attention of a significant part of the physics community on nuclear phenomena and on the new possibilities for fruitful research in that field, possibilities which were expanded yet further with the development, availability and increasing-

Berkeley cyclotrons. Lefthand photo shows chamber of the 11-inch cyclotron. In early 1932, Ernest Lawrence and M. Stanley Livingston achieved a 10^{-9}-ampere, 1.22-MeV proton beam; later experiments with this chamber confirmed the artificial disintegration of lithium that Cockcroft and Walton had observed at lower energies. The 60-inch Berkeley cyclotron (right) was built for medical applications. This 1938 photograph shows Luis Alvarez astride the magnet-coil tank, Edwin McMillan on the ''D'' stem casing and, standing (left to right), Don Cooksey, Dale Corson, Lawrence, Robert Thornton, John Backus and Winfield Salisbury. (Photos: Lawrence Radiation Laboratory.)

ly productive use of particle accelerators. These instruments became central to experimental work at a number of new research centers that began to flourish during the period.

Special role of Berkeley

Because Berkeley, and particularly Lawrence's radiation laboratory there, played such a major role in these developments, let us take a brief glimpse into the Berkeley scene in the 1930's. Clearly the Berkeley work was very important in 1932, and it accounted for a large part of the field's subsequent productivity in the US. Throughout the 1930's Berkeley not only produced more nuclear-physics papers and PhD's than other US institutions but also had the lion's share of funded nuclear research. These statistics, however, are only a part of the story, for Berkeley also played a key social role in developing the entire field of nuclear physics internationally.[24]

Berkeley was the home of the cyclotron, the instrument that became central to nuclear physics research as it took root in more and more institutions throughout the world in the 1930's. The early 11-inch model, which first accelerated protons to energies of 1.2 million electron volts by the beginning of 1932 and achieved nuclear disintegrations later that year, had been made possible by a grant of $500 from the National Research Council of the National Academy of Sciences in the spring of 1931. By the spring of 1940 Lawrence had obtained a grant from the Rockefeller Foundation for more than one million dollars toward the cost of creating a 100-million volt cyclotron. In the intervening years—aided by grants from the University of California, the Research Corporation, the Chemical and Macy Foundations, the US Works Progress Administration (WPA), as well as individual donors—several generations of cyclotrons of steadily increasing energy and wide applications had been developed at Berkeley by Lawrence and the team he had assembled there.

The cyclotron had proved to be an excellent instrument for particle-scattering experiments and an unsurpassed producer of powerful neutron sources that could make a large variety of new isotopes, thus providing previously unavailable data essential for a fuller understanding of nuclear structure. These unstable isotopes were also used for therapeutic medical applications and as tracers in pioneering studies of chemical and biological processes. The unique role of the cyclotron as a producer of isotopes began in 1934, after the Joliot-Curies discovered artificially induced radioactivity. Later that year Fermi's group in Rome demonstrated induced radioactivity by neutron bombardment. The Berkeley cyclotron was soon at work systematically producing artificially radioactive isotopes of a number of elements. Lawrence's production of a radioisotope of sodium in 1934 was especially significant because of its potential application to medical therapy.

The potential biological and medical applications helped to create interest in and financial support for the subsequent development of cyclotrons at Berkeley and at other places. During 1935 a number of institutions started to build cyclotrons because they recognized that it was a major tool for nuclear studies. At several of these places—for example, Bohr's institute in

Copenhagen where George de Hevesy was pursuing his tracer studies, the University of Rochester where the physics department was headed by Lee DuBridge and the University of Michigan where Harrison Randall was department chairman—the cyclotron projects were proposed and financed as part of planned collaborative research efforts involving the physics, medicine and biology departments. At Berkeley, such joint efforts were wholeheartedly pursued and were immensely strengthened when the physician John Lawrence arrived from Yale in the mid-1930's to collaborate with his brother and others in a full medical program involving not only isotopes but also experiments in the use of neutron beams for cancer therapy. Radiochemistry also blossomed at Berkeley where strong ties existed between the physics and chemistry departments.

Recognition of the role of the cyclotron in physics, chemistry, biology and medicine resulted in a proliferation of the instruments at institutions throughout the world in the late 1930's, and almost all of these projects depended on assistance from the Berkeley experts. Detailed technical information and advice was communicated through a lively network of personal letters, circulation of unpublished technical memoranda and progress reports, personal visits, and exchange of personnel. Don Cooksey, who played a key role as the Berkeley hub of this international informal communication network, jokingly referred to it in June 1938 as the "Cyclotron Union of the World."[25] At that time Berkeley-trained physicists were building cyclotrons in Copenhagen, Stockholm, Paris, Cambridge, Liverpool, Tokyo, and at more than a dozen US institutions. The Berkeley radiation laboratory played a key role as an international information center, a training school, a supplier of cyclotron-produced radioactive materials for use in other laboratories, and a source of skilled physicists who were available to help other institutions enter the cyclotron field. Thus the impact of Lawrence's laboratory transcended the important results being obtained in Berkeley and had a tremendous multiplier effect on the entire field in the 1930's.

I have described some of the events of 1932 and the immediate responses of some of the participants. It was clear to them that the new developments would open up an exciting period for fruitful research in nuclear physics. The field did flourish in the following years and by the mid-1930's was firmly established in a number of new centers of nuclear research.

In March 1972 champagne toasts were drunk in Batavia, Illinois to celebrate the achievement of accelerating protons to record energies of 200 GeV through the four-mile circumference of the giant new accelerator there. It was a fitting observance of the 40th anniversary of the "annus mirabilis" of 1932, and makes one wonder how soon we might see another "marvelous" year and what its impact may be on physics and society in the decade that follows.

* * *

The location and study of historical materials used in this article have been supported by grants from the National Science Foundation and the John Simon Guggenheim Memorial Foundation, and have been greatly facilitated by the information resources of the AIP Niels Bohr Library. Permission to use and quote archival materials and oral history interviews was kindly granted by the appropriate institutions and individuals cited. The author is grateful for this assistance.

References

1. J. Boyce to J. Cockcroft, 8 January 1932, Sir John Cockcroft Papers, Churchill College Library, Cambridge, UK.

2. C. D. Anderson to R. A. Millikan, 3 November 1931, Robert A. Millikan Papers, California Institute of Technology Archives, Pasadena. For Millikan's account of his talks in Europe about the photographs, see R. A. Millikan, *Electrons (+ and −), Protons, Photons, Neutrons and Cosmic Rays* (Chicago, 1935), pages 327–330. The reaction of some European physicists to these talks is documented and analyzed in N. R. Hanson, *The Concept of the Positron* (Cambridge, UK, 1963), page 139–142, 216–217.

3. C. D. Anderson, Science **76**, 238–239, 9 September 1932.

4. Some of the 1932 nuclear events are also discussed in C. Weiner, "Institutional Settings for Scientific Change: Episodes from the History of Nuclear Physics," in A. Thackray and E. Mendelsohn, eds., *Science and Values* (Humanities Press, N.Y., 1972).

5. J. Chadwick, "Some Personal Notes on the Search for the Neutron," *Proceedings* of the 10th International Congress of the History of Science, Ithaca, 1962 (Paris, 1964), page 161.

6. J. Chadwick, "On the Possible Existence of the Neutron," Nature **129**, 312, 27 February 1932.

7. J. Chadwick to N. Bohr, 24 February 1932, and N. Bohr to J. Chadwick, 25 March 1932, Niels Bohr Papers, Niels Bohr Institute, Copenhagen. The Bohr Papers have been microfilmed by The American Physical Society–American Philosophical Society project on Sources for History of Quantum Physics, and the films are deposited at the American Philosophical Society, Philadelphia; at the Bancroft Library, University of California, Berkeley and at the AIP Niels Bohr Library, New York.

8. Information on the conferences is available in the Niels Bohr Institute administrative archive, Copenhagen.

9. Translation by Barbara Gamow in George Gamow, *Thirty Years That Shook Physics*, Doubleday, New York (1966), page 214.

10. For a full discussion of Heisenberg's treatment of the neutron, see J. Bromberg, "The Impact of the Neutron: Bohr and Heisenberg," in *Historical Studies in the Physical Sciences*, Vol. 3 (1971), page 307–341.

11. S. Goudsmit to N. Bohr, 4 November 1932. Niels Bohr Institute administrative archive, Copenhagen.

12. N. Bohr to S. Goudsmit, 28 December 1932, Niels Bohr Institute administrative archive, Copenhagen.

13. Interviews conducted in connection with the joint AIP-American Academy of Arts and Sciences conferences on the history of nuclear physics held in May 1967 and May 1969. The proceedings and abstracts of the interviews are in *Exploring The History of Nuclear Physics*, AIP Conference Proceedings 7 (1972).

14. E. Rutherford to N. Bohr, 21 April 1932, Niels Bohr Papers, Niels Bohr Institute, Copenhagen.

15. N. Bohr to E. Rutherford, 2 May 1932, Rutherford Papers, Cambridge University Library, Cambridge, UK.

16. Interview with J. Cockcroft by C. Weiner, 28 March 1967, Oral History Collection, AIP Niels Bohr Library, New York.

17. E. Lawrence to J. Cockcroft and E. T. S. Walton, 20 August 1932, Cockcroft Papers, Churchill College Library, Cambridge, UK.

18. Interview with M. S. Livingston by C. Weiner, 21 August 1967, Oral History Collection, AIP Niels Bohr Library.

19. E. Rutherford to W. Tisdale, 6 March 1933, Cockcroft Papers, Churchill College Library, Cambridge, UK.

20. E. Lawrence to J. Cockcroft, 2 June 1933, Cockcroft Papers, Churchill College Library, Cambridge, UK.

21. E. Rutherford to G. N. Lewis, 30 May 1933, G. N. Lewis Papers, Bancroft Library, University of California, Berkeley.

22. F. Spedding to G. N. Lewis, 1 December 1934, Lewis Papers, Berkeley.

23. The data are drawn from the statistical study of the physics journal literature conducted by H. Small with the assistance of D. Schreibersdorf at the AIP Center for History of Physics under a National Science Foundation grant. Work-in-progress reports were presented by Small at the History of Science Society annual meetings in 1970 and 1971 and are in the AIP archives.

24. The brief sketch here is based on archival materials from the Lawrence Radiation Laboratory, the Cavendish Laboratory and the Bohr Institute in Copenhagen; physics-department files at several US universities; Herbert Childs's biography of Lawrence, *An American Genius* (Dutton, New York, 1968); and on historical accounts of the cyclotron such as those by M. Stanley Livingston and Edwin M. McMillan in PHYSICS TODAY, October 1959, 18–34, and Livingston's *Particle Accelerators: A Brief History* (Harvard, Cambridge, 1969).

25. D. Cooksey to M. S. Livingston, 8 June 1938, E. O. Lawrence Papers, Bancroft Library, University of California, Berkeley. □

The idea of the neutrino

To avoid anomalies of spin and statistics Pauli suggested in 1930 that a neutral particle of small mass might accompany the electron in nuclear beta decay, calling it (until Chadwick's discovery) the *neutron*.

Laurie M. Brown

PHYSICS TODAY / SEPTEMBER 1978

During the 1920's physicists came to accept the view that matter is built of only two kinds of elementary particles, electrons and protons, which they often called[1] "negative and positive electrons." A neutral atom of mass number A and atomic number Z was supposed to contain A protons, all in the nucleus, and A negative electrons, $A - Z$ in the nucleus and the rest making up the external electron shells of the atom. Their belief that both protons and negative electrons were to be found in the nucleus arose from the observations that protons could be knocked out of light elements by alpha-particle bombardment, while electrons emerged spontaneously (mostly from very heavy nuclei) in radioactive beta decay. Any other elementary constituent of the atom would have been considered superfluous, and to imagine that another might exist was abhorrent to the prevailing natural philosophy.

Nevertheless, in December 1930 Wolfgang Pauli suggested a new elementary particle that he called a *neutron,* with characteristics partly like that of the nucleon we now call by that name, and partly those of the lepton that we now call *neutrino* (more precisely the electron antineutrino, but this distinction is not needed here). Pauli's neutron–neutrino idea became well-known to physicists even before his first publication of it, which is in the discussion section following Heisenberg's report on nuclear structure at the Seventh Solvay Conference,[2] held in Brussels in October 1933.

Shortly after attending this conference, Enrico Fermi published his theory of beta decay, which assumes that a neutrino always accompanies the beta-decay elec-

tron, and that both are created at their moment of emission. Perhaps because of the rapid acceptance of Fermi's theory and the tendency to rethink history "as it should have happened," the true nature of Pauli's proposal has been partly overlooked and its radical character insufficiently emphasized. Contrary to the impression given by most accounts, Pauli's "neutron" has some properties in common with the neutron James Chadwick discovered in 1932 as well as with Fermi's neutrino.

Flaws in the model

By the end of 1930, when our story begins, quantum mechanics had triumphed not only in atomic, molecular and crystal physics, but also in its treatment of some nuclear processes, such as alpha-particle radioactivity and scattering of alpha particles from nuclei (including the case of helium, in which quantum-mechanical interference effects are so important). However, the situation regarding electrons in the nucleus was felt to be critical. The main difficulties of the electron–proton model of the nucleus were:

▶ The symmetry character of the nuclear wave function depends upon A, not Z as predicted by the model; when $A - Z$ is odd the spin and statistics of the nucleus are given incorrectly. For example, nitrogen ($Z = 7$, $A = 14$) was known from the molecular band spectrum of N_2 to have spin 1 and Bose–Einstein statistics.

▶ No potential well is deep enough and narrow enough to confine a particle as light as an electron to a region the size of the nucleus (the argument for this is based on the uncertainty principle *and* relativistic electron theory).

▶ It is hard to see how to "suppress" the very large (on the nuclear scale) magnetic moments of the electrons in the nucleus,

which conflict with data on the hyperfine structure of atomic spectra.

▶ Although both alpha and gamma decay show the existence of narrow nuclear energy levels, the electrons from a given beta-decay transition emerge with a broad continuous spectrum of energy.

The strong contrast between the successes and the failures of quantum mechanics applied to the nucleus are nowhere more evident than in a book by George Gamow.[3] In it, all the passages concerning electrons in the nucleus are set off in warning symbols (skull and crossbones in the original manuscript).

Some physicists (among them Niels Bohr and Werner Heisenberg[4]) took these difficulties to indicate that a new dynamics, possibly even a new type of space–time description, might be appropriate on the scale of nuclear distances and energies, just as quantum mechanics begins to be important on the atomic scale. These physicists were impressed by the similarity of the nuclear radius to the value e^2/mc^2, the classical electron radius of H. A. Lorentz. At this distance it had been anticipated that electrodynamics would probably fail (and maybe, with it, the special theory of relativity). Bohr was willing to relinquish the conservation of energy, except as a statistical law, in parallel with the second law of thermodynamics. At the same time Heisenberg was considering the introduction of a new fundamental length into the theory. It seemed that anything might be considered acceptable as a way out of the dilemma—or perhaps anything except a new elementary particle.

Pauli's proposal

It was in this context of ideas that Pauli dared to suggest the existence of a new neutral particle. His proposal, intended to rescue the quantum theory of the nu-

Laurie M. Brown is a professor in the Department of Physics and Astronomy, Northwestern University, Evanston, Illinois.

PAULI ON THE WAY TO PASADENA, 1931

cleus from its contradictions, was presented in good humor as a "desperate remedy," although it was a serious one. (The Viennese version would have been, according to the old joke: desperate, but not serious.) During the next three years he lectured on what he called the "neutron" at several physics meetings and he discussed it privately with colleagues.

Pauli's first proposal was put forward only tentatively, as he recalled in a lecture he delivered in Zürich in 1957, after receiving news of the experiments confirming parity violation in beta decay.[5] Invited to a physics meeting in Tübingen, Germany, which he was unable to attend (because of a ball to be held in Zürich, at which he declared he was "indispensable"), he sent a message with a colleague as an "open letter," although it was intended mainly for Hans Geiger and Lise Meitner. An English translation of this letter is given in the Box on page 27. Pauli was anxious for their expert advice as to whether his proposal was compatible with the known facts of beta decay.

In the 1957 lecture Pauli also tells how he became convinced of a crisis associated with beta decay. During the decade that followed the discovery by Chadwick in 1914 of beta rays with a continuous energy spectrum, it became established that *these* were the true "disintegration electrons," rather than those making up discrete electron line spectra, which were later shown to arise from such causes as photoelectric effects of nuclear gamma rays, internal conversion and Auger processes. Because a continuous spectrum seemed to disagree with the presence of discrete quantum states of the nucleus (as indicated by alpha and gamma emission), some workers, including Meitner, thought that the beta rays were radiating some of their energy as they emerged through the strong electric field of the nucleus.[5,6,7]

This led C. D. Ellis and William Wooster at the Cavendish Laboratory in Cambridge, England, who did not believe in the radiation theory, to perform a calorimetric experiment with radium E (bismuth) as a source. Their result, later confirmed in an improved experiment by Meitner and W. Orthmann,[8] was that the energy per beta decay absorbed in a thick-walled calorimeter was equal to the mean of the electron energy spectrum, and not to its maximum (endpoint). Furthermore, Meitner showed that no gamma rays were involved. According to Pauli (in 1957), this allowed but two possible theoretical interpretations:

▶ The conservation of energy is valid only statistically for the interaction that gives rise to beta radioactivity.

▶ The energy theorem holds strictly in each individual primary process, but at the same time there is emitted with the electron another very penetrating radiation, consisting of new neutral particles. To the above, Pauli adds, "The first possibility was advocated by Bohr, the second by me."[5]

But although the conservation of energy, and possibly other conservation laws in beta decay were very much in Pauli's mind at this time, this was not his *only* reason for proposing the neutrino. He makes this point (already obvious from his Tübingen letter) quite explicit in his 1957 Zürich lecture. After pointing out one of the major difficulties with the nuclear model containing only protons and electrons (the symmetry argument mentioned above), Pauli says:

"I tried to connect this problem of the spin and statistics of the nucleus with the other of the continuous beta spectrum, without giving up the energy theorem, through the idea of a new neutral particle."

Neutrinos—ejected or created?

It is often overlooked in discussing the history of the neutrino idea that Pauli suggested his particle as a constituent of the nucleus, with a small but not zero mass, together with the protons and the electrons. (Chien-Shiung Wu, for example, emphasizes the non-conservation of statistics that would occur in beta decay without the neutrino.[6,7,9] However, Pauli refers rather to the spin and statistics of *stable* nuclei such as lithium 6 and nitrogen 14.) This point is of some significance; had Pauli proposed in 1930 that neutrinos were created (like photons) in transitions between nuclear states, and that they were otherwise not present in the nucleus, he would have anticipated by three years an important feature of Fermi's theory of beta decay. Pauli did not claim to have had this idea when he wrote the Tübingen letter, but he did say (in his Zürich lecture) that by the time he was ready to speak openly of his new particle, at a meeting of The American Physical Society in Pasadena, held in June of 1931, he *no longer* considered his neutrons to be nuclear constituents. It is for this reason, he says, that he no longer referred to them as "neutrons"; indeed, that he made use of no special name for them. However, there is evidence, as we shall see, that Pauli's recollections are incorrect; that at Pasadena the particles *were* called neutrons and *were* regarded as constituents of the nucleus.

I have not been able to obtain a copy of Pauli's Pasadena talk or scientific notes on it; he said later that he was unsure of the matter and thus did not allow his lecture to be printed. The press, however, took notice. For example, a short note in *Time,* 29 June 1931, headed "Neutrons?", says that Pauli wants to add a fourth to the "three unresolvable basic units of the universe" (proton, electron and photon); adding, "He calls it the *neutron.*"

Upon examining the program of the Pasadena Meeting, I discovered that Samuel Goudsmit spoke at the same session as Pauli (and even upon the same announced subject—hyperfine structure). I wrote to Goudsmit and received a most interesting reply, from which I should like to quote:

"Pauli accompanied my former wife and me on the train trip across the US. I forgot whether we started in Ann Arbor or arranged to meet in Chicago. We talked little physics, more about physicists. Pauli's main topic at the time was that he could imitate P. S. Epstein and he insisted that I take pictures of him while doing that. We spent a couple of days in San Francisco, where we almost lost him in Chinatown. He'd suddenly rush ahead and around a corner while we were window shopping ... He may have talked about the "neutron" on that trip, but I am not at all certain ..."

Goudsmit does not now recall exactly what Pauli said at Pasadena, except that he mentioned the "neutron"; however, he sent me a copy of his report at the Rome Congress on what Pauli had said four months earlier in Pasadena. To continue, then, with Goudsmit's letter:

"Fermi was arranging what was probably the first nuclear physics meeting. It was held in Rome in October 1931 ... It was the best organized meeting I ever attended, because there was very much time available for informal discussions and get-togethers ... Fermi had arranged marvelous leisurely sightseeing trips for the group. There were about 40 guests and 10 Italians.

"Fermi ordered the then 'young' participants, namely [Nevill] Mott, [Bruno] Rossi, [George] Gamow (who could not leave Russia but sent a manuscript) and myself, to prepare summary papers for discussion ... As you know, I don't use and don't keep notes. But I have a clear picture of Pauli lecturing [at Pasadena] and his mention of the 'neutron' ... Pauli was supposed to attend the Rome meeting, but he arrived a day or so late. In fact, he entered the lecture hall the very moment that I mentioned his name! Like magic! I remarked about it and got a big laugh from the audience."

Goudsmit's Rome report

At Fermi's request, then, Goudsmit reported at the Rome Conference on Pauli's talk in Pasadena. Here is what he said:[10]

"At a meeting in Pasadena in June 1931, Pauli expressed the idea that there might exist a third type of elementary particles besides protons and electrons, namely 'neutrons.' These neutrons should have an angular momentum $\frac{1}{2} h/2\pi$ and also a magnetic moment, but no charge. They are kept in the nucleus by magnetic forces and are emitted together with beta-rays in radioactive disintegration. This, according to Pauli, might remove present difficulties in nuclear structure and at the same time in the explanation of the beta-ray spectrum, in which it seems that the law of conservation of energy is not fulfilled. If one would find experimentally that there is also no conservation of mo-

GOUDSMIT (MIDDLE) AND FERMI (RIGHT) WITH UNIDENTIFIED MAN

mentum, it would make it very probable that another particle is emitted at the same time with the beta-particle. The mass of these neutrons has to be very much smaller than that of the proton, otherwise one would have detected the change in atomic weight after beta-emission."

Goudsmit added that Pauli believed "neutrons may throw some light on the nature of cosmic rays."

It does appear clear from this passage (to which Pauli evidently made no objection at the time) that at Pasadena the neutron was intended to be a particle that could be bound in the nucleus by magnetic forces. In his letter to me Goudsmit also said, "It was Maurice Goldhaber who some time ago pointed out that I was the first to put Pauli's idea on paper and in print."

After leaving Pasadena Pauli remained in the United States until the fall, when he went to Rome. He gave a seminar at the Summer Session of the University of Michigan at Ann Arbor (probably at one of their Symposia on Theoretical Physics, where Fermi had, the previous summer, given his famous lectures on the quantum theory of radiation). At the seminar Pauli spoke, according to the Berkeley theorists J. F. Carlson and J. Robert Oppenheimer,[11] about "the elements of the theory of the neutron, its functions and its properties."

Tracks in the cloud chamber

Carlson and Oppenheimer wondered whether Pauli's "neutrons" could be used to solve yet another puzzle: the appear-

ance of certain lightly ionizing cloud-chamber tracks from cosmic rays that had been reported.

The complex problem of the energy loss of relativistic charged particles was crucial to the interpretation of the various components of the cosmic rays observed in the atmosphere, and had attracted the attention of many theorists. Carlson and Oppenheimer were unable to account for cloud-chamber tracks that appeared thinner than those of an "ordinary radioactive" beta particle. Their calculations of energy loss (which agreed in a general way with independent calculations by Heisenberg and Hans Bethe, and with an older classical estimate by Bohr) showed that charged particles should have a relativistic increase of ionization with energy. The particles leaving light tracks were very penetrating (and thus probably relativistic) and it was concluded that they could not be electrons or protons. (These quarklike tracks have not, to my knowledge, been explained. Perhaps they were examples of old and "faded" tracks, which often plagued cloud chambers of the untriggered variety.)

Carlson and Oppenheimer decided therefore to make a theoretical investigation, as they said,[11] of the "ionizing power of the neutrons which were suggested by Pauli to salvage the theory of the nucleus. These neutrons, it will be remembered, are particles of finite proper mass, carrying no charge, but having a small magnetic moment ..."

Could thin tracks, like those in the cosmic rays, be seen from beta decays? "If they were found, we should be cer-

The participants in the Seventh Solvay Conference, where Pauli presented his neutrino idea, included, in the **first row**, E. Schrödinger, I. Joliot, N. Bohr, A. Joffé, M. Curie, O. W. Richardson, P. Langevin. Lord Rutherford, T. De Donder, M. de Broglie, L. de Broglie, L. Meitner, J. Chadwick, and in the **second row**, E. Henriot, F. Perrin, F. Joliot, W. Heisenberg, H. A. Kramers, E. Stahel, E. Fermi, E. T. S. Walton, P. A. M. Dirac, P. Debye, N. F. Mott, B. Cabrera, G. Gamow, W. Bothe, P. M. S. Blackett, M. S. Rosenblum, J. Errera, E. Bauer, W. Pauli, J. E. Verschaffelt, M. Cosyns (in back), E. Herzen, J. D. Cockcroft, C. D. Ellis, R. Peierls, A. Piccard, E. O. Lawrence, L. Rosenfeld. The photograph is by Benjamin Couprie.

tain that the neutrons not only played a part in the building of nuclei, but that they also formed the cosmic rays."

The calculations of Carlson and Oppenheimer were published[12] almost a year later, in September 1932; by that time they no longer believed that "neutrons" might leave observable cloud-chamber tracks. In addition, the situation in nuclear physics had changed profoundly, as it also was about to in cosmic-ray physics: Chadwick's study of "the penetrating radiation produced in the artificial disintegration of beryllium" had revealed the existence of the neutron, announced the previous February; Anderson's discovery of the positron in cosmic rays was announced in August.[13] Certainly one could no longer speak of the proton as synonymous with positive electricity, and one might suppose that now a new particle like Pauli's would be acceptable; but this was not the case:

For one thing, the positron was thought to be only the *absence* of a negative electron of negative energy, a *hole* in the vacuum. For another, the neutron of Chadwick, the *heavy* neutron, was generally regarded as a composite object (it was not thought to be unstable when free), a kind of tightly bound hydrogen atom or neutral nucleus made of a proton and an electron, like other nuclei. It was perhaps thought to be elementary only by Ettore Majorana in Rome who (according to Emilio Segrè) called it the *neutral proton*.

For our purposes, the Carlson–Oppenheimer article is significant in what it tells us about the view held by Pauli, in that summer of 1931, about his neutral particle, which, following the Berkeley authors, we will now call the *magnetic neutron* to distinguish it from Chadwick's neutron and Fermi's neutrino. Carlson and Oppenheimer state that the neutral particle of spin $\frac{1}{2}$, satisfying the exclusion principle, was introduced by Pauli not only to resolve the difficulties in nuclear theory, but "on the further ground that such a particle could be described by a wave function which satisfies all the requirements of quantum mechanics and relativity . . . The experimental evidence on the penetrating beryllium radiation suggests that neutrons of nearly protonic mass do exist; and since our calculations may be carried through without specifying the mass or magnetic moment of the neutron, we shall consider the most general particle which satisfies the wave equation proposed by Pauli. It is important to observe that there may very well be other types of neutral particles, which are not elementary, and to which our calculations do not apply . . ."

Thus we find, surprisingly, that there were thought to be also purely *theoretical* grounds for considering a neutral particle with a magnetic moment; it is one of the few simple types of elementary particles that are allowed by relativistic quantum theory. In the wake of Chadwick's neutron discovery, Carlson and Oppenheimer in 1932 redefined Pauli's particle to be one whose wave function obeys a certain relativistic wave equation. We should not, however, assume that the Berkeley theorists were soft on new particles. On the contrary, the final paragraph of their lengthy article reads, "We believe that these computations show that there is no experimental evidence for the existence of a particle like the magnetic neutron."

Pauli's wave equation for the neutral particle, given at Ann Arbor, is a variant of the linear Dirac equation for the electron, containing an additional term (*Zusatzglied*) called the "Pauli anomalous magnetic moment" term.[14] This equation describes a spin-$\frac{1}{2}$ particle that may be either charged or neutral; the extra term makes a contribution to the charge-current four-vector, which need not vanish for a neutral particle.

Fermi is positive

Carlson and Oppenheimer derived a general formula for the collision cross section of magnetic neutrons and examined the result for small velocities. (They were well aware of the perils involved in pushing this highly singular interaction to excessive energies.) For the collision of a neutron against a particle of equal mass, they found a large probability, nearly independent of velocity and proportional to the square of the magnetic moment. The average energy loss per collision was relatively large, and they deduced that such a particle "will never produce ion traces in a cloud chamber, since it tends to lose an appreciable fraction of its energy, and suffer an appreciable deflection at every impact." For targets much lighter or heavier than the neutron, smaller energy losses occur; cloud-chamber tracks might result in this case, but the collision probabilities are small unless the magnetic moment of the neutron is assumed to be improbably large. The concluded (correctly) that there is no evidence for magnetic neutrons. (The heavy neutron, with a magnetic moment only one thousandth of a Bohr magneton, leaves no tracks.) At the Seventh Solvay Conference in 1933, Pauli no longer felt the magnetic neutron to be "well-founded."

Let us return now to the Rome Congress of 1931, which Pauli considered important in the development of the neutrino concept, for there he had the opportunity to discuss it with Bohr and especially with Fermi, with whom he had a number of private conversations. While Fermi's attitude toward the neutrino was very positive, Bohr was totally opposed to it, preferring to think that within nuclear distances the conservation laws were breaking down.[15]

"From the empirical point of view," said Pauli, "it appeared to me decisive whether the beta spectrum of the electrons showed a sharp upper limit" or, instead, an infinitely falling statistical distribution. Pauli felt that if the limit were sharp, then *his* idea was correct, and Bohr's was wrong.

In mid-1933, Ellis and Mott suggested that the beta-ray spectrum has indeed a sharp upper limit, corresponding to a unique energy difference between parent and daughter nucleus.[16] Furthermore, they added,

"According to our assumption the β-particle may be expelled with *less* energy than the difference of the energies . . . of the two nuclei, but not with *more* energy. We do not wish in this paper to dwell on what happens to the excess energy in those disintegrations in which the electron is emitted with less than the maximum energy. We may, however, point out that if the energy merely disappears, implying a breakdown of the principle of energy conservation, then in a β-ray decay energy is not even statistically conserved. Our hypothesis is, of course, also consistent with the suggestion of Pauli that the excess energy is carried off by particles of great penetrating power such as neutrons of electronic mass."

The question of the upper limit of the beta spectrum, although not easily resolved, is of some importance, for the shape of the upper end of the spectrum is sensitive to the neutrino mass. This was discussed again by Ellis at an international conference in London, held in the fall of 1934, where he referred to accurate magnetic spectrograph measurements of W. J. Henderson that strongly suggested a neutrino of zero mass.[17] Fermi's theory of beta decay had already been published,[18] and Ellis assumed it in his analysis, but an energy-nonconserving theory, that of Guido Beck and Kurt Sitte, shared equal time with Fermi's at the conference.

Fermi spoke at the London Conference, but his subject was the neutron-activation work of the Rome experimental nuclear physics group. He also had attended the Seventh Solvay Conference, held in October, 1933, where he heard Pauli present his first suggestion *for publication* of the existence of a neutrino. The complete Solvay remarks of Pauli are given in English translation in the Box on page 28; we leave it to the reader to decide whether Pauli still thought that the neutrino or the

Pauli proposes a particle

The letter in which Pauli proposed the neutrino, translated from the German of reference 5, reads as follows:

Zürich, 4 December 1930
Gloriastr.

Physical Institute of the
  Federal Institute of Technology (ETH)
Zürich

Dear radioactive ladies and gentlemen,

As the bearer of these lines, to whom I ask you to listen graciously, will explain more exactly, considering the "false" statistics of N-14 and Li-6 nuclei, as well as the continuous β-spectrum, I have hit upon a desperate remedy to save the "exchange theorem" * of statistics and the energy theorem. Namely [there is] the possibility that there could exist in the nuclei electrically neutral particles that I wish to call neutrons, which have spin $\frac{1}{2}$ and obey the exclusion principle, and additionally differ from light quanta in that they do not travel with the velocity of light: The mass of the neutron must be of the same order of magnitude as the electron mass and, in any case, not larger than 0.01 proton mass.—The continuous β-spectrum would then become understandable by the assumption that in β decay a neutron is emitted together with the electron, in such a way that the sum of the energies of neutron and electron is constant.

Now the next question is what forces act upon the neutrons. The most likely model for the neutron seems to me to be, on wave mechanical grounds (more details are known by the bearer of these lines), that the neutron at rest is a magnetic dipole of a certain moment μ. Experiment probably requires that the ionizing effect of such a neutron should not be larger than that of a γ ray, and thus μ should probably not be larger than $e.10^{-13}$ cm.

But I don't feel secure enough to publish anything about this idea, so I first turn confidently to you, dear radioactives, with the question as to the situation concerning experimental proof of such a neutron, if it has something like about 10 times the penetrating capacity of a γ ray.

I admit that my remedy may appear to have a small *a priori* probability because neutrons, if they exist, would probably have long ago been seen. However, only those who wager can win, and the seriousness of the situation of the continuous β-spectrum can be made clear by the saying of my honored predecessor in office, Mr. Debye, who told me a short while ago in Brussels, "One does best not to think about that at all, like the new taxes." Thus one should earnestly discuss every way of salvation.—So, dear radioactives, put it to the test and set it right.—Unfortunately I cannot personally appear in Tübingen, since I am indispensable here on account of a ball taking place in Zürich in the night from 6 to 7 of December.—With many greetings to you, also to Mr. Back, your devoted servant,

W. Pauli

* In the 1957 lecture, Pauli explains, "This reads: exclusion principle (Fermi statistics) and half-integer spin for an odd number of particles; Bose statistics and integer spin for an even number of particles."

CHADWICK

electron were constituents of the nucleus. (That a massless neutrino could be *created* at the moment of its emission with the electron was clearly proposed that year[19] by Francis Perrin, who also attended the Seventh Solvay Conference.) There was, in any case, no doubt that a light or massless neutral particle of spin $1/2$ has to be emitted with the beta-decay electron in order to save the conservation laws, and *that* is surely the idea of neutrino!

Fermi's theory of beta decay is in many ways still the standard theory. Called by Victor Weisskopf "the first example of modern field theory,"[20] it eventually caused Bohr to withdraw[21] his doubts concerning "the strict validity of the conservation laws." A radical generalization of quantum theory was *not* required, though new particles and new interactions were. Within a few months of Fermi's theory, positron beta decay was seen (the first example of artificial radioactivity); and beta decay was to be the prototype of a larger class of weak interactions.

The neutrino can be regarded as one of the first (if not the first) of the new particles that made the new physics of the 1930's, even though it took two more decades to observe the first neutrino-capture event. The weak interactions have been notorious for their capacity to flout the expectations of physicists with regard to symmetries and conservation laws. Although Bohr was too willing, in his 1931 Faraday Lecture,[15] "to renounce the very idea of energy balance," the conclusion of that lecture is probably still appropriate today: "... notwithstanding all the recent progress, we must still be prepared for new surprises."

* * *

This work was supported in part by a grant from the National Science Foundation. I would like to express my sincere appreciation to Arthur L. Norberg of The Bancroft Library, University of California, Berkeley, and to Judith Goodstein of the Robert A. Millikan Memorial Library of the California Institute of Technology. I am much obliged to Samuel Goudsmit for his letter and for his kind permission to quote from it.

Pauli becomes bolder

The discussion comments in which Pauli presented the idea of the neutrino at the Seventh Solvay Conference, ref. 2. The text is based on the translation from the French original by Chien-Shiung Wu, ref. 9, with corrections by Laurie Brown noted in brackets.

The difficulty coming from the existence of the continuous spectrum of the β-rays consists, as one knows, in that the mean lifetimes of nuclei emitting these rays, as that of the resulting radioactive bodies, possess well-determined values. One concludes necessarily from this that the state as well as the energy and the mass, of the nucleus which remains after the expulsion of the β particle, are also well-determined. I will not persist in efforts by which one could try to escape from this conclusion for I believe, in agreement with Bohr, that one always stumbles upon insurmountable difficulties in explaining the experimental facts.

In this connection, two interpretations of the experiment present themselves. The interpretation supported by Bohr admits that the laws of conservation of energy and momentum do not hold when one deals with a nuclear process where light particles play an essential part. This hypothesis does not seem to me either satisfying or even plausible. In the first place the electric charge is conserved in the process, and I don't see why conservation of charge would be more fundamental than conservation of energy and momentum. Moreover, it is precisely the energy relations which govern several characteristic properties of beta spectra (existence of an upper limit and relation with gamma spectra, Heisenberg stability criterion). If the conservation laws were not valid, one would have to conclude from these relations that a beta disintegration occurs always with a loss of energy and never a gain; this conclusion implies an irreversibility of these processes with respect to time, which doesn't seem to me at all acceptable.

In June 1931, during a conference in Pasadena, I proposed the following interpretation: the conservation laws hold, the emission of beta particles occurring together with the emission of a very penetrating radiation of neutral particles, which has not been observed yet. The sum of the energies of the beta particle and the neutral particle (or the neutral particles, since one doesn't know whether there be one or many) emitted by the nucleus in one process, will be equal to the energy which corresponds to the upper limit of the beta spectrum. It is obvious that we assume not only energy conservation but also the conservation of linear momentum, of angular momentum and of the characteristics of the statistics in all elementary processes.

With regard to the properties of these neutral particles, we first learn from atomic weights [*of radioactive elements*] that their mass cannot be much larger than that of the electron. In order to distinguish them from the heavy neutrons, E. Fermi proposed the name "neutrino." It is possible that the neutrino proper mass be equal to zero, so that it would have to propagate with the velocity of light, like photons. Nevertheless, their penetrating power would be far greater than that of photons with the same energy. It seems to me admissible that neutrinos possess a spin $1/2$ and that they obey Fermi statistics, in spite of the fact that experiments do not provide us with any direct proof of this hypothesis. We don't know anything about the interaction of neutrinos with other material particles and with photons: the hypothesis that they possess a magnetic moment, as I had proposed once (Dirac's theory induces us to predict the possibility of neutral magnetic particles) doesn't seem to me at all well founded.

In this connection, the experimental study of the momentum difference [read *balance*] in beta disintegrations constitutes an extremely important problem; one can predict that the difficulties will be quite insurmountable [read *very great*] because of the smallness of the energy of the recoil nucleus.

References

1. R. A. Millikan, in *Encyclopedia Britannica*, 14th edition, volume 8, page 340 (1929).

2. *Rapports du Septième Conseil de Physique Solvay*, 1933, Gauthier-Villars, Paris (1934), page 324. Pauli's remarks are in French.

3. G. Gamow, *Constitution of Atomic Nuclei and Radioactivity*, Oxford U.P. (1931).

4. J. Bromberg, Hist. Stud. Phys. Sci. **3**, 307 (1971).

5. W. Pauli, *Aufsätze und Vorträge über Physik und Erkenntnistheorie*, Braunschweig (1961); *Collected Scientific Papers*, volume 2, Interscience (1964), page 1313.

6. C. S. Wu, S. A. Moszkowski, *Beta Decay*, Interscience (1966).

7. C. S. Wu, in *Trends in Atomic Physics* (O. R. Frisch *et al*, eds.), Interscience (1959), page 45; C. S. Wu, in *Five Decades of Weak Interactions* (N. P. Chang, ed.), New York Acad. Sciences, New York (1977), page 37; A. Pais, Rev. Mod. Phys. **49**, 925 (1977).

8. C. D. Ellis, W. A. Wooster, Proc. Roy. Soc. (London) A **117**, 109 (1927); L. Meitner, W. Orthmann, Zeit. f. Phys. **60**, 413 (1930).

9. C. S. Wu, in *Theoretical Physics in the Twentieth Century* (H. Fierz, V. F. Weisskopf, eds.), Interscience (1960), page 249.

10. S. A. Goudsmit, in *Convegno di Fisica Nucleare*, Reale Accademia d' Italia, Atti, Rome (1932), page 41.

11. J. F. Carlson, J. R. Oppenheimer, Phys. Rev. **38**, 1737 (1931).

12. J. F. Carlson, J. R. Oppenheimer, Phys. Rev. **41**, 763 (1932).

13. C. Weiner, PHYSICS TODAY, May 1972, page 40.

14. W. Pauli, in *Handbuch der Physik*, Band 24/1 (1933), page 233; ref. 12, page 778.

15. N. Bohr, in ref. 10, page 119; J. Chem. Soc. (London), page 349 (1932).

16. C. D. Ellis, N. F. Mott, Proc. Roy. Soc. (London), A **141**, 502 (1933).

17. C. D. Ellis, in *International Conference on Physics, London, 1934, Vol. I, Nuclear Physics*, Cambridge (1935); W. J. Henderson, Proc. Roy. Soc. (London) A **147**, 572 (1934).

18. E. Fermi, Z. f. Phys. **88**, 161 (1934); English translation: F. L. Wilson, Amer. J. Phys. **36**, 1150 (1968).

19. F. Perrin, Compt. Rend. **197**, 1625 (1933).

20. V. F. Weisskopf, in *Exploring the History of Nuclear Physics* (C. Weiner, ed.), Amer. Inst. of Physics, N.Y. (1972), page 17.

21. N. Bohr, Nature **138**, 25 (1936). □

The birth of elementary-particle physics

PHYSICS TODAY / APRIL 1982

In the 1930s and 1940s physicists significantly revised their views on the elementary constituents of matter, which during the 1920s they had assumed to be only the electron and the proton.

Laurie M. Brown and
Lillian Hoddeson

By 1930, relativity and quantum mechanics were established, yet the excitement of the new physics was far from over. Indeed, the next half-century was characterized by startling experimental and theoretical discoveries and by new puzzles that appeared wherever one looked.

In the late 1920s all matter was thought to be made up of protons and electrons. There were, of course, many difficulties with this view, and the effort to revise it led to new problems—and to the birth of the field of modern elementary-particle physics. Three currents flowed together to make particle physics: nuclear physics, cosmic rays and quantum field theory. By the mid-1930s, there was conflict and apparent paradox where these fields overlapped, and although some of the conflict was resolved by the end of the 1940s, the resolution raised new and urgent problems.

Today there is increasing interest in this historical process. An international symposium was held recently at Fermilab to study the history of particle physics through lectures by important participants and through discussions among physicists and historians. An earlier symposium, at the University of Minnesota,[1] considered the role of nuclear physics in the origins of particle physics. This article is an outgrowth of the Fermilab meeting, which concentrated mainly on the parts played by cosmic rays and quantum field theory in the emergence of the

new field. In the discussion of the origins of particle physics with which we begin this article, we retain that emphasis: We will mention the role of the atomic nucleus, but we shall concentrate on the roles of cosmic rays and theory.

The nucleus and cosmic rays

There were many problems in treating the nucleus as a quantum mechanical system of protons and electrons.
▶ The nucleus was supposed to contain A protons and $A - Z$ electrons. But when the latter number is odd, as for lithium-6 or nitrogen-14, the spin and statistics are incorrect.
▶ Moreover, unpaired electron spins in the nucleus implied a hyperfine splitting of atomic spectral lines on a scale about a thousand-fold larger than is observed.
▶ In the relativistic quantum theory of the electron it was impossible to confine the light electron within the small nucleus.
▶ Finally, there was the continuous spectrum of β-decay electron energies, which called into question even the conservation of energy.

Physicists seriously considered radical suggestions for modifying the mechanics, the electrodynamics and even the conservation laws. But the resolution was to hinge on new particles: the neutron, discovered by James Chadwick in 1932, and the neutrino, proposed by Wolfgang Pauli in 1930 and incorporated in a theory of β decay by Enrico Fermi in 1934. These two neutral particles permitted the banishment of electrons from nuclear models. Soon after Carl David Anderson's 1932 discovery of the positron in cosmic rays, Irene Curie and Frederic Joliot produced artificially radioactive light elements that decayed by positron emission, and the picture of nuclear β decay was complete.

Cosmic rays were discovered as a result of post-1900 investigations of

fine-weather "atmospheric electricity," that is, ionization in the absence of an electrical thunderstorm. After one had accounted for all known sources of ionization, there remained a "residual" conductivity, even in closed vessels that were heavily shielded. This phenomenon implied the existence of a penetrating radiation of unknown origin.

Researchers—notably Victor F. Hess in Austria—conducted balloon flights, mainly in central Europe, to investigate altitude dependence of atmospheric conductivity. The manned balloons carried sealed electrometers whose rates of discharge first decreased with altitude, but then (above 2 km) began a marked increase. This pattern of ionization suggested the existence of an extraterrestrial source for the penetrating radiation, so that by the late 1920s one spoke of the *cosmic rays* (see box on page 39, *Discovery*). Until 1930, their specific ionization (ions per cm³ per sec) was the only property systematically observed.

The focus changed at the end of the 1920s when researchers used two methods, coincidence counting and the cloud

Laurie M. Brown is professor of physics and astronomy at Northwestern University in Evanston, Illinois. Lillian Hoddeson is historian of physics at Fermilab and in the physics department of the University of Illinois at Urbana–Champaign.

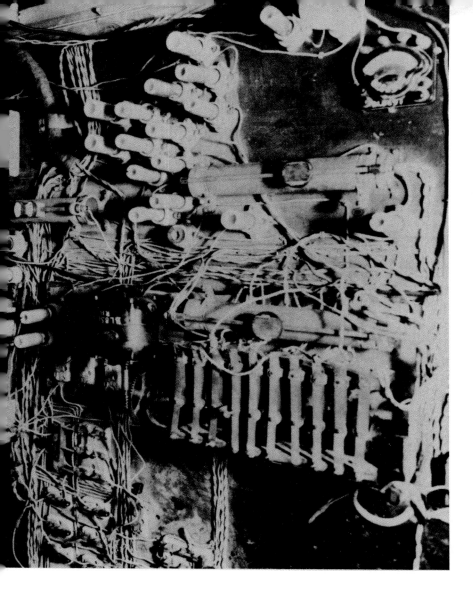

Carl Anderson and control panel for cloud chamber in trailer on Pike's Peak, 1935. (Courtesy of Carl Anderson.)

curved, with a radius of curvature directly proportional to the particle's momentum and inversely proportional to the magnetic field. Skobeltzyn noted that the tracks appeared to be associated with each other, to a degree difficult to account for by the scattering processes known at that time.[3] His was the first method for studying the interactions of particles of energies higher than those available from radioactive sources.

Skobeltzyn's counterpart in California was Carl David Anderson, who had been using a cloud chamber to study photoelectrons produced by x rays. Anderson wanted to move on to study Compton collisions of nuclear γ rays, but in 1930, at the urging of his boss, Robert A. Millikan, he began tooling up a cloud chamber and a strong magnetic field to observe cosmic-ray interactions. Anderson was to discover two new particles in cosmic rays: the positron and the muon.[4]

The other major step forward was Patrick M. S. Blackett and Giuseppe P. S. Occhialini's invention and use in 1932 of the counter-controlled cloud chamber.[5] In such a chamber, both the expansion and camera are activated by an electronic pulse from a counter array that selects a class of events, so that the incident particle "takes its own picture." Soon after Anderson had discovered what he referred to as "easily deflectable positives," Blackett and Occhialini used their new instrument to observe electron pair production and cascade showers. By 1930, therefore, the technical framework had been established for two decades of spectacular cosmic-ray and new-particle discoveries, made using counter and cloud-chamber techniques.

Theory

Relativistic electron theory, which led to the "prediction" of the positron, and the quantum theory of fields were both on the agenda of theoretical physics after Werner Heisenberg and Erwin Schrödinger invented quantum mechanics in 1925 and 1926. Both theories emerged from the fertile brain of Paul A. M. Dirac. In a pioneering work of February 1927 on quantum electrodynamics (QED), Dirac proposed a solution to the problem of the wave–particle duality, which had puzzled physicists since Albert Einstein hypothesized the light-quantum in 1905.[6] At the end of his paper, Dirac summarized its contents as follows:

The problem is treated of an assembly of similar systems satisfying the Einstein–Bose statistical

chamber with magnetic field, to study the individual behavior of the charged particles produced by collisions of primary cosmic rays with air molecules. They adapted both methods from techniques used to study x rays and radioactivity. The two methods were flexible, permitting a variety of experiments to be performed; and they could be combined. Their descendants are the principal tools used today to study the interactions of elementary particles, whether the source be cosmic rays or accelerators. The pioneers in this enterprise were Walter Bothe and Werner Kolhörster in Berlin and Dmitry Skobeltzyn in Leningrad.

Improved detectors. Kolhörster, a colleague of Bothe at the Physikalisch–Technische Reichsanstalt in Charlottenburg, outside Berlin, and an experienced cosmic-ray worker, pointed out in 1928 that by aligning two point-counters in a vertical array, one could use Bothe's counting technique of coincidence to make a γ-ray telescope for cosmic rays. Bothe and Kolhörster then implemented a similar scheme, using the far-more-efficient Geiger–

Müller tube counter. By mid-1929 they established that a 4.1-cm-thick gold block placed between the counters reduced the coincidence rate by only 24%, and they concluded from this that the primary rays had a "corpuscular nature."[2] Until then the rays had been thought to be high-energy photons and had been called (by Hess, for example) "ultra γ rays."

Bruno Rossi, at the physics laboratory of the University of Florence in Arcetri, Italy, soon found a way to improve the technique. By using a vacuum-tube circuit to detect the coincident discharges of the tube counters, he achieved greater flexibility and time resolution. With three out-of-line counters, he discovered that there was a great abundance of secondary radiation—later identified as "cascade showers."

Meanwhile in Leningrad, Skobeltzyn, who had been studying γ radiation from radioactive materials, began using the Wilson cloud chamber to observe the trajectories of cosmic-ray particles in a magnetic field. In such a field a charged particle's track is

mechanics, which interact with another different system, a Hamiltonian function being obtained to describe the motion. The theory is applied to the interaction of an assembly of light-quanta with an ordinary atom, and it is shown that it gives Einstein's laws for the emission and absorption of radiation.

The interaction of an atom with electromagnetic waves is then considered, and it is shown that if one takes the energies and phases of the waves to be q-numbers satisfying the proper quantum conditions instead of c-numbers, the Hamiltonian function takes the same form as in the light-quantum treatment. The theory leads to the correct expressions for Einstein's As and Bs.

(The As and Bs are light-quantum emission and absorption probability amplitudes.) From this we can see that Dirac treated the electromagnetic field as a Bose–Einstein gas of light-quanta. The following year, Pascual Jordan and Eugene Wigner gave the analogous treatment for a Fermi–Dirac gas, applicable to electrons.[7] The Jordan–Wigner type of quantization, designed to prohibit more than one electron from occupying a given state, was just what Dirac needed to formulate the theory of holes and the notion of antimatter.

In his 1927 papers on the quantum theory of the electromagnetic field, Dirac quantized only the radiation part of the field, consisting of transverse waves. The Coulomb interaction was considered a part of the energy of the "matter" system, that is, the charged particles. This separation is conven-

ient and often is a calculational necessity. However, as Gregor Wentzel has remarked, it "not only appears contrary to the spirit of Maxwell's theory, but also raises questions from the viewpoint of relativity theory... the splitting is not [relativistically] invariant."[8] Thus, in 1929, Heisenberg and Pauli took up a task whose completion would require the best theoretical efforts of the next two decades:

... to connect, in a contradiction-free manner, mechanical and electrodynamic quantities, electro-magneto–static interaction, on the one hand, and radiation-induced interactions on the other, and to treat them from a unified viewpoint. Especially [to take] into account in a correct manner the finite propagation velocity of electromagnetic forces.[9]

In the course of this work they discovered that the self-mass of the point electron was infinite, just as in the classical theory. (See box on page 42.) It was not until the postwar period that a more self-consistent QED was achieved. Nevertheless, the admittedly imperfect QED could still be fashioned into an effective tool for analyzing the high-energy cosmic rays.

QED and cosmic rays

Some disturbing experiments at moderate energies—energies some-

Paul A. M. Dirac (at right) in Ann Arbor, Michigan, in 1929. Leon Brillouin is in the background.
Robert A. Millikan and G. Harvey Cameron (below) with early cosmic-ray electroscopes. In this photo, taken about 1925, Millikan (left) is holding some lead shielding and Cameron an electroscope. (Photos courtesy AIP Niels Bohr Library.)

what larger than twice the rest mass energy of the electron—showed a much greater energy degradation and scattering of high-energy γ rays than was predicted by a Compton-effect calculation based on Dirac's relativistic electron theory.[10] By 1933, the excess absorption was found to be due to the production of electron–positron pairs, and the excess "scattering" was traced to photons produced by pair annihilation. This resolved the "doubts at $2mc^2$,"

which, however, moved then to $137mc^2$. The existence of doubts about the validity of QED at energies of the order of $137mc^2$ is corroborated by Anderson, who says that in 1934 members of the Caltech group spoke among themselves of " 'green' electrons and 'red' electrons—the green electrons being the penetrating type, and the red the absorbable type." But the green electrons did not behave like electrons. Although the formulas giving the ionization energy loss for very fast charged particles were considered to be accurate, there seemed to be a problem with the radiation formulas, even in 1934. Referring to Anderson's analysis of cloud chamber photographs, Hans Bethe and Walter Heitler said that *the theoretical energy loss by radiation is far too large to be in any way reconcilable with the experiments of Anderson.*[11] For the particles of energy 300 MeV (the assumed green electrons), Anderson found an energy loss of 35 MeV per centimeter of lead, whereas Bethe and Heitler concluded that "it seems impossible that the theoretical energy loss can be smaller than about 150 million volts per centimetre lead for Anderson's electrons."

Instead of suggesting that these strangely behaving electrons might be some other particles, Bethe and Heitler proposed a possible explanation that reveals the spirit of the time:

This can perhaps be understood for electrons of so high an energy. The de Broglie wave-length of an electron having an energy greater than $137mc^2$ is smaller than the classical radius of the electron $r_0 = e^2/mc^2$. One should not expect that ordinary quantum mechanics which treats the electron as a point-charge could hold under these conditions. It is very interesting that the energy loss of the fast electrons really proves this view and thus provides the *first instance in which quantum mechanics apparently breaks down for a phenomenon outside the nucleus.* We believe that the *radiation of fast electrons will be one of the most direct tests for any quantum-electrodynamics to be constructed.*[11]

QED proves indispensable. The problem was not with QED but with the assumption that Anderson's penetrating high-energy "green" electrons were electrons. They were, in fact, mesotrons (now called muons), about 200 times as massive. But about three years had to pass before anyone had the courage, or the faith in QED, to ascribe the discrepancy to new particles.

It was tempting at the time to explain away discrepancies between observed high-energy phenomena and theoretical expectations by appealing to a breakdown of QED at small distances, at a "fundamental length" or at the corresponding large momenta. But this became impossible by 1937; by that time, through a complex series of steps, QED showed itself to be not only useful after all, in spite of its menacing infinities, but also the indispensable means for understanding the nature of the cosmic rays. Although the electrodynamics of such energetic particles were questioned, Evans James Williams showed in 1933 that the important momentum transfers involved are small and that in a suitably chosen reference frame the collisions are gentle ones that do not involve high energies or small distances.[12]

In another step, taken in 1934, Bethe and Heitler calculated the relativistic formulas for bremsstrahlung (x-ray production) and electron pair creation. As noted earlier, they found significant disagreement with Anderson's results when they assumed Anderson was looking at electrons. Also, Williams and Carl-Friedrich von Weizsäcker showed that no disagreement with theory was to be expected, *even if QED were to break down at $137mc^2$.* Again, the argument was based upon looking at the collisions in a suitable rest frame.[12] As Williams said in his 1935 article, "We find that the quantum mechanics which enter into the existing treatments really concerns energies of the order of mc^2 however big the energy of the electron or photon."

By 1937 QED had also demonstrated its usefulness by explaining the behavior of the "soft component" of the cosmic rays, the cascade showers. Many physicists contributed to the solution of this problem; the first successes were by Homi J. Bhabha and Heitler, and by J. F. Carlson and J. Robert Oppenheimer.[13] However, the infinities of QED remained, and to obtain useful results they had to be ignored or thought of as corrections that would be "small," were they calculable in finite terms.

Particles envisioned, particles seen

The particle discoveries of the early 1930s (if we can call the neutrino proposal a discovery) permitted the banishment of electrons from the nucleus. On the heels of the discovery of the neutron, Heisenberg made a model of the nucleus as a non-relativistic quantum-mechanical system of neutrons and protons in which the neutron was to some extent treated as an elementary particle, the neutral counterpart of the proton. However, within this scheme Heisenberg tried to model the neutron as a tightly bound compound of proton and electron, in which the electron loses most of its properties—notably its spin, magnetic moment, and fermion character. The dominant nuclear force was to consist of the ex-

---

Development of cosmic-ray physics

In successive periods there was always at least one change that was so significant that it required a totally new interpretation of the previous observations.

Prehistory (to 1911, especially from 1900):
▶ "Atmospheric electricity" during calm weather
▶ Conductivity of air measured by electrometers
▶ Connection with radioactivity of earth and atmosphere
▶ Geophysical and meteorological interest

Discovery (1911–1914) **and exploration** (1922–1930):
▶ Observers with electrometers ascend in balloons and measure the altitude dependence of ionization, showing that there is an ionizing radiation that comes from above
▶ Such measurements begin in 1909 and continue (at interval) to about 1930, in the atmosphere, under water, underground
▶ The primaries are assumed to be high enough photons from outer space
▶ Search for diurnal and annual intensity variations
▶ Study of energy homogeneity

Early particle physics (1930–1947):
▶ Direct observation of the primaries is not yet possible, but the "latitude effect" shows they are charged particles
▶ Trajectories of secondary charged particles are observed with cloud chambers and counter telescope arrays, and momentum is measured by curvature of trajectory in a magnetic field
▶ Discoveries of positron and pair production
▶ Soft and penetrating components
▶ Radiation processes and electromagnetic cascades
▶ Meson theory of nuclear forces
▶ Discovery of mesotron (present day muon)
▶ Properties of the muon, including mass, lifetime and penetrability
▶ Two-meson theory and the meson "paradox"

Later particle physics (1947–1953):
▶ Particle tracks observed in photographic emulsion
▶ Discovery of pion and π–μ–e decay chain
▶ Nuclear capture of negative pions
▶ Observation of primary cosmic-ray protons and fast nuclei
▶ Extensive air showers
▶ Discovery of the strange particles
▶ The strangeness quantum number

Astrophysics (1954 and later):
▶ Even now the highest energy particles are in cosmic rays, but such particles are rare
▶ Studies made with rockets and earth satellites
▶ Primary energy spectrum, isotopic composition
▶ X-ray and γ-ray astronomy
▶ Galactic and extragalactic magnetic fields

change of this much abused electron.

After Fermi's successful theory of β-decay gave the neutrino a more legitimate status than it had previously enjoyed, there were attempts (although not by Fermi) to incorporate electron–neutrino pair exchange into the Heisenberg nuclear picture—the so-called Fermi-field model. However, it was shown to be impossible to fit simultaneously the range and strength of nuclear forces together with nuclear β decay. In an attempt to resolve this conflict, Hideki Yukawa, in Japan, made a bold imaginative stroke by introducing a new theory of nuclear forces that required the existence of a new type of particle, a fundamental massive boson.[14] The particle was to carry either the positive or negative electronic unit charge, and its exchange was to be the agent of Heisenberg's charge-exchange nuclear force. From the range of nuclear forces its mass was determined to be about 200 electron masses. Furthermore, Yukawa's meson (as it later became known) was to be capable of decaying into an electron and neutrino, in accord with Yukawa's proposed mechanism for nuclear β decay. Finally, it was predicted to be a part of the cosmic-ray flux.

In 1937 Anderson and others discovered in cosmic rays both positive and negative charged particles with masses about 200 times that of the electron. Some researchers greeted this as a fulfillment of Yukawa's prediction.[15] A number of properties of these particles, including mass, charge and lifetime, were determined before or during World War II; properties such as spin and parity, and the characteristics of interactions, were not determined unambiguously until the large accelerators came into use at the turn of the 1950s.[18] The fact that the known properties, other than the charge, did not provide a satisfactory match between the meson observed in cosmic rays and Yukawa's postulated meson of nuclear force stimulated new field theories that went beyond QED.

Because these new field theories had even worse divergence difficulties than QED, and because their strong interactions made perturbation methods far more questionable, there again arose practical as well as esthetic demands for curing or circumventing "the infinities" of field theory. The theoretical struggle was double-pronged: One effort was to find a version of meson theory that agreed with the cosmic-ray meson's behavior; another was to find a meson theory to fit the nuclear forces, whose complicated behavior came to be better known. An important success of the second approach was Nicholas Kemmer's symmetric meson theory of nuclear forces, which established the

utility of the concept of isospin and called for the existence of a charged triplet of positive, negative and neutral mesons.[16] The neutral meson, whose two-photon decay initiates the majority of cascade showers in the cosmic rays, was not observed until it was artificially produced in 1950.

Admitting a new particle. The cosmic-ray meson (muon) is the main component of the hard or penetrating cosmic rays. The penetrating rays were seen as early as 1929 in the first absorption measurements made on individual cosmic-ray particles, and perhaps were suggested by even earlier measurements. But it was not until 1937 that Seth Henry Neddermeyer and Anderson claimed these cosmic-ray mesons to be new charged particles—neither electrons nor protons—on the basis of their ability to penetrate a 1-cm thickness of platinum.[4] Even though scientists elsewhere, in England and France for example, were making similar observations, the preferred interpretation was that QED breaks down at high energies. That was the view of Blackett, who called the particles electrons and considered that a modification of the radiation formulas was in order.[17] Two French cloud-chamber groups emphasized that there were "two species of corpuscular rays" (like Anderson's red and green electrons) differing in their penetrating power; however, they did not insist on any *new* particles.[18] Two observations of mesons stopping in the gas of a cloud chamber permitted a determination of their masses sufficient to show them to be roughly 200 times the electron mass, or about one-tenth of the proton mass.[19]

The next step was to deal with the problem posed by the false identification of the muon with Yukawa's nuclear meson: If the cosmic-ray meson were Yukawa's strongly interacting particle, why did it not seem to interact at all? The remaining story of the muon—the determination of its mass, lifetime and interaction properties, and the growing sense of bewilderment and paradox in the confrontation between experiment and theory—was climaxed when an Italian group proved that negative muons stopping in carbon decay before they can be captured by the nucleus.[20]

The grand finale came when a group at Bristol University in England used a new nuclear photographic emulsion technique to reveal the pion, Yukawa's nuclear meson, and its decay into the muon, the cosmic-ray meson.[21] However, the solution of the π–μ paradox produced a new one, the "muon puzzle": making sense of the evidence that the muon was a heavy version of the electron—in modern terms, a second-generation lepton. The observation of the complete decay chain, pion→muon

→electron, together with the long muon lifetime, strongly suggested this similarity. Today this is known as the puzzle of the "generations" of quarks and leptons.

Unification and diversification

Many physicists today believe that we are approaching a new synthesis in our view of matter, in which the world will be seen as made up of a few types of elementary particles that interact by means of a small number of forces, with both particles and forces being aspects of a few or perhaps even a single quantum field. An important reason for this confidence in unification is the apparent success of the theory of the unified electroweak field. The 1979 Nobel lectures in physics deal with this subject and with speculative theories of a more advanced type, having names such as "electronuclear grand unification" and "extended supergravity."

The mood of those lectures is one of barely qualified optimism.[22] Sheldon Glashow, for example, while cautioning against the adoption of a "premature orthodoxy," contrasts the present with 1965, when he began theoretical physics and when "the study of elementary particles was like a patchwork quilt." He continues:

Things have changed. Today we have what has been called a "standard theory" of elementary particle physics in which strong, weak, and electromagnetic interactions all arise from a local symmetry principle. It is, in a sense, a complete and apparently correct theory, offering a qualitative description of all particle phenomena and precise quantitative predictions in many instances. There are no experimental data that contradict the theory. In principle, if not yet in practice, all experimental data can be expressed in terms of a small number of "fundamental" masses and coupling constants. The theory we now have is an integral work of art: The patchwork quilt has become a tapestry.

These remarks are reminiscent of other far-reaching syntheses: not only the "mechanical philosophy" of the eighteenth century and the "electromagnetic synthesis" at the end of the nineteenth century, but also physics as it appeared about 50 years ago. Then it was believed that there were only two fundamental material particles (electron and proton), only two fundamental forces (gravitation and electromagnetism), and that the fundamental laws were known (relativity and quantum mechanics). Accordingly, as Stephen Hawking reports, shortly after Dirac published his relativistic wave equation for the electron, Max Born said that "Physics, as we know it, will be

Lunches at the Niels Bohr Institute in Copenhagen. These photos, taken in 1934, show (at right) Walter Heitler with Leon Rosenfeld and (below) Werner Heisenberg and Niels Bohr. (Photos courtesy Paul Ehrenfest Jr)

over in six months.[23]

Although the positron discovery of August 1932 was a validation of Dirac's theory, that particle (and the neutron, neutrino and meson) totally destroyed the synthesis that appeared to be at hand in 1930. As Millikan said: "Prior to the night of 2 August 1932, the fundamental building-stones of the physical world had been universally supposed to be simply protons and neg-ative-electrons."[24] Progress in the 1930s and the next few decades would lie not in unification of forces and reduction in the number of elements but rather in diversification—the dis-covery of new particles, the enlarge-ment of the particle concept and the recognition of new nuclear forces, both strong and weak. During the 1930s and 1940s there were discovered the first antiparticle (the positron), the second baryon (the neutron), the second lepton (the muon), a neutral massless lepton (the neutrino, although actually first detected in 1953), the first massive field quanta, both charged and neutral (the pions), and the strange particles. By 1950, the modern idea of families of particles and the distinction between hadrons and leptons had already emerged. The idea of the universal

weak interaction was also in the air. For hadrons there was the beginning of what Victor Weisskopf has called the "third spectroscopy" (that is, after those of atoms and nuclei), although all three cases involve not only spectrosco-py but also structures. Thus the path toward unification, which looked at-tainable for a few years after resolution of the π-μ paradox, now seemed to twist through a minefield of the most diverse phenomena.

Particles and human attitudes

Because of their fundamental and universal character, elementary parti-cles (and their unexpected properties such as indeterminacy, complementar-ity, strangeness and spin) both influ-ence and are influenced by our general world outlook, from our primitive per-ceptions to our most advanced philoso-phical conceptions. Space limitations allow us only a glance at these issues, which we explore more fully in the symposium volume on which this arti-cle is based.

Some of the greatest battles occurred in the 1930s and 1940s over the en-largement of the concept of elementary particles far beyond the Newtonian mass point. Two of these battles in-

volve the neutron and the neutrino and belong also to nuclear physics. At the Minnesota symposium, Maurice Gold-haber recalled:

I remember being quite shocked when it dawned on me [in 1934] that the neutron, an "elementary particle" as I had by that time already learned to speak of it, might decay by β-emission with a half-life that I could roughly esti-mate ... to be about half an hour or shorter ...[1]

The battles over the positron and over the two mesons illustrate the psy-chological resistance of physicists to admit new particles to their cherished scheme. Dirac, in his first paper on the positron, and Anderson, in tune with what he called the "spirit of conserva-tism," both initially identified this new particle as a proton. Dirac even tried to make an argument for increasing the positron mass to the size of the proton mass; he realized that the new particles could not be protons only after Her-mann Weyl proved mathematically that the holes had to have the same mass as electrons.

Yukawa had virtually no support outside Japan for his proposed nuclear meson until the mu meson was ob-served. Bohr's response to the proposal by the Kyoto group that there is a neutral meson in addition to the charged one was "Why do you want to create such a particle?" And the tanta-lizing π-μ paradox during 1937–1947 arose out of the reluctance to admit that there could be a second particle, having a mass similar to that of the Yukawa particle but in other respects behaving differently.

Researchers in the 1930s and 1940s were strongly affected by the over-whelming economic, social and politi-cal upheavals of that period. To list but a few:

▶ the economic depression, which took away jobs and financial security
▶ the rise of fascism in Europe, which displaced many physicists (including

Development of quantum field theory

Prehistory
Classical (19th century):
▶ Electromagnetism (Faraday, Maxwell, Hertz, Lorentz)
Quantum (1900–1927):
▶ Blackbody radiation (Planck, 1900)
▶ Photon hypothesis (Einstein, 1905)
▶ Stationary states of atom (Bohr, 1913)
▶ Atomic emission and absorption coefficients (Einstein, 1916)
▶ Bose and Fermi statistics (1924)
▶ Electron waves (de Broglie, 1924)
▶ Exclusion principle and spin (Pauli, Goudsmit, and Uhlenbeck, 1925)
▶ Quantum mechanics of atoms and molecules (Heisenberg, Schrödinger, Dirac, Born, 1925–1926)
▶ General transformation theory (Dirac, 1927)

Birth and early development (1927–1929):
▶ Quantum electrodynamics (QED) (Dirac, 1927)
▶ Second quantization (Jordan and Klein, 1927, and Jordan and Wigner, 1928)
▶ Relativistic electron theory (Dirac, 1928)
▶ Relativistic QED (Heisenberg and Pauli, 1929)
▶ Theory of holes (Dirac, 1929)

Developments, difficulties and doubts (1929–1934):
▶ Applications of QED and Dirac theory (Klein and Nishina, 1929; Oppenheimer *et al.*; Bethe and Heitler, 1934)
▶ Experimental tests (Meitner and Hupfeld, 1930, Tarrant, Gray, Chao)
▶ Specter of infinite energy shifts (Oppenheimer, 1930)
▶ Specter of infinite vacuum polarization (Dirac, 1932)

New fields (1934–1946):
▶ Scalar field theory (Pauli and Weisskopf, 1934)
▶ Beta decay theory (Fermi, 1934)
▶ Meson theory of nuclear forces (Yukawa, 1935)
▶ Relativistic spin-one theory (Proca, 1936)
▶ "Infrared" radiation (Bloch and Nordsieck, 1937)
▶ S-matrix (Wheeler, 1937, and Heisenberg, 1943)
▶ Developments of meson theory (Fröhlich, Heitler, Kemmer, Yukawa, Sakata, Taketani, Kobayasi, 1938)

Renormalization (1947 and later):
▶ Lamb shift (Lamb and Retherford, 1947)
▶ Calculation of Lamb shift (Bethe, 1947)
▶ Electron magnetic moment (Foley and Kusch, 1948)
▶ Renormalized relativistic QED (Tomonaga, Schwinger, Feynman, Dyson, 1948–1949)

pation in 1945–1951 slowed nuclear-physics research by explicitly prohibiting experimental nuclear physics. Yet at the same time the occupation helped to establish the institutional basis for Japan's rapid progress in nuclear physics during the 1950s and 1960s.

In the postwar period, particle physics grew very rapidly, as did other subfields of physics. Many factors contributed to this postwar boom:
▶ the greater internationalism of science resulting from the war
▶ new experimental techniques developed as part of the weapons programs
▶ new funding mechanisms that emerged from the wartime support for research, resulting in, for example, the National Science Foundation and the Atomic Energy Commission
▶ the new widespread appreciation of the value of science for national security
▶ the sudden reentry into physics of graduate students and other researchers who, after about four years away, were anxious to make up for lost time
▶ the closer relationship between theory and experiment resulting from the experience of the large wartime projects such as building the bomb and developing radar for defense.

These and other influences need to be illuminated in detailed scholarly studies, for such larger issues are inseparable from the intellectual development of physics. Scholars will need to probe them deeply to understand fully the birth of particle physics.

* * *

This article is an abridged version of the introductory essay to the proceedings of the International Symposium on the History of Particle Physics, held at Fermilab 28–31 May 1980. The Proceedings were published in 1983 as The Birth of Particle Physics *(Cambridge U.P., New York).*

Weisskopf, Bethe, Fermi, Rossi and Rudolf Peierls) from their homes in Germany and Italy, and at the same time dissolved much of the research establishments in those countries
▶ the political controls on philosophy (including physics) in certain countries
▶ the brutal war, with its diversion from research to defense work
▶ its bombings and destruction
▶ the death camps

▶ the economic shortages
▶ the breakdown of communications between countries
▶ the occupations.

Other authors have dealt with these developments, but their effects on physics have not yet been fully examined. Many vital social issues have not been considered. For example, the postwar occupations had a definite impact on physics. In Japan, the American occu-

Robert Millikan (center) visits Seth Neddermeyer (right) and Carl Anderson on the summit of Pike's Peak, where Anderson set up his cloud-chamber experiment. The photograph was taken in 1935. (Courtesy of Carl Anderson.)

Hideki Yukawa and Richard Feynman during Feynman's visit to Kyoto, Japan, in the summer of 1955. Left to right: Mrs. Yukawa, Satio Hayakawa, Feynman, Yukawa, Koichi Mano, Minoru Kobayasi. (Courtesy of Satio Hayakawa.)

References

1. H. Steuwer, ed., *Nuclear Physics in Retrospect; Proceedings of a Symposium on the 1930s*, U. of Minnesota P., Minneapolis (1979).
2. W. Kolhörster, Naturwiss. **16**, 1044 (1928); W. Bothe and W. Kolhörster, Naturwiss. **16**, 1045 (1928).
3. D. Skobeltzyn, Z. f. Phys. **43**, 354 (1927); **54**, 686 (1929).
4. C. D. Anderson, Science, **76**, 238 (1932); S. H. Neddermeyer, C. D. Anderson, Phys. Rev. **51**, 884 (1937).
5. P. M. S. Blackett, G. P. S. Occhialini, Proc. Roy. Soc. **A139**, 699 (1933)
6. P. A. M. Dirac, Proc. Roy. Soc. (London) **A114**, 243 (1927).
7. P. Jordan, E. Wigner, Z. f. Phys. **47**, 631 (1928).
8. Gregor Wentzel, in *Theoretical Physics in the Twentiety Century*, M. Fierz, V. F. Weisskopf, eds., Interscience, New York (1960).

9. W. Heisenberg, W. Pauli, Z. f. Phys. **56**, 1 (1929); **59**, 168 (1930), Part II.
10. O. Klein, Y. Nishina, Z. f. Phys. **52**, 853 (1929).
11. H. Bethe, W. Heitler, Proc. Roy. Soc. (London) **A146**, 83 (1934). (Italics of Bethe and Heitler.)
12. E. J. Williams, Proc. Roy. Soc. (London) **A139**, 163 (1933); Phys. Rev. **45**, 729 (1934); K. Danske Vid. Selskab (Math.-Phys. Meddelelser) **13**, No. 4, 1 (1935); C. F. von Weizsäcker, Z. f. Phys. **88**, 612 (1934).
13. H. J. Bhabha, W. Heitler, Proc. Roy. Soc. (London) **A159**, 432 (1937); J. F. Carlson, J. R. Oppenheimer, Phys. Rev. **51**, 220 (1937).
14. H. Yukawa, Proc. Phys.-Math. Soc. Japan **17**, 48 (1935).
15. J. R. Oppenheimer, R. Serber, Phys. Rev. **51**, 1113 (1937); E. C. G. Stueckelberg, Phys. Rev. **52**, 41 (1937).
16. N. Kemmer, Proc. Camb. Phil. Soc. **34**, 354 (1938).

17. P. M. S. Blackett, J. G. Wilson, Proc. Roy. Soc. (London) **A160**, 304 (1937).
18. J. Crussard, L. Leprince-Ringuet, Compt. rend. **204**, 240 (1937); P. Auger, P. Ehrenfest Jr, Journ. de Phys. **6**, 255 (1935).
19. J. C. Street, E. C. Stevenson, Phys. Rev. **51**, 1005 (1937); Y. Nishina, M. Takeuchi, T. Ichimiya, Phys. Rev. **52**, 1198 (1937).
20. M. Conversi, E. Pancini, O. Piccioni, Phys. Rev. **71**, 209 (1947).
21. C. M. G. Lattes, H. Muirhead, G. P. S. Occhialini, C. F. Powell, Nature **159**, 694 (1947).
22. S. Weinberg, Rev. Mod. Phys. **52**, 515 (1980); A. Salam, Rev. Mod. Phys. **52**, 525 (1980); S. L. Glashow, Rev. Mod. Phys. **52**, 539 (1980).
23. S. Hawking, *Is the End in Sight for Theoretical Physics*, Cambridge U. P., New York (1980).
24. R. A. Millikan, *Electrons*, Cambridge U. P., New York (1935), page 320. □

the Discovery of
ELECTRON TUNNELING
into SUPERCONDUCTORS

By *Roland W. Schmitt*

PHYSICS TODAY / DECEMBER 1961

IN August 1960, Ivar Giaever published a discovery about electron tunneling into superconductors [1]; the discovery was elegant and had the esthetic simplicity that makes a scientist wonder why it had not been made before. It is too early to assess the importance of the discovery; it may be recorded as only a small but neat strand of science, or the train of work it has set off may produce a web of new knowledge about solids. Regardless of the final assessment that science makes of it, the discovery was surrounded by novel circumstances that dramatize the unexpected course of discovery.

Other physicists had come close to making the discovery or seemed on the verge of doing so: some had been doing similar experiments, but missed the discovery; some were looking for the wrong effect because of mistaken ideas; some were experimenting in the same field and, though not looking for a particular effect, could have stumbled on it. The experiment could have been done with equipment and techniques that were common a decade ago; it was not blocked by inadequate techniques and did not have to wait for the development of new research tools. Only a simple vacuum system for evaporating thin metallic films, a voltmeter, ammeter, and liquid helium were needed. The discovery was technically an easy one. Why, then, did the experiment remain undone during the previous decade while many physicists were working on superconductivity, including thin films? In spite of being simple, of being unblocked by technical complexities, of being in the arena of attention of many physicists, the discovery remained unsought and undetected until it was looked for and found by a young mechanical engineer just changing to a career in physics.

This story is the story of the discovery and the discoverer. I have only two reasons—other than the appeal of an entertaining story about research—for writing about the details of this microcosm in the history of science. Occupying an administrative post close to the people who played roles in the discovery, I had an intimate, but detached, view of the events that occurred. Also, in this story it is reasonably clear what was discovered and when it was discovered; what was new did not emerge slowly through the hazy fringes

Roland W. Schmitt is a physicist in the Metallurgy and Ceramics Research Department of the General Electric Research Laboratory, Schenectady, N. Y.

of discovery nor was it clouded by almost indistinguishable parallel discoveries. Goudsmit's fear that "when we try to look at a recent event with a microscope, the resolving power may often be insufficient" [2] does not hover too ominously in the background of this story. Except for these particular reasons, I make no claim that this story ought to be told any more than the stories of hundreds of other discoveries that go unreported.

THE history of superconductivity is a checkered one; it is characterized by long lapses between the major experimental discoveries and by an extraordinary hiatus between the original discovery and the first acceptable, fundamental theory of the phenomenon. Kammerlingh Onnes, in 1911 at Leiden, discovered superconductivity and found the characteristic property of zero resistance; he also learned that a high magnetic field would destroy superconductivity so that the state existed only at very low temperatures and in low magnetic fields. Another bulk property of superconductors remained hidden until 1933, when Meissner in Germany found it: in low magnetic fields, superconductors are perfect diamagnetics and expel all magnetic flux from their interior. The fundamental theory of the phenomenon still could not be developed in spite of intense efforts, but in 1950 the discovery of the isotope effect—a variation in the superconducting transition temperature with isotopic mass—confirmed an emerging suspicion of several theoreticians: that the interaction of electrons with lattice vibrations played the key role in producing superconductivity. Nevertheless, not until 1957, forty-six years after the original discovery, did Professor John Bardeen and two of his associates, Leon Cooper and J. Robert Schrieffer, develop a satisfactory theory of superconductivity.

One feature of this theoretical development is especially interesting for the story of electron tunneling into superconductors. The BCS theory, as it has come to be known, showed that a small but nonzero energy difference separated the first excited state of a superconductor from the ground state. Translated into the usual one-electron picture that physicists use when thinking about metals, this feature becomes a forbidden energy gap centered at the Fermi energy; in a superconductor, no electrons can have energies in this forbidden range.

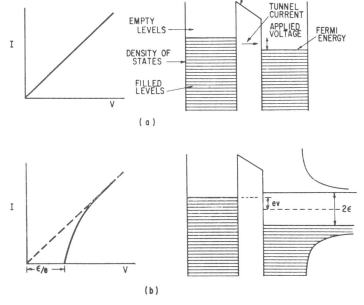

(a) The tunneling current between two normal metal films varies linearly with voltage at low voltages. As the voltage is increased, more and more filled levels of the metal film with negative bias are exposed (through the thin insulating barrier) to empty levels in the opposed metal film. This permits more and more electrons to tunnel through the barrier into empty states.

(b) The discovery. The forbidden energy gap at the Fermi level in superconductors prevents electrons from tunneling through the barrier into the superconductor until the biasing voltage exceeds half the gap width. The shape of the current-voltage curves measures the gap width and the density of states near the gap. The curve in this figure corresponds to $T = 0$.

Speculations about this energy gap reach back twenty years into the history of superconductivity, and experiments to detect it engaged physicists both before and after the BCS theory. The most convincing evidence for the gap came from studies of the way infrared radiation passed through or was absorbed by very thin films of superconductors.[3] These studies, carried out by Professor M. Tinkham and his students at Berkeley, demanded the most skillful experimental techniques; they needed talented experimentalists for their success.

The presence of the forbidden gap in superconductors means that if one tries to inject electrons with the forbidden energies into a superconductor, they will be rejected by it. Giaever showed this to be true with his experiment; it gave the most simple, direct evidence for the existence of the energy gap in superconductors and also gave information about the behavior of electrons with energies near the gap. The experiment is to inject electrons into a superconductor by letting them tunnel through a very thin, insulating barrier. Such a barrier allows one to vary the potential difference between the metal from which electrons are drawn and the metal into which they are injected and, therefore, makes it possible to vary the injection energy. Furthermore, the barrier prevents the free flow of electrons from the metal into the superconductor, as would occur with direct contact, but still allows single electrons to move through it one at a time. The original experiment used a thin, evaporated, aluminum film, coated with its own oxide and topped by another thin, evaporated film of lead. At the boiling point of helium, the lead, but not the aluminum, is superconducting. At very low voltages, almost no current flows through the junction because the energy of the injected electrons is in the forbidden energy range of electrons in the superconductor, but the current grows rapidly as the voltage

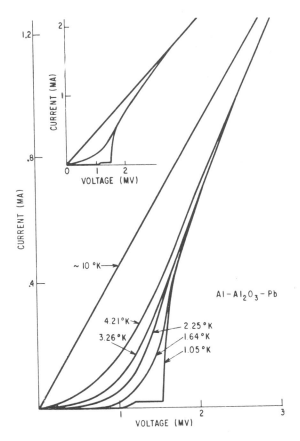

Experimental results showing tunneling current between aluminum and lead at various temperatures. At 10°K neither metal is superconductive, between 4.2°K and 1.3°K only lead is superconductive, and below 1.3°K both are superconductive.

reaches a value equal to half the width of the forbidden gap, for then the injection energies are equal to the allowed energy values in the superconductor. The current-voltage curve reveals directly the existence of the energy gap in superconductors and permits a simple measure of the size of this gap. With this elementary experiment, Giaever not only opened a new realm of experimental work on superconductors, but also created the hope of further discoveries about tunneling into metals, semimetals, and semiconductors.

The story behind the discovery begins in 1957. John C. Fisher became interested in the electronic properties of thin films; he talked about experimental possibilities with several people in our research group, but, because his main interest was different, there was no further activity until the latter part of 1958 when Giaever joined the section.

GIAEVER was born and educated in Norway; in 1954 he emigrated to Canada as a mechanical engineer. There he worked for a while as an architect's aide, but soon joined the Canadian General Electric Company. In 1956, he came to Schenectady in order to follow an advanced training program for engineers. During this period he had one assignment of six months at the General Electric Research Laboratory and worked on a problem of heat flow—a problem in applied mathematics associated with an applied-research project. During this time Giaever noticed that there were solid-state physicists at the Laboratory who were working on problems that seemed to be more interesting to him than the problems of engineering. Near the end of his assignment he asked if he could switch fields and try to become a physicist.

He joined our group, a group devoted to solid-state physics research, in September 1958 and began work under John Fisher. At the same time, Giaever began taking advanced courses in physics at Rensselaer Polytechnic Institute in Troy, N. Y. These studies were to prove critical in the discovery.

Fisher and Giaever began their work on thin films with Langmuir films; they tried, by various techniques, to put metallic electrodes on opposite sides of monomolecular layers and to measure electrical conductance through them. This technique proved so cumbersome and unreliable that after a few months they abandoned it and turned to evaporated-film junctions of aluminum-aluminum oxide-aluminum. With these films they did a series of experiments measuring the relation of electrical current through the oxide film with film thickness, voltage, and temperature, and showed that electron tunneling caused the current through the barriers [4]. During the year occupied with this work, Giaever learned both physics and experimental techniques, and by the end of 1959 he was carrying most of the work forward while Fisher's main efforts remained with other problems; nevertheless, Fisher continued to be the main source of stimulation, ideas, and criticism other than Giaever himself.

Aluminum is a superconductor if cooled below $1.2°$ K, and it may have been because of this fact alone, and for no better reason, that it was first suggested that the $Al-Al_2O_3-Al$ junctions be cooled to see what effect, if any, superconductivity would have on the tunneling current. The origin of the question, "Why don't you cool them to superconducting temperatures?" is lost; the question is of a type continually being asked in an active research group, and several people asked it at one time or another. Each time Giaever rejected the suggestion, because, he argued, most of the junction resistance was in the barrier itself and a vanishing resistance of the metal films would make no important difference to the junction current. In the light of subsequent events this argument may seem astonishing, yet no one in the preceding decades had joined a conception of the experiment with a reason for doing it, and it is not surprising that Giaever did not at first do so. In any case, he could not at that time have seen the real reason for doing the experiment, for he did not know of the energy gap at the Fermi level in superconductors! He had not, in one year as a physicist, learned all of the things that a person with conventional training would be expected to know, and none of the solid-state physicists among whom he worked had mentioned the superconducting energy gap in a way that had caught his attention.

Early in the spring of 1960, the question about cooling the junctions to superconducting temperatures was asked again, and this time it almost coincided with the study of superconductivity in a course at RPI. There Giaever learned of the energy gap; he recognized that this gap could have an effect on the tunneling current and suggested this possibility to John Fisher, Charles Bean, and Walter Harrison. The first reaction of all three was that probably the gap would not be noticeable. It was, after all, quite small and was only a crude representation of a more complex, many-electron effect; one could not take the simple picture, so like the

Ivar Giaever, Walter Harrison, Charles Bean, John Fisher.

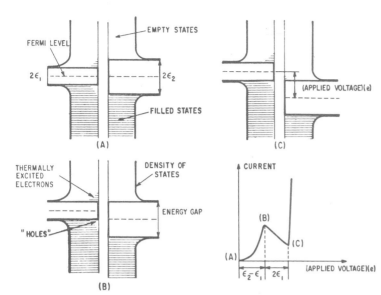

Tunneling between two different superconductors. (**A**) The two superconductors with no voltage applied. Thermally excited electrons above and holes below the gaps are shown. (**B**) When a voltage is applied, the thermally excited electrons in the left superconductor can tunnel into empty levels above the gap of the right superconductor. (**C**) When the voltage is increased further, only the same number of electrons may flow and they now face a lower, less favorable density of states in the right superconductor. The current decreases as a function of voltage until the electrons below the gap of the left superconductor are lifted enough to flow into the levels above the gap of the right superconductor.

picture of a semiconductor, too literally. Nevertheless, they all urged Giaever to try the experiment, and he soon calculated the width of a gap in units one would use in the experiment—in volts. Until then, none of us had noticed (at a time when it would have been meaningful) that superconducting gap widths are in the millivolt range, yet this simple fact was critical at that delicate moment when the experimenter had to decide whether to go ahead or not.

Giaever chose an aluminum-aluminum oxide-lead junction but failed to get definitive results in the first few trials. But, by this time, the conviction that there should be an effect was strong enough to carry the work on, and these efforts were shortly rewarded with success. Within a day or two of this success, Giaever and Charles Bean, who recognized the possible import of the experiment and began to work with Giaever, noticed that a simple model of the electron tunneling allowed them to deduce the density of states near the gap in a superconductor from the shape of the current-voltage curves. This observation suggested that electron-tunneling experiments might yield the density of states near the Fermi energy in normal metals and semimetals. Also, Giaever quickly recognized that tunneling between two superconductors should yield dynamic negative-resistance regions in the voltage-current characteristics. The second of these predictions has proved to be correct, and, following Giaever's publication of the original discovery[1], scientists at Arthur D. Little Company also recognized the possibility, and they, as well as Giaever, proved it to be true[5, 6]. The hope that tunneling experiments could measure the density of states in normal metals and semimetals became dim after detailed theoretical studies of Walter Harrison gave results different from the first, intuitive model. Subsequent experiments have failed to show interesting behavior, but all hope for some effects has not been abandoned.

By now, this discovery has firmly entered the science of superconductivity. It has also broadened the possibilities of other work on tunneling—technological as well as scientific—beyond the realm of semiconductors.

IN the end it is not possible to answer the question asked at the beginning of this story: why did this elegant experiment, one that is so easy to do, remain undone during the previous decade? Sir C. G. Darwin has said of the discovery of atomic numbers that it was an "easy" discovery, meaning that "when discovered, it is so easy to understand that it is difficult afterwards to see how people had got on without it"[7]. This kind of discovery, with its birth, destroys unrecognized barriers to the discovery that cannot subsequently be recreated or imagined. In this sense, Giaever's discovery was also an easy one.

Some of the ingredients that led to success are apparent in the story of the discovery: there was a question asked, catalyzing the reaction of knowledge about superconductors with experiments on electron tunneling; there was a delicate balance between theoretical knowledge and naiveté; there was a predisposition for working with simple, uncomplicated equipment; there was the permissive attitude of more senior research people. Chance played a role in arranging these factors, but to no greater extent than it plays a daily role in the research of every scientist; in spite of these ingredients the discovery could have been missed. The final key was that Giaever deliberately tried to make the discovery, and, in the end, knew why he wished to do the experiment and what he was looking for. This fact is probably the crucial fact that caused him to succeed while other scientists in a position to make the discovery did not.

Other discoveries have been made in other ways and the story of this one is not a prescription. But it is a reminder that even in this age of complexity there remain simple, but important, discoveries to be made.

References

1. Giaever, I., Phys. Rev. Letters **5**, 147 (1960).
2. Goudsmit, S. A., Physics Today, June 1961, p. 18.
3. Glover, R. E., III and Tinkham, M., Phys. Rev. **108**, 243 (1957).
4. Fisher, J. C. and Giaever, I., J. Appl. Phys. **32**, 172 (1961).
5. Nicol, J., Shapiro, S. and Smith, P. H., Phys. Rev. Letters **5**, 461 (1960).
6. Giaever, I., Phys. Rev. Letters **5**, 464 (1960).
7. Darwin, Charles, Proc. Roy. Soc. **A236**, 285 (1956).

Victor F. Weisskopf

PHYSICS TODAY / NOVEMBER 1981

Victor Weisskopf is Institute
Emeritus Professor of physics
at MIT. From 1961–1965 he
was Director General of CERN.

The development of field theory in the last 50 years

This article is devoted to the development of quantum field theory, a discipline that began with quantum electrodynamics,[1] which was born in 1927 when P. A. M. Dirac published his famous paper "The Quantum Theory of the Emission and Absorption of Radiation." Figure 1 reproduces the first page. Note that it was communicated by Niels Bohr himself. Also note the second and third sentences. The latter is an understatement indeed: *Nothing* had been done up to this time on quantum electrodynamics.

The pre-Dirac time

Classical electrodynamics started in 1862 when James Clerk Maxwell created his equations connecting the electric field **E** and the magnetic field **B** with the charge density ρ and the current density **j**. Together with the expression of the Lorentz force acting on a system carrying charge and current in an electromagnetic field, it led to an understanding of light as an electromagnetic wave, of the radiation emitted by moving charges and of the effects of radiation upon charged bodies. The results were splendidly verified by Heinrich Hertz in 1885 for radiations emitted and absorbed by antennas.

The application to atomic radiation was stymied by two facts: First, ρ and **j** in atoms were unknown to them; second, they faced a fundamental difficulty when the statistical theory of heat was applied to the radiation field. The number of degrees of freedom of a radiation field in a unit volume is infinite, and if each degree is supposed to get an energy $kT/2$ according to the equipartition theorem, the total energy density becomes infinity; empty space would be an infinite sink of radiation energy. Furthermore, apart from this distressing result, the classical theory of light had no explanation of the daily experience that incandescent matter changes its color with rising temperature—from red to yellow and then to white. The physicists must have felt before 1900 much as the neurophysiologists of today feel without any explanation of what memory is.

Then came quantum theory. It developed with increasing speed within a quarter century beginning with Max Planck's insight into the nature of blackbody radiation in 1900, followed by Albert Einstein's revolutionary idea of the existence of a photon in 1905, by Niels Bohr's atomic model in 1913, and by Louis DeBroglie's daring hypothesis of the wave–particle duality of particles in 1924. It reached its peak with the formulation of quantum mechanics by Werner Heisenberg, Erwin Schrödinger, Dirac, Wolfgang Pauli and Bohr in 1925.

The difficulties of the classical theory disappeared with one stroke—not without bringing about other difficulties about which much more will be said soon. Of course, the problem of heat radiation was immediately solved and the reasons for the sharp characteristic spectral lines of each atomic species became evident.

Atomic stabilities, sizes and excitation energies could be derived from first principles: The chemical forces turned out to be a direct consequence of quantum mechanics; chemistry became part of physics.

However, before the publication of Dirac's 1927 paper, it was not possible to derive the expressions for ρ and **j** within the atoms for the purpose of calculating the emission of light quanta.

Actually, the Schrödinger equation allowed the calculation of transitions under the influence of an external radiation field, that is the absorption of light and the *forced* emission of an additonal photon in the presence of an incident radiation. The field of an incident light wave could be considered as a perturbation on the atom in the initial state; it was possible by means of the Schrödinger equation to calculate the probability of a transition, which turned out to be proportional to the intensity of the incident light wave. However, the emission by a transition from a higher to a lower state in a field-free vacuum could not be treated. It was assumed at that time the matrix elements $\langle a|\rho|b\rangle$ and $\langle a|\mathbf{j}|b\rangle$ between two stationary states a, b of the atom play the role of charge and current density responsible for the radiation connected with the quantum transition from a to b or vice versa. The atom was considered as an "orchestra of oscillators," and the matrix elements determined the strengths of those oscillators ascribed to each pair of states. To determine the intensity

Title page of paper *(below) by*
P. A. M. Dirac (left) on radiation
theory (from Proceedings of the
Royal Society ***114****, 243,*
1927). Figure 1

AIP NIELS BOHR LIBRARY

The Quantum Theory of the Emission and Absorption of Radiation.

By P. A. M. DIRAC, St. John's College, Cambridge, and Institute for Theoretical Physics, Copenhagen.

(Communicated by N. Bohr, For. Mem. R.S.—Received February 2, 1927.)

§ 1. *Introduction and Summary.*

The new quantum theory, based on the assumption that the dynamical variables do not obey the commutative law of multiplication, has by now been developed sufficiently to form a fairly complete theory of dynamics. One can treat mathematically the problem of any dynamical system composed of a number of particles with instantaneous forces acting between them, provided it is describable by a Hamiltonian function, and one can interpret the mathematics physically by a quite definite general method. On the other hand, hardly anything has been done up to the present on quantum electrodynamics. The questions of the correct treatment of a system in which the forces are propagated with the velocity of light instead of instantaneously, of the production of an electromagnetic field by a moving electron, and of the reaction of this field on the electron have not yet been touched. In addition, there is a serious difficulty in making the theory satisfy all the requirements of the restricted

of spontaneous emission, one had to use either the oscillator model and equate the emission with the classical radiation of these oscillators, or one had to use the Einstein relations, from which it follows that the probability of spontaneous emission from b to a is equal to the absorption probability from a to b when the light intensity per frequency interval $d\omega$ is put equal to a certain value I_0:

$$I_0 d\omega = \frac{\hbar\omega^3}{4\pi^2 c^2} d\omega \qquad (1)$$

This happens to be the light intensity when each degree of freedom of the radiation field contained one photon. According to this rule the probability of spontaneous emission is equal to the probability of a *forced* emission by a fictitious radiation field of the intensity 1.

But why? According to the Schrödinger equation, any stationary state should have an infinite lifetime when there is no radiation present.

Quantization of the radiation field

Dirac's fundamental paper in 1927 changed all that. Quantum mechanics must be applied not only to the atom via the Schrödinger equation, but also to the radiation field. Dirac made use of the old idea of Paul Ehrenfest (1906) and Peter Debye (1910), to describe the electromagnetic field in empty space as a system of quantized oscillators. In the presence of atoms or of other systems of charged particles, the coupling between the charged particles and the field is expressed by an interaction energy

$$H^1 = e\int \mathbf{j}\cdot\mathbf{A}dx^3 \qquad (2)$$

where **j** is the current density of the particles. The value e of the particle charge is inserted here as an explicit factor and **A** is the vector potential. Both magnitudes are operators in the quantized system of the atom and the field oscillators. Expression 2 is a direct consequence of Maxwell's equations. The Hamiltonian of the combined system then has the form

$$H = H_0 + H^1$$
$$H_0 = H_{field} + H_{atom} \qquad (3)$$

where H_{field} is the Hamiltonian of the isolated field oscillators and H_{atom} is the Schrödinger Hamiltonian of the atom isolated from the electromagnetic fields.

The Hamiltonian H_0 describes field and atom without interaction. The effects of H^1 are treated as a perturbation upon the system H_0. The stationary states of H_0 are characterized by

$$(\ldots n_i \ldots; a) \qquad (4)$$

Here n_i are the occupation numbers of the radiation oscillators (the numbers of photons present in each oscillator i) and a indicates the stationary state of the atom.

The states 4 are no longer stationary when the perturbation energy H^1 is taken into account. The theory yields simply and directly the laws of emission and absorption of light. Indeed, the state $(\ldots 0,0, \ldots; a)$ of an atom in an excited state a without any radiation present is not stationary according to the Hamiltonian 3. A first-order perturbation calculation gives a probability $P_{ab} d\Omega$ per unit time for a transition from a to a lower state b, accompanied by the emission of a photon of a frequency $\omega = (\epsilon_a - \epsilon_b)/\hbar$ into the solid angle $d\Omega$ and with a polarization vector $\mathbf{s}$:

$$P_{ab} d\Omega = \frac{e^2}{\hbar c} \frac{(2\pi)^2}{\hbar \omega^2} I_0 |\mathbf{s}\mathbf{j}_{ab}|^2 d\Omega \qquad (5)$$

I_0 is given by the expression 1. The matrix element is determined by (for a one-electron system)

$$\mathbf{j}_{ab} = \int \psi_a^* \, \mathbf{j} \, \exp(i \, \mathbf{k}_{ab} \mathbf{x}) \psi_b \, dx^3$$

where $\mathbf{j}$ is the operator of the current, and $\mathbf{k}_{ab}$ the wave vector of the emitted quantum. The effect of the size of the system compared to the wavelength is taken into account by the exponential; it was neglected in the oscillator picture (dipole approximation). According to equation 5 spontaneous emission appears as a forced emission caused by the zero-point oscillations of the electromagnetic field, which are always present,

even in a space without any photons.

This was the start of an interesting development in theoretical physics. After Einstein had put an end to the concept of aether, the field-free and matter-free vacuum was considered as a truly "empty space." The introduction of quantum mechanics changed this situation and the vacuum gradually became "populated." In quantum mechanics an oscillator cannot be exactly at its rest position except at the expense of an infinite momentum, according to Heisenberg's uncertainty relation. The oscillatory nature of the radiation field therefore requires zero-point oscillations of the electromagnetic fields in the vacuum state, which is the state of lowest energy. The spontaneous emission process can be interpreted as a consequence of these oscillations.

Dirac's theory produced all results regarding the absorption and emission of light by atoms that previously were obtained by unreliable arguments. The results followed from the Hamiltonian 3 when the interaction energy 2 was treated as a first-order perturbation. Some other radiation phenomena such as photon scattering processes, resonance fluorescence and nonrelativistic Compton scattering of photons by electrons, appear in the second order of the perturbation treatment. The theory gave excellent account of all radiation phenomena in that order of perturbation in which they first appear. The higher approximations give rise to difficulties, which will be discussed later on.

Coupling to relativistic systems

In 1928 Dirac published two papers on a new relativistic wave equation of the electron. It was his third great contribution to the foundations of physics; the first was the reformulation of quantum mechanics, the "transformation theory,"[2] the second was the theory of radiation. The Dirac equation was supposed to replace Schrödinger's equation for cases where electron energies and momenta are too high for a nonrelativistic treatment. It immediately gave rise to four great triumphs:

▶ The spin $\hbar/2$ of the electron appeared to be a natural consequence of the relativistic wave equation. (It turned out later that there exist relativistic wave equations for particles with different spin. Dirac's equation for a spin $\hbar/2$ is distinguished by the fact that the energy operator appears linearly.)
▶ The g-factor of the electron necessarily has the value $g = 2$. The value of the magnetic moment of the electron followed directly from the equation.
▶ When applied to the hydrogen atom, the equation yields the correct Sommerfeld formula for the fine structure of the hydrogen spectrum.

The coupling of the quantized radiation field with the Dirac equation made it possible to calculate the interaction of light with relativistic electrons. The most important results were the derivation of the Klein–Nishina formula for the scattering of light by electrons, the Møller formula for the scattering of two relativistic electrons, and the emission of photons when electrons are scattered by the Coulomb field of nuclei.

In spite of these amazing successes a number of serious difficulties turned up immediately and it took a long time to solve them. The difficulties came from the existence of states of negative kinetic energy or negative mass. There was no way to get rid of them. If one tried to exclude them from the Hilbert space of the electron, the space becomes incomplete; furthermore, the Klein–Nishina formula could not be derived without them. Taken at face value, the existence of those states would imply that the hydrogen atom is not stable because of radiative transitions from the ordinary states to the states of negative energy. The properties of those impossible states were constantly in the center of discussion during those years. George Gamow referred to electrons in these states as "donkey electrons" because they tend to move in the opposite direction to the applied force.

Triumph and curse of the filled vacuum

It was again Dirac who proposed a way out of the difficulty in

1929. As it happens with ideas of great men, it was not only "a way out of a difficulty" but it was a seminal idea that led to the recognition of the existence of antimatter and ultimately to the development of field theory with all its concomitant insights into the nature of matter. He made use of the Pauli principle and assumed that, in the vacuum, all states of negative kinetic energy are occupied. This was the second step in the development of "populating" the vacuum. Later on this step was somewhat mitigated by eliminating the notion of an actual presence of those electrons, but the fluctuations of matter density in the vacuum remained as an additional property of the vacuum besides the electromagnetic vacuum fluctuations.

Dirac's daring assumption had most disturbing consequences, such as an infinite charge density and infinite (negative) energy density of the vacuum. Some of these impossible consequences were circumvented later, as is reported in the next section. However, the assumption not only solved most of the problems of the negative energy states but led to an impressive and unexpected broadening of our views about matter.

First of all, the transitions from positive to negative energy states were excluded, and the stability of the atoms was assured. Furthermore, Dirac's assumption required the existence of processes in which one particle from the "sea" of filled negative states is lifted to a state of positive energy, if the necessary energy is supplied by absorption of photons or by other means. A hole in the sea and a normal particle would be created. The hole would have all the properties of a particle of opposite charge. Moreover, a particle could fall back into a hole with the emission of photons of the right amount of energy and momentum. This, of course, would be a process of particle–antiparticle annihilation. Thus Dirac's assumption led to the recognition of the existence of antiparticles and of the existence of two new fundamental processes: pair creation and annihilation.

In the beginning these ideas seemed incredible and unnatural to everybody. No positive electron was ever seen at that time; the asymmetry of charges, positive for the heavy nuclei, negative for the light electrons, seemed to be a basic property of matter. Even Dirac shrank away from the concept of antimatter and tried to interpret the positive "holes" in the sea of the vacuum electrons as being protons. It was soon recognized, however, by Hermann Weyl, Robert Oppenheimer and by Dirac himself, that this interpretation would again lead to an unstable hydrogen atom and that the holes must have the same mass as the particles. Antimatter ought to exist. Indeed the positron was found by Carl Anderson in 1932; the antiproton was discovered 25 years later because its production needed energy concentrations several thousand times higher than were available before the invention of the synchrocyclotron. (The possibility of antiparticles was already mentioned by Pauli[3] and Einstein.[4] More about this can be found in a review by A. Pais.[5])

Once the idea of the filled vacuum took hold, it was relatively easy to calculate the cross section for the annihilation of an electron and a positron into two photons and the cross section for pair creation by photons in the Coulomb field of atomic nuclei. It is astonishing that it took more than three years after the identification of the holes with positrons, before the pair creation in a Coulomb field was calculated, although it was a very simple determination of a transition probability. It illustrates the wonder and incredulity that those ideas encountered during the first years.

Today it is hard to realize the excitement, the skepticism and the enthusiasm aroused in the early years by the development of all the new insights that emerged from the Dirac equation. A great deal more was hidden in the Dirac equation than the author had expected when he wrote it down in 1928. Dirac himself remarked in one of his talks that his equation was more intelligent than its author. But it was Dirac who found most of the additional insights himself.

The formulas derived for the creation of pairs and for radiative scattering (Bremsstrahlung) also gave an excellent account of the development of cosmic-ray cascade showers in matter, once the incoming energy is transformed into electrons and photons. It is interesting to observe how this success was interpreted. First it was considered as proof that radiation theory and pair creation are valid even at very high energy. Then, when it turned out that a part of the cosmic rays do not form showers (the part consisting of the then-unknown muons), doubts were expressed as to the validity of radiation theory at high energies. But it was shown by Enrico Fermi[6] and then by C. F. Von Weizsäcker[7] and E. J. Williams[8] that the effect of a Coulomb field on a fast-moving electron can be expressed as the effect of light quanta whose energy is only a few mc^2, when a suitable system of reference was used (the system in which the electron is at rest). This analysis of the production of cascade showers showed clearly that only energies and momenta of the order mc^2 and mc are exchanged in the relevant processes. Hence the shower production does not test the theory at high energies, nor could any deviation from the expected showers be explained by a breakdown of the theory at high energies.

Indeed, electron accelerators of many GeV were needed to test the theory at large energies. Recent measurements with electron–positron colliders have shown radiation theory to be valid at least up to energy exchanges of 100 GeV.

How unreasonable the idea of antimatter seemed at that time may be illustrated by the fact that many of us did not believe in the existence of an antiparticle to the proton because of its anomalous magnetic moment. The latter was measured by Otto Stern in 1933 and could be interpreted as an indication that the proton does not obey the Dirac equation. The fundamental character of the matter–antimatter symmetry and its independence of the special wave equations was recognized only very slowly by most physicists.

The following conclusions must be drawn from the new interpretation of the negative-energy states in the Dirac equation. There are no real one-particle systems in Nature, not even few-particle systems. Only in nonrelativistic quantum mechanics are we justified to consider the hydrogen atom as a two-particle system; not so in the relativistic case, because we must include the presence of an infinite number of vacuum electrons. Even if we consider the filled vacuum as a clumsy description of reality, the existence of virtual pairs and of pair fluctuations shows that the days of fixed particle numbers are over.

Furthermore, relativity requires that time and space be treated equivalently. In nonrelativistic quantum mechanics, time is a parameter, whereas the space coordinates of the particles are considered as operators. In relativistic quantum mechanics the particles appear as quanta of a field, just as the photons are quanta of the electromagnetic one. The fields assume the role of operators and the coordinates are parameters indicating the space- or time-dependence of the field operators. The theory of the interaction of charged particles with the radiation field becomes a field theory in which two (or more) quantized fields interact: the matter field and the radiation field.

The field amplitudes are expressed as linear combinations of creation and destruction operators that increase or decrease the number of particles in the quantum states of the system. It is a direct generalization of the quantization of the electromagnetic field as decomposed into oscillator amplitudes. The operator of an oscillator amplitude contains matrix elements only between states that differ by one unit of excitation. The corresponding operator either adds (creates) or subtracts (destroys) a quantum of the oscillator.

There are essential differences between a field of particles with spin ½ and the radiation field. The former describes the behavior of fermions, whereas the latter is an example of a boson field. In the classical limit, the boson fields are classical fields whose field strength is a well-defined function of space and time (radio wave). The fermion fields cannot have a classical limit because no more than one fermion can be put into one wave; its classical limit is a particle with a well-defined momentum and position. So far, the constituents of matter have all been shown to be fermions interacting by means of boson fields.

Furthermore, the interaction between fermion and boson fields in its simplest form necessarily is bilinear in the fermion fields and linear in the boson fields. This is indicated by the fact that the current density is a bilinear expression of the particle wave functions. One cannot construct a Lorentz-invariant expression that is linear or cubic in the spinor wave functions. Boson field (vector or scalar), however, may appear linearly in the interaction.

When the fields are expressed in terms of creation and annihilation operators, the form of the interaction can be interpreted in the following way: The fundamental interaction between fermions and bosons consists of the product of

James Clerk Maxwell

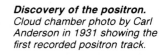

Discovery of the positron.
Cloud chamber photo by Carl Anderson in 1931 showing the first recorded positron track.

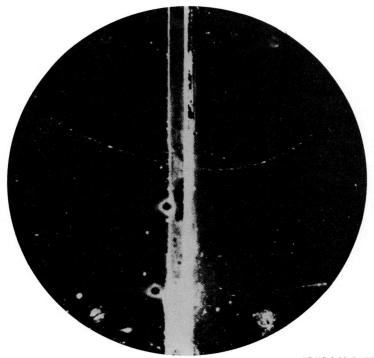

two fermion creation or destruction operators $b^\dagger$ and b, and one boson operator a or $a^\dagger$: $b^\dagger ba$ or $b^\dagger ba^\dagger$. It is interpreted as a change of state of a fermion "destroyed" in one state and "created" in another) accompanied with either an emission or an absorption of a boson.

The fight against infinities: elimination of the vacuum electrons

In spite of all successes of the hole theory of the positron, the infinite charge density and the infinite negative energy density of the vacuum made it very difficult to accept the theory at its face value. A war against infinities started at that time. It was waged with increasing fervor by the developers of quantum electrodynamics when more intricate infinities appeared besides those mentioned before, as will be described in the subsequent sections.

There is a rather primitive way to take care of the infinite charge density, by a slight change in the definition of charge and current. It amounts to the following argument: Because the theory is completely symmetric in regard to electrons and positrons, it would be equally valid to construct a theory in which the positrons are the particles and the electrons are the holes in a sea of positrons that occupy negative energy states. The actual theory then could be considered as a superposition of these two theories, one with an infinite negative charge density and the other with infinite positive one. This combination also serves to emphasize the symmetry between matter and antimatter. The vacuum charge densities cancel; the corresponding expressions for charge and current indeed give a more satisfactory description of the phenomena.

It was recognized in 1934 by Heisenberg[9] and by Oppenheimer and Wendell Furry[10] that the creation and destruction operators are most suitable for turning the liability of the negative energy states into an asset, by interchanging the role of creation and destruction of those operators that act upon the negative states. This interchange can be done in a consistent way without any fundamental change of the equations. The consequences are identical to those of the filled-vacuum assumption, but it is not necessary to introduce that disagreeable assumption explicitly. Particles and antiparticles enter symmetrically into the formalism, and the infinite charge density of the vacuum disappears. One even can get rid of the infinite negative-energy density by a suitable rearrangement of the bilinear terms of the creation and destruction operators in the Hamiltonian. After all, in a relativistic theory the vacuum must have vanishing energy and momentum. There remains, however, the unpleasant fact of the existence of vacuum fluctuations[11] without any energy.

The fundamental interaction between charged fermions and photons now contains three basic processes: the scattering of a fermion with the emission or absorption of a photon, the creation and the annihilation of a fermion–antifermion pair with the emission or absorption of a photon. All electrodynamic interaction processes are combinations of these fundamental steps.

Surprisingly enough, it took many years before the physicists realized the great advantages of this new formalism. One still reads about the "hole theory" of positrons in papers written in the late 1940s, when renormalization was the topic of the day.

An interesting episode in the fight for the elimination of vacuum electrons was the quantization of the Klein–Gordon relativistic wave equation for scalar particles. It seemed to be a rather academic activity because no scalar particle was known at that time. In that theory, the charge density $(\dot\phi^*\phi - \phi\dot\phi^*)$ and the wave intensity $|\phi|^2$ are not identical. Therefore, it seemed posssible that, under the influence of external electromagnetic fields, the total intensity $\int |\phi|^2 dx^3$ may change in time, although the total charge remains conserved. It smelled of a creation or annihilation process of oppositely charged particles. The problem attracted the attention of Pauli and myself[11] because we saw that the quantized Klein–Gordon equation gives rise to particles and antiparticles and to pair creation and annihilation processes without introducing a vacuum full of particles. Note that at the time the method of exchanging the creation and destruction operators (for negative energy states) was not yet in fashion; the hole theory of the filled vacuum was still the accepted way of dealing with positrons. Pauli called our work the "anti-Dirac paper;" he considered it as a weapon in the fight against the filled vacuum, which he never liked. We thought that this theory only served the purpose of an unrealistic example of a theory that contained all the advantages of the hole theory without the necessity of filling the vacuum. We had no idea that the world of particles would abound with spin-zero entities a quarter of a century later. This was the reason why we published it in the venerable but not widely read *Helvetica Physica Acta*.

Our work on the quantization of the Klein–Gordon equation led Pauli to formulate the famous relation between spin and statistics. Pauli demonstrated in 1936 the impossibility of quantizing equations of scalar or vector fields that obey anticommutation rules. He showed that such relations would have the consequence that physical operators do not commute at two points that differ by a spacelike interval. This lack of commutativity would contradict causality because it would require that mea-

Hideki Yukawa

surements interfere with each other when no signal can pass from one to the other. Thus Pauli concluded that particles with integer spin cannot obey Fermi statistics. They must be bosons. During the days of the hole theory it was obvious that particles with spin ½ cannot obey Bose statistics because it would be impossible to "fill" the vacuum. Four years later Pauli proved the necessity of Fermi statistics for half-integer spins, also on the basis of the same causality arguments.

The fight against infinities: infinite self mass

The infinities of the filled vacuum and of the zero-point energy of the vacuum turned out to be rela-

Wolfgang Pauli in 1931

tively harmless compared to other infinities that appeared in quantum electrodynamics when the coupling between the charged particles and the radiation field was considered in detail. No difficulties appeared as long as only the first terms of the perturbation treatment were taken into account, that is those terms in which the phenomena under consideration appear in the lowest order. It soon turned out that the higher terms always contain infinities, as Oppenheimer[12] had pointed out for the first time.

In 1934 Pauli asked me to calculate the self energy of an electron according to the positron theory. It was a modern repetition of an old problem of electrodynamics. In classical theory the energy contained in the field of an electron of radius a (neglecting the inside) is $4\pi e^2/a$ and would diverge linearly if the radius goes to zero. The corresponding calculation in the positron theory is much more complicated. One had to calculate the difference between two infinite amounts: the energy of the vacuum and the energy of the vacuum plus one electron. The result was equivalent to the statement that the electric field inside one Compton wave length $\lambda_c = h/mc$ from the electron is not e/r^2 but $(e/r^2)(r/\lambda_c)^{1/2}$. When r goes to zero it increases only as $r^{-3/2}$. The self energy then becomes[12]

$$E = m_0 c^2 + (3/2\pi) m_0 c^2 \times (e^2/\hbar c)\log(\lambda_c/a) \qquad (6)$$

where m_0 is the intrinsic or "mechanical" mass of the electron, which appears in the Hamiltonian of the electron when it is decoupled from the electromagnetic field. It diverges only logarithmically.

(This brings back one of the dark moments of my professional career. I made a mistake in the first publication that resulted in a quadratic divergence of the self-energy. Then I received a letter from Furry, who kindly pointed out my rather silly mistake and the fact that actually the divergence is logarithmic. Instead of publishing the result himself, he allowed me to publish a correction quoting his intervention. Since then the discovery of the logarithmic divergence of the electron self-energy is wrongly ascribed to me instead of to Furry.)

A consistent relativistic theory

requires a point electron, that is $a \rightarrow 0$. It is worth noting, however, that the value of a for which the second term of 6 becomes half of the first is as small as 10^{-72} cm! Even the Schwarzschild radius of the electron is only 10^{-55} cm. This value means that the deformation of the space around the electron is strong enough to prevent the electron from interacting with photons of that wave length, thus providing a natural cut-off long before the electromagnetic self-energy becomes important. Unfortunately, no consistent calculation of this effect has ever succeeded.

Another somewhat more benign type of infinities appeared in quantum electrodynamics when emissions of photons of very low frequencies were considered. Such emissions take place, for example, when electron beams are scattered by static electric fields. Classical theory predicts that the emitted energy does not vanish in the limit of zero frequencies. The quantum result ought to be identical with the classical one at that limit; it would indicate that the number of emitted quanta goes to infinity. This trouble, called "infrared catastrophe," can be avoided by describing this limit with the help of classical fields, as Bloch and Arnold Nordsieck[14] have shown in their important paper of 1937. It put an end to any worries about this kind of infinity.

The fight against infinities: infinite vacuum polarization

The virtual pairs endow the vacuum with properties similar to a dielectric medium. We may ascribe a dielectric coefficient ϵ to the vacuum. A direct calculation of this dielectric effect leads to a dielectric coefficient that consists of a constant part ϵ_0 and an additional part that depends upon the electromagnetic fields and their derivatives in time and space.

$$\epsilon = \epsilon_0 + \epsilon(\text{field}) \qquad (7)$$

The constant part ϵ_0 cannot have any physical significance because it serves only to redefine the unit of charge. Any charge Q_0 would appear as $Q = Q_0/\epsilon$. The actual value of ϵ_0 turns out to be logarithmically divergent (it goes as $\log(\Lambda/m)$ where Λ is the highest momentum considered in the calculation). The additional field-dependent term, however, turns out to be finite and therefore should have physical significance.

Let us now consider what happens to a charge Q_0 when placed in a vacuum with a dielectric coefficient of the form 7. At large distances r the effective charge will be Q_0/ϵ_0. When r becomes of the order $\lambda_c = \hbar/(mc)$ or less the second term of 7 becomes important. Calculations of this term for a Coulomb field were carried out by Robert Serber[15] and E. Uehling.[16] They found that $\epsilon(r)$ decreases with r when r becomes smaller than the Compton wave length λ_c. This is so because, for smaller r, only those virtual pairs contribute whose energy is larger than $\hbar c/r$. This decrease is finite and calculable. The infinite value of ϵ_0 was interpreted as an indication that the intrinsic "true" charge Q_0 is infinite so that the observed charge becomes finite and equal to $e = Q_0/\epsilon_0$ for $r \to \infty$. The decrease of ϵ with decreasing r when $r < \lambda_c$ would then amount to an increase of the effective charge Q_{eff} at those small distances.

This increase of Q_{eff} for $r < \lambda_c$ over the value e at large distances is rather small; it is of the order of $e/137$. A strong increase occurs only at very small distances $r \sim \lambda_c \exp(-\hbar c/e^2)$; these are the same distances as the ones we discussed in connection with the self-energy, at which the theory most likely is inapplicable. We then get a dependence of Q_{eff} on the distance as shown in figure 2. It is the first example of a "running coupling constant," which plays an important role in quantum chromodynamics.

The fight against infinities: renormalization

The appearance of infinite magnitudes in quantum electrodynamics was noticed in 1930. Because they only occurred when a certain phenomenon was calculated to a higher order of perturbation theory than the lowest one in which it appeared, it was possible to ignore the infinities and stick to the lowest-order results that were good enough for the experimental accuracy at that period. However, the infinities at higher order indicated that the formalism contained unrealistic contributions from the inter-

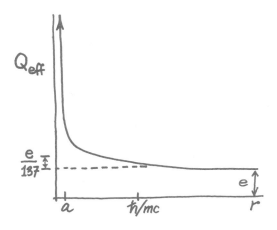

Running coupling constant in QED. The effective charge Q_{eff} as a function of the distance r. The distance a, the distance at which Q_{eff} is about 137 e, is very much smaller than indicated in this drawing.   Figure 2

Willis Lamb in 1947

Julian Schwinger

Richard Feynman *(photo by Sylvia Posner, courtesy of the CalTech Archives)*

action with high-momentum photons.

Already in 1936 the conjecture was expressed[17,18] that the infinite contributions of the high-momentum photons are all connected with the infinite self mass, the infinite intrinsic charge Q_0 and with nonmeasurable vacuum quantities such as a constant dielectric coefficient of the vacuum. Thus it seemed that a systematic theory could be developed in which these infinities are circumvented. At that time nobody attempted to formulate such a theory, although it would have been possible then to develop what is now known as the method of renormalization.

There was one tragic exception and that was E. C. G. Stueckelberg.[19,20] He wrote several important papers in 1934–38, putting forward a manifestly invariant formulation of field theory. This could have been a basis of developing the ideas of renormalization. Later on (in 1947) he actually formulated the complete renormalization procedure quite independently of the efforts of other authors. Unfortunately, his writings and his talks were rather obscure and it was very difficult to understand them or to make use of his methods. Had the theorists been capable of grasping his ideas they may well have calculated the Lamb shift and the correction to the magnetic moment of the electron at a much earlier time.

A new impetus to such attempts came from an experimental result. Willis Lamb and R. C. Retherford[21] were able to measure reliably the difference in energy between the $2S_{1/2}$ and $2P_{1/2}$ state of hydrogen (Lamb shift). The two states should have been exactly degenerate according to the Dirac equation applied to the hydrogen problem. Already in the 1930s the degeneracy of these two levels was in doubt from spectroscopic measurements, but Lamb and Retherford, using newly developed microwave methods, definitely established the splitting and measured it with great precision.

It had been conjectured long ago that such a splitting should be caused by the coupling of the radiation field with the atom, but early attempts to calculate it ran into difficulties because the infinite mass and vacuum polarization appeared in the same approximation. It was

H. A. Kramers who pointed out[22] that one ought to be able to calculate the effect by carefully subtracting the infinite energy of the bound electron from that of the free one and thereby separating the parts that contribute to the mass and charge from those of real significance. Infinities are always difficult to subtract in an unambiguous way. After the Lamb shift had been measured, Bethe had made an attempt to estimate the effect of the radiation coupling, simply by omitting the coupling with photons of an energy larger than mc^2. This attempt was successful because most of the effect comes from the coupling with photons of lower energy, which can be treated nonrelativistically.

An exact calculation to the lowest order in $(e^2/\hbar c)$ was then performed by Norman M. Kroll and Lamb[23] and by J. B. French and myself[24] (1949) and resulted in good agreement with the experiment. However, the methods used by those authors of subtracting two infinities were clumsy and unreliable. Subsequently, a formidable group of physicists, including Julian Schwinger, Richard Feynman, Freeman Dyson and Sin-Itiro Tomonaga, developed a reliable way to deal with the infinities. They introduced a method of renormalization in which the initial parameters were eliminated in favor of those with immediate physical significance. In any computation of an electrodynamical result, the effects of the mass and charge redefinitions had to be incorporated. Infinite "counterterms" are introduced into the Hamiltonian in such a manner that they compensate for the infinite mass and charge. In order to make this procedure unambiguous it was necessary to keep the expressions in a manifestly relativistic and gauge-invariant form throughout the calculations.

The results were most encouraging. Schwinger found that the magnetic moment of the electron should indeed be larger by the factor $1 + \alpha/(2\pi)$ than the Bohr magneton, a result that was observed shortly before by I. I. Rabi and his disciples and then more accurately by Henry Foley and Polykarp Kusch. The Lamb-shift results were recalculated in a much simpler way, radiative corrections of higher order in $e^2/\hbar c$ to scattering processes were unambiguous-

ly determined, and the vacuum polarization effects were worked out in detail; the latter found an impressive experimental confirmation in the measurements of the spectrum of muonic atoms (the electron replaced by a muon); the muon moves in the region $r \sim (\hbar/m_e c)$ where the vacuum polarization is a one-percent effect. Another remarkable test of the new methods was the agreement between the predicted and observed properties of positronium—the atom consisting of an electron and a positron, discovered and investigated for the first time by Martin Deutsch.

The war against infinities was ended. There was no reason any more to fear the higher-order terms. The renormalization took care of all infinities and provides an unambiguous way to calculate with any desired accuracy any phenomenon resulting from the coupling of electrons with the electromagnetic field. It was not a complete victory, because infinite counter-terms had to be introduced to remove the infinities. Furthermore, the procedure of eliminating infinities could be carried out only by renormalizing successively at each step of the perturbation expansion in powers of the coupling parameter. It still is not clear whether this method leads to a convergent series. It is like Hercules's fight against Hydra, the many-headed sea monster, which grows a new head for every one cut off. But Hercules won his fight and so did the physicists. Sidney Drell characterized the situation most aptly as "a peaceful coexistence with the infinities."

Here are the signs of victory in the war against infinities:

▶ **Lamb shift** (about 10% is due to vacuum polarization; most of the rest is the interaction with the zero-point oscillations of the electromagnetic field):

$$\Delta\nu(2S_{1/2} - 2P_{1/2}) = \frac{1057.862\ (20)\ \text{MHz (exp.)}}{1057.864\ (14)\ \text{MHz (theor.)}}$$

▶ *g*-**factor of the electron** $(a = \tfrac{1}{2}(g-2)) \times 10^3$

$$a = \frac{1.15965241\ (20)\ \text{(exp.)}}{1.159652379\ (261)\ \text{(theor.)}}$$

▶ **Vacuum polarization.** 90% of the Lamb shift in muonic helium (α particle + muon) is caused by vacuum polarization:

$$\Delta E(2S_{1/2} - 2P_{3/2}) = \frac{1.5274\ (0.9)\ \text{eV (exp.)}}{1.5251\ (9)\ \text{eV (theor.)}}$$

In spite of these victories there remain nagging problems in quantum electrodynamics. There are definite indications that we understand only a partial aspect of what is going on. As was mentioned before, the elimination of infinities is possible only in a perturbation approach; it is contingent upon the smallness of $e^2/\hbar c$. But the effective coupling constant at very small (indeed incredibly small) distances becomes larger than unity. Will there be a theory that avoids renormalization by using nonperturbative methods? Or will a future unification of electrody-

Sin-Itiro Tomonaga

Steven Weinberg

E. B. BOATNER

Abdus Salam

Sheldon L. Glashow

E. B. BOATNER

namics and general relativity heal the disease of divergencies because of the fact that the dangerous distances are smaller than the Schwarzschild radius of the electron?

Moreover, there is no way to understand and derive the mass of the electron within today's electrodynamics. This problem has become even more acute since heavier electrons such as the muon and the τ-electron have been discovered. There is not the slightest indication why electrons with different masses should exist. In present-day field theories the masses are arbitrary parameters that may assume any values.

Quantum electro-weak dynamics

The tremendous quantitative success of renormalized quantum electrodynamics (QED) has elevated this theory as an (almost) spotless example of a physical theory dealing with the interactions of electrically charged particles with fields. No wonder that the physicists tried to apply similar methods whenever interactions between fermions and bosons occurred. The first well-known use of QED as an example was the attempt of Hideki Yukawa (1935) to describe the nuclear force between protons and neutrons as an emission and subsequent absorption of a virtual boson. He had to ascribe a mass to that boson, because the nuclear force has a short range r_0 of the order of 10^{-13} cm. Any field theory modelled after QED would give an exponential force between fermions of the form $r^{-1} e^{-rMc/\hbar}$, with M the mass of the boson. The observed range of nuclear forces leads to a mass of about 200 MeV. No such bosons were known at that time, but he predicted the existence of them. His prediction was confirmed ten years later—an impressive success of a simple idea. Actually the nuclear force turned out to be the effect of somewhat more complicated processes; it does not detract from the beauty of his prediction.

The second early attempt to use QED as an example is a little known contribution by Oskar Klein.[25] He suggested a model for the weak interactions in which massive charged vector bosons mediated processes such as β de-

cay. He even called them by the currently used letter W. He was the first to propose that the neutron decay: $n \rightarrow p + e + \bar{\nu}$ be split into two consecutive steps:

$$n \rightarrow p + W^-, \quad W^- \rightarrow e + \bar{\nu} \quad (8)$$

He even went as far as to assume that the coupling constant for such processes is $e^2/\hbar c$, the same as for electromagnetic events. He attributed the smallness and the short range of the weak interactions to a large mass of the W, as it is done today, and he arrives at a W mass of about 100 GeV. This was 20 years before Schwinger independently took up this idea again. Schwinger initiated a development that brought forward the present unified quantum electro-weak dynamics, referred to as QEWD, a development in which a large number of theorists took part, including Martinus Veltman, Gerard 't Hooft, P. W. Higgs, R. Brout, Sheldon Glashow, Steven Weinberg, Benjamin W. Lee and Abdus Salam. An excellent historical survey has been written by Sidney Coleman.[26]

Before entering the discussion of those new ideas it is necessary to modernize the relations 8. We assume today that the proton and the neutron are not elementary but are made up of three quarks, the proton being the combination uud, the neutron ddu. Here u and d stand for the two most important quark types; u carries the charge $\frac{2}{3} e$ and d carries $-\frac{1}{3} e$. They represent an isotopic doublet. Thus the transitions 8 and their inverse are pictured today as transitions between the two doublet states:

$$
\begin{aligned}
d &\rightarrow u + W^- \\
W^- &\begin{array}{l} \nearrow e + \bar{\nu}_e \\ \rightarrow \mu + \bar{\nu}_\mu \\ \searrow \tau + \bar{\nu}_\tau \end{array} \qquad (9) \\
W^+ &\begin{array}{l} \nearrow \bar{e} + \nu_e \\ \rightarrow \bar{\mu} + \nu_\mu \\ \searrow \bar{\tau} + \nu_\tau \end{array}
\end{aligned}
$$

The bar denotes the antiparticle. (There is a refinement that we will not treat in any detail. In the fundamental weak interaction process d is replaced by a linear combination $d' = ad + bs$, where s is the so-called strange quark. This refinement allows a weak transition in which the strangeness changes. These effects are smaller than 9 because $b < a$. Similar mixtures between quark types in weak interactions appear between the higher quark types.)

C. N. Yang and R. L. Mills[27] provided the key idea that was necessary in order to apply field theory to weak and later to strong interactions. It is a generalization of the field concept that underlies QED. In the latter the source of the field is a scalar magnitude, the charge of the particles. The field does not carry any charge; the charge always stays with the particles. Such theories are called "abelian" theories. Nonabelian field theories, as the ones introduced by Yang and Mills, contain two new features:

▶ The source of the field is not a *scalar* charge, but an internal quantum number of the source particle, for example a *spinor* charge, such as the isotopic-spin quantum number (called "up" or "down" in the case of proton and neutron).

▶ The source particle can exchange its "charge" (the isospin) with the field in the interaction process.

In such theories the field itself carries charge and, therefore, acts as a source of fields; there is a direct interaction process between field quanta. Whereas the fundamental diagram of QED is the coupling of the charged particle with the field (see figure 3a) the nonabelian theories also contain another fundamental diagram denoting the coupling between field quanta. The mathematical formulation of nonabelian field theories is based upon a generalization of gauge invariance; we will not enter here into these formal, though essential, arguments, except by noting that they require the field quanta to be massless vector bosons.

To come closer to an understanding of the present view regarding electro-weak dynamics, we start by discussing the theory at very high energies, much higher than the mass of the W, that is much higher than 100 GeV. In that region the weak interactions and the electric interactions are neatly separated. Let us first discuss the former ones. We introduce the so-called weak isodoublets, consisting of the u-d quark pair (actually u − d'; see parenthetical remark on page 80), and the three neutrino–electron pairs:

Doublet (left-handed)	u	ν_e	ν_μ	ν_τ
	d	e	μ	τ
Hypercharge	η'	η	η	η

Only the left-handed particles form these isodoublets. The right-handed ones have no weak interactions. These doublets emit or absorb three types of bosons according to the scheme:

$$\begin{aligned} a &\rightleftarrows b + W^+ \\ b &\rightleftarrows a + W^- \\ a &\rightleftarrows a + W^0 \\ b &\rightleftarrows b + W^0 \end{aligned} \quad (10)$$

Here a − b stands for any isodoublet of the table above; the coupling constant for each process is g. The process corresponds to the diagram of figure 3a with a coupling constant g. The basic gauge invariance of this formalism requires that the three processes 8 have the same probabilities and that the three W's are massless vector bosons.

In addition to the "SU(2)-type" couplings of equation 10 we also introduce a "hyper-electromagnetic" coupling. It is analogous to the ordinary electromagnetic one ("U(1) coupling"), but the two members a and b carry the same scalar "hypercharge" η' or η, depending on whether we consider the quark pair or the lepton pairs. This coupling does *not* distinguish right- and left-handed particles; it applies to both. We therefore get the processes (with coupling constants η' or η)

$$\begin{aligned} a &\rightleftarrows a + B^0 \\ b &\rightleftarrows b + B^0 \end{aligned} \quad (11)$$

where B^0 is the massless quantum (vector boson) of the hyper-electromagnetic field. At very high energies we then expect the quarks and leptons to be coupled to the W field in a nonabelian way because, according to equation 10 the iso-spinor charges are transferred to the field and vice versa; but they are coupled to the B field in an abelian way via the scalar hypercharge η or η'.

This picture can be right only at very high energies. The mass of the W would show up at a lower energy. We also find there that the electromagnetic field is coupled to different charges in each isodoublet. How does Nature achieve these deviations from the symmetric theory at high energies? The current theories postulate something that is called "spontaneous symmetry breaking" at lower energies. It is caused by a new isotopic spinor field—the Higgs field. It has the following remarkable property: Its energy is

such that it has a minimum not when the field is zero but when it has a finite value given by the spinor $\{\phi_0, 0\}$. That would mean that the vacuum has a certain fixed direction in isospace, namely the direction of the spinor ϕ_0. At high energy this is no longer true because there the energy gained by choosing ϕ_0 instead of zero is negligible. The situation is like that of a ferromagnet, in which a direction in real space is determined as long as the energy transfers are smaller than the Curie energy. Thus at low energies the Higgs field destroys the symmetric situation described before. The effects of this destruction by the finite expectation value of the Higgs field are as follows:

▶ The hyper-electromagnetic field **B** and the W^0 field get mixed by an arbitrary mixing angle θ_W, called the Weinberg angle. The two emerging linear combinations are

$$\begin{aligned} \mathbf{Z} &= \cos\theta_W W^0 + \sin\theta_W \mathbf{B} \\ \mathbf{A} &= -\sin\theta_W W^0 + \cos\theta_W \mathbf{B} \end{aligned} \quad (12)$$

▶ The Higgs field is coupled with the other field in such a way that W^+ and W^- acquire a mass M_W; Z gets a different mass M_Z, whereas the field A remains massless and becomes the electromagnetic field (photons).

▶ The fact that W^+ and Z have large masses reduces the weak interaction effects compared to the electric ones, at low energies.

▶ The coupling of the quarks and leptons to the electromagnetic field **A** is different from the coupling to the hyper-electromagnetic field **B**. Indeed it is such that the members of an isospin pair acquire the different electric charges, the ones that we usually ascribe to them.

▶ The bosons $W^\pm$ acquire an electric charge $\pm e$ that couples them to the field **A**.

▶ The weak transitions mediated by Z (no charge transfer, "neutral currents") are different from those transmitted by the $W^\pm$. The latter ones are characterized by a maximum parity violation because only the left-handed leptons and quarks are coupled to them. The Z, however, contains not only the W^0, which is coupled to left-handed particles, but also the hyper-electromagnetic field **B** that does not distinguish the handedness in its coupling.

So much for the description of

C. N. Yang

© ALAN W. RICHARDS.

Fundamental diagrams
(below). (a) shows the
fundamental diagram of QED.
The straight lines are electron
states; the wavy line is a
photon state. (b) shows the
three fundamental diagrams of
QCD. The straight lines are
quark states; the wavy lines are
gluon states.   Figure 3

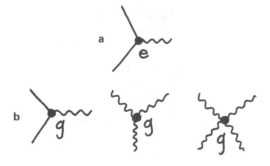

Running coupling constant in
QCD. The effective "charge"
Q_{eff} as a function of the
distance. The distance r_0,
where $Q_{eff} = 1$, is of the order
of the proton radius.   Figure 4

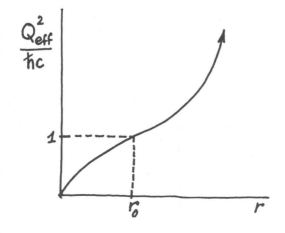

quantum electro-weak dynamics. The experiments have borne out the predicted consequences as far as they are accessible to today's experimentation. In particular the mixing of equations 12 could be verified and the angle θ_W determined. Several different experiments lead to the same result: $\sin^2\theta_W = 0.23 \pm 0.02$.

The most important experimental verification is still outstanding: the observation of the intermediate bosons. It is a similar situation to the one of Maxwell's theory of unification of electric and magnetic fields before Hertz's experiments. Woe to the theory if the bosons are not seen when the necessary energy and intensity for their production is reached at some of the accelerators under construction!

A questionable feature of this theory is the introduction of the Higgs field and its somewhat arbitrary couplings with other fields that are adjusted such that they produce the correct masses. The theory also requires the existence of Higgs-field particles of undetermined mass that have not yet been identified. It is hoped that a future formulation of the theory produces the effects of the Higgs field in a more elegant way and gets rid of it, as QED got rid of the vacuum filled with electrons of negative mass!

Quantum chromodynamics

The second theory that was structured as a parallel to quantum electrodynamics was "quantum chromodynamics (QCD)." It deals with the strong interactions. Since the discovery of the quark structure of hadrons one understands by "strong interaction" the forces between quarks. The nuclear force between nucleons was the previous candidate for that name. Today the nuclear force is considered as a weaker derivative of the quark–quark forces, just like the forces between atoms are weaker derivatives of the Coulomb forces between the atomic constituents.

Considering the successes of field-theoretical approaches, it is no surprise that present attempts to describe the interquark forces are also structured according to the model of quantum electrodynamics. Here is a dictionary of the analogies:

QED	QCD
electron	quarks
charge	color
photon	gluons (massless)
positronium	$\rho^0, \omega, \phi, J/\psi, \Upsilon$

Five analogs to positronium exist in QCD because five different types of quarks have been discovered up to now. Actually QED also predicts the existence of two more "positroniums," made of each of the two heavy electrons (μ, τ) and their antiparticles.

There are important differences between these two field theories, which mainly come from the different nature of the charge. In QED the charge is a scalar and remains with the fermions. The field is uncharged. In QCD, what acts as the charge is a "trivalent" magnitude ascribed to the quarks, referred to as "color." It is trivalent in the same sense in which the isotopic spin is a bivalent magnitude.

The color was introduced because three quarks were often found to be in the same quantum state. Because quarks are supposed to obey the Pauli principle, they must possess an internal quantum number capable of assuming three different values. There is a historic parallel to this: The fact that two electrons are found in the ground state of helium has contributed to the discovery of a two-valued internal quantum number—the spin.

QCD assumes that the color is the source of the field. Thus, we again face a nonabelian situation, but here the source is a trivalent "spin," whereas in quantum electro-weak dynamics we had the isotopic doublets of the pairs in the table on page 81 as sources. The consequences of QCD are also derived from a general gauge invariance with respect to the abstract "directions" of the trivalued spin. We obtain again a vector boson field whose massless quanta are the gluons. The properties of this field are analogous to the electromagnetic field. We may use terms such as "gluo-electric" and "gluo-magnetic" fields. There is one essential difference: The fields carry color charge in a similar sense as described in expressions 10. Because now we have three quark colors a, b, c, we find eight different types of gluons, arising from the following emission processes in which the quark colors may change:

$$a \rightarrow b + G_{a\bar{b}} \qquad a \rightarrow c + G_{a\bar{c}}$$
$$b \rightarrow c + G_{b\bar{c}} \qquad b \rightarrow a + G_{b\bar{a}} \qquad (13)$$
$$c \rightarrow a + G_{c\bar{a}} \qquad c \rightarrow b + G_{c\bar{b}}$$

$$a \rightarrow a + G_0 \qquad a \rightarrow a + G_0'$$
$$b \rightarrow b + G_0 \qquad b \rightarrow b + G_0' \qquad (14)$$
$$c \rightarrow c + G_0 \qquad c \rightarrow c + G_0'$$

Here $G_{a\bar{b}}$, etc., stands for the emitted gluon that carries double color (a, anti-b). There are eight different gluon colors. The transitions 14 give rise to colorless gluons, but invariance considerations show that there are only two: G_0, G_0', just as there is only one W^0 in equation 10. The fact that the gluons carry color charge leads to the typical nonabelian diagram in figure 3b, which indicates that gluons interact among each other.

A detailed description of QCD goes beyond the aims of this article. It may be important, however, to stress two surprising consequences of this theory, of which the second is not yet established with certainty. The first is called "asymptotic freedom." In contrast to electrodynamics, the effective coupling constant decreases when the distance decreases or when the momentum transfer increases. The coupling decreases as the inverse of the logarithm of the distance and, therefore, vanishes at infinitely close distances. For increasingly larger distances, however, the effective coupling constant does not remain finite as in QED, but seems to increase steadily. Here again we encounter an example of a "running" coupling constant but the dependence of the effective charge Q_{eff} on r is very different from the one in QED that was shown in figure 2. The situation in QCD is sketched in figure 4. The potential energy, say, between a quark and an antiquark (the analog to the Coulomb energy $- e^2/r$ between two opposite charges) probably increases linearly as ar at large distances and goes to infinity for $r \rightarrow \infty$.

The consequences of these relations are most unusual. It follows that single quarks cannot exist as free particles. Because the effective charge would become infinite at large distances, the energy necessary to isolate a quark from its partners in a hadron would be infinite. An isolated quark would be surrounded by a field that does not decrease with the distance. Obviously, no isolated quarks (or gluons) can exist in Nature if these conclusions are con-

firmed. Only systems whose total color charge is zero can exist in isolation. In the spin analogy to color, it would mean that the spins of the constituents must be opposed to each other and form a state of zero spin (singlet). In the trivalent case, three quarks are needed so that their colors add up to zero, or a quark–antiquark pair. Hence hadrons consist of either three quarks or of a quark–antiquark pair, because the antiquark has the complementary color to the quark. (This property justifies the use of the term "color". The three fundamental colors add up to white, and so do a color and its complementary one.)

The fact that hadrons carry no net color charge emphasizes the previously mentioned parallel between the nuclear force and the forces between atoms. Atoms are electrically neutral but when they approach each other, their structure is sufficiently altered that attraction occurs through resonance (Van der Waals forces) or through formation of new quantum states (chemical force). The same would happen when color-neutral nucleons approach each other.

Here we encounter a new situation: The elementary constituents—quarks and gluons—can only exist in bound states, never as single free particles. It should be noted that this paradoxical situation most probably follows (it has not yet been proved beyond a doubt) from a field theory that is a generalization of QED. In the latter, of course, fermions and bosons do exist as free particles; moreover, the system of free particles is the natural limit reached when the coupling constant goes to zero. This limit does not exist in QCD except for very small distances, the opposite situation to that of free particles.

One may ask why a similar situation—the impossibility of isolated particles—does not occur in the case of the weak interaction, which is also a nonabelian field theory. The answer lies in the fact that the symmetry of the iso-spin space is broken by the Higgs field at low energies (which means low momentum transfers and large distances) whereas the symmetry of the color space does not seem to be broken. Indeed, the mass M of the field quanta (a consequence of the Higgs field) prevents the fields from spreading

over distances larger than h/M. Isolated particles do not have infinitely strong fields in QEWD.

Unsolved problems

The development of quantum field theory since its inception half a century ago is most impressive. Today we have the means to calculate electromagnetic effects with incredible accuracy; two new field theories were created that seem reasonably appropriate to deal with the strong and weak interactions, the new forces of nature that were discovered during this half century. These forces are more complicated than the electromagnetic ones and exhibit different properties, such as charge-carrying fields, symmetries broken by vacuum fields, and forever confined particles. The fact that they nevertheless can be described by field theories is an indication that the concepts of those theories play an important role in natural phenomena. Certainly the language of field theory is used by Nature. There exist today attempts to bring together into one unified theory not only the weak and electromagnetic interactions but also the strong ones. These attempts use quantum electroweak dynamics as a model, to bring the SU(2) doublets of the weak forces and the SU(3) triplets of the color variety into one super group with new types of intermediate bosons. They are encouraged by the fact that the strong coupling constant decreases towards higher energies so that one might imagine a very high energy (10^{15} GeV) at which the electro-weak and strong coupling constants merge to one universal parameter. The differing values at lower energies are again caused by symmetry-breaking fields of the Higgs type.

It is by no means clear as to whether these attempts will turn out to be successful or not. In this so-called "grand unification" scheme, the Weinberg angle is no longer arbitrary and seems to come out close to the observed value. It also predicts transitions between quarks and leptons. For example, the u quarks, each having the charge $2/3$, end up as a positron (charge 1) and an anti-d quark (charge $1/3$). Thus a proton (a uud combination) can decay into a π^0 (a d$\bar{d}$ combination) and a positron. The proton would have a finite lifetime! Such transitions would be very slow because they would be mediated by some of those new intermediate bosons that are supposed to have masses near the characteristic energy of 10^{15} GeV. The lifetime of the proton should be of the order of 10^{32} years. If the numerous ongoing experiments to measure such lifetimes turn out to be successful, the ideas of field theory would win a new victory, and a unification of the three forces of Nature would be in sight. This still would leave gravity alone. The characteristic energy at which quantum effects become important in gravity is given by the mass of the particle pair, whose gravitational potential energy at a distance r is equal to the quantum energy $\hbar c/r$. It is of the order of 10^{19} GeV. This is about 1000 times higher than the characteristic energy of the grand unification attempt.

There are many indications that we understand only a partial aspect of what is going on. Here is an incomplete list of questions that are still unanswered:

▶ Is the renormalization procedure sound? So far it can only be carried out in successive perturbation steps. Can it be applied to a theory with an arbitrarily large coupling constant? The answer to this question may save or condemn field theory. A better understanding of the strong coupling limit (small distances in QED, large distances in QCD) may result in a satisfactory solution to the problems of infinities and of confinement or it may reveal fundamental shortcomings.

▶ The large value of the effective coupling constant of quantum chromodynamics at small momentum transfers causes serious problems as to the nature of the vacuum itself. The field fluctuations may turn out to be very large and may require new conceptions of the nature of the vacuum.

▶ Is the present interpretation of the electro-weak interactions correct? Do the intermediate bosons and the Higgs field really exist? These are questions that will soon be answered by experiments.

▶ The present theories contain arbitrary constants. In QED it is the coupling constant $e^2/\hbar c$ at large distances and the masses of the different electrons. Today three such electrons are known, but there may be more. There is

no way visible at present to explain how their mass values may emerge from the field theories. Moreover, the question remains why there is only one value of the electric charge (the quark charges are simple rational fractions of it) but several mass values seemingly without any simple relations.

In the electro-weak interaction there are two coupling constants between fermions and intermediate bosons, both of the order $e^2/\hbar c$. The Weinberg angle determines the ratio between the two. Furthermore, we find arbitrary coupling constants with the Higgs field that are chosen in order to yield the correct mass for the particles.

In QCD the situation is worse in respect to the mass problem because we deal with many different types of quarks, each having its own mass value. The coupling constant problem, however, is less difficult in QCD, if it turns out for sure that we deal with a running coupling from 0 at very small distances to infinity at large ones. Such a theory does not contain a fixed value at large distance, like $e^2/\hbar c$. But it contains length r_0 (of the order of 10^{-13} cm) at which the running coupling constant is near unity. We expect the composite quark systems to be of that size, and their masses to be of the order $\hbar/r_0 c$, in particular when the masses of the constituent quarks can be negligible compared to that mass. This is indeed the case for those hadrons that are made of u and d quarks. Therefore QCD has the advantage of containing the proton mass as a basic ingredient. (In our description of Nature we expect three intrinsic magnitudes to appear that determine the units of our measuring system. Their values do not require any explanation. These units may well be h, c, and the length r_0 as defined above.) But there is no indication whatsoever how the masses of the heavier quarks are determined by field theory. The theory does not even allow us to hope that the mass problem may be answered by strong coupling effects at small distances. Asymptotic freedom excludes any such effects.

The importance of the mass problem may be illustrated as follows. We have no explanation for the mass of the electron, that is for smallness of the ratio $(1836)^{-1}$ between the electron mass and the proton mass. (The latter may

be considered as the natural unit defined by QCD.) The small value of this ratio determines the properties of everything we see around us. It is the precondition of molecular architecture, of the fact that the positions of atomic nuclei are well defined within the surrounding electron clouds. Without it there would be no materials and no life. We have no idea about the deeper reasons for the smallness of that important ratio.

▶ Our present view of elementary particles is plagued by the following problem: Nature as we know it consists almost exclusively of u and d quarks (the constituents of protons and neutrons), and of ordinary electrons; all important interactions are mediated by photons, intermediate bosons and gluons. But there definitely exist higher families of particles, such as the heavier quarks and the heavier electrons. These additional particles are very short-lived or give rise to short-lived hadronic entities. They appear only under very exceptional circumstances that are realized during the early instances of the big bang, perhaps in the center of neutron stars, and at the targets of giant accelerators. What is their role in Nature, why do they exist? Rabi exclaimed when he heard of the first of those "unnecessary" particles, the muon: "Who ordered them?" Again, field theory does not seem to contain the answer to this question. Are they, perhaps, an indication of a deeper internal structure within the quarks and leptons? Are they the excited states of systems made of more elementary units held together by more elementary forces? Will the quantum ladder, the progression from atoms to nuclei, to nucleons and to quarks, ever reach an end?

We will find out sooner or later whether field theory is able to clear up some of these outstanding problems. It may be that a very different approach will be required to solve the questions for which field theory so far has failed to provide answers. Nature's language may be much wider than the language of field theory. We have not yet been able to make sense of much of what Nature says to us.

Looking back over a lifetime of field theory, it seems obvious that we have learned much since 1927, but there is a great deal more that is still shrouded in darkness. New ideas and new experimental facts will be needed to shed more light upon the deeper riddles of the material world.

* * *

Parts of this article appeared in the proceedings of a symposium on the history of particle physics held in May 1980 at Fermilab and also in the 1979 Bernard Gregory Lectures, CERN Report No. 80-03 (1980).

References

1. There exist two interesting studies about this subject: A. Pais, The Early History of the Electron 1897–1947 in Aspects of Quantum Theory, A. Salam, E. Wigner, eds., Cambridge University Press, 1972; S. Weinberg, Notes for a History of Quantum Field Theory, Daedalus, Fall 1977.

2. P. A. M. Dirac, Proc. Roy. Soc. **109**, 642 (1926); **114**, 243 (1927).

3. W. Pauli, Phys. Z. **20**, 457 (1919).

4. A. Einstein, Physica **5**, 330 (1925).

5. A. Pais, Rev. Mod. Phys. **51**, 861 (1979).

6. E. Fermi, Zeits. f. Phys. **29**, 315 (1924).

7. C. V. von Weizsäcker, Z. Phys. **88**, 612 (1934).

8. E. J. Williams, Phys. Rev. **45**, 729 (1934).

9. W. Heisenberg, Z. Phys. **90**, 209 (1934).

10. J. R. Oppenheimer, W. Furry, Phys. Rev. **45**, 245 (1934).

11. W. Pauli, V. F. Weisskopf, Helv. Phys. Acta **7**, 709 (1934).

12. J. R. Oppenheimer, Phys. Rev. **35**, 461 (1930).

13. V. F. Weisskopf, Zeits. f. Phys. **89**, 27; **90**, 817 (1934).

14. F. Bloch, A. Nordsieck, Phys. Rev. **52**, 54 (1937).

15. R. Serber, Phys. Rev. **48**, 49 (1935).

16. E. Uehling, Phys. Rev. **48**, 55 (1935).

17. H. Euler, Ann. d. Phys. V **26**, 398 (1936).

18. V. F. Weisskopf, Kgl. Dansk. Vid. Selsk. **14**, no. 6 (1936).

19. E. C. G. Stueckelberg, Ann. d. Phys. **21**, 367 (1934).

20. E. C. G. Stueckelberg, Helv. Phys. Acta **9**, 225 (1938).

21. W. Lamb, R. Retherford, Phys. Rev. **72**, 241 (1947).

22. H. A. Kramers, Nuovo Cim. **15**, 108 (1938).

23. N. Kroll, W. Lamb, Phys. Rev. **75**, 388 (1949).

24. J. B. French, V. F. Weisskopf, Phys. Rev. **75**, 1240 (1949).

25. O. Klein in *New Theories in Physics*, Conf. Proc. (Warsaw, 1938), Institut International de la Cooperation Intellectuelle, ed., M. Nijhoff, The Hague (1939), page 77.

26. S. Coleman, Science **206**, 1290 (1979).

27. C. N. Yang, R. Mills, Phys. Rev. **96**, 190 (1954).

For Further Reading

The history of physics has been growing so rapidly that the bibliographical notes in many of the articles in this book are out of date. We therefore want to suggest some ways the interested reader can venture further into this fascinating field.

There are many doorways to the history of modern physics. Several journals regularly publish scholarly articles in the field. Among these, *Historical Studies in the Physical Sciences* (University of California Press, Berkeley) carries by far the most articles, all of exceptional quality, and of particular interest to physicists. For the history of science in general, the central journal is *Isis*, issued by the History of Science Society (E.F. Smith Hall, University of Pennsylvania, Philadelphia). Anyone interested in the field is urged to join the Society. Members of The American Physical Society should also join their Division of History of Physics, at no charge, and receive its newsletter, which carries much information on current activities such as historical sessions at meetings, grants, and recent books. The Center for History of Physics at the American Institute of Physics, New York City, also issues a newsletter, available free whether or not one makes a contribution to the Friends of the Center; this newsletter carries information about current journal articles and projects, archival repositories, and other news. The Center's staff and its Niels Bohr Library are always glad to answer inquiries on historical matters. For information on a particular historical personage, the first place to look is the *Dictionary of Scientific Biography*, a fine multivolume work available at most libraries. A thorough bibliography is in John L. Heilbron, Bruce R. Wheaton, *et al.*, *Literature on the History of Physics in the 20th Century* (University of California Office for History of Science and Technology, Berkeley, 1981). An excellent selected list is in Lars Rodseth and Stephen G. Brush, "Library Checklist of Books and Periodicals in the History of Science," 1981 edition; copies are available from the AIP Center for History of Physics. We list below some more recent books which make good reading. This is only a sample of a large and rapidly growing body of work in the history of modern physics.

Badash, Lawrence, *Radioactivity in America: Growth and Decay of a Science* (Johns Hopkins University Press, Baltimore, 1979).

Bohr, Niels, *Collected Works*. Volumes 1-4 issued to date. (North Holland, Amsterdam, 1972–   ).

Bromberg, Joan Lisa, *Fusion: Science, Politics, and the Invention of a New Energy Source* (MIT Press, Cambridge, MA, 1982).

Brown, Laurie M., and Lillian Hoddeson, eds., *The Birth of Particle Physics* (Cambridge University Press, Cambridge, 1983).

Brush, Stephen G., *Statistical Physics and the Atomic Theory of Matter, from Boyle and Newton to Landau and Onsager* (Princeton University Press, Princeton, 1983).

Bunge, Mario, and William R. Shea, eds., *Rutherford and Physics at the Turn of the Century* (Dawson, London; Science History, London, 1979).

Dyson, Freeman J., *Disturbing the Universe* (Harper & Row, New York, 1979).

French, A. P., ed., *Einstein: A Centenary Volume* (Harvard University Press, Cambridge, MA, 1979).

Frisch, Otto R., *What Little I Remember* (Cambridge University Press, Cambridge, 1979).

Goldberg, Stanley, *Understanding Relativity: Origin and Impact of a Scientific Revolution* (Birkhauser, Cambridge, MA, 1984).

Hankins, Thomas L., *Sir William Rowan Hamilton* (Johns Hopkins University Press, Baltimore, 1980).

Hartcup, Gary, and Allibone, T. E., *Cockcroft and the Atom* (Adam Hilger, Bristol, 1984).

Hendry, John, *The Creation of Quantum Mechanics and the Bohr–Pauli Dialogue* (Reidel, Dordrecht, 1984).

Holton, Gerald, and Yehuda Elkana, eds., *Albert Einstein: Historical and Cultural Perspectives* (Princeton University Press, Princeton, NJ, 1982).

Kargon, Robert H., *The Rise of Robert Millikan: Portrait of a Life in American Science* (Cornell University Press, Ithaca, NY, 1982).

Kevles, Daniel, *The Physicists: The History of a Scientific Community in Modern America* (Knopf, New York, 1978).

McCormmach, Russell, *Night Thoughts of a Classical Physicist* (Harvard University Press, Cambridge, MA, 1982).

Mott, Nevill, ed., *The Beginnings of Solid State Physics: A Symposium...* (The Royal Society, London, 1980).

Oppenheimer, Robert, *Robert Oppenheimer: Letters and Recollections*, edited by Alice Kimball Smith and Charles Weiner (Harvard University Press, Cambridge, MA, 1980).

Pais, Abraham, *"Subtle is the Lord...": The Science and Life of Albert Einstein* (Clarendon Press, Oxford, 1982).

Segrè, Emilio, *From X-rays to Quarks: Modern Physicists and Their Discoveries* (Freeman, San Francisco, 1980).

Shea, William R., ed., *Otto Hahn and the Rise of Nuclear Physics* (Reidel, Boston, 1983).

Smith, Robert W., *The Expanding Universe: Astronomy's "Great Debate," 1900-1931* (Cambridge University Press, New York, 1982).

Sopka, Katherine Russell, *Quantum Physics in America, 1920-1935* (Arno, New York, 1980).

Stuewer, Roger, *Nuclear Physics in Retrospect: Proceedings of a Symposium on the 1930s* (University of Minnesota Press, Minneapolis, 1979).

Szilard, Leo, *Leo Szilard: His Version of the Facts*, edited by Spencer Weart and Gertrud Weiss Szilard (MIT Press, Cambridge, MA, 1978).

Truesdell, Clifford, *The Tragicomical History of Thermodynamics, 1822-1854* (Springer-Verlag, New York, 1980).

Weart, Spencer, *Scientists in Power* (Harvard University Press, Cambridge, MA, 1979).

Wheaton, Bruce R., *The Tiger and the Shark: Empirical Roots of Wave–Particle Dualism* (Cambridge University Press, Cambridge, 1983).